Biomass Utilization

NATO Advanced Science Institutes Series

A series of edited volumes comprising multifaceted studies of contemporary scientific issues by some of the best scientific minds in the world, assembled in cooperation with NATO Scientific Affairs Division.

This series is published by an international board of publishers in conjunction with NATO Scientific Affairs Division

A	Life Sciences	Plenum Publishing Corporation
B	Physics	New York and London
C	Mathematical and Physical Sciences	D. Reidel Publishing Company Dordrecht, Boston, and London
D	Behavioral and Social Sciences	Martinus Nijhoff Publishers The Hague, Boston, and London
E	Applied Sciences	
F	Computer and Systems Sciences	Springer Verlag Heidelberg, Berlin, and New York
G	Ecological Sciences	

Recent Volumes in Series A: Life Sciences

Volume 61—Genetic Engineering in Eukaryotes
    edited by Paul F. Lurquin and Andris Kleinhofs

Volume 62—Heart Perfusion, Energetics, and Ischemia
    edited by Leopold Dintenfass, Desmond G. Julian, and Geoffrey V. F. Seaman

Volume 63—Structure and Function of Plant Genomes
    edited by Orio Ciferri and Leon Dure III

Volume 64—Gene Expression in Normal and Transformed Cells
    edited by J. E. Celis and R. Bravo

Volume 65—The Pineal Gland and Its Endocrine Role
    edited by J. Axelrod, F. Fraschini, and G. P. Velo

Volume 66—Biomagnetism: An Interdisciplinary Approach
    edited by Samuel J. Williamson, Gian-Luca Romani, Lloyd Kaufman, and Ivo Modena

Volume 67—Biomass Utilization
    edited by Wilfred A. Côté

Biomass Utilization

Edited by
Wilfred A. Côté

State University of New York
College of Environmental Science and Forestry
Syracuse, New York

Plenum Press
New York and London
Published in cooperation with NATO Scientific Affairs Division

Proceedings of a NATO Advanced Study Institute on
Biomass Utilization,
held September 26-October 9, 1982,
in Alcabideche, Portugal

Library of Congress Cataloging in Publication Data

NATO Advanced Study Institute on Biomass Utilization (1982: Alcabideche, Portugal)
  Biomass utilization.

  (NATO advanced science institutes series. Series A, Life sciences; v. 67)
  "Proceedings of a NATO Advanced Study Institute on Biomass Utilization, held September 26-October 9, 1982, in Alcabideche, Portugal"—T.p. verso.
  "Published in cooperation with NATO Scientific Affairs Division."
  Includes bibliographical references and index.
  1. Biomass energy—Congresses. I. Côté, Wilfred A. II. North Atlantic Treaty Organization. Scientific Affairs Division. III. Title. IV. Series.
TP360.N28 1983   660.'6   83-9585
ISBN 0-306-41376-0

©1983 Plenum Press, New York
A Division of Plenum Publishing Corporation
233 Spring Street, New York, N.Y. 10013

All rights reserved. No part of this book may be reproduced, stored in a retrieval system, or transmitted in any form or by any means, electronic, mechanical, photocopying, microfilming, recording, or otherwise, without written permission from the Publisher

Printed in the United States of America

PREFACE

This proceedings volume represents the culmination of nearly three years of planning, organizing and carrying out of a NATO Advanced Study Institute on Biomass Utilization. The effort was initiated by Dr. Harry Sobel, then Editor of Biosources Digest, and a steering committee representing the many disciplines that this field brings together. When the fiscal and logistical details of the original plan could not be worked out, the idea was temporarily suspended.

In the spring of 1982, the Renewable Materials Institute of the State University of New York at the College of Environmental Science and Forestry in Syracuse, New York revived the plan. A number of modifications had to be made, including the venue which was changed from the U.S.A. to Portugal. Additional funding beyond the basic support provided by the Scientific Affairs Division of NATO had to be obtained. Ultimately there were supplementary grants from the Foundation for Microbiology and the Anne S. Richardson Fund to assist student participants. The New York State College of Forestry Foundation, Inc. provided major support through the Renewable Materials Institute.

The ASI was held in Alcabideche, Portugal from September 26 to October 9, 1982. Eighty participants including fifteen principal lecturers were assembled at the Hotel Sintra Estoril for the program that was organized as a comprehensive course on biomass utilization. The main lectures were supplemented by relevant short papers offered by the participants.

The course was introduced by discussion of the basic concepts of biomass utilization. Then the raw material (forest biomass, agricultural resources, aquatic resources and municipal solid waste) was considered from the point of view of its availability, assessment, preparation and general suitability. The structure and chemical composition of the biomass were addressed by a number of speakers before the conversion methods were presented.

Biological and thermochemical routes for conversion of biomass to energy, chemicals or food were discussed for several days as this is the main thrust of biomass utilization today. Finally the engineering aspects and the economics of biomass utilization were taken up in order to examine the feasibility of the various elements that comprise this multidisciplinary field.

One of the original objectives of this ASI was to provide young professionals in research, engineering and teaching, exposure to the ideas and experience of a cadre of veterans in this field. It was judged by the majority of participants to have been achieved. The papers offered in this volume contain the principal lectures and a number of supplementary papers which represent the substance of the presentations at the ASI. In this way, it is hoped that many individuals who could not be accommodated at the conference will have access to the material that was offered during the course.

Many individuals contributed to the success of this ASI. Those who helped in the planning, Dr. Harry Sobel, Dr. Irving Goldstein and Dr. Roscoe Ward deserve special credit. I owe a debt of gratitude to Dr. Eugene G. Kovach for his cooperation and advice on the management of ASI's. The administrative phases of the ASI could not have been handled without the outstanding efforts of Dr. John Yavorsky. All who presented principal lectures or supplementary papers and those who participated through active discussion made this conference a worthwhile undertaking. Finally, these proceedings would not have appeared without the diligent efforts of Mrs. Linda Boshart and Miss Judy Barton who assisted with the mechanical details of manuscript preparation.

February 1983
Wilfred A. Côté

ACKNOWLEDGMENTS

This Advanced Study Institute was co-sponsored by the Scientific Affairs Division of the North Atlantic Treaty Organization and the State University of New York College of Environmental Science and Forestry. Major financial support was provided by NATO and the NYS College of Forestry Foundation, Inc. The Foundation for Microbiology, the Anne S. Richardson Fund, and the U.S. National Science Foundation provided support for travel and associated expenses for several students.

Support for the planning and management of this Advanced Study Institute was provided by several offices of the College of Environmental Science and Forestry, with the Renewable Materials Institute and the School of Continuing Education most prominent.

On behalf of all the participants we express our thanks and appreciation to all of the organizations and individuals who contributed to the success of the ASI on Biomass Utilization.

CONTENTS

BIOMASS UTILIZATION -- THE CONCEPT

Biomass for Energy -- Fuels Now and in the Future 1
   D. O. Hall

Food, Chemical Feedstocks and Energy from Biomass 23
   R. F. Ward

Maximizing the Use of New Biomass with Population Limits 51
   H. Sobel

THE RAW MATERIAL AND ITS PREPARATION

Forest Inventories as the Basis for a Continuous Monitoring
   of Forest Biomass Resources 63
   T. Cunia

Aerial Photo Biomass Equation 103
   J. Kasile

Forest Biomass Utilization in Greece 109
   G. Tsoumis

Biomass -- The Agricultural Perspective 117
   W. J. Sheppard and B. Young

The Aquatic Resource 137
   M. Indergaard

Municipal Solid Waste -- A Raw Material 169
   R. F. Vokes

The Potential Role of Densification in Biomass Utilization 181
   J. J. Balatinecz

Mass Propagation of Selected Trees for Biomass by Tissue
    Culture 191
  S. Venketeswaran, V. Gandhi, E. J. Romano and R. Nagmani

The Production of Microalgae as a Source of Biomass 205
  E. W. Becker

Hydrogen from Water -- The Potential Role of Green Algae as
    Solar Energy Conversion System 227
  B. Mahro and L. H. Grimme

Utilization of Aquatic Biomass for Wastewater Treatment 233
  G. Lakshman

The Importance of Spartina maritima in the Recycling of Metals
    from a Polluted Area of the River Sado Estuary, Portugal . 241
  F. Reboredo

THE STRUCTURE AND CHEMICAL COMPOSITION OF BIOMASS

The Anatomy, Ultrastructure and Chemical Composition of Wood ... 249
  W. A. Côté

Cellulose: Elusive Component of the Plant Cell Wall 271
  E. J. Soltes

Some Structural Characteristics of Acid Hydrolysis Lignins 299
  J. Papadopoulos

CONVERSION METHODS -- BIOLOGICAL

General Concepts of Waste Utilization for the Production of
    Microbial Biomass and Bio-energy 309
  B. Pekin

The Fermentation of Biomass -- Current Aspects 365
  D. E. Eveleigh

Biotechnological Upgrade of Rice Husk: I. The Influence of
    Substrate Pretreatments on the Growth of Sporotrichum
    pulverulentum 393
  J. M. C. Duarte, A. Clemente, A. T. Dias and M. E. Andrade

Food from Biomass 411
  M. T. A. Collaço

CONTENTS

Potential Substrates for Single Cell Protein Production 443
    J. C. Royer and J. P. Nakas

Enzymatic Hydrolysis of Starch-containing Crops --
    Status and Possibilities in the Coming Years 461
    P. B. Poulsen

Biofilm Reactors: A Reaction Engineering Approach to
    Modelling 503
    A. E. Rodrigues

Methane Generation from Livestock Waste -- A Review 511
    S. Kirimhan

High Temperature Composting as a Resource Recovery
    System for Agro-industrial Wastes 529
    J. M. Lopez-Real

CONVERSION METHODS -- THERMOCHEMICAL

Thermochemical Routes to Chemicals, Fuels and Energy from
    Forestry and Agricultural Residues 537
    E. J. Soltes

Pyrolysis of Wood Wastes 553
    J. L. Figueiredo, J. M. Orfao, J. C. Monteiro and S. Alves

Hydrolysis of Cellulose by Acids 559
    I. S. Goldstein

Efficient Utilization of Woody Biomass: A Cellulose-
    Particleboard-Synfuels Model 567
    R. A. Young and S. Achmadi

Methanol from Wood, A State of the Art Review 585
    A. A. C. M. Beenackers and W. P. M. Van Swaaij

ENGINEERING AND ECONOMICS IN BIOMASS UTILIZATION

Energy Balances for Biomass Conversion Systems 611
    R. Katzen

Municipal Solid Waste -- Process Technology 623
    R. F. Vokes

Chemicals from Biomass 635
    W. J. Sheppard and E. S. Lipinsky

Energy Efficiency in the Process of Ethanol Production
    from Molasses 659
  A. Beba

Integrated Processes for Chemical Utilization of Biomass 669
  I. S. Goldstein

Biomass Utilization: Economic Analysis of a Systems
    Approach 675
  M. Dicks and J. P. Doll

Economics of Municipal Solid Waste as a Raw Material 687
  R. F. Vokes

Technical and Economic Evaluation -- Biomass Energy/
    Chemicals Integrated Systems 695
  R. Katzen

Participants...713

Contributors...719

Index 721

BIOMASS FOR ENERGY - FUELS NOW AND IN THE FUTURE

D. O. Hall

University of London
King's College
68 Half Moon Lane
London SE24 9JF U.K.

INTRODUCTION

Today about 14 percent of the world's primary energy is derived from biomass, equivalent to 20 m barrels of oil per day. Predominant use is in the rural areas of developing countries where half of the world's population lives. For example, Kenya derives three-fourths, India one-half, China one-third, and Brazil one-quarter of their total energy from biomass. A number of developed countries also derive a considerable amount of energy from biomass, e.g. Sweden nine percent and the U.S.A. three percent. Worldwide expenditure on biomass programmes is over two billion dollars per year. However, let me start by indicating what I am not going to advocate. I do not suggest that biomass will solve the energy problems of the world. I am not going to propose cutting down all the trees in the world or to reforest the world; and I am not advocating that we all become vegetarians. What I hope to clarify is that biomass already contributes a significant part of the world's energy; it is an important provider of energy to very many people. But how much biomass will contribute in the future will depend very much on decisions that are made both at the local level and at the national level, in addition to international policy making. Decisions that are made over the next few years will significantly influence the level of biomass energy use in the future.

The oil/energy problem of the 1970's has had three clear effects on biomass energy and development. Firstly, in a number of developed countries large research and development programmes have been instituted which have sought to establish the potential, the costs and the methods of implementation of energy from biomass. The prospects look far more promising that was thought even three years ago. Demonstrations, commercial trials and industrial projects are being implemented. Estimated current expenditure is over a billion dollars per annum in North America and Europe. Secondly, in at least two countries, viz. Brazil (which currently spends over half of its foreign currency on oil imports) and China, with over 7 million biogas digesters, large scale biomass energy schemes are being implemented - the current investment is about 1.3 billion dollars per annum in Brazil. Thirdly, in the developing countries as a whole there has been an accelerating use of biomass as oil products have become too expensive and/or unavailable.

BIOMASS

Biomass is a jargon term used in the context of energy for a range of products which have been derived from photosynthesis; you will recognize the products as waste from urban areas and from forestry and agricultural processes, specifically grown crops like trees, starch crops, sugar crops, hydrocarbon plants and oils, and also aquatic plants such as water weeds and algae (Table 1). Thus everything which has been derived from the process of photosynthesis is a potential source of energy. We are talking essentially about a solar energy conversion system, since this is what photosynthesis is. The problem with solar radiation is that it is diffuse and it is intermittent, so that if we are going to use it we have to capture a diffuse source of energy and need to store it - this is what plants solved a long time ago.

The process of photosynthesis embodies the two most important reactions in life. The first one is the water-splitting reaction which evolves oxygen as a byproduct; all life depends on this reaction. Secondly is the fixation of carbon dioxide to organic

Table 1. Sources of Biomass for Conversion to Fuels (Coombs 1980).

Water & Residues	Land Crops	Aquatic Plants
Manures	Ligno-cellulose (trees)	Algae (uni-cellular)
Slurry	Eucalyptus	Chlorella
Domestic rubbish	Poplar	Scenedesmus
Food wastes	Firs, Pines	Navicula
Sewage	Leucaena	Algae (Multi-cellular)
Wood residues	Casuarina	Kelp
Cane tops	Starch crops	Water weeds
Straw	Maize	Water hyacinth
Husks	Cassava	Water reeds/rushes
Citrus peel	Sugar crops	
Bagasse	Cane	
Molasses	Beet	

Table 2. Photosynthetic Efficiency and Energy Losses (UK-ISES, 1976).

	Available light energy
At sea level	100%
50% loss as a result of 400-700 nm light being the photosynthetically usable wavelengths	50%
20% loss due to reflection, inactive absorption, and transmission by leaves	40%
77% loss representing quantum efficiency requirements for CO_2 fixation in 680 nm light (assuming 10 quanta/CO_2)*, and remembering that the energy content of 575 nm red light is the radiation peak of visible light	9.2%
40% loss due to respiration	5.5%
	(Overall photosynthetic efficiency)

compounds. All our food and fuel is derived from this CO_2 fixation from the atmosphere. When looking at an energy process we need to have some understanding of what the _efficiency_ of this process will be; one needs to look at the efficiency over the entire cycle of the system, and in the process of photosynthesis we mean incoming solar radiation converted to a stored end-product. Most people agree that the practical maximum efficiency of photosynthesis is between five and six per cent. It might not seem very good, but I should remind you that this represents _stored energy_ (Table 2).

The photosynthetic efficiency will determine the biomass dry weight yields. For example, in the UK with 100 Wm^{-2} incoming radiation a good potato crop growing at one percent efficiency (usually not higher than this in temperate regions), will yield 20-25t dry weight per hectare per annum. Obviously if we can grow and adapt plants to increase the photosynthetic efficiency the dry weight yields will increase and of course alter the economics of the crop.

---

* If the minimum quantum requirement is 8 quanta/CO_2 then this loss factor becomes 72% instead of 77%, giving the final photosynthetic efficiency of 6.7% instead of 5.5%.

very interesting areas of research is to try to understand what the limiting factors are in photosynthetic efficiency in plants both for agriculture and for biomass energy (Coombs and Hall 1982).

There is another aspect of photosynthesis that we should all appreciate. That is, the health of our biosphere and our atmosphere is totally dependent on the process of photosynthesis. Every three hundred years all the CO_2 in the atmosphere is cycled through plants. Every two thousand years all the oxygen, and every two million years all the water. Thus three key ingredients in our atmosphere are dependent on cycling through the process of photosynthesis. (References for this section: Coombs 1980; Experientia 1982; Hall 1979; UK-ISES 1976; Zaborsky, 1981).

WORLD ENERGY USE

How much photosynthesis actually occurs on the earth? Table 3 shows that the world's total annual use of energy is only one tenth of the annual photosynthetic energy storage, i.e. photosynthesis already stores ten times as much energy as the world needs. The problem is getting it to the people who need it (Tables 4 and 5). Secondly, the energy content of stored biomass on the earth's surface today, which is 90% in trees, is equivalent to our proven fossil fuel reserves. In other words the energy content of the trees is equivalent to the commercially extractable oil, coal and gas. Thirdly, during the Carboniferous Era quite large quantities of photosynthetic products were stored, but in fact they only represent one hundred years of net photosynthesis. The overall photosynthetic efficiency during the Carboniferous Era was less than 0.0002% (Moore 1983). Our total possible fossil fuel reserves thus only represent one hundred years of net photosynthesis. Fourthly, the problem of CO_2 cycling in the atmosphere. Many people are rightly concerned about the problem of build-up of CO_2 in the atmosphere if we continue to burn fossil fuels. It is a problem of cycling between two or three pools of carbon. The amount of carbon stored in the biomass is approximately the same as the atmospheric CO_2 and the same as the carbon as CO_2 in the ocean surface layers; there are three equivalent pools. The problem is how is the CO_2 distributed between these pools, and how fast does it equilibrate into the deep ocean layers. However, we should appreciate that increasing CO_2 concentrations in the atmosphere may be good for plants (since CO_2 is a limiting factor in photosynthesis and plants have better water use efficiency at higher CO_2 concentrations) (Wittwer 1982). Plants could also act as CO_2 sinks if photochemical means for fixing CO_2 One of the were not available to alleviate the problem (Connolly 1981) (Reference for this section: Hall 1979).

Table 3. Fossil Fuel Reserves and Resources, Biomass Production and CO_2 Balances (Hall 1979).

1. Proven reserves   <u>Tonnes coal equivalent</u>

    Coal 5×10^{11}
    Oil 2×10^{11}
    Gas $\underline{1 \times 10^{11}}$

               8×10^{11} t = 25×10^{21} J

2. Estimated resources

    Coal 85×10^{11}
    Oil 5×10^{11}
    Gas 3×10^{11}
    Unconventional gas
      and oil $\underline{20 \times 10^{11}}$

               113×10^{11} t = 300×10^{21} J

3. Fossil fuels used so far (1976) 2×10^{11} t carbon = $\underline{6 \times 10^{21} \text{ J}}$

4. World's annual energy use $\underline{3 \times 10^{20} \text{ J}}$

               (5×10^9 t carbon from fossil fuels)

5. Annual photosynthesis
   (a) net primary production 8×10^{10} t carbon
               (2×10^{11} t organic matter)

               = $\underline{3 \times 10^{21} \text{ J}}$

   (b) cultivated land only 0.4×10^{10} t carbon

6. Stored in biomass
   (a) total (90% in trees) 8×10^{11} t carbon = $\underline{20 \times 10^{21} \text{ J}}$

   (b) cultivated land only (standing mass) 0.06×10^{11} t carbon
7. Atmospheric CO_2 7×10^{11} t carbon
8. CO_2 in ocean surface layers 6×10^{11} t carbon
9. Soil organic matter $10\text{-}30 \times 10^{11}$ t carbon
10. Ocean organic matter 17×10^{11} t carbon

Table 4. Annual Biomass Production (tonnes).

Net primary production (organic matter)	2×10^{11}
Forest production (dry matter)	9×10^{10}
Cereals (as harvested)	1.5×10^9
as starch	1×10^9
Root crops	5.7×10^8
as starch	2.2×10^8
Sugar crops	1×10^9
as sugar	9×10^7

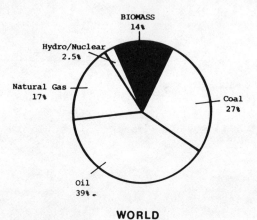
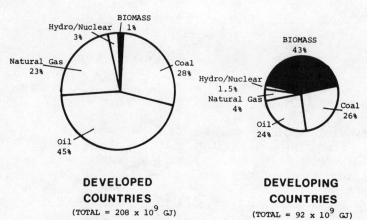

Figure 1. Percentage Of Energy Use Provided By Biomass.

Table 5. Total Production (10^6 t) and yields (t/ha) of the major sugar and starch crops (Coombs 1980).

	a) Sugar				b) Starch grains						c) Starch roots and tubers						
	Cane		Beet		Cereals		Maize		Rice		Wheat		Roots & Tubers		Potatoes		Sweet Potatoes		Cassava	
	Total	Yield	Total	Yield	Total	Yield	Total	Yield	Total	Yield	Total	Yield	Total	Yield	Total	Yield	Total	Yield	Total	Yield
World	737	56	290	32	1459	1.9	349	3.0	386	2.6	366	1.7	570	11.0	272	15.0	138	9.6	110	8.8
Developed countries	70	79	263	32	765	2.5	236	4.7	26	5.4	261	1.9	225	15.7	222	15.7	2	14.7	--	--
Developing countries	667	54	26	30	687	1.6	113	1.7	340	2.5	125	1.3	345	9.2	70	10.4	136	9.5	110	8.8
U.S.A.	25	82	23	46	273	4.1	161	5.7	5	4.9	55	2.0	17	28.0	16	29.2	11	12.4	--	--
Canada	--	--	1	39	42	2.3	48	5.9	--	--	19	1.9	3	22.4	3	22.4	--	--	--	--
Europe	0.3	63	143	38	249	3.5	49	4.2	1.5	3.8	82	3.3	114	18.7	83	10.3	--	--	--	--
U.S.S.R.	--	--	93	25	187	1.5	11	3.2	2.2	4.0	92	1.5	83	11.8	83	11.8	--	--	--	--
Asia	305	52	25	31	603	1.8	54	2.0	335	2.6	108	1.3	224	10.3	61	11.2	128	9.9	33	11.2
S. America	187	57	--	--	64	1.7	31	1.8	13	1.9	9	1.1	44	11.0	9	9.5	3	10.4	32	11.6
Africa	60	64	16	32	66	0.9	26	1.3	8	1.8	8	0.9	73	6.8	4	7.8	5	6.3	44	6.6
Oceania	26	75	--	--	16	1.0	0.4	4.5	0.5	5.3	10	0.9	2	10.3	1	23.0	0.6	5.4	2	11.0

These figures are on a crop basis excluding the considerable amounts of crop wastes, straw, etc. The cereal figures exclude such crops grown for forage silage (FAO 1978, 1979).

DEVELOPING COUNTRIES

There is another very important aspect of biomass energy use in the world - at least half the world's population are primarily dependent on biomass as their main source of energy. This realization has only widely occurred in the last few years - it has been called the "fuelwood crisis" or the "second energy crisis." There is no doubt that half the people in the world have a much more serious energy problem than most of us who are primarily concerned about the price of petrol and whether we should turn the heat up or down. About 14% of the world's annual fuel supplies are currently derived from biomass (Figure 1). The average person in the rural areas of the developing world uses the equivalent of about one tonne of wood. This is mainly for cooking and heating but also for small-scale industry, agriculture, food processing, and so on. The use of wood and charcoal in urban areas and for industry is also often much greater than is realized. Fourteen percent of the world's energy use represents the equivalent of twenty million barrels of oil a day, which is nearly equal to the total USA oil use. I don't think many people realized the importance of this because the statistics were not available to show this significance, and the consequences of biomass overuse were not readily evident. Until a few years ago world energy supply statistics listed biomass at 3-5%, if at all. It is now known that about half of all the trees cut down in the world today are used for cooking and heating. The problem of deforestation with its consequent flooding, desertification and agricultural problems is not solely due to over-cutting of trees for cooking and heating. There are obviously other factors involved such as commercial and illegal cutting, absence of replanting, and so on.

Recently there have been a few very good papers published on studies in Southern and East Africa which show that in an average family of six or seven, one person's sole job is to collect firewood and they will often have to walk great distances, which of course has other deleterious consequences. In urban environments, households can spend up to forty percent of their income on fuelwood and charcoal. Another aspect which has been highlighted in Tanzania is the curing of tobacco; for each hectare of tobacco you need to burn the wood from one hectare of savannah woodland. There are many examples to show that it is not only domestic fuelwood use but also agricultural, urban and small-scale industrial uses which are having serious long-term consequences. Serious attempts are being made by a number of international and national groups to try and reverse this problem of deforestation by vigorously promoting reforestation, village fuelwood lots or community forestry, agro-forestry, and so on. One study which has just been published by the US National Academy of Sciences (1981a) is a long overdue manual for tree species especially suited for fuelwood in the humid tropics, the arid tropics and temperate regions. Another study from ICRAF in

Nairobi (Nair 1980; Huxley 1982) promotes the concept of agroforestry where one can derive both food and fuel from the reforestation schemes. The World Bank (1980) concluded that if the deforestation problem was to be reversed, one would need to spend 6.75 billion dollars over the next five years in order to start reforesting fifty million hectares.

There is little hope that this will happen, but it was what realistically was thought to be needed. There are a number of reasons why this won't be possible, but I will mention only one here. It is the very low status that foresters have in developing (and also developed) countries. Consequently it is all very well promulgating reforestation schemes to help solve the energy crisis in various parts of the world, but unless you have the people on the ground with the experience and the knowledge, there is no way that these schemes can be implemented. (References for this section: Earl 1975; Hall et al. 1982; Moss & Morgan 1981; Unasylva 1981.)

BIOMASS ATTRIBUTES

Biomass as a source of energy has problems and it has advantages. Like every other energy source one must realize that it is not the universal panacea. Some advantages and disadvantages are listed in Table 6. I wish to emphasize one that is very interesting to me, namely the large biological and engineering development potential which is available for biomass (Experientia 1982; Rabson and Roger 1981). Presently we are using knowledge and experience which has been static for very many years; the efficiency of production and use of biomass as a source of energy has not progressed the way agricultural yields for food have increased. Thus there is an undoubted potential to increase biomass yields. The most obvious problems that immediately come to mind are land use in competition with food production (Brown 1980; Hall 1982b; Kovarik 1982; Pimentel and Pimentel 1979). Existing agricultural, forestry and social practices are also certainly a hindrance to promoting biomass as a source of energy whether in a developing or a developed country.

BIOMASS PROJECTS

Here follows some examples of biomass programmes around the world.

Brazil

The largest programme is in Brazil (Trindade 1981; Lima Acioli 1981) which is currently spending about 1.3 billion dollars of government money on subsidizing the production of alcohol, primarily from sugar-cane. The reason why so much money is being spent,

Table 6. Some Advantages and Problems Foreseen in Biomass for Energy Schemes.

Advantages	Problems
1. Stores energy	1. Land and water use competition
2. Renewable	2. Land areas required
3. Versatile conversion and products; some products with high energy content	3. Supply uncertainty in initial phases
4. Dependent on technology already available with minimum capital input; available to all income levels	4. Costs often uncertain
	5. Fertilizer, soil and water requirements
5. Can be developed with present manpower and material resources	6. Existing agricultural, forestry, and social practices
6. Large biological and engineering development potential	7. Bulky resource; transport and storage can be a problem
7. Creates employment and develops skills	8. Subject to climatic variability
8. Reasonably priced in many instances	9. Low conversion efficiencies
9. Ecologically inoffensive and safe	10. Seasonal (sometimes)
10. Does not increase atmospheric CO_2	

having been started in 1975, is due to the fact that currently Brazil spends over half of its total foreign income, about $11 billion on importing oil, in order to provide the petroleum to run its transport system. Brazil does not have an extensive railway system and is very much dependent on the internal combustion engine. Interestingly enough, it still derives about 25% of its energy from biomass; in fact its steel industry uses large amounts of charcoal. What Brazil is trying to do is to reverse this great dependence on imported petrolum by the production of alcohol; currently they are producing about 4 billion litres, and this is proposed to be expanded up to about 11 billion litres by 1985. At present all the petrol sold in Brazil is blended to 20% with alcohol. On Sundays such gasohol is not sold but one can buy alcohol - about a million cars now run on hydrated alcohol which is 95% by volume. The price of alcohol-engine cars is less than ordinary cars that run of gasohol, and the hydrated alcohol is about 10-20% cheaper than the gasohol. Thus there are subsidies to the producers, to the car buyer and to the car user. One can argue that there are advantages and disadvantages to this system, but the Brazilians' argument is that they have to "get off the oil hook." They calculate that each barrel of oil does not cost them $34, it eventually costs them $100 a barrel when taking into account financing charges and opportunity costs.

In 1987 production is anticipated to be 14 billion litres when they hope to replace half their petroleum imports with alcohol. Everything is not perfect with this programme. In fact the Brazilians have recently published an assessment of the programme which shows their great awareness of the problems themselves (Min. Ind. 1981). But as someone in Brasilia has said, if any government has to spend $6bn in six years it can make mistakes! One of the problems which was immediately evident is pollution in rivers. Most of the alcohol currently being produced comes from sugar-cane plantations in the South West of Brazil, where often the stillage which is a byproduct from the distillation of the alcohol has been put into the rivers. For every litre of alcohol produced about 8-10 litres of stillage is produced and this has a very high chemical and biological oxygen demand. There are certainly ways of countering this pollution problem. It has a high protein content and it can be a valuable food when dried. It can also be fermented to methane or be put into lakes to grow water hyacinths and algae to be fed to cattle or even directly fermented.

Zimbabwe

You don't have to be a country the size of Brazil in order to have a successful alcohol programme. In 1980 Zimbabwe opened an alcohol distillery in the Triangle area of the South East lowlands which will save them $10-12 million of foreign currency per annum. They produce 40 million litres of ethanol a year from sugarcane which is blended into petrol by the oil companies at a 15% blend; it is planned to increase this up to a 25% blend when a new refinery/distillery is opened. The yields of sugarcane are high at about 120 t per hectare per annum. Using the stillage from the fermentation as a fertilizer on the sugar plantations increases the yields by about 6% and this increase allows for the extra installations to be paid for within one year. (References for this section: Grundy 1980; Hall 1982.)

USA

The biomass resources of the US are indeed quite large. Currently they derive about 3% of their total energy requirements from biomass, which is equivalent to one million barrels of oil a day. The energy content of the standing forests of the US are at least 50% greater than the oil reserves and about equivalent to the gas reserves. The Americans have a gasohol programme which blends 10% alcohol, primarily derived from corn (maize), with unleaded gasoline to provide a high octane fuel. You can buy gasohol in over eleven thousand gasoline stations at a recent count. A gasoline pump may have an advertisement on it as follows: "Gasohol: the home grown fuel, the high octane fuel that is good for your car and good for your country." It is a bit of useful propaganda because it often costs a few cents more to buy the gasohol than it does to buy unleaded gasoline.

It is significant that this gasohol programme is relevant to the discussion that we are having about lead in petrol in Europe. The use of alcohol in gasoline has been known for a long time; cars were sold in 1902 which ran on ethanol (Kovarik 1982)! In the intervening years of the world wars and even after WWII there were cars running on blends of alcohol with anywhere from five percent up to about thirty-five percent alcohol. Recent studies (Alexander 1981 at the National Parks Service, Colorado) looked at a 25% ethanol blend using unmodified vehicles, besides boats and small equipment, which had been manufactured in the last 10 years and used either leaded or unleaded gasoline. The vehicles were driven 170,000 miles as of October 1980. The results indicated that vehicles using this blend had significantly lower emissions, fuel economy changes were not measurable, reliability was not changed and drivability had not been substantially affected, alcohol had not adversely affected any metal or plastic parts, engine maintenance had not been affected, and so on.

The question is often asked as to what happens when ethanol is added to gasoline. A report from the US Office of Technology Assessment (1980) based on a study of the Union Oil Company shows that for each 1% ethanol that is added to gasoline the octane rating is raised by an average of 0.24 for the motor octane number and 0.36 in research octane number. What this implies is that if 10% alcohol is added to gasoline the octane number increases by about 3-4 points. In the United States they have tried not very successfully, to get small-scale ethanol production going, especially in rural and local communities in the Midwest and the far West. The US Department of Energy provides a book free: 'Fuel from farms, a guide to small-scale ethanol production' (1980). It explains "how to do it" and the book has forms to fill out which are sent to the Revenue Service.

One of the serious discussions about using biological material for producing alcohol is whether you obtain more energy out than you have put into the systems; and secondly whether food is diverted from the food market and from world trade. There are no easy answers to these questions. With regard to energy ratios, it seems obvious that if you use more energy to manufacture the alcohol than obtained at the end you should not manufacture alcohol. But the problem is not as simple as that, because it depends on what kind of energy is used; if low value energy is used to fuel the system but high value liquid fuel is obtained at the end, it implies that the energy ratio is not significant. The analysis done by the US Office of Technology Assessment describes three different overall systems using corn (maize): one which results in a negative energy ratio and two which give positive energy ratios. One such system shows that if you use coal or biomass as a boiler fuel and the ethanol is used as an octane booster there is a definite positive benefit from turning corn into alcohol (See also TRW 1980 and Perez-Blanco and Hannon 1982).

Another aspect of the US energy programme is called Silviculture Biomass Plantations, which are essentially forests! What has been done is the selection of species which grow in specific sites, with close spacing, short rotation, and coppicing ability, i.e. after cutting down they immediately resprout (Steinbeck 1981). One is trying to obtain the maximum amount of energy per hectare per annum in such demonstration schemes. The species which have been looked at are pines, willows, oaks, maples, sycamores, etc. (References for this section: BioEnergy Council 1980; Lipinsky 1981; OTA 1980; US-DOE 1980.)

Europe

Nearly all the countries in Europe have energy-from-biomass schemes. Why in Europe should we look seriously at biomass as a source of energy? Currently the EEC derives over half its total energy requirements from imported oil. For Europe as a whole some substitution of imported liquid fuels is what makes biomass look so interesting. The EEC programme examines the use of agricultural resources, the use of forestry resources, the use of algae, the digestion of biological materials to produce methane, and thermochemical routes such as gasification to produce methanol. We in Europe are fairly crowded, so how much land do we have to produce energy from biomass? A recent study indicated that the EEC could produce about 75 million tonnes of oil equivalent, which is $1\frac{1}{2}$ million barrels of oil a day, providing about 6% of our estimated 1985 energy demand (Palz and Chartier, 1980). This is more than we use for agriculture. In developed countries agriculture generally uses about 3-4% of that country's energy requirements. Agriculture unfortunately often has a bad name because it is thought to be energy intensive, but there is no doubt that if ever there is an energy shortage this 4% is the most important. The study also showed that if there was a maximum disturbance to agriculture and forestry where a crash programme was required, we could possibly achieve a 20% provision of our energy requirement in the EEC. It is highly unlikely that this will happen. Thus a 6% energy provision is there for the taking, with minimal disturbance to current agricultural and forestry practices.

Some points about the land area, afforestation and agriculture in the EEC countries are required. Italy, Germany, France and the UK each have the same population - about 55 million. Germany is 28% forested, France 27%, Italy 21% and UK 8%. France has twice the land area as the UK, Germany and Italy (which are equal). The EEC currently spends about $5 billion (2/3 of its budget) on the Common Agricultural Policy and most of this goes to subsidizing animals

(Pearce 1981). To provide the food for these animals we devote a very large percentage of our land area for growing grains, we also import large quantities of grains from the United States and other countries. We produce milk, butter and cheese - it is a problem getting rid of these mountains, so very often they are fed back to animals. Since 90% of the energy in the food is lost every time you go through the animal, it seems rather an unusual way of doing things.

Even two years ago it was not thinkable to mention the possibility that we could use some of our agricultural land and our agricultural surpluses to produce something different from them, such as energy. The EEC is also the world's second largest exporter of sugar. We produce about 1.5 million tonnes of surplus sugar every year (from 1980 data); we also import 1 million tonnes from Africa, Caribbean and Pacific countries, and then we turn round and dump the 2.5 million tonnes at a cost of $1 b a year on to the world market. We also are the second biggest meat exporters in the world, have the biggest wine lakes, and have the biggest olive oil lakes in the world. The problem in Europe is not food shortages, it is overproduction (Hall 1982b).

The French renewable energy programme, which is now over $200 million dollars a year, devotes about 40% of its budget to production of "petrol from the earth" (COMES 1980). Various schemes are being implemented from methanation to straw burners, to running trucks on charcoal, to "carburol" blending into petrol, etc. "Carburol" can be acetone-butanol mixtures derived from fermentation processes, or other alcohols; trials are now starting in various parts of France on the blending of gasoline with "carburol." Ireland has a short-rotation forestry scheme, primarily using peat bogs in order to grow rapid-growing trees like willows and alder (Neenan and Lyons 1980). A UK Department of Energy study indicated that Britain could derive about 9% of its total energy requirements from biofuels. The study shows that this is an upper limit which depended on the implementation of a number of studies which were in the process of being completed; these studies will come to fruition rather soon and it is hoped that the UK will make a positive commitment to biofuels (King 1982). The energy is there for the taking especially from urban, agricultural and forestry wastes; the question is, can it be done economically and can it be done in the present circumstances. Studies by the Volkswagen Company in Germany, who are also very much involved in the Brazilian biofuels programme, predict that by the end of the century only half the cars in the world will run on petrol, 15% on diesel, 3.5% on biomass ethanol, 23% on coal methanol, some off LPG, and a little off electrical drive systems (Bernhart 1981). They predict that cars will be much more adaptable in their fuel use in the future. Sweden currently derives about 9% of its total energy requirements from biomass. They have a very interesting series of advanced biomass trials

(Heden 1982). Fast-growing willows have been selected; every year in Sweden they ask people to submit the tallest-growing shoot of willow that they have found that year and prizes are awarded. It is very successful because they now have nine clones which they call "super willows" which have twice the yield of those previously used. The Swedes say that if it was so desired they could derive half their energy requirements from biomass. They have the land and a low population. (References for this section: Chartier and Palz 1981; Moss and Hall 1982; Palz et al. 1981).

VEGETABLE OILS

It has been known for a long time that vegetable oils can be used in diesel engines. In fact in 1911 Diesel wrote an article saying that he advocated the use of vegetable oils in his engines in agricultural regions of the world, and he predicted that it would become important in the future. It has been known for a long time that you can put all types of vegetable oils into diesel engines. But studies in Zimbabwe, South Africa, Australia, Brazil, Philippines, United States, Austria, Germany, and others, show for example that in the sunny countries if a maize farmer devoted 10% of his land area to growing sunflowers or peanuts he could run all the diesel-powered machines that he uses in his farming operation. It is not generally advocated that these vegetable oils are used pure. Probably a blend of from 10-30% is preferable; the question is whether to spend money on refining the oil or to use it unrefined - if used unrefined, you have to make sure that oil filters and jets are cleaned frequently. There is also a lot of very interesting work on the esterification of sunflower oil as methyl or ethyl esters; the esterified oil has fuel properties very close to those of diesel and the esterification can be done on the farm. The Brazilians are devoting most of their effort to extraction of oil from peanuts and carefully ascertaining the use of palm oil as a 6% blend into diesel. Oils from soyabean, castor seeds, and indigenous plants such as malmeleiro and babasssu nut are also being investigated. (References for this section: BioEnergy Council 1981; Hall 1981; IUCEM 1981; Lipinsky et al. 1981).

HYDROCARBON PLANTS

Proposals to use plants directly to produce gasoline have been around for quite some time with the main recent proponent being Calvin of the University of California. What he advocates is growing _Euphorbia_ _lathyrus_ for the extraction of hydrocarbons which have molecular weights very close to those of petroleum (Calvin 1980). There are large trials, mostly in Arizona, financed by the Diamond Shamrock Corporation and others, to establish whether this is economically viable. Unfortunately the initial claims of high yields were not substantiated, but the recent studies show yields of about 10 barrels (1.5 tonnes) of oil per hectare per annum under irrigation (Johnson and Hinman 1980; Kingsolver 1982). The question is whether such yields are sustainable in arid environments? There are at least five other trials in various parts of the world to see if this is economically viable.

There is another requirement from oil, and that is for the manufacture of synthetic rubber. Guayule, Parthenium argentatum, which grows naturally in Northern Mexico and the southern USA, can be used as a source of rubber with properties which are indistinguishable from that of the rubber tree (CIQA 1978). By 1910 Rockefeller and Vanderbilt made fortunes supplying a large part of the world's rubber from such guayule bushes. There is now a pilot plant of one tonne a day in Saltillo, Mexico, and a 50 tonne a day plant is now being proposed in Northern Mexico.

An alga, Botryococcus braunii has been shown in Australia to yield 70% of its extract as a hydrocarbon liquid closely resembling crude oil. This has led to the work and ideas in France of immobilizing these algae in solid matrices such as alginates and polyurethane and using a flow-through system to produce hydrocarbons (Casadevall 1981). A green alga called Dunaliella discovered in the Dead Sea (Israel) produces glycerol, beta-carotene, and also protein. This alga does not have a cell wall and it grows in these very high salt concentrations; thus to compensate for the high salt externally it produced glycerol internally. A recent publication showed that if glycerol, beta-carotene and protein are produced this can be an economically viable system (Ben Amotz & Avron 1981). (References for this section: Buchanan et al. 1978; Hall 1980; Sonalysts 1981; Stewart et al. 1982).

BIOGAS

There are many energy problems which cannot be solved just by technology. For example, cow-dung is gathered in the surrounding areas of cities and towns in India and then sold as a fuel. This has serious consequences to agriculture. There are however means of obtaining energy from cow-dung, and at the same time producing fertilizer as a by-product (since one does not want to remove the manure fertilizer from the soil). The process is fermentation in biogas digesters which could be very useful in many parts of the world (Table 7). There is unfortunately however no universal prescription in advocating biogas as an energy source. In China they have a biogas digester which is a single family unit which was built at about the rate of one million a year mostly in Szechwan Province in Southern China; 7 million have been constructed so far. They are cheap, not very efficient, but they do work. In India they have about 90,000 biogas digesters, which are efficient, with steel domes and concrete bases, but they require a minimum of about five cows in order for them to work efficiently (Ghate 1979). Unfortunately there are not a great percentage of families in India with five cows. Thus ideally such biogas digesters would be best suited to a community or village; but there are social problems which must be understood before putting this type of system into an Indian village and in other parts of the world (Reddy 1981). (References for this section: Barnett et al. 1978; Chen 1981; Hughes et al. 1982; Experientia 1982; NAS 1981b; Smil 1982; Stafford et al. 1980; TERI 1982; UN-ESCAP 1981.)

PHOTOCHEMISTRY AND PHOTOBIOLOGY

I shall conclude with an area of research with which we are concerned in our laboratory, namely, how we might use photobiological and photochemical systems, not in the near future but maybe twenty years or more hence. We know that plants fix CO_2 and that they require soil, fertilizers, and water, and they must be protected from all types of predators. Hence, is it possible to short circuit the plant and construct systems which can directly fix CO_2, or split water, or fix nitrogen, without the intervention of the plant? The path of photosynthetic carbon fixation from CO_2 in the atmosphere to sugars, carboyhydrates, fats, etc. is well known.

Table 7. Alternative Uses of Biogas.

Application	Output from 1 cubic metre of Biogas
Lighting	Illumination equivalent to a 60-100 watt bulb for 6 hours
Cooking	Cooks 3 meals for a family of 5 - 6 people
Stationary Power	Runs a 1 horsepower motor for 2 hours
Road Vehicles	Runs a 3 tonne truck for 2.8 kilometres
Electricity Generation	Generates 1.25 kWh of electricity

Source: Chinese Biogas Manual (A. van Buren, Ed. 1979) IT Publ., 9 King St., London, WC2, UK.

Note: Typical Chinese household - size digesters produce 1 - 2 cubic metres of biogas per day.

Unfortunately, we do not know how the most crucial initial step works. The enzyme ribulosebisphosphate-carboxylase, which fixes all the CO_2 in the atmosphere which has gone into our oil, coal, gas, and currently goes into our food and into our trees, is the most abundant enzyme on earth. Half the protein in the lettuce leaf that you eat comprises this single enzyme. Unfortunately, we do not know how it works even though we are now learning quite a lot about it. Can we mimic the CO_2 fixation process? There are two reports from Israel and Japan on the photochemical fixation of CO_2 to methanol, formic acid, and formaldehyde. These are photoelectrochemical systems which work, however, only in UV light. It is early days but significantly these are the first reports on fixing CO_2 in the atmosphere into fuels.

With regard to the splitting of water, Jules Verne said in 1874: "I believe that water will one day be employed as fuel, that hydrogen and oxygen which constitute it used singly or together will furnish an inexhaustible source of heat and light" - unfortunately he did not tell us how to do it and we still do not know exactly how to do it! There have been recent reports on photochemical systems from a number of laboratories that use colloidal particles incorporating titanium and ruthenium which can split water and produce hydrogen and oxygen. Experimental problems still exist with such systems, but they are the first reports of purely chemical systems that appear as if they might function. The only system which actually does function in visible light on a continuous basis is the biological system. In mimicking such biosystems we use the water-splitting reaction of the chlorophyll membrane, throw away all the CO_2 fixation components, add in enzymes from bacteria, shine light on the system, and produce hydrogen gas (plus oxygen as a byproduct). This is all very well, but unfortunately it is a biological system and it eventually dies. We and other groups have tried to immobilize the system so that it can continuously produce hydrogen and oxygen, because if the system is ever going to be practically useful it has to work for years and not hours. A system immobilized in alignate on a steel grid produces hydrogen and oxygen in the light, but it dies after about twelve hours. We have been able to substitute the enzymes using platinum or synthetic iron-sulphur catalysts. What we have not been able to do is to substitute for the water-splitting reaction which is the crux of the whole problem.

Can we mimic, or can we construct, a system which will split water using visible light so that the electrons and protons from water can be used to produce high energy compounds like fixed carbon, hydrogen or ammonia? At present the answer is "No." What we can do quite positively is to use waste materials like organic compounds or sulphur (where you don't have to split water) to give continuous production of hydrogen gas upon illumination. In our laboratory we have had such a system running for fifteen days continuously producing pure hydrogen from compounds such as ascorbic acid. But we have not solved the ultimate problem of splitting water. (References for this section: Bolton and Hall 1979; Calvin 1980; Connolly 1981; Experientia 1982; Hall and Palz 1981; Hall et al. 1980).

CONCLUSION

What I have tried to say is that the process of photosynthesis is really quite marvellous. We start with water and light and use chlorophyll-containing membranes to produce an energy-rich compound which is usually an iron-sulphur catalyst. Normally the energy in that compound, which has a very high energy content, is used to fix carbon dioxide into a whole range of organic compounds. But we can also use this ultimately to fix nitrogen, reduce sulphur, reduce oxygen, and also use it to produce hydrogen. Plants and their derived systems can do nearly everything!

REFERENCES

Alexander, G. 1981. High percentage gasohol fleet reliability tests. See BioEnergy Council, 1981, p. 1023.
Bernard, G.W. and Hall, D.O. 1982. Energy from renewable resources: Ethanol fermentation and anaerobic digestion. In "Biotechnology Vol. III"., ed. H. Dellweg, Verlag Chemie, Weinheim (In press).
Barnett, A., Pyle, D.L. and Subramanian, S.K. 1978. Biogas technology in the Third World: A multidisciplinary review. Intl. Devel. Res. Corp. (IDRC), Ottawa, K1P 5YF, Canada.
Ben-Amotz, A. and Avron, M. 1981. Glycerol and -carotene metabolism in the halotolerant alga Dunaliella: A model system for biosolar energy conversion. Trends Biochem. Sci. TIBS 6, 296-298.
Bernhart, W. 1981. Alternative fuels from biomass and their use in Transport. In: Energy from Biomass, Palz, W., Chartier, P. and Hall, D.O., Eds. Applied Sci. Publ., London. pp. 815-825.
BioEnergy Council. 1980. BioEnergy '80 Proceedings. The BioEnergy Council, Washington, D.C. 20006, USA.
BioEnergy Council. 1981. International BioEnergy Directory. The BioEnergy Council, Washington, D.C. 20006, USA.
Bolton, J.R. and Hall, D.O. 1979. Photochemical conversion and storage of solar energy. Ann. Rev. Energy 4 353-401.
Brown, L. 1980. Food or fuel - new competition for the world's cropland. Paper no. 35, Worldwatch Institute, Washington, D.C. 20036, U.S.A.
Buchanan, R.A., Cull, I.M., Otey, F.H., and Russell, C.R. 1978. Hydrocarbon and rubber-producing crops. Econ. Bot. 32, 131-153.

Calvin, M. 1980. Petroleum Plantations and Synthetic Chloroplasts. Energy 4, 851-869; 1982. Plants can be a direct fuel source. Biologist 29, 145-148.

Casadevall, E. 1981. Renewable hydrocarbon production by cultivation of the green alga Botryococcus braunii. See Chartier & Palz, 1981, pp. 95-102.

Chartier, P. and Palz, W. 1981. Energy from biomass. Series E. Vol. 1. D. Reidel Publ., Dordrecht, Holland.

Chen, Ruchen. 1981. The development of biomass utilization in China. Biomass 1, 39-46.

CIQA. 1978. Guayule. CIQA Press, Aldamo Ote 351, Saltillo, Mexico.

COMES, 1980. Etude et Recommandations pour l'exploitation de l' energie-vert. Comite Biomasse-Energie. Commissariat a l' energie-solaire. 208 rue Raymond-Losserand, 75014 Paris, France.

Connolly, J.S., ed. 1981. Photochemical conversion and storage of solar energy. Academic Press, N.Y.

Coombs, J. 1980. Renewable sources of energy (carbohydrates). Outlook Agric. 10,235-245.

Coombs, J. and Hall, D.O. 1983. Energy and biotechnology, in Biotechnology - Principals and Applications. I.J. Higgins, Ed. Blackwell Sci. Publ., Oxford (In press).

Earl, D.E. 1975. Forest energy and economic development. Clarendon Press, Oxford.

Experientia (Special Issue). 1982. New trends in research and utilization of solar energy through biological systems. Vol. 38, pp. 1-66 & and 145-228.

Ghate, P.B. 1979. Biogas: A Decentralised Energy System. A Pilot Investigation Project. Economic and Political Weekly (Bombay) July 7, 1979, pp. 1132-6.

Grundy, T. 1980. Put the Alcohol in your Tank - Zimbabwe Style. Africa Business, 23 July 1980. pp. 18-19, London, U.K.

Hall, D.O. 1979. Solar energy use through biology - past, present and future, Solar Energy 22, 307-318; 1981. Solar energy through biology: Fuel for the Future. In: Advances in Food Producing Systems for Arid and Semi-arid Lands. Manassah, J.T. and Briskey, E.J., eds. Academic Press, N.Y., pp. 105-137; 1982, J. Roy. Soc. Arts 130, 457-471.

Hall, D.O. 1980. Renewable Resources (hydrocarbons). Outlook Agric. 10, 246-254.

Hall, D.O. 1981. Put a sunflower in your tank. New Scientist 89, 524-526.

Hall, D.O. 1982(a). Alcohol and Hot air in Zimbabwe. Earthscan Bull. (2), 9. Earthscan, London W1P ODR, U.K.

Hall, D.O. 1982(b). Food versus fuel, a world problem? In: Energy from Biomass; 2nd. EC Conference. Strub, A., Chartier, P. and Schleser, G., Eds., Applied Sci. Publ., London (In press).

Hall, D.O., Adams, M.W.W., Gisby, P.E. and Rao, K.K. 1980. Plant power fuels hydrogen production. New Scientist 86, 72-75.

Hall, D.O., Barnard, G.W. and Moss, P.A. 1982. Biomass for energy in the developing countries. Pergamon Press, Oxford.
Hall, D.O. and Palz, W. 1982. Photochemical, photoelectrochemical and photobiological processes. Series D. Vol. 1. D. Reidel Publ., Dordrecht, Holland.
Heden, K. 1982. Swedish energy forestry. Biomass 2, 1-3.
Hughes, D.E., Wheatley, B.I., Stafford, D.A., Baader, W., Lettinga, G., Nyns, E.J. Verstraete, W. and Wentworth, R.L. Eds., 1982. Anaerobic Digestion 1981. Elsevier, Amsterdam.
Huxley, P.A. 1982. Agroforestry - a range of new opportunities? Biologist 29, 141-143.
IEA. 1976 et seq. Biomass Conversion Technical Information Service. Nat'l Board Sci. Technol., Dublin 4, Ireland.
IUCEM. 1981. Beyond the Energy Crisis, Opportunity and Challenge. Vol. 2. Third Intl. Conf. on Energy Use Management, Berlin (West). Pergamon Press, Oxford.
Johnson, J. and Hinman, H.E., 1980. Oil and rubber from arid lands. Science 208, 460-464; 1981, see BioEnergy Council, p. 156.
King, G. 1982. Biofuels. Renewable Energy News, No.4. ETSU, Harwell, Didcot, U.K.
Kingsolver, B.E. 1982. Euphorbia lathyrus reconsidered - its potential as an energy crop in arid lands. Biomass 2, 281-298.
Kovarik, B. 1982. Fuel alcohol--energy and environment in a hungry world. Earthscan, London W1P ODR, U.K.
Lima Acioli, J. 1981. The alternative energy program in Brazil. Renewable Energy Rev. J. 3, 1-10.
Lipinsky, E.S. 1981. Chemicals from biomass: petrochemical substitution options. Science 212, 1465-1471.
Lipinsky, E.S., McClure, T.A., Kaesovich, S., Otis, J.L., Wagner, C.K., Trayser, D.A., and Applebaum, H.R. 1981. Systems studies of animal fats and vegetable oils for use as substitute and emergency diesel fuels. Report to US-Dept. of Energy, Washington, D.C. 20545, USA.
Ministry of Industry and Commerce, 1981. Assessment of Brazil's Alcohol Program, Ministry of Industry & Commerce, Secretariat of Industrial Technology, Brasilia, Brazil.
Moore, P.D. 1982. Plants and the paeleoatmosphere. J. Geol. Soc. London, Vol. 140, (In press).
Moss, P.A. and Hall, D.O. 1982. Biomass for energy in Europe. Intl. J. Solar Energy 1 (In press).
Moss, R.P. and Morgan, W.B. 1981. Fuelwood and Rural Energy, Production and Supply in the Humid Tropics. Natural Resources and Environment Series, Vol. 4, UN University Press, Tokyo.
Nair, P.K.R. 1980. Agroforestry species - a crop sheets manual. Intl. Council Research Agroforestry, P.O. Box 30677, Nairobi, Kenya.
NAS. 1981(a). Firewood crops; Shrub and Tree Species for Energy Production. Natl. Acad. Sci. USA, Washington, DC 20418, USA.
NAS. 1981(b) Food, fuel and fertilizer from organic wastes. Natl. Acad. Sci. USA, Washington, DC 20418, USA.

Neenan, M. and Lyons, G., eds. 1980. Production of Energy from Short Rotation Forestry. Oak Park Research Centre, Carlow, Ireland.
OTA, 1980. Energy from Biological Processes. Office of Technology Assessment, U.S. Congress, Washington, DC 20510, USA.
Palz, W. and Chartier, P. 1980. Energy from Biomass in Europe. Applied Sci. Publ., London.
Palz, W., Chartier, P. and Hall, D.O. eds., 1981. Energy from Biomass. Applied Sci. Publ., London.
Pearce, J. 1981. The Common Agricultural Policy. Routledge & Kegan Paul, London.
Perez-Blanco, H, and Hannon, B. 1982. Net energy analysis of methanol and ethanol production. Energy 7, 267-280.
Pimentel, D. and Pimentel, M. 1979. Food, Energy and Society. Edward Arnold Publ., London.
Rabson, R. and Rogers, P. 1981. The role of fundamental biological research in developing future biomass technologies. Biomass 1, 17-38.
Reddy, A. K. N. 1981. An Indian village agricultural ecosystem-- case study of Ungra village. II. Discussion. Biomass 1, 77-88.
Smil, V. 1982. Chinese biogas program sputters. Soft Energy Notes 5, 88-90.
Sonalysts, Inc. 1981. Assessment of Plant Derived Hydrocarbons. Report to U.S. Dept. Energy, Washington, D.C. 20545, USA.
Stafford, D.A., Wheatley, B.I. and Hughes, D.E. 1980. Anaerobic Digestion. Applied Sci. Publ., London.
Steinbeck, K. 1981. Short rotation forestry as a biomass source. In: Energy from Biomass, Palz, W., Chartier, P. and Hall, D.O., Eds., Applied Sci. Publ., London. pp. 163-171.
Stewart, G.A. et al. 1982. Potential for Production of hydrocarbon fuels from crops in Australia. CSIRO Div. Chem. Tech., Canberra, Australia.
TERI. 1982. Biogas Handbook. Tata Energy Res. Inst., Bombay, 400023, India.
Trindade, S.G. 1981. Energy crops - the case of Brazil. See Palz et al., 1981, pp. 59-74.
TRW. 1982. Energy balances in the production and end-use of alcohol derived from biomass, Report to U.S. Dept. Energy, Washington, D.C. 20545, USA.
UK-ISES. 1976. Solar energy: a UK assessment. Chap. 9. UK-ISES Publ. 19 Albemarle St., London WIX 4BS, UK.
Unasylva. 1981. Vol. 33, issue no. 131. Wood Energy Special Edition 1, FAO, Rome 00100, Italy.
UN-ESCAP. 1981. Renewable Sources of Energy. Vol. II. Biogas. ESCAP/TCDC, UN Bldg., Bangkok 2, Thailand.
Wittwer, S.H. 1982. Carbon dioxide and crop productivity. New Scientist 95, 233.
World Bank, 1980. Energy in the developing countries. World Bank, Washington, DC 20433, USA.
Zaborsky, O.R. ed. 1981. Handbook of Biosolar Resources. Vol. II Resource Materials. CRC Press, Boca Raton, Florida.

FOOD, CHEMICAL FEEDSTOCKS AND ENERGY FROM BIOMASS

Roscoe F. Ward

United Nations
Interregional Advisor-Biomass Energy
U.S.A.

INTRODUCTION

The Oxford Concise Dictionary (1976) defines biomass as "the total quantity or weight of organisms." The root of the word "bio" comes from the Greek <u>bios</u> which means life and <u>mass</u> has it origin from the old French and Latin meaning "a lump." The term biomass has been used during the last few years to refer to all materials of recent plant and animal origin. The definition of biomass can encompass oil, natural gas, tar sands and coal, but these materials are usually excluded.

We, as humans, depend totally on biomass to provide us with food which gives us our basic metabolic energy. In addition we use biomass in one of its many forms to cook our food and to provide us with heat for our homes. Our homes in many cases are entirely built from biomass. Biomass provides us with fiber for our clothing and provides us with our organic chemical feedstocks. Nothing is as important or serves as many functions as biomass. This paper will review the sources and uses of biomass.

FOOD

Food is a basic necessity of man. The daily needs of man for food vary depending upon the age, size, climate and degree of activities. An adult requires approximately 8 KJ/day of food. The 8 KJ/day is net energy and excludes the energy that may be used to produce and prepare the food. The energy required to cook the food man eats is about double the amount required as food intake. In addition when one includes the energy of the food residues the 8 KJ increase to 40 KJ (Stout 1979).

When man began to practice primitive agriculture with the assistance of animals the quantities of energy required increased to 50KJ/capita/day. With advanced agricultural systems the energy requirements increased to 300KJ/day/capita for low technology systems and up to an estimated 900 KJ/day/capita for advanced systems.

The land areas required to sustain man have been changing rapidly with time. Early hunters under favorable conditions required about 1.5 sq. km. to provide them with food for one person and in the harsher environments as much as 80 to 100 sq. km.

The first forms of agriculture were the shifting cultivation. A hectare of land under shifting cultivation can provide an all-crop diet for one person. Shifting agriculture is still widely used in the world and an estimated 250 million persons depend upon this system to provide them their food needs.

As pressures have increased on land the need for annual cropping has become a necessity. This type of agriculture requires more labor and has lower yields per unit of energy but one-half hectare or less can sustain one person. The additional labor to grow the crops can be supplied by animal power or some external energy source to power a tractor or other mechanical equipment. An advantage in using animals is that they can be used to provide food. Animals have the disadvantage that they need pasture or to be fed some crops which increases the land requirements. Mechanized agriculture requires an energy source, i.e. a fuel.

Developed countries have been able to achieve substantial increases in the yields of biomass. This increase in yield has come at the price of additional energy and chemical inputs as shown in Tables 1 and 2 for rice and corn production. Not only have they been able to increase the yields per unit area but also the yield per worker as shown in Table 3.

The United Nations in 1980 estimated that the population of the world was about 4.4 billion. The annual growth rate in 1980 was 1.7 percent or approximately 75 million persons per year (see Table 4). The growth rate has decrease since 1960 when it was about two percent. Most of the decline in growth rate has been in the developed countries and East Asia. However, some of the developing countries are still growing at over four percent per year. The current projections are that the earth's population will be over six billion persons in the year 2000.

At the beginning of the 20th Century estimates were that the earth could only support six billion persons. The sustainable population of six billion appears to be low when one examines the estimates by FAO (1980) on land availability to grow biomass. FAO's estimates are that the earth contains three to four billion hectares

FOOD, CHEMICAL FEEDSTOCKS AND ENERGY

of potentially arable land and that only about 1.5 billion hectares are currently developed. The rate of development of land in the past has been only about one half percent per annum. In 1980 FAO estimated that to keep pace with population growth rates the land development rates would have to increase at about two percent per year. In the development of land not only does the utilization change but also the intensity changes. As an example, by the addition of irrigation systems the same land in some areas can be more intensively used producing more crops.

With only one half or less of the arable land developed does not mean that the world does not face problems. The developed countries in general either produce excess or adequate food to sustain their population growth. If they do not produce adequate food they produce goods and services that they can use to purchase additional food that they require.

Developing countries, however have not only large populations and high growth rates but they often have poorly developed agricultural systems which are unable to sustain their populations. Some countries lack energy or chemicals which might allow them to better use their land resources. Figure 1 attempts to relate agricultural and energy self-sufficiency for a number of countries. Those countries which are neither food nor energy sufficient and are not industrialized face real problems in the future. FAO has projected

Table 1. Energy Required For Agriculture Rice.

Inputs	Modern Quantity/ha	Modern Energy	Traditional Quantity	Traditional Energy	Tradit. Energy
Labor					0.173
Machinery		4.2		0.335	
Fuel	224.7 L	8.98	40 L	1.6	
Nitrogen Fert.	134.4 kg	10.752	31.5 kg	2.52	
Phosphate Fert.					
Potassium	67.2 kg	.605			
Seeds	112	3.36	110 kg	1.65	
Irrigation	683.4 L	27.336			
Insecticide	5.6 kg	.56	1.5 kg	0.15	
Herbicide	5.6 kg	.56	1 kg	0.1	
Drying		4.6			
Electricity		3.2			
Transport		.724		0.03	
Total		64.885		6.386	0.173
YIELD (kg/ha)	5800		2700		1250

Energy Joules x 10^9/Hectare
Ref: FAO 1977

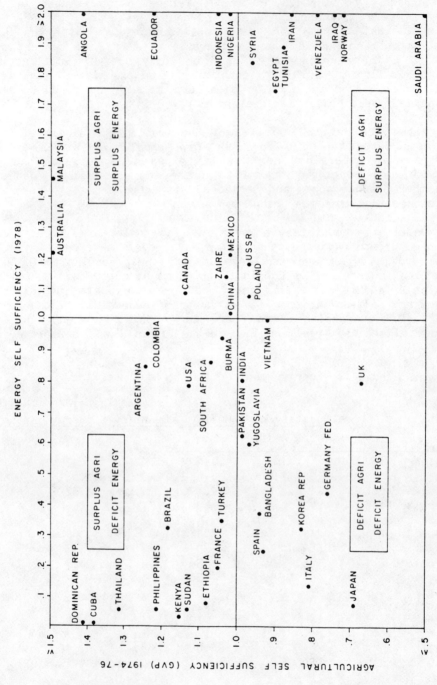

Figure 1. Energy and agricultural self sufficiency.

that the inability of a number of developing countries to support their populations is expected to worsen. By the year 2000 as much as 60 percent of the population of developing countries will live where arable land expansion opportunities will have been largely exhausted.

ENERGY

Earlier it was shown that as the world has developed, man has used more energy per capita to grow and prepare the food that is required to support the growing population.

The early source of energy to cook food was from biomass primarily wood. Wood is still the major fuel for about one half of the world's population (U.N. 1981). Even in developed countries like the United States biomass is used in rural regions to provide heat and to do the cooking. As development has taken place, coal, gas, kerosene and electricity have replaced wood as the fuel for cooking. These sources of energy are easier to use and in many cases less costly than wood.

Energy to produce the additional food was first provided by animals which were used to both work the soil and to provide food. As mechanization has taken place other sources of energy have been

Table 2. Commercial Energy For Corn Production.

Inputs	Modern Methods		Traditional Methods	
	Quantity	Energy	Quantity	Energy
Labor				0.173
Machinery		4.2		
Fuel		8.42		
Nitrogen Fert.	125 kg	10.		
Phosphate Fert.	34.7 kg	0.586		
Potassium Fert.	67.2 kg	0.605		
Seed	20.7 kg	0.621	10.4 kg	
Irrigation		0.351		
Herbicide	1.1 kg	0.11		
Insecticide	1.1 kg	0.11		
Drying		1.239		
Electricity		3.248		
Transport		0.724		
Total		30.034		0.173
YIELD (kg/Ha)	5,083		950	

Energy = Joules x 10^9/Hectare
Ref: FAO 1977; Pimentel 1974

used. First steam energy and then gasoline or diesel engines were used in agriculture.

Energy is required for industrialization. Early factories were located along streams and used water power. Development has used many other sources of energy including coal, gas, oil and nuclear. In examing the ratios of energy used per capita in the developed countries versus the world's average one will notice that the United States uses about six times the world's average while areas like Asia (excluding Japan) use only about one-tenth the average (Table 5). Table 6 compares the use of biomass-based fuel to alternative currently used energy sources.

CHEMICAL FEEDSTOCKS

Intensive agricultural techniques require fertilizers, insecticides, herbicides and fungicides which could be suppied by biomass. Nitrogen fertilizers are primarily produced using a hydrogen source and nitrogen from the atmosphere, resulting in ammonia or urea. The hydrogen to produce fertilizers is now obtained from natural gas, fuel oil, coal, or naphtha, but it can be produced by the gasification of biomass. The quantities of nitrogen applied depend upon the nitrogen content of the soils, the crops that are being grown and the cost of the fertilizer.

Table 3. Commercial Energy Used and Cereal Output.

Region	Energy/ Hectare	Energy/ Worker	Output/ Hectare	Output/ Worker
Developed Countries	24.8	107.8	3,100	10,508
N. America	20.2	555.8	3,457	67,882
W. Europe	27.9	82.4	3,163	5,772
Oceania	10.8	246.8	976	20,746
Other Dev. C.	19.4	19.1	2,631	2,215
Developing Countries	2.2	2.2	1,255	887
Africa	0.8	0.8	829	538
Latin America	4.2	8.6	1,440	1,856
Near East	3.8	4.4	1,335	1,386
Far East	1.7	1.4	1,328	781
Centrally Planned	5.9	6.8	1,744	1,518
Asia	2.4	1.7	1,815	911
E. Europe & USSR	9.3	28.5	1,682	4,109
World	7.9	9.9	1,821	1,671

Ref: FAO 1977

Table 4. Population Size and Rate of Increase.

Region	Population Millions					Average Annual Rate of Growth				
	1980	1985	1990	2000	2005	1985	1990	1995	2000	2025
World (Total)	4432	4826	5242	6119	8195	1.7	1.65	1.60	1.5	0.96
Dev. Regions	1131	1170	1206	1272	1377	0.68	0.61	0.58	0.48	0.24
Developing Reg.	3301	3656	4036	4847	6818	2.04	1.98	1.89	1.77	1.1
Africa	470	546	635	853	1542	3.	3.02	2.99	2.9	1.91
East Africa	134	156	183	250	478	3.09	3.17	3.17	3.12	2.09
Middle Africa	53	61	70	91	162	2.67	2.74	2.75	2.72	1.9
North Africa	109	126	144	186	296	2.87	2.77	2.62	2.45	1.47
South Africa	33	38	44	58	101	2.87	2.87	2.82	2.71	1.76
West Africa	141	166	195	267	505	3.18	3.23	3.2	3.13	2.01
Latin America	364	410	459	566	865	2.38	2.28	2.15	2.02	1.48
Caribbean	31	34	37	43	62	1.82	1.74	1.71	1.64	1.19
Central Amer.	93	107	122	156	243	2.83	2.72	2.51	2.31	1.43
Temp. S.Amer.	41	44	47	52	62	1.29	1.20	1.09	0.00	0.55
Tropic S.Amer.	199	226	254	315	498	2.46	2.36	2.23	2.1	1.66
North America	248	261	274	299	343	1.04	0.95	1.05	0.7	0.42
East Asia	1175	1250	1327	1475	1712	1.24	1.2	1.09	1.02	0.38
China	995	1060	1128	1257	1469	1.27	1.24	1.13	1.05	0.39
Japan	117	120	123	129	131	0.62	0.49	0.48	0.49	-0.07
Other East Asia	63	70	76	88	111	1.92	1.76	1.54	1.35	0.69
South Asia	1404	1565	1731	2075	2819	2.17	2.02	1.9	1.72	0.95
Europe	484	492	499	512	522	0.34	0.30	0.27	0.24	0.03
East Europe	110	113	116	121	131	0.61	0.49	0.43	0.43	0.23
North Europe	82	82	82	83	81	0.04	0.03	0.05	0.02	-0.10
South Europe	139	143	147	154	161	0.61	0.53	0.46	0.4	0.12
West Europe	153	153	154	155	150	0.04	0.07	0.08	0.05	-0.16
Oceania	23	25	26	30	36	1.44	1.36	1.29	1.19	0.61
USSR	265	278	290	310	355	0.93	0.84	0.7	0.64	0.5

Ref: United Nations 1981

Table 5. Estimated Total and Agricultural Use of Commercial Energy 1972-73. Joules x 10^{15}.

Region	Total Use	Agricultural Use	Percent Agricultural
Developed Countries	135.	4.6	3.4
N. America	76.9	2.1	2.8
W. Europe	42.9	2.1	4.9
Oceania	2.4	0.13	5.6
Other Dev. Countries	13.3	0.25	1.8
Developing Countries	19.3	0.9	4.8
Africa	1.6	0.07	4.5
Latin America	8.1	0.3	3.8
Near East	2.6	0.17	6.4
Far East	7.0	0.37	5.3
Centrally Planned	64.	2.0	3.2
Asia	14.3	0.4	2.9
E. Europe & USSR	49.8	1.6	3.3
World	219.	7.6	3.5

Ref: FAO 1977.

Table 6. Comparison of Biomass Derived Fuels to Selected Alternative Conventional Fuels.

Category	Biomass	Alternative
HEAT	Wood	Coal, Oil, Natural Gas
	Low Joule Value Gas	Crude Oil, Natural Gas
	Oils-Pyrolysis	Crude Oil
	Oils-Liquefaction	Crude Oil
CLEAN HEAT	Low Joule Value Gas	Coal
	Med. Joule Value Gas	Coal, Natural Gas
	Synthetic Natural Gas	Coal, Natural Gas
	Methanol	Natural Gas
TRANSPORTATION FUELS	Ethanol	Gasoline
	Methanol	Gasoline
	Synthetic Gasoline	Gasoline

FOOD, CHEMICAL FEEDSTOCKS AND ENERGY

Herbicides, insecticides and fungicides protect the crops and reduce the quantities of hand labor that are necessary to control the weeds, insects and fungus. These chemicals are energy intensive but since the application rates are low when compared to fertilizers the benefits per unit of energy used can be much larger.

Other chemicals can be manufactured from biomass that satisfy the domestic and/or industrial markets. Alcohol is one which will have considerably more attention devoted to it at this Institute. Ethanol had an equivalent of 12.2 million tons of 95 percent alcohol produced in 1980 and methanol about 12.5 to 13 million tonnes. However, about three quarters of the ethanol equivalent was used as a beverage with the other quarter sold for industrial and energy uses (Ward 1981). Fermentation also produces a number of other products:

1. Foods such as vinegars and single celled proteins.
2. Industrial fermentaton products including acetic acid, acetone-butanol, butane diol and critic acid.
3. Pharmaceuticals which represent high value products of limited quantities.

Since most of these items can also be produced from other feedstocks the plants must be inexpensive and operate efficiently.

ABILITY OF BIOMASS TO SATISFY THE DEMAND FOR FOOD, CHEMICAL FEEDSTOCKS AND ENERGY

It has been postulated earlier in this paper that even though some countries currently or in the future will be unable to satisfy the basic demand for food, on a world wide basis adequate food should be available.

In the case of energy there are many sources other than biomass that can be used. The issue is whether biomass can compete on an economic basis with other alternatives.

FOOD VERSUS FUEL

In meeting man's basic need for food, agricultural residues and food processing wastes are produced which can be used as fuels or chemical feedstocks. Table 7 lists some typical crop residues production factors. The quantities of residues in the table include only the amounts which are left above ground as most consider it necessary to leave the stem root system to supply organic matter to the soil and to prevent erosion. For each tonne of wheat produced approximately 1.8 tonnes of straw and chaff are produced. Straw and chaff are used in some areas as feed supplements for animals while in other regions it can be used as an energy or chemical feedstock source.

Table 7. Residues Yields Per Tonne of Crop at Harvest (Using Harvest Weight and Including Moisture.*)

Crop	Residue Factor
Barley	1.28
Beans and Peas	0.66
Corn (Maze)	0.6 (*)
Cotton	1.2
Oats	1.25
Peanuts	1.15
Potatoes	0.12
Rice	1.15 (*)
Rye	1.80
Sorghum (Milo)	0.63 (*)
Soy Beans	1.83
Sunflowers	5.15
Sugarbeets	0.1
Sugarcane	0.13
Wheat	1.83

Trash and Hulls

Almonds	1.0
Cotton Gin Trash	0.47
Rice	0.17
Sugarbeet Pulp	0.06
Sugar Cane Begasse	0.19

(*) For Tropical varieties the quantities of residues per tonne of product will likely be higher.

Ref: Alich 1978
* (Based on harvest weight including moisture.)

Food products have been used as an energy source. For example, Brazil has undertaken a major programme to produce ethanol for use as a chemical feedstock or as a fuel from sugarcane. Surpluses of sugar can be removed from the market providing Brazil with a fuel or ethanol which they can export. At times of shortages of sugar the raw sugar and molasses are worth more as foods on the world market than as fuels. Table 8 shows the current prices of some commodities used to make ethanol in the U. S. and the resulting materials cost per liter of ethanol produced. If an average of $0.08/liter conversion cost is added to the raw material prices some of the materials can be used to produce ethanol competitively with synthetic ethanol. The ethanol is _not_ competitive with the world prices of gasoline and has been used as a fuel supplement only where it is heavily subsidized. (Wall Street Journal, Sept. 7, 1982).

Table 8. Ethanol Costs From Various Feedstocks.

Raw Materials	Prices 29/VI/81 tonne $ (1)	29/VI/82 tonne $ (1)	Raw Material Cost per liter $/ 1981	Cost ethanol $/ 1982
Corn (Maize) Central Illinois	118.	98.	0.33	0.26
Milo (Sorghum) Gulf Coast U.S.	123	104.	0.33	0.28
Potatoes (reds)	220	242	2.04	2.40
Raw Sugar	333	158	0.60	0.29
Wheat No. 2 (soft)	118	120	0.34	0.35

Notes: (1) Ethanol sells at the plant (no taxes) (2) $0.44 lt.
       (2) Processing Cost (including capital and indirect costs) est. $0.08 lt.
       (3) Gasoline Sells at Port of Entry (no taxes) (1) $0.24 lt.
       (4) If an energy equivalent to gasoline, ethanol should be $0.16 lt.

Ref: (1) Wall Street Journal 30/VI/1982
      (2) Wall Street Journal 7/IX/1982

Vegetable oils have been proposed for use in place of diesel oil. Table 9 lists the current market prices for vegetable oils and compares them to diesel oil. The cost of the oils are up to two times that of diesel fuels per GJ (Lipinsky 1981). Not only are they more expensive, but recent tests have indicated the difficulties of using these oils in internal combustion engines (Chem. Engr. 1982).

RESOURCE BASE

A more systematic approach to examine the potential of biomass would be to look at the Biomass Resource Base. For purposes of this paper the resource base can be classified into three major headings: (1) Wastes, (2) Residues and (3) Materials grown specifically for energy and chemical production. Wastes have been classified as materials which have cost of disposal associated with them. Others classify wastes as materials for which the producer of the material has no further use. Residues are materials which have a traditionally recognized value associated with them. For example, animal manure is sold for its fertilizer value. When one examines the major waste streams, municipal and solid wastes the quantities vary depending upon a number of factors some of which Alich (1978) identified (See Table 10).

The quantities of biomass collected from the agricultural sector depend upon the types of crops, the methods of harvesting the crop, the quantities that are uncollectable, and the quantities that must be left in the field to provide for organic matter in the soil, nutrients, and erosion control. Some factors that must be considered which affect the supply are shown in Table 11.

Municipal Solid Wastes

The municipal solid wastes includes all those materials which are generated as a consequence of the activities within the munici-

Table 9. Price of Vegetable Oils and Diesel Fuel.

Item	Location of Item	Price $/GJ (1)
Coconut Oil Crude	New Orleans	14.
Corn Oil	Chicago	14.5
Cotton Seed Oil	Mississippi	12.5
Linseed Oil	Minneapolis	17.
Palm Oil	New Orleans	13.
Peanut Oil	S.E. U.S.A.	15.5
Soybean Oil	Decature, Illinois	12.
Diesel Oil (2)	East Cost Port (imported)	7.

Notes: (1) $/G.J. Based on Specific Gravity of 0.92
Heat Content Assumed to be 90% of Diesel Oil/Unit Volume
(2) Based on World Oil Price of $42/bbl. Current Price is $34/bbl.

Ref: Wall Street Journal 29/VI/82.

Table 10. Factors Affecting Quantities of Materials From Municipal and Industrial Wastes.

        Climatic conditions
        Regional Distribution of Populations
        Urban and Rural Population Distributions
        Standards of Living and Growth Rates
        Water Availability and Use by Sector
        Water Pollution Problems
        Existing and Planned Sanitation Services
        Population Served by Various Sanitation Systems
        Wastes Produced by Sectors

Table 11. Factors Affecting Supply of Biomass From Residues.

Level of primary crop, livestock and forestry production
Residues generated per unit pof primary production
Harvesting and collection technology
Conventional energy feedstock market prices
Actual market prices or perceived values of residues in
   alternative uses

Factors Affecting Supply of Biomass From Crops.

Alternative resource use technology
Production technology including yields
Weather fluctuations
Conventional energy feedstock market prices

pality. Cointreau (1982) reviewed a series of projects with respect to waste collection transfer and disposal. The quantities and characteristics of the municipal solid wastes vary greatly according to the country and the social-economic conditions of the population (See Table 12). The quantities and characteristics also vary greatly according to the levels of collection service provided (Table 13).

Industrial Wastes

Industries generate both liquid and solid wastes. The characteristics and quantities vary greatly depending upon the type of process used. However, if the type of industry, the process used, and the production are known, the literature can provide more detailed information about the waste streams (Nemerow 1971). Some of the wastes are toxic and if added to the biological systems can result in serious problems.

Human Wastes

Human wastes represent the major problem from a health point of view. They must be either properly used or disposed of to reduce the transmission of disease. Kalbermatten et al. (1980) reviewed the technical and economic options that are available to collect, transport and treat these wastes. It is very likely that any community will have a number of types of collection systems. One may find households served by night soil collectors, other sections served by pit latrines, septic tanks and/or water carriage systems. It is important in assessments that the sources, quantities, and characteristics of the wastes are identified.

Table 12. Urban Refuse Generation Rates (Cointreau 1982).

City or Country	Waste Generation Rate
Industrialized Countries:	
New York, New York, U.S.A.	1.80 kg/cap/day
Hamburg, Germany	.85
Rome, Italy	.69
Middle-Income Countries	
Singapore	.87
Hong Kong	.85
Tunis, Tunisia	.56
Medellin, Columbia	.54
Kano, Nigeria	.46
Manila, Philippines	.50
Cairo, Egypt	.50
Low-Income Countries	
Jakarta, Indonesia	.60
Surabaya, Indonesia	.52
Bandung, Indonesia	.52
Lahore, Pakistan	.55
Karachi, Pakistan	.60
Calcutta, India	.51
Kanpur, India	.50

Note: For those cities where the total refuse mix was subdivided into major categories of waste, data indicate that the residential portion of the total refuse was between 60 and 80%.

Agricultural and Forest Residues

Agricultural and forest residues are broken down further to high moisture, low moisture, woody and manure categories. The different characteristics suggest different processing techniques. Forecasts of these materials requires first determining residue factors for each residue-generating crop or activity (Alich 1977; 1978). Farmers and foresters rarely can tell you the quantities of residues produced, but they can tell you the quantities of crops or trees they sell. Therefore they developed some factors which have been used in the United States. These factors were based on intensive field investigation at a number of sites as well as extensive data collection from crop and animal production experts throughout the U. S. The values are dry weight factors excluding the stem below the cut and the roots which much be left in the field.

FOOD, CHEMICAL FEEDSTOCKS AND ENERGY 37

Table 13. Municipal Refuse Compositional Data
(In percentage by weight)
(Cointreau 1982)

Countries	Paper	Glass, ceramics	Metals	Plastics	Leather, rubber	Textiles	Wood, bones straw	Vegetative putrescible	Misc. inerts	Total
Industrialized										
London, England	37	8	8	2	–	2	–	28	15	100
Brooklyn, New York	35	9	13	10	–	4	4	22	4	100
Rome, Italy	18	4	3	4	–	–	–	50	21	100
Middle Income										
Singapore	43	1	3	6	–	9	–	5	32	100
Hong Kong	32	10	2	6	–	10	–	9	31	100
Medellin, Colombia	22	2	1	5	–	4	–	56	10	100
Lagos, Nigeria	14	3	4	–	–	–	–	60	19	100
Kano, Nigeria	17	2	5	4	–	7	–	43	22	100
Manila, Philippines	17	5	2	4	2	4	6	43	17	100
Low Income										
Jakarta, Indonesia	2	1	4	3	–	1	4	82	3	100
Lahore, Pakistan	4	3	4	2	7	5	2	49	24	100
Karachi, Pakistan	1	1	1	–	1	1	1	56	40	100
Lucknow, India	2	6	3	4	–	3	1	80	2	100
Calcutta, India	3	8	1	1	–	4	5	36	42	100
Bandung, Indonesia										
residential	10	1	2	6	–	4	1	72	6	100
market	8	1	1	2	–	1	1	84	5	100
commercial	12	1	1	7	–	3	1	69	7	100
Colombo, Sri Lanka										
residential	8	6*	2	1	–	1	1	80	1	100
market	8	2*	1	1	–	1	0	88	0	100
commercial	28	8*	1	2	–	1	2	58	0	100

Note: The above values have been rounded to the nearest whole number, unless the amount is less than 1.0

Residue factors for logging and the primary wood products were developed by Howlett (1977) for the U. S. Logging residues were defined as the above ground residues of growing stock and harvested rough, rotten and dead commercial species, tops and branches. Excluded are the stumproot systems, trees of non-commercial species and trees of commercial species less than 12.5 cm in diameter at breast height. Logging and primary wood products factors may need to be developed for other regions of the world.

Energy Crops

The concept of growing energy crops involves the utilization of unused or under-utilized lands which are managed as energy farms or forests. The land varies from unused crop lands to potentially productive bodies of water. Energy crops which have been proposed include sugarcane, sweet sorghum, cassava, soybeans, sunflowers, Euphorbia, trees, woody plants, Napier grass, water hyacinths and many others.

One concept of an energy farm is to use short-rotation trees where the production objective is to optimize the biomass output taking into account the cost of the operation. Trees in such energy farms are grown closer together and for shorter periods than normal forestry practices; hence the term "short rotation forestry" is used for the products of the system. For example, fast growing species, grown at high density, can be harvested at frequent intervals (e.g. 3-4 years) allowing the species to regrow from the stump (i.e. coppicing during a 20-30 year period). The possibility of high yields, up to 20 dry tonnes per hectare per year on a sustainable basis have offered significant changes on the economics of energy farming systems. Already short rotation forests are competitive with traditional systems.

Experimental work using the synergistic effects of planting mixed species or clones or varieties of the same species appear to reduce the risk of biological hazards due to monocultures. These systems, in addition, appear to allow for maintenance of wildlife, assure greater diversity of products and make easier the insertion of such energy farms in rural land use systems.

In the case of forestry there are considerable advantages in multipurpose planning and use of the products for timber, pulpwood, and energy.

A variety of crops can be grown on herbaceous energy farms. The basic difference between woody and herbaceous crops is that the herbaceous crops offer an advantage over woody crops in either production or conversion, i.e., sugar crops for fermentation ethanol, or oil crops for use as fuel substitutes.

Aquatic Biomass

With the rapid growth of the population of the world, competition for land will increase in many countries. As a result, there will be less marginal land available on which to produce biomass. Therefore, it is only natural that aquatic plants are being considered as energy sources (Ashare 1978). Water plants are already used for food and some species of seaweeds are used commercially to produce chemicals. Research has been conducted using fresh water, brackish water and marine micro and macro algae and aquatic plants as potential biomass feedstocks (Goldman 1977). Aquatic plants appear to offer greater potential than algae since they can be harvested easier (Kresovich 1981). Some species of water plants that are receiving considerable interest include water hyacinths, duck weed and <u>Hydrialla</u> (Ryther 1981).

A major disadvantage of biomass over coal is that it has a low bulk density and a wide range of moisture contents. These characteristics preclude the transportation of biomass over long distances. Wood can be used in a number of forms which include round wood, chips and powdered wood. Commercial systems are available which can improve biomass properties, reduce it moisture content and increase it Joule/weight ratio. These systems can incorporate drying, briquetting and/or pelletization. The systems can use agricultural or forest residues which are ground to a suitable size, dried and compressed. The processes may or may not require binders depending upon the feedstock. Difficulties have been encountered in the storage of the briquets or pellets as they can absorb moisture and disintegrate. If this occurs, dry storage is required which would increase further the cost of handling of the biomass. Reports are available on a numer of processes which have been widely used. Some of the reports include costs and energy balances (Janczak 1980, Kinderman et al. 1980).

DIRECT USE OF BIOMASS

Man and all other life require energy. For plants the energy is obtained from the sun. For animals and man the energy is obtained from the breaking down of food obtained from plants and other animals. Food can be in the forms which can be eaten directly and broken down in the digestive system or from food that needs to be partly broken down by cooking to make the organic matter digestible.

Early man used caves to provide shelter from adverse climatic conditions. He did not worry about clothing. Later he found that he could build structures utilizing trees and grasses to provide shelter and that he could manufacture cloth from plant materials.

Man has used fire to provide himself with warmth and to cook food. The early source of fuel was wood. Later man found that he could use other stored sources of hydrocarbons such as coal, oil and natural gas. All of these materials had their source of energy from plants and animals.

INDIRECT USES OF BIOMASS

Man found that he could manage biological systems. For example, he used meat as an energy source. To provide for a more uniform supply of meat he raised animals, first by allowing the animals to graze and later he found that he could grow grasses and grain which would produce a better meat product. The livestock served as a conversion process breaking down the cellulose and producing protein. There are a number of other conversion systems which can break biomass down into end use forms or intermediate forms. These systems are grouped into two areas: biological conversion and thermochemical conversion.

Biological Conversion

Biological conversion utilizes the enzymatic breakdown of biomass by microorganisms to produce useful end products. Three types of processes will be discussed. (1) Ethanol fermentations, (2) Biomethanation (anaerobic digestion) and (3) other microbiological processes.

Ethanol Fermentation. The production of ethanol by biological conversion is the oldest known biological conversion process. The sugars obtained from carbohydrates, starch and lignocellulosic materials are converted by microorganisms like yeasts, bacteria or fungi to ethanol and carbon dioxide. The technology of ethanol production from sugars and starches by fermentation is well established. Facilities of different sizes, using different processes and producing up to 700,000 liter per day are in use. In most cases the carbohydrates and starches used as food for humans or animals are expensive and these make the ethanol produced from them expensive. Because of this high price for the feedstocks synthetic ethanol processes were developed. In 1930 one process using ethylene as a raw material was developed. The ethylene is converted by indirect hydration to ethanol. Ethylene was so low in price that more than 95% of the industrial ethanol was produced from it. Since the ethanol could only be distinguished from fermentation ethanol by carbon dating, laws were passed in some countries which prohibited the use of it as a beverage ethanol. The capacity of synthetic ethanol plants in the United States as recently as 1978 was twice the capacity of all natural fermentation plants. In the United States today over 95 percent of the industrial ethanol is still produced from ethylene and it is only recently because of the low grain and sugar prices that fermentation ethanol is competitive (Business Week, 26, July, 1982).

The traditional fermentation process used is a batch process. This means that the sugars and yeasts are placed in a fermentor where the sugars are converted into ethanol and byproducts. If one could convert all the sugar placed in the fermentator, carbon dioxide and two moles of ethanol would be produced for each mol of C6 sugar used. In practice yeast cells and other end products including fusel oil are produced. Therefore only about 1.8 mol of ethanol is obtained per mol of C6 sugar with the balance going to yeast and fusel oil. The fractions of the products vary according to a number of conditions. One of the arguments for the use of continous fermentation processes is that the quantities of yeasts are reduced by up to 50 percent and more ethanol can be obtained. Kinderman et al. (1980) noted that both the capital cost and the operating costs for continuous fermentation process are less than for batch processes.

In examining the production of ethanol from biomass the major sugar crops used are sugarcane and sugar beets. Experiment work has been conducted on the use of sweet sorghum which can be grown in colder climates and areas with shorter growing seasons than sugar cane. Sugar beets serve only as a sugar source in North America and in Europe where the industry is heavily subsidized. Table 14 shows the distribution and consumption of sugar in 1977 throughout the world.

Sugarcane processing recovers about 90 percent of the sugar juice from the fibrous stalks leaving a residue called bagasse. The bagasse was originally burned to get rid of the waste and to produce some energy for the processing of the sugar. Bagasse has also been used as a source of fiber for paper or composite building board. The more sugar that is removed from the fibers the more the fibers are worth. A new separator called the Tilby Separator permits the separation of the sugarcane into four major components: (1) epidermal wax cells from the surface to the cane, (2) rind fiber, (3) pith, and (4) sugar juice. This separator has been estimated to have

Table 14. 1977 World Production and Consumption of Sugar.

	Production 10^6 T/yr	Cane Sugar Percent	Consumption 10 m^6 T/yr
Europe	30.48 n	2	31.74
North America	4.74	31	11.29
Central America	13.59	100	4.49
South America	12.72	89	9.07
Asia	18.46	88	20.98
Oceania	4.76	100	1.05

Ref: International Sugar Organization 1977.

similar investment and operating costings but has a higher extraction efficiency and allows for separation of the fractions (Kinderman et al. 1980).

In conventional sugar crystallation facilities molasses is a byproduct. The molasses is widely used as a cattle feed supplement, to produce rum or to produce industrial ethanol.

Grains have been used traditionally as a sugar source to produce ethanol. Corn (maize) or grain sorghum (milo) are the two major grains used for fermentation. In processing corn, two routes are used. The first is the whole kernel route in which the corn is cooked, subjected to hydrolysis and fermentation. The byproducts are distillers' dried grains. Wet milling is the second route, where addition byproducts such as corn oil, gluten meal, corn bran, germoil meal, and steepwater are byproducts (Kinderman et al. 1980).

Mid- and long-term research and development efforts have been directed toward the hydrolysis of agricultural and forest residues and new crops grown for conversion into ethanol. Non-food biomass usually has a low price and can be used to provide an assured supply of inexpensive raw materials to a conversion facility. A number of techniques have been developed which can be used to produce ethanol from glucose and other C6 sugars derived from lignocellulosic materials. The cost of most of these systems developed so far has been unattractive except for managed economies like that of the USSR (Bungay 1980). These conversion processes depend upon one of three different groups of hydrolyzing agents to release the sugars from the lignocellulose:

1. Dilute mineral acid processes using sulfuric acid.

2. Concentrated mineral acid processes using sulfuric or hydrochloric acid.

3. Enzymatic hydrolysis processes.

O'Neil (1978) and Bungay (1980) have presented very comprehensive reviews of the various hydrolyzing systems.

When one examines the chemical composition of a typical hardwood (see Table 15) one notes the sizeable amounts of hemicellulose, lignins and extractives which are available from the process for use in byproduct operations. Unfortunately the demand for most of the byproducts is limited.

Biomethanation

Biomethanation, which is often referred to as anaerobic digestion or biogas, is a familiar process because of its wide application in wastewater treatment for the stabilization of settleable

Table 15. Composition of Hardwood (Percent, Cry Basis).

Cellulose (all Hexosan)	48
Hemicellulose	25
Pentosans 20	
Hexosans 5	
Total Hexosans	53
Lignin	20.5
Extractives	6
Ash	0.5

solids and recycling of nutrients. The process produces a gaseous product of methane and carbon dioxide (a fuel gas). Normally the quantity of methane gas produced is larger than that of the carbon dioxide. The process works best on organic sustances with high moisture contents. The process is very slow at ambient temperatures in septic tanks and land fills but is accelerated at higher temperatures. Recent research and full scale facilities have led to an improvement of methane generation through the use of either microorganism accumulation, thermophilic micro-organisms, or through a better adaptation of the process to environmental conditions. Selection of appropriate environmental parameters can also lead to higher productivity of methane. The gas can be burned directly or it can be upgraded to a high joule value gas or synthetic natural gas (SNG) by the removal of the carbon dioxide.

With most feed stocks a humus like material which contains the nutrients is left. This material has been used as animal feed or soil conditioner. The liquid effluent also contains nutrients that can be recycled, used in crop irrigation or in algae-fish ponds where it will serve as a food supply.

There are three types of digester systems that have been used. The first is a small scale unit which is for the individual home or farm. This type of system uses human wastes, animal manures, crop residues or special grown crops as feeds. The second type of digesters is the upflow sludge blanket type system which can be used for soluble or low strength wastes. The third type of system is the large scale unit serving a community or an industry.

In examining small scale systems we can see several approaches that have been used to design the systems. The Chinese approach has been to use a sealed septic tank which will oeprate at ambient temperatures producing a fuel gas and a soil conditioner, and reduce the cost of the system. Gas storage is incorpated within the digester vessel.

The Indiana type digester has been designed on a empirical basis scaling down the large digesters which have been successfully used at sewage treatment plants. The design uses a concrete tank with a steel gas holder which makes the system very expensive and places the unit out of the reach of most individuals.

The third type of small scale system is a bag digester which operates on a plug flow basis. This type of unit has been widely used in Taiwan and a version of the bag digester has been developed for use in the colder climates of the United States on dairy farms.

The upflow anaerobic sludge blanket digesters allow the low strength wastes to flow in an upward direction through a dense blanket of anaerobic sludge. In the sludge layer the gas is produced and the solids are retained. The applications of this type of digester have been primarily for industries.

The large scale digester system is usually a conventional stirred tank reactor which is fed as high a solids concentration as can be pumped into the vessel. To optimize gas production the thermophilic temperature range of 50 to 65 degrees C has been used which allows for the use of small tank thus reducing some of the capital costs. Lower temperatures have been used but these require longer retention times and thus large vessels. The heat for the digesters is provided by using a portion of the gas produced and external heat exhangers can be used to recover the sensible heat of the effluent streams. To increase the digestibility of the residues, the feed can be pre-treated or post-treated prior to going to a secondary digester.

Thermochemical Conversion

Thermochemical conversion denotes technologies that use elevated temperatures to convert the fixed carbon of the biomass materials by (1) direct combustion to produce heat, (2) pyrolysis to produce gas, pyrolytic liquids, chemical and char, (3) gasification to produce low or intermediate joule value gas. (The intermediate joule value gas can be subjected to indirect liquefaction to produce ammonia, methanol, Fisher-Tropsch liquids or upgraded to synthetic natural gas.) And (4) liquefaction to produce heavy fuel oil or with upgrading, lighter boiling products used as distillates, light fuel oil or gasoline.

There are three major issues in addition to supply that must be considered. They are 1) Commercial availability of the systems, 2) Cost and 3) Efficiency of the process. Table 16 illustrates the commercial status of the thermochemical conversion processes and the expected time of commercialization of those processes which are under development. The relative efficiencies of the processes are shown in Table 17.

Table 16. Commercial Status of Selected Thermochemical Conversion Processes.

Process	Produce	Status	Commercial Availability
Direct Combustion	Heat for Cooking and Heating	Commercial	1982
	Steam	Commercial	1982
	Power	Commercial	1982
	Steam & Power (Cogeneration)	Commercial	1982
Gasification	Low Joule Value Gas-Raw	Commercial	1982
	Low Joule Value Gas-Clean	Commercial	1982
	Medium Joule Value Gas	Lab/Pilot	1988
	Fluidized Bed-SNG	Concept	1988
	Shaft Conversion	Demonstration	1988
	Methanol	Concept	1988
	Methanol Gasoline	Pilot	1988
	Polymerization to Gasoline	Laboratory	1990
Pyrolysis	Liquefaction to Oil	Commercial	1982
Liquefaction	Oil	Pilot	1990

Note: Date Commercial assumes that the process is available at a fixed price with guaranteed performance and that the costs are competitive with the competition.

Table 17. Conversion Efficiency of Thermochemical Processes.

Process	Efficiency
Gasification-Low Joule Value Gas-Raw	81
Gasification-Med. Joule Value Gas	76
Combustion-Steam	70
Gasification-Low Joule Value Gas-Clean	69
Combustion-Steam and Electricity	68
Liquefaction	63
Gasification-Synthetic Natural Gas	61
Gasification-Methanol	57
Gasification-Synthetic Gasoline	45
Liquefaction-Pyrolysis	29
Combustion-Electricity	20
Polymerization-Gasoline	19

Ref: Kinderman et al. 1980.

Direct Combustion

Today direct combustion of biomass provides energy for cooking and heating to the majority of the rural population of the world. The biomass source may be wood, crop residues and/or manure. These materials are gathered and used with essentially no financial cost to the user. The combustion systems used in these areas include open fires, simple cook stoves and Dutch ovens. Biomass provides process heat and/or electricity throughout the pulp and forest products industry (Howlett 1977). In addition, utilities throughout the world are producing electricity from wood fired power plants. For industrial purposes cyclone burners, suspension firing and fluidized bed combustion systems are used to burn biomass to produce heat, steam and or electricity.

Many of the wood burning systems now use wood chips to provide a uniform feedstock which can be handled by mechanical equipment and will result in improved biomass utilization. The moisture content of the biomass and particle size influence the type of direct combustion systems used.

Pyrolysis

Pyrolysis is a thermochemical process which operates in the absence of oxygen to produce approximately equal quantities of char, organic liquid (oil) and low Joule value gaseous fuel that must be used on site. For the process to work, either external heat must be supplied or one of the products must be used. Although the process has received a great deal of attention it is used only to produce charcoal.

Gasification

In many instances it has been considered more desirable to use a gaseous or liquid fuel rather than a solid fuel. Biomass, because of its bulk density and high moisture content can be transported only short distances. Direct combustion of biomass often results in low energy release, tar and smoke. Depending upon the process used either low, medium or high Joule value gas can be produced. The low Joule value gas must be used at the production site while the medium and high value gas can be transmitted over longer distances before being used. The gases can be burned in conventional oil and gas-fired units.

Gasification-Indirect Liquefaction to Produce Methanol

To produce methanol from biomass three steps are involved: steam gasification, shift of the product gas to syngas, and methanol synthesis. The first two steps can be combined in a single reaction.

In the first step the moist biomass under moderate temperature and pressures in a gasifier through an exothermic process produces methane, hydrogen, carbon monoxide, carbon dioxide and water. This gas is then heated in the presence of catalysts and shifted in composition to yield a ratio of hydrogen and carbon monoxide which can be used for methanol synthesis. The heat is recovered from the syngas, the carbon dioxide is stripped and methanol is produced in a catalyzed reactor. The residual gas can provide enough additional heat for the process through combustion.

Liquefaction

High temperature high pressure catalytic processes to convert biomass to oil are being developed. One of the liquefaction processes is the Lawrence Berkely Process (LBL). In this process the biomass is reduced to pumpable slurry through acid hydrolysis at elevated temperatures and pressures. The slurry and syngas, carbon monoxide and hydrogen, are added to a reactor where biomass is converted to an oil product. The off gas containing unreacted syngas is used as a fuel for the process.

CONCLUSIONS

Biomass will have its greatest value when used as food for humans. Other uses in descending order of value are food for animals, chemical feedstocks and the lowest value will be for energy. Under existing conditions the quantities of biomass required to satisfy the World's needs are the least for chemical feedstocks, with increasing quantities required for humans and animals and the largest usage is for energy. Through the use of integrated systems the needs for food, chemical feedstocks and fuels of many regions can be satisfied. For each site, region and country there will be alternatives which may solve specific needs. It is important that the policy and decision makers be aware of the alternative and their economic and social benefits.

REFERENCES

Alich, J., Schooley, F.A., Ernest, R.K., Hamilton, R., Louks, B.M., Miller, K.A., Veblen, T.C. and Witwen, J.G. 1977; 1978. An Evaluation of the Use of Agricultural Residues as an Energy Feedstock - A 10 Site Survey, Vol. 1&2, U.S. Dept. of Energy Report No. TID-27904/1&2, SRI International, Menlo Park, Calif.

Ashare, E., Hruby, T. and Nyhart, J.D. 1978. Cost Analysis of Aquatic Biomass Systems, U.S. Dept. of Energy Report No. HCP/ET-4000-78/1, Dynatech R/D Co., Cambridge, Mass., July 1978.

Bungay, H.R. 1981. Energy the Biomass Option, Wiley-Interscience, New York.
Business Week. 1981. Come Back for Grain Alcohol, Business Week, July 26, 1982.
Concise Oxford Dictionary of English. 1976. 6th Edition, Oxford.
Chemical Engineering News. 1982. pp. 18, June 14.
Cointreau, S.J. 1982. Environmental Management of Urban Wastes in Developing Countries, World Bank, Washington, D.C.
FAO. 1977. The State of Food and Agriculture, 1976, Rome.
FAO. 1980. Energy Cropping Versus Agricultural Production, presented 16th Regional Converence for Latin America, Doc. LARC/80/8, Havana, Cuba, June 1980.
Goldman, J.C., Ryther, J.H., Waaland, R. and Wilson, E.H. 1977. Topical Report on Sources and Systems for Aquatic Plant Biomass as an Energy Resource, U.S. Dept. of Energy Report No. DSJ-4000-77/1, Dynatech R/D Co., Cambridge, Mass., Oct. 1977.
Howlett, K., Gamache, A. 1977. Silviculture Biomass Farms, Volume VI, MITRE Corp., McLean, Virginia, May, 1977.
International Sugar Organization. 1977.
Janczak, J. 1980. Compendium of Simple Techniques for Agglomeration and/or Densifying Wood, Crop and Animal Residues, UN Conference on New and Renewable Sources of Energy, Tech. Panel on Fuelwood and Charcoal Paper, Dec. 1980.
Kalbermatten, J.M., Julius, D.S. and Gunnerson, C.G. 1980. Appropriate Technology for Water Supply and Sanitation-Technical and Economic Options, World Bank, Washington, D.C., Dec. 1980.
Kinderman, E.M., Yam, A.Y., Kohan, S.M., Jones, J.L., Bomberger, D., Capener, E.E., Chatterjee, A., Ravenscroft, P.H., Schooley, F.A., Semrau, K.T., Stallings, J.W. and Wilhelm, D.J. 1981. Technical and Economic Evaluations of Biomass Utilization Processes, U.S. Dept. of Energy Report No. ET/20605-T-4, SRI International, Menlo Park, Calif.
Kresovich, S., Wagner, D.A., Scantland, D.A. and Lawhon, W.T. 1981. Emergent Aquatic Plants: Biological & Economic Perspective. Submitted to Biomass Journal, Nov. 1981.
Lipinsky, E.S., McClure, T.A., Kresovich, S., Otis, J.L., Wagner, D.A., Trayser, D.A. and Applebaum, H.R. 1981. System Study of Vegetable Oils and Animal Fats for Use as Substitutes & Emergency Diesel Fuels. U.S. Dept. of Energy Report No. DOE/NBB-0006, Battelle Columbus Laboratory, Columbus, Ohio Oct. 1981.
Nemerow, N.L. 1971. Liquid Waste of Industry: Theory, Practice and Treatment. Addison-Wesley, Reading, Mass.
O'Neil, D., Colcord, A.R., Bery, M.R., Day, S.W., Roberts, R.S., El-Barbary, I.A., Havlieek, S.C., Anders, M.E. and Sondhi, D. 1979. Design, Fabrication and Operation of Biomass Fermentation Facility. First Quarterly Report, Georgia Institute of Tech., Atlanta, Ga., Jan.

Pimentel, D.L., MacReynolds, W.R., Hewes, M.T. and Rusk, S. 1974. Workshop on Research Methodology for Studies of Energy, Food, Man and Environment, Cornell University, Ithaca, N.Y.

Ryther, J.C. 1981. Wastewater Nutrient Recycling by Multispecies Aquaculture Systems, presented to World Mariculture Society, Venice, October, 1981.

Stout, B.A. 1979. Energy for World Agriculture, FAO, Rome.

Ward, R.F. 1981. Alcohols as Fuels, the Global Picture. Solar Energy Vol. 26, pp. 169-173.

United Nations. 1981. Report of the Technical Panel on Fuelwood and Charcoal, Report No. A/CONF/100/PC/34, New York, Feb.

MAXIMIZING THE USE OF NEW BIOMASS

WITH POPULATION LIMITS

>Harry Sobel
>
>Editor, Biosources Digest
>P. O. Box 5820
>Sherman Oaks, California 91403
>U.S.A.

INTRODUCTION

The attention of mankind has historically been directed to increasing security and survivorship. For most of mankind today, this is still the major goal. More recently in man's story, there has been a drive for improving the quality of life for those populations which have the resources to opt for this. Bare survivorship or high quality of life depend on the availability of nature's bounty and the development of needed technology to use this bounty.

The needs are, of course, for food, clothing, structural products, chemicals and energy. With adequate technology, boundless areas for growing food and an unlimited supply of resources, the ultimate goals should be easily achievable. However, for at least 75 years, it has been apparent that that which was thought to be boundless and unlimited was indeed limited -- that the drives for improving the quality of life resulted in enormous increases in population with increased need for these resources, thus endangering the very same quality of life by the consumption of irreplaceable resources and the development of enormous environmental pollution problems. These now limit the ultimate goals and indeed threaten mankind as a whole. The requirements for food, which demand an ever improving use of land, availability of water, recovery of eroded soils, etc., are now limiting the progress that mankind has sought.

The major proportion of mankind includes people with marginal levels of existence, who are further corroding crop lands, grazing lands, and consuming on-hand fuels. The enormity of the problem can be grasped with the understanding that, at the present time, the world population is doubling every 35 years and the supply of resources of all kinds, food particularly, cannot keep up with this demand.

THE RESOURCES

The resources of Earth relate to its origenic components: The elemental composition, which permitted the processing of life, the internal fiery heart, radioactive elements and the gravitational forces, as well as the radiation from the hydrogen fusion furnace, the sun. The particular orbital qualities of Earth about the sun and periodic rotational attributes played enormous roles in the evolution of life and life processes.

The energy which is used in processing events which occur on Earth are stored as (I) geothermal processes and solar energy generated processes, manifested by movements of air and water, direct heating and photovoltaic effects and as (II) chemically stored energy, generated almost entirely since life began, by photosynthetic bioorganic processes. The latter storage materials constitute biomass. The biomass materials are in the form of fossil products, modified plant materials stored as oil, tar, coal, peat and natural gas or newly generated photosynthetic materials. The possibility that origenic methane and more complex hydrocarbons has been considered, but it has been held to be inconceivable that large amounts of hydrocarbons could remain in the mantle. The theory that great natural gas resources lie deep in the Earth and are being outgassed suggest a very long term resource. This gas is abiogenic in origin, and if it is still there, there is still no real measure of its extent (Haggin 1982). Most observers, therefore, consider methane to be of biogenic origin.

The use of fossil biomass energy results in the rapid release of CO_2 and other pollutants stored slowly over the eons of time or in the case of new biomass, in the release of CO_2 and other products which are equilibrated with their removal during synthesis. It has been said of fossil fuels that man consumes in a year what nature took a million years to create.

THE PROBLEM

The problem posed by the above relates to maximizing the use of renewable non-pollutional aspects of Earth, while minimizing the use of pollution producing nonrenewable fossil energy and chemical sources as an enormous increase of population is occurring.

POPULATION

The most limiting factor in the resolution of the problem relates to population growth. The well known principle announced by T. R. Malthus in 1798 emphasized the disparity between the potential rates of expansion of population and food supply. He argued that mortality would intervene to curb overpopulation. His essay was written at the initial phase of the industrial revolution. The success of the revolution itself and ensuing confidence in the genius of technology, modern economics, and medical progress in solving all problems relegated this principle to an historical curiosity, although not all social and demographic historians ignored it. However, as we shall see, mankind is now facing a new version of the Malthusian dilemma.

The present world's population is approximately 4.3 billion. It is expected to reach 6.4 billion by the year 2000 A.D. and 10 billion by 2030 A.D. Another estimate by the International Institute for Applied Systems Analysis, suggests that the world's population will reach 9 billion by the middle of the next century. It is expected to reach 30 billion by the end of the 21st century. This last corresponds to the estimate by the U.S. National Academy of Sciences, as the maximum carrying capacity of the Earth (Barney 1980). It is difficult for the writer to accept this figure without an extreme restraint on the illdefined term "quality of life". Rocks and Runyon have postulated this limit to be 10 billion (Rocks and Runyon 1972).

An additional dilemma is now rapidly developing as a result of progress in an area which has been a wish of mankind throughout the ages - the extension of life expectancy. From a life expectancy of 26 years during the Roman period, approximately 1% reached 65 in Bismarck's time, while 4% achieved this in the U.S.A. by 1900 A. D. Now, 26 years have been added to life expectancy, and it has been estimated that 20% of all Americans will be over 65 by the year 2020 (Butler 1982). The implications are staggering. Today, life expectancy in the developing countries is 15 to 20 years less than it is in Europe and the U.S. This actually represents a rapid increase in life expectancy, which has taken place since World War II and is one of the main causes of population growth. This rapid growth seriously aggravates economic and social problems. Best current estimates suggest that the population of the third world will triple in the next century.

In the U.S.A., life expectancy in 1981 reached an all time high of 70.3 years for men and 77.9 years for women. Aside from fantastic claims of extension of life span, most reasonable estimates to result from the consequences of understanding the aging process and progress in geriatric health care, suggest that a large segment of the population will live into the mid-eighties, largely through the control of heart disease, stroke and cancer.

The activities of the international programs such as those sponsored by the World Health Organization are greatly reducing childhood mortality. The gap in life expectancy is being narrowed, as it should be, but ultimate worldwide life expectancy cannot even be guessed, if progress in life extension continues, as is expected.

Population growth must be kept to a minimum and, finally, zero population growth must be achieved, or a new Malthusian spectre will arise, first limiting quality of life to the level of third world countries today, and then once more increasing mortality.

The well-known paradox of the starving cats in the Roman Forum, who when fed, increase their breeding and thus more starving cats are produced, is apparent. In the U.S.A., there has recently been a similar example involving starving deer, which resulted from a natural program of breeding, overgrazing, and then starvation, which resulted in a requirement of partial extermination to reduce their number.

Mankind is less than 75 years away from the new Malthusian spectre, even less, if we consider a reasonable quality of life. Worldwide population growth must be reduced dramatically and voluntarily, or enormous increases in mortality from natural or man-made causes, is to be anticipated. Furthermore, stabilization of population must occur at a level of equilibrium with resource production and suitable environmental safety. Clearly, mankind must decide how many people can comfortably be accommodated by Earth's resources and establish an adequate quality of life.

ENERGY ANALYSIS

The composite of industrial resources, labor and goods has been equated through a common set of values under the general heading of economics. Energy resources was, until recently, not treated as a separate entity, and was assumed to be replaceable with adequate investment of labor and capital. However, with the growing importance of energy as a resource class which cannot be replaced, there has relatively recently developed the discipline of energy analysis as a supplement to economics (Anderson 1979). Energy analysis has been officially defined as "the study of the energy, free energy, or availability of any other thermodynamic quantity sequestered in the provision of goods or services" (IFIAS 1975). An example of the application of economics with energy analysis is provided by the calculation of an approximate conversion factor, the ratio of economic and energy fluxes. For the U.S. in 1973, $1.4 -- trillion circulated per year within the economy. Energy consumption during the same year was 3.5×10^{16} cal. The conversion factor is then 25,000 cal. per dollar. This allows a rough calculation of per capita energy consumption as a function of annual income. This ratio may have some usefulness in establishing a quality of life, but it is limited by the variable value of the unit of money. What is needed is a reference based on a biological value for minimal energy requirements for survivorship. It requires a ratio of energy expended in industrial activity to this energy unit.

QUALITY OF LIFE

Terms such as "quality of life" or of similar inference, such as living standard, gross national product, etc., cannot be adequately defined. They usually imply the comparison of actual or desired conditions with those required for bare subsistence, in needs for food, housing, clothing and warmth. Gross national product (GNP) is a rough and inadequate measure of social and economic welfare. One way of comparing quality of life between countries is to determine the number of hours of work required to purchase some basic object, like a loaf of bread.

Rocks and Runyon (1972) ingeniously used power expenditure to permit comparisons. They point out that the average human operates at 100 watts (2000 Cal/day). This represents metabolically derived energy from biochemical processes. The U.S.A. industrial power expenditure represents 10,000 watts per capita, or a hundred times the biochemical power of a person. It is 1500 watts per capita as a world average, and 500 watts per capita in India. While such calculations are useful, it is difficult to express certain of man's needs, such as living space, etc., on a wattage equivalent.

The question of the desired quality of life, as expressed in wattage, is arbitrary. What is desired? Is it to be set at 3000 watts per capita worldwide, or 20,000 watts per capita in the U.S.A.? Even if it is set between these values worldwide, the maximum number of persons that the world can nurture must be set at a lower value than the limits set today. With tripling of population and GNP, the size of energy systems required to carry the world population could increase up to tenfold. However, Columbo (Columbo 1982) found that the energy consumption/GNP in industrialized countries decreases precipitously after an increase resulting from growth of base industries and infra structures with society's demands concentrated on material goods. Later, there is stabilization and increased efficiency. However, there are other scenarios. Haefele and co-workers (Haefele 1982) consider a three-to-fivefold increase in world primary energy consumption over the next fifty years, while Lovins and coworkers (Lovins 1977) indicate that it will be possible to survive, even with an improved standard of living, 50 years hence on a little as half the energy consumed in the world today, notwithstanding doubling of the world population.

WORLD ENERGY NEEDS

As suggested above, the various needs for living (money, food, water, clothing, housing, etc.) while diverse, can be conjoined through energy or energy equivalents. Money will buy, at any time and place, a certain amount of workman's labor, or his product which he created with his own labor. So can an item of clothing or place to live be expressed as energy equivalents, whether as power, BTU or barrels of oil, with its equivalents in energy.

The energy used by the farmer, who sows and reaps by hand, is purely biochemically generated energy, which he obtains from what he eats. The excess of what he produces can be sold for other energy derived products. He can use it to buy a form of stored energy not generated by his own body, and increase the efficiency in what he can sow and reap. He can rent machinery, ultimately produced from stored energy, and greatly expand his efficiency, etc. So some expression for energy or its equivalents will be useful in relating need to productivity, and ultimately to compare economies and GNP. We deal in the unit of biochemical (photosynthetically derived) energy, plus energy ultimately of origenic or solar origin, the latter stored as fossil energy.

Consider an adult carrying out daily a somewhat above average basal activity, which is to be used as a unit of human intake and energy expenditure, say 2000 Cal. This is acquired in the form of food, and assumes a concomitant intake of required amounts of protein, minerals and vitamins. A person carrying out very heavy work will exceed 4000 Cal. a day and infants and children much less. The average unit population energy expenditure could be calculated from a knowledge of feeding habits and weight and age distribution. Since 2000 Cal. a day corresponds roughly to a rate of 100 watts, which is familiar as the rate of energy required to light an electric bulb, this is a good standard for comparison, In the U.S.A., with a population of 220 million, the daily biochemical expenditure is 2000 Cal. x 220 million (4.4×10^{11} Cal.). This is derived from food which contains photosynthetically derived energy. However, it must be emphasized that this basic biochemical unit does not include the cost in farming the food which requires planting, soil prepara- soil preparation, cultivation, fertilizer, pesticides, machinery, water delivery, processing, storage and delivery of the food itself. The delivery of water and the water shortage is rapidly threatening the world's food supply. It has been estimated that in 1973, in the U.S., 40% of the total U.S. energy consumption was used for growing, harvesting, transportation and preparation of food.

The national industrial usage of energy was 75 quadrillion BTU's (quads) a year in 1973 or 205×10^{12} BTU's per day, resulting in an industrial/individual biochemical ratio of 116. A BTU is approximately one-quarter of a calorie.

Today, worldwide, the industrial energy consumption is 20 quads per year or 657×10^{12} BTU's per day. With a population of 4.3 billion, the individual biochemical daily expenditure is 34.4×10^{12} BTU's, or an industrial/biochemical expenditure of 19. It is evident from the above that most of this must be given to the food economy.

NONRENEWABLE RESOURCES

The world resources of nonrenewable energy sources is thought to be as follows (Barney 1980):

petroleum	9,634 Quads BTU
natural gas	8,663 " "
solid fuel	120,852 " "
tar sands	(5,600-9,000) "
uranium	1,960 " "

The total resources add to 166,250 Quads. Thus with a world population of 12 billion, this would be consumed in 94 years, assuming that all of it can be made available and can be utilized, if the desired industrial/individual biochemical value is set at about 50. If the ability to extract these resources is lower than anticipated, if the world population becomes greater than 12 billion, this period could be greatly shortened.

The environmental consequences in deriving and consuming this energy from these sources may itself be cataclysmic, even without regard to a need to feed, clothe and house a world population above 12 billion. The consequences of the increase in atmospheric CO_2 resulting from converting fossil materials alone, could have profound effects on world climate and land mass (Revelle 1982). The release of nitrates, sulphates, etc. is amply threatening today. The consequences of radiopollution in the use of radioactive sources is amply being discussed as a major area of concern.

PHOTOSYNTHESIS

The Earth's surface receives annually approximately 5.5×10^{23} Cal., or about 100,000 Cal./Cm^2/year (Galston et al. 1980). About one third of this total is expended in evaporating water, leaving roughly 67,000 Cal./Cm^2/year for photosynthesis and other purposes. Each year, photosynthesis by green plants converts 200 billion tons of carbon from atmospheric CO_2 to sugar. Photosynthesis is relatively inefficient in utilizing the sun's radiant energy. Annual photosynthesis for the entire Earth averages only about 33 Cal. Cm^2, which means that photosynthesis converts only 1/2000 of the available energy. This figure presents a somewhat inaccurate picture of photosynthetic efficiency, however, for much solar radiation reaches Earth at locations devoid of vegetation. If we calculate only the radiation absorbed by green plants, the overall efficiency of photosynthesis (radiant energy stored/radiant energy absorbed) rises to several percent.

It is generally thought that there are 32 billion acres of total world ice free land areas, half of which are given to mountains and deserts and are unavailable for agriculture. 8 billion acres are potentially arable and one half of this is now being farmed. 8 billion acres are potentially grazable and one half is now being grazed.

NEW BIOMASS

10 to 30 tons of new biomass per acre per year can be harvested. One pound of dry plant tissue, when harvested and burned, can produce as much as 7,500 BTU's of heat (Fowler and Fowler 1977), a little more than half of the heat available from a similar amount of coal. With some improvements, a ton of dry biomass, can proceed to yield 1.25 barrels of oil, 1,200 cu. ft. of medium BTU gas, and 750 lbs. of solid residue, roughly equivalent to coal in heat value.

The U.S.A. produces 5 billion tons of new biomass a year. At 7,500 BTU's/lb. of dry material, this results in more than the 75 Quad, BTU's consumed a year. Worldwide, 150 billion tons are produced a year. If converted to energy, this would result in an industrial/biological value of 70 for 12 billion people. It has been estimated that arable land will increase by only 4% by the year 2000 A.D. (Barney 1980), but increased food output will occur through higher yields, involving great improvement in agricultural productivity, improvements in preparation, storage and processing of food. Farm productivity has increased three fold in the last 40 years. Current developments in biomass technology and genetic engineering (Chemrawn II 1982) will result in vastly greater plant biomass yields and better energy crops. It has been stated that most of the products derived from petroleum can be produced directly from biomass. Future possibilities for new biomass have been discussed at this conference and elsewhere. At the present time, it seems that water shortage may be the limiting factor in meeting the future growth and needs.

In spite of the exuberant discourse on the food and energy role of biomass in meeting the needs of an approaching finite world population, it must not be assumed that the writer believes that biomass can or should supply all of the needs for food, energy and chemicals. Rather, the quantity and usage of biomass must be maximized and interwoven with other renewable nonpolluting systems, which, except for biomass, are sources only of energy. We must maximize these and minimize the use of fossil sources. The latter, such as remains, may be needed to build the machinery and technology for the former. The technology for using new biomass and other renewable energy sources must be developed on a regional basis and related to regional availability.

NEW BIOMASS/SOLAR TECHNOLOGY

It must be emphasized that the cost of transportation consumes a considerable proportion of the content of substances being transported. Solids must be moved by trucks, liquids by trucks and pipe lines, and gas, by pipe lines. Furthermore, such channels leave control with centralized industry and not only consume a large proportion of the material delivered, but also irreplaceable minerals and metals. In a country with a large area, centralization becomes formidable and limits the future development to the interest of those who control centralized distribution. Furthermore, it has been pointed out that centralized distribution points are vulnerable against military attack (Holden 1981). Finally, the expectation that coal technology will replace petroleum in the U.S.A. has resulted in momentous disappointment. Coal chemistrty is not a variation of petroleum chemistry. Coal is not oil. New biomass can replace oil and coal, is cleaner and may be grown with local technology and utilized with local technology. It has in fact been suggested that biomass is suitable for deriving coal and oil based chemicals, but may be too costly as a source of energy (Lipinsky 1981). The energy needs may be met by renewable nonpolluting origenic energy sources.

An example of a site specific, low pollution program is under development in Hawaii (Shupe 1982), which is now tied in with the national energy delivery program consuming 37×10^6 barrels of oil a year (211×10^{12} BTU's) which must be delivered over long distances. Since Hawaii is geologically of recent origin, it does not have its own fossil sustances. This is used for electricity, heat and transportation. Hawaii will become self-sufficient with the use of its locally generated sources of energy and new biomass.

1. Direct solar energy will be utilized through collectors, solar ponds and photovoltaic technology.
2. Ocean thermal energy conversion.
3. Wind energy.
4. Geothermal energy.
5. Hydroelectric power.
6. Municipal waste residues.
7. Biomass from bagasse, molasses, forest products, tree farming, sugar cane, pineapple residues and algae.

We return to the theme that renewable biomass must be expanded and maximized to supply food and other of its benefits to a world population whose number is being rapidly delimited, while the use of fossil materials is minimized and perhaps preserved for specific purposes. The use of renewable biomass should be interwoven with the use of renewable nonpolluting origenic sources of energy.

CONCLUSIONS

1. The population explosion which has resulted from man's humanitarianism and the ingenuity of his mind, threatens his very existence. A new Malthusian spectre now looms.

2. Based on the present scenario, a population limit must be set which is bounded by the ability to produce sufficient food, land and living space, energy availability, agricultural and potable water at relatively low costs, and minimum ecological pollution. This problem must be completely resolved in less than 75 years, or natural forces or man himself, using nonhumanitarian means, will set the limit.

3. With present usage, nonrenewable energy resources will last less than 100 years. The use of these resources is leading to ecological disaster.

4. Mankind must include the concept of a "quality of life" which goes beyond survival. A crude estimate of "quality of life" may be measured by the expression:

daily industrial energy expenditure/daily individual biochemical expenditure (2,000 Cal.).

This value is now roughly 116 in the U.S.A., and is only approximately 19 worldwide. A reasonable value of 50 worldwide will set a population limit of approximately 12 billion persons. This population will be reached in less than 60 years at the present rate of population growth.

5. New biomass alone, if expanded by new technology, can generate the amount needed to produce a quality of life of 50 units.

6. Fossil energy sources should be reserved for developing regional technologies directed toward self-sufficiency.

7. In order to maximize the use of new biomass for food, chemicals, and energy and to develop alternate sources of energy, using renewable nonpolluting resources, it will be necessary to develop new technologies. Such an effort must be international and be initiated at once.

8. Photosynthetic and origenic processes can produce regional self-sufficiency in many areas of the world. Control and delivery of energy over large distances is wasteful and vulnerable to destructive means.

REFERENCES

Anderson, R. E. 1979. Biological Paths to Energy Self Reliance. A Guide to Biological Solar Energy Conversion. Van Nostrand-Reinhold Co., N.Y. 367 pp.
Barney, G. O. Study Director. 1980. The Global 2000 Report to the President: Entering the Twenty-first Century. Three volumes. Govt. Printing Office, Washington, D. C. Reprinted by Penquin Books, Great Britain, 1982.
Butler, R. N. 1982. Quoted in Science 217: 616.
Chemrawn II. 1982. Alternate Energy Futures: The Case for Electricity. Science 217: 705-709.
Fowler, K. and Fowler, J. M. 1977. Factsheet. National Science Teachers Association. I. Fuels from Plant. Bioconversion.
Galston, A. W., Davies, P. J. and Satter, R. L. 1980. The Life of the Green Plant. Third Edition. Prentice-Hall, Inc., N.J. 464 pp.
Haefele, W. 1982. World Energy Production and Productivity. In: Proceedings of International Energy Symposium I, Clinard, L. A., English, R. A. and Bohm, R. A., Eds. Ballinger, Cambridge, Mass.
Haggin, J. 1982. Energy R & D: New strategies begin to emerge. Chem. Eng. News, June 21: 7-14.
Holden, C. 1981. Energy: Security and War. Science 211: 683.
IFIAS. 1975. Energy Analysis and Economics. A Workshop Report. Solna, Sweden. International Federation of Institutes for Advanced Study.
Lipinsky, E. S. 1981. Chemicals from Biomass: Petrochemical substitution options. Science 212: 1465-1471.
Lovins, A. B. 1977. Soft Energy Paths. Harper and Row, New York. 240 pp.
Revelle, R. 1982. Carbon Dioxide and World Climate. Sci. Amer. 247: 35-43.
Rocks, L. and Runyon, R. P. 1972. The Energy Crisis. Crown Publishers, Inc., New York. 189 pp.
Shupe, J. 1982. Energy Self-sufficiency for Hawaii. Science 216: 1193-1199.

FOREST INVENTORIES AS THE BASIS FOR A CONTINUOUS

MONITORING OF FOREST BIOMASS RESOURCES

T. Cunia

School of Forestry
State University of New York
College of Environmental Science and Forestry
Syracuse, New York 13210 - U.S.A.

INTRODUCTION

The forest biomass constitutes an important part of the total biomass resources. According to Milescu and Axente (1969), it is estimated that approximately 4 out of the 13 billion hectares of land on earth (excluding the arctic and antarctic regions) is occupied by forest; that is, about 30 percent of the land is forest land. This is an average percent applicable to the entire earth. If a breakdown by large geographical areas is desired, it is estimated that the percent of forest land is about 20 for Asia (excluding Siberia), 25 for Africa, 30 for Europe (excluding Russia) and Oceania (including Australia), 34 for USSR (including Siberia) and about 42 percent for the North and South America combined.

There is no reliable data on green or oven-dry weight of the total forest biomass. To get an idea, however, of the relative importance of the biomass by large geographical areas, we can use the estimates given by Milescu and Axente (1969) that out of a total of some 240 billion cubic meters of merchantable wood volume, about 12 are to be found in Europe (excluding Russia), 80 in USSR (including Siberia), 17 in Asia (excluding Siberia), 4 in Africa, 4 in Oceania (including Australia) and about 120 billions in America (North and South).

Most of the temperate countries, and this is particularly true for the countries from Europe and North America, have a long tradition in forest management, as their forests have been under intensive management regimes for many years. The various land uses have reached relatively stable conditions and there seems to be little changes occurring in the state or equilibrium of the various ecosystems of the temperate lands.

On the other hand, it appears that the tropical regions go through a stage, long passed in most temperate countries, where land uses change relatively fast. The greater needs for agricultural products, and more importantly, the present indiscriminate practices of harvesting may lead to changes in land uses from forest to non-forest or from forest to waste or desert land. According to Lanly (1982), the clearing of closed and open tree formations in the tropical areas involves, on the average, some 11 million hectares per year. As about 1 million hectares of these lands are reforested by planting every year, it appears that there is a net loss of some 10 million hectares of forest land every year. In view of this rapid depletion of forest biomass resources, there seems to be a need for a continuous system of monitoring the world forests.

This need is also important from another point of view. The international trade in forest resources and products made therefrom has increased tremendously since the end of World War II. Often, the wood based industry of one country must rely on wood supply from another country. For better decision making, the management of the wood based industry must be able to forecast the future availability of wood supply at various points in time, its quantity by quality classes and its geographical distribution. The present wood suppliers must be considered together wih the potential alternative suppliers. When the inventory of the forest resources is adequate, the management of the wood based industry is in a good position to make long range plans. But when the inventory is not adequate, the forecast would generally be poor. And the situation is further complicated by the fact that the control of the forest resources is in the hands of people other than those using these resources.

The continuous monitoring of the world forest resources can be viewed as having two main aspects. There is first the need for monitoring the changes in land classes from forest to non-forest uses and then there is the need for knowing what happens in these forests. If the main concern is the need of knowing the trends of forest depletion it may not be necessary to have a complex system of forest inventory. To estimate changes in the land shifts from forest to non-forest uses, it is sufficient to record whether randomly selected permanent sample units fall within or outside forested areas over successive points in time. This can be done very efficiently by remote sensing techniques. With the present technology, the forest area of the entire globe can be covered with high resolution by Landsat type, multispectral scanners at relatively short time intervals with a relatively low cost. This would provide a permanent record of the forest conditions over time, which, from a conceptual point of view at least, could be used to continuously estimate changes in forest conditions.

Real world problems are seldom that simple since estimation of the forest depletion is only a part of the overall picture of monitoring the forest resources broadly defined. It is generally recognized that in addition to the traditional merchantable wood resources, the unmerchantable biomass and other forest resources such as water, air and soil polution, wildlife, fish and insects, etc. are often of great importance. Changes in the basic ecology of the forest environment may greatly affect the quality of life for generations to come. We live in a small world, and ecological changes in one part of the world may greatly affect the well-being of people living in different parts of the world. In this sense, it seems advisable to monitor the entire environment of the forest land, not only the extent by which the forest land shifts from forest to non-forest uses or desert. What to actually measure in detail is rather difficult to summarize, and more will be said later.

It can be generally stated that to better manage and monitor forest biomass resources over space and time, policy makers and forest land managers need, among many other things, a continuous assessment of the conditions of the biomass resources and the land on which they grow. They need information about the <u>quantity</u> and <u>quality</u> of the resource, the <u>current values</u> (at given points in time) as well as the <u>rates of change</u> (with time) in these values. They also need information about the <u>distribution</u> of the biomass over geographical areas and its <u>composition</u> with respect to species, age, site quality, possible utilization, etc.

It is impossible to have <u>perfect</u> knowledge about biomass resources; the forest is too large and the trees are too numerous for that. Furthermore, the measurement of the tree biomass, in particular that of the oven-dry weight, requires destructive sampling. Finally, even if the knowledge about past performance of the forest is perfect, the future performance is a random variable whose value cannot be exactly predicted. The best one can hope for is to estimate its <u>expected</u> value and its probable error. There are many biological, <u>social</u> and economic factors that come into play and affect the delicate interrelationship between forest evolution and its environment. And the impact of accidents (such as fires, blowdowns, and insect or disease epidemics) on growth components (such as mortality, growth on surviving trees or biomass utilization) is extremely difficult to assess with any degree of accuracy.

Therefore, to obtain some of the information needed for the control of the future development of the forest, one must rely on forest inventory by sampling. It is by sampling rather than complete enumeration and measurement that one obtains estimates of the present conditions, rates of change in these conditions, as well as the factors affecting these changes. It is these estimates that provide most of the basis for the mathematical models used in making predictions about the future evolution of the forest as affected by the inherent forces of nature and the past decisions of the management.

It is customary to define forest inventory as a systematic procedure used to: 1) collect mensurational data on forest biomass and the land on which it grows; (2) process and analyse these mensurational data, and 3) present for management use estimates (with their precision) of quantity of biomass by quality class as distributed over geographical areas. In this sense, the forest inventory is defined as both, the methodology used in the derivation of the biomass estimates and the estimates themselves, the "inventory".

It is important to realize that in forest management, the biomass inventory is a relatively new concept referring to the weight, usually the oven-dry weight, of the biomass components of the forest. Because of the rather limited use in the past of estimates other than those related to the merchantable tree bole, no inventory of the forest was normally contemplated for any biomass component other than merchantable volume of the main tree bole. It is only recently that forest managers showed any interest in the biomass of additional tree components, such as biomass of the leaves, crowns or roots. This is mainly due to (1) the increasing needs for raw material of the wood-using industry and (2) the new perception of the biomass of the entire tree as a possible new source of energy. To my knowledge, the first biomass inventory, an isolated event, was made in Maine, USA, only a few years ago. As reported by Young, Hoar and Tryon (1976), the inventory included that of the biomass (above and below ground) of all the woody shrubs and trees taller than 15 centimeters. The main objective of this inventory was to show that the biomass inventory is a natural extension of the usual merchantable volume inventory which can be made with little additional costs.

Although the right time for the complete biomass inventory of the forest is yet to come, we are living in a transitional period where we have to prepare ourselves for the time when we would be required to take such an inventory. Because (1) the merchantable tree bole is the most important component of the total forest biomass, (2) the volume is a relative measure of the biomass, and (3) we have a long experience with the methodology of taking the inventory of the merchantable wood volume, we can refer back to that experience when we will be faced with the problem of devising future systems of forest biomass inventory. It is the main purpose of the present paper to review and discuss the basic concepts and methodology of sampling as presently applied in volume inventory, viewed from the point of view of their possible application to future biomass inventories.

BASIC TYPES OF FOREST INVENTORIES

The inventory is not an end in itself; the availability of good inventory data is essential to good forest management. The decision making function of management at all its levels of activity requires quantitative information about biomass resources. It is hard to conceive good decision making without good informational data.

At one extreme, the top management requires information for long-term planning about (1) the current state of the forest resources (quantity of biomass by quality classes) and (2) the rates by which these resources change. The knowledge of the geographical distribution of the biomass is only needed in very general terms and over large area units. How much there is and how it changes with time is much more important to know for long-term planning than where the resource is presently located. At the other extreme, the local, low-level management requires information about (1) the current state of the resources and (2) the exact location of the biomass within small geographical areas. As the inventory data are ordinarily used for short-term planning (as, for example, completing natural regeneration by additional planting or constructing forest roads for harvesting operations), the precise knowledge of the rate of forest growth is not critical. And in between these two extreme management levels, there are many intermediate levels with their own specific informational needs that should be satisfied by a variety of estimates derived by forest inventory.

Because the type of inventory used in any particular situation is a direct consequence of the informational needs of the decision level of forest management, it may be useful to classify the inventory systems into three, rather broad classes. As outlined by Cunia (1978), the three classes can be defined as follows:

(1) Operational Inventory is an intensive, in-place inventory primarily designed to provide estimates of the current values of the biomass resources as growing on specific, relatively small forest areas that may be as small as a few hectares or as large as several hundred hectares in size. While the estimates of the forest growth may sometimes be of some use, they are seldom of any primary importance. Ordinarily of the one-shot type, the operational inventory provides local management with informational data that are ordinarily used for specific purposes and for short-term planning.

For example, the timber sales inventory is of this type. It is a common practice to sell individual trees for sawlogs or veneer production to the highest bidder, with the buyer being responsible for their removal from the forest. Even though they may be sold in groups, the trees are ordinarily measured and marked for sale in the field on an individual basis. The main objectives òf this type of inventory are to measure the size and quality of the trees marked for sale. Together with some assessment of the logging difficulties and marketing conditions, these measurements are the basis for the determination of the total sale value.

The logging inventory is also of this type. One of the main activities of the low and mid-level forest management is that of planning for harvesting operations and the subsequent execution of these plans. Because the harvesting is to take place in the very near future, knowledge of forest growth or mortality is of little use. What is needed most is knowledge about the quantity and quality of merchantable volume as is currently distributed over relatively small areas. This knowledge is used for better decision making; what logging equipment and system provides maximum efficiency in a given forest land, where to locate the logging camps and the network of logging roads so as to minimize logging costs, how long the harvesting will take or how much one should pay the woodcutters, since the harvesting time and pay scale of woodcutters are both functions of logging conditions, etc.

Operational inventories are also the basis for estimating the value of forest lands when contemplating their sale or acquisition, for planning silvicultural treatments such as thinning or fertilization, for assessing the damage caused by insect or disease epidemics, fire or windstorm, etc.

(2) <u>Management</u> <u>Inventory</u> is an extensive inventory, primarily designed to insure a continuous flow of information about the general conditions of the biomass resources; the current values and the rates of change in these values due to growth, mortality or harvesting, the approximate location of the mature stands that may be ready for harvesting in the near future, etc. It provides estimates that apply to large forest areas such as entire forest ownerships or large subdivisions of these ownerships, ordinarily known as management units. Its main results are expressed as general statements about species composition, tree diameter distribution, average site quality, past trends in forest evolution, etc. It cannot provide accurate estimates applicable to small forest areas such as forest stands, logging areas, etc.; these estimates are derived when needed by the one-shot type operational inventory.

The management inventory data are primarily used for medium and long-term planning by the top and mid-level management. In particular they are used in the calculation and scheduling of the annual cuts, in the planning for the increase in the overall forest production through either stand improvement or shifting to different species composition or merchantable tree size, in the making of stand projections into the future as functions of the various possible courses of action by the management, in the identification of areas of applied research that are promising in terms of future financial returns, etc. The inventory will provide data justifying answers to questions of the following type. How much of the growing stock can be harvested every year to insure a steady supply of biomass? What types of trees or stands should the harvesting consist of? What can one do to increase the forest production to the maximum potential of the land? Is it worth fertilizing the entire forest area on a continuous basis, or is it better to concentrate stand improvement operations to the best sites only?

In short, the inventory data are being used in the preparation of a general management working plan over a medium and long-range time horizon. In addition, the inventory system must also provide the management with data for a continuous monitoring of the forest biomass resources. It is important to know whether past stand projections were accurately made, whether the biomass supply remains stable or changes over time, whether the consequence of the management decisions are as predicted, etc.

(3) <u>National forest inventory</u> is an extensive inventory which covers very large forest areas, usually the forest area of an entire country or that of a suitably defined geographical, economical or political subdivision or region of a given country. The primary objectives and needs for forest data are generally similar to those of the management inventory since estimates of both current values and rates of change of biomass resources are normally required.

The basic differences between management and national inventories can be traced back to the fact that the forest resources of a country, both public and private, are ordinarily viewed as part of the natural resources belonging to the entire nation. The management inventory covers only a forest ownership (or some suitable subdivision) and its main purpose is to provide management with specific data to be used in the selection of a specific course of action that seems optimum in terms of the main objectives of the forest landowner. The effect of this course of action on neighboring forest lands are of little, if any concern. On the other hand, the national forest inventory concerns itself with the forest land of an entire country (or some suitable subdivision) belonging to a variety of ownerships. The biomass resources are ordinarily combined for all land ownerships together, even though the land ownership itself may be one characteristic that may ordinarily be measured. In addition, the inventory data are much more general in nature and they are ordinarily used to solve general problems of national concerns. National courses of action are sought that seem optimum in terms of the well being of the nation taken as a whole.

The data from National Forest Inventory are used for better decision making at the highest possible level, ordinarily federal or state governments. They are specifically used to (1) define a national forest policy, (2) express this policy as a set of national programs and laws, and (3) create the necessary organizational structure to carry out and execute these programs. The data may also be used to monitor the evolution of the forest resources at the national level; to identify areas of research for the better management of the national forest lands; to protect the forest resources by controlling the damage due to fire, windstorm or insect or disease epidemics; to locate areas that are not utilized to their full potential, and thus, plan for their reforestation, stand improvement or shift from forest to agricultural uses, etc. In short, the data are being used at the national level to (1) make decisions about forest management, (2) carry out these decisions, and (3) monitor forest evolution by verifying whether the consequences of these decisions are as predicted.

Finally, the data from national inventories may also be used in the private sector in the deployment of new forest industry or the management of the existing one. For example, in the developing countries with important biomass resources, a national forest inventory of a very low intensity (of the type usually known as a reconnaissance inventory) may be the first step in the long process of developing a national forest industry. To establish, and later maintain a sound forest industry, one must be as certain of the present and future availability of the basic wood supply, its quality and quantity, its accessibility and logging conditions, the rate of growth and mortality, etc., as he is of the marketing conditions for the finished products.

To the above three basic types, one may add a last type, that of the global forest inventory. At the national level, the objectives of the inventory are those of monitoring the forest resources of a given nation with the implicit assumption that the informational data provided by the inventory are to be used to determine, and then take those courses of action that seem best for the nation as a whole. In a similar way, the global forest inventory is designed to monitor the evolution of the global resources. It can determine what happens to these resources and make the world community aware of any developing trends in the forest evolution that may endanger the existing delicate balance among the many physical and biological components of the ecosystems of the world.

Contrary to national forest inventories which assume the existence of national bodies that can take courses of action benefiting nations as a whole, no international body exists that can make a nation change its policy, when this policy is viewed as detrimental. All one can hope for is to have an international body which, on the basis of the results from some global inventory system can determine and recommend for adoption good courses of action. It cannot force the adoption of these courses of action by any nation not willing to do it. It can only exert international pressure by the world community on nations to change those policies that are viewed as detrimental to a sound management of the forest resources, whenever detrimental trends are shown to exist.

There is no doubt that we have the technical capability of devising a sound and efficient global system of taking the continuous inventory of the forest biomass resources on the global level. The current national forest inventory systems can serve as basic models that can be easily adapted to the global level. What may be lacking, however, is (1) a sufficiently strong perception of the fact that such a system is critically needed, and (2) a sufficiently strong desire to commit the necessary human and financial resources to its establishment. Because a statistically sound system would ordinarily require ground sample units randomly distributed among a variety of countries governed by a variety of political regimes, the political problems associated with the establishment and the maintenance of a global monitoring system may become critical. If such a system is to be established, the political and financial problems must be solved first; the technical problems are easy to handle.

Lacking a sound, statistically designed system for monitoring forest resources, one can use another approach. It can sum up the forest information from the national inventories of various countries. Although far from satisfactory (many countries have no sound national systems to start with, or if they have, the specifications will vary from country to country) this approach may be the only approach presently available.

As an example of an elementary global inventory system consider the FAO and UNEP project of making an assessment of the present and current development of the tropical forest resources. Done within the framework of the Global Environmental Monitoring System (GEMS) the project took place between 1978 and 1981. The findings of the survey are presented in four technical reports, the first three dealing with the tropical areas of America, Africa and Asia respectively, and the fourth report summarizing the findings of the previous three. For more details, the interested reader is referred to the summary report by Lanly (1982) and the associated three detailed reports.

It may be interesting to note that there were two main factors that contributed to this undertaking. There was first the realization that unwise exploitation of the tropical forests has led to the depletion and degradation of many tropical forest lands. And then, there was the perception that the global forest biomass resources represent part of the resources of the mankind as a whole. Major changes in the forest biomass in one part of the globe are viewed as affecting the biological, economical and social environment of many other parts of the globe. What is being done in one country may have a marked effect on the well being of the populations of other countries.

We have briefly discussed the basic classes of forest inventory systems. Without any doubt the informed reader may know of inventories which may be quite different from the typical ones described above. Some systems may present features which are common to more than one type, and features which are so different that additional classes may be needed to include them. Our aim, however, was not to define a foolproof classification system; rather to give an idea of some basic types.

GENERAL STRUCTURE OF THE SAMPLING DESIGN FOR FOREST INVENTORY

The general sampling design of a forest biomass inventory consists of two major phases. In the first phase, a relatively large sample of trees, hereby called the <u>main sample of trees</u>, is selected by some random procedure. All, or a subsample of trees are measured for a variety of (1) tree characteristics such as species, diameter at breast height, total or merchantable height, quality class, possible utilization, etc., (2) surrounding stand characteristics such as species composition, age class, stand history, etc., and (3) characteristics of the ground on which the trees grow, as for example, slope, aspect, soil depth, etc. The trees of this phase are not measured for their biomass. In the second phase, a relatively small sample of trees is selected by some random procedure and the trees are measured for their biomass (volume or weight of various tree components, such as merchantable stem with or without bark, crown, leaves, stump-root system, etc.) in addition to a variety of tree, stand and ground characteristics of the type measured on the sample trees of the first phase.

The data from the first phase sample are used to derive estimates of the tree frequency by various characteristics (normally by species and diameter) within geographical areas. The trees of the second phase are used to derive estimates of the relationships between biomass components and tree characteristics. These relationships are usually expressed as regression functions or biomass tables giving estimates of some kind of average tree biomass by tree, and possibly stand characteristics such as species, diameter, height, site quality, age class, etc. The estimates of the first and second phase are then appropriately combined to yield estimates of average biomass per unit area by species, tree quality, tree size, possible tree utilization, etc.

To illustrate these concepts, let us describe a basic design applied to a forest area containing a pure stand of a single species. The first phase sample consists of a set of n fixed area sample plots selected by a random (or systematic) procedure, with all of the trees contained in these plots measured for their diameter d. Let d_{ij} denote the diameter of the j-th tree in the i-th plot, $i = 1,2$ --, n and $j= 1,2$ ---, m_i, where m_i represents the number of trees for the plot i. The second phase consists of a random sample of trees, selected independently of the first phase sample. These trees are measured for their diameter d and some biomass component, say merchantable volume y, and these data are used to determine, usually but not necessarily by the least squares method, a regression function of y on d of the form

$$y = b_1 + b_2 d + b_3 d^2$$

This regression function can be used, if so desired, to construct a <u>one-way tree biomass table</u> giving the estimate of the average value of y for any given diameter d. Symbolically this table can be expressed as

$$y = b_1 + b_2 d + b_3 d^2 \text{ for } d = 1, 2, \text{---}, d_M$$

where d denotes a diameter class (to the nearest inch or centimeter) and d_M represents some maximum value of d.

To combine the data from the two phases, the regression function (or the biomass table) is first applied to each individual tree diameter d_{ij} to determine the corresponding volume y_{ij}. The various values y_{ij} are then added over the m_i trees of a given plot i to find v_i, the estimate of the biomass of plot i. We continue with the calculation of the sample average biomass per plot $\bar{v}$ by summing up the values v_i over all plots and then dividing by the total number n of sample plots. Finally, the estimates v_a and v_t of the average biomass per unit area (acre or hectare) and the volume of the entire forest area respectively, are obtained by multiplication of $\bar{v}$ with appropriately selected converting factors.

It can be shown that it is not necessary to calculate the individual tree volumes y_{ij}, since $\vec{v}$ can be determined by the formula

$$\vec{v} = b_1 \bar{z}_1 + b_2 \bar{z}_2 + b_3 \bar{z}_3$$

where $\bar{z}_1$ = average "number of trees" per plot,

$\bar{z}_2$ = average "sum of tree diameters" per plot, and

$\bar{z}_3$ = average "sum of squared tree diameters" per plot

The procedure for the calculation of $\bar{z}_1$, $\bar{z}_2$ and $\bar{z}_3$ is similar to that used in the calculation of $\bar{v}$, where v_i is replaced respectively by

$z_{1i} = m_i$ = number of trees in plot i

z_{2i} = sum of diameter values d_{ij} in plot i, and

z_{3i} = sum of squared diameter values d_{ij} in plot i.

The last expression of $\bar{v}$ is extremely useful when one wishes to estimate the standard error of $\bar{v}$, and include in this estimation the statistical errors of the samples of the first and second phase.

When previously constructed regression functions or <u>standard, one-way biomass tables</u> are available, and it is thought that they are applicable to the forest area being currently inventoried, the second phase sample is no longer necessary. In this case one should note that it is implicitly assumed that the sample trees from which the standard biomass tables were constructed are similar, on the average, to the trees of the forest area to which the tables are currently applied. This is a big assumption to make, since it is well known that the average tree biomass by diameter (and species) may vary considerably from one to the next forest area.

It is also known that the average tree biomass by diameter and height (and species) varies much less. Then, the above procedure can be modified in the sense that (1) it is assumed the ready availability of a <u>two-way standard tree biomass table</u> by diameter and height (and species); these tables having been constructed from the data of a relatively small sample of trees, previously selected by some random procedure so as to represent a wide range of forest conditions, tree diameters and heights, (2) a subsample of the trees of the first (main) phase sample are measured for their tree height in addition to their diameter (and species); these trees are used to determine a relationship between tree diameter and height (by species) and (3) the standard two-way biomass table and the diameter-height relation are combined to yield a <u>one-way local biomass table</u> by diameter (and species) alone that is applicable to the forest area being currently inventoried.

There are many ways by which the biomass tables of (1) above can be combined with the data of the subsample of trees of (2) above, so as to construct the local one-way biomass tables. One procedure widely used in America is to (1) take the tree measurements of the subsample of trees (of the first phase sample) and estimate the regression function of the tree height on tree diameter (by species), usually but not necessarily by the least square method and, (2) use the average height estimates by diameter (and species) as determined by this regression function, in the two-way biomass tables by diameter and height (and species) to derive the one-way biomass table by diameter (and species) alone. For example, if in the forest area being currently inventoried, the estimate of the average height for a given 10-inch diameter tree is 57 feet, and if the average biomass of a 10-inch diameter and 57-foot height tree is estimated by the two-way biomass table as 23.7 cubic feet, then, the estimate of the average biomass of a 10-inch diameter tree as given by the newly constructed one-way local volume table will be 23.7 cubic feet.

A second procedure, also currently used in America, is to (1) use the two-way biomass table to estimate the biomass y of every tree of the subsample of trees of the first phase (already measured for diameter and height) and (2) apply the least squares method to the trees of the subsample to calculate the regression of the estimated biomass y on diameter d. It is then this regression function that is used to construct the one-way local biomass tables. Many other procedures exist, mostly in Europe, for the construction of local one-way biomass tables, but they are not described here.

It may by worthwhile to state here that one can go one step further and extend this inventory sampling method in the sense that (1) the standard biomass table of the second phase is constructed so as to yield estimates of average biomass by tree diameter, tree height, species, tree form factor, crown diameter, etc., and (2) the trees of the subsample from the first phase are measured for the same characteristics, namely diameter, height, species, form factor, crown diameter, etc. but not measured for biomass. The data from (1) and (2) above are then combined to construct one-way local biomass tables by diameter (and species) alone. These extensions, however, are not discussed here.

THE ERROR OF INVENTORY ESTIMATES AND ITS MAIN SOURCES

Because the biomass estimates of a given forest area are based on sample data, they have errors. Statistically, these errors are ordinarily expressed as confidence limits of the true values of the parameters being estimated or standard errors of the estimates. Because confidence limits require use of a probability level (usually 95 percent), standard errors are to be preferred. However, it is more convenient to work with the square of the standard error, the variance of the estimates; it can be better manipulated mathematically, and many times it can be divided into additive components, one variance component associated with a specific source of error.

The statistical error has ordinarily two parts, a random and a systematic error or bias. When working with confidence limits, standard errors or variances it is implicitly assumed that the error of the biomass estimates have only the random part and have no bias; that is, in the long run and on the average the estimates are equal to the parameters they try to estimate. This is seldom true. Sometimes, when the sampling design and the measurement techniques are sound, the bias is ordinarily small and its effect can be ignored. But many times the bias is too large and cannot be ignored. Means must be found to evaluate it.

For the basic two-phase sampling design outlined in the previous section, the error of the biomass estimates (including its random and systematic part) has several sources, and with each source one can define an error component.

There is first the error component associated with the inherent variation between trees as selected by the first phase sampling design. It is intuitively clear that successive applications of the same sampling procedure do not generally yield the same biomass estimates, even though the procedure is applied to the same forest area, using the same sampling frame and with the same biomass tables (of the second phase) used every time. This is because a different set of trees is obtained every time the sampling procedure is applied and these sample trees have different dimensions from one to the next time. For instance, if we refer back to the illustrative example of the previous section where the first phase sample data consisted of a set of sample plots, a different set of plots would normally be obtained every time the sampling procedure is being applied, the trees of these plots will be different in size, and the different plot values v_i would result in different estimates $\bar{v}$, v_a and v_t.

The size of this error component is affected by, (1) the sampling design of the first phase used in the selection of the sample trees; it makes a difference whether the trees are selected by a systematic, stratified, cluster, multiphase or multistage procedure, (2) the sample size; the larger the sample the smaller the error component, and (3) the inherent variation between clusters of trees (plots) as determined by the sampling frame of the first phase; the larger the cluster, the smaller (in general) the variation between the values v_i.

The second error component is associated with the biomass table and is due to the difference between the true (unknown) biomass table and the biomass table as constructed from the sample trees of the second phase. Similar to the first error component, various applications of the sampling design of the second phase would result in a different set of sample trees, and a different set of measurement values and biomass tables. The size of the error is also affected by the same type of factors, (1) the sampling design used in the selection of the sample trees, (2) the sample size and (3) the inherent variation of the tree biomass about its true regression function of biomass on tree diameter or on diameter and height.

A third component is the measurement error. It is customary to define this error as the difference between the true and the measured value of a sample unit. For example, the true diameter value of a sample tree is not necessarily equal to the value found by the physical measurement of the tree diameter. The difference between these two diameter values (the true and the measured one) is the measurement error of the specific tree diameter. As with any error it may have a bias due, among other things, to poorly calibrated instruments or poorly defined (or applied) measurement procedures. It may also have a random component; different measurements of the same tree diameter by the same individual and with the same instrument would normally result in different values.

One should not confuse the measurement with the sampling error. The first refers to one's inability of determining the true value of a given sample unit, the second to the particular set of sample units (and their values) one happens to select in a given case. If one is able to measure exactly the true value of a sample unit, the measurement error is completely eliminated while the sampling error is still present. On the other hand, if all sample units of a population are measured one is not certain of finding the true values of the parameters of interest. The sampling error is zero since the entire population is included in the sample, but the measurement error may still be present.

Although not necessary, it has been found convenient to include the random, but not the systematic part (the bias) of the measurement error (the third component) with the first and second error components as defined above. This is because the random part is easily accounted for by the usual statistical formulae used in the calculation of the error. We have preferred discussing it separately only to draw attention to the fact that the sampling error of the first two components is essentially different from the error of measurement. The sampling design has little effect, if any, on the error of measurement, and the sample size has only a limited effect on the random but not on the systematic part of the measurement error. The main way to control the size of the error is to carefully design the measurement procedures and calibrate the measuring instruments.

There are also other sources of error and they define additional error components. For example, there are errors associated with the measurement of the forest area when v_t is calculated or the measurement of the area of the various forest strata when stratified sampling is used. Or there are errors associated with the statistical models used. For example, when calculating the error of the biomass estimates by the systematic sampling design one would normally use the simple random sampling formulae which are only approximately right. Or when the biomass tables are constructed, one must make a reasonable assumption about the form of the regression function. As there are many reasonable assumptions to make in any given case, different statisticians working with the same sample data (of the second phase) would arrive at different biomass tables and, thus, would arrive at different biomass estimates.

The current practice in forest inventory is to ignore all sources of error except that associated with the sample of the first phase. This means that the only component accounted for, when the standard errors of the biomass estimates are calculated, is the first component, which, for convenience, is made to include the random part, but not the systematic bias of the measurement error. This case will be discussed here. Although not of general practice, we shall also describe means for taking into account the error component of the second phase sample data, in addition to the first component. The measurement bias and all other sources will no longer be considered here.

It will be an extremely difficult undertaking, if at all possible to show in general terms, and for the general case of the two-phase sampling design outlined in the previous section, how to express in a suitable form (1) the error due to the first source (the sample of the first phase), 2) the error due to the second source (the sample of the second phase), and (3) how to combine the error components from these two sources. Instead, we shall prefer referring back to the illustrative example of the previous section where (1) the first phase sample consists of a randomly selected set of sample plots (clusters of trees) and (2) the one-way biomass table (constructed from the randomly selected trees of the second phase sample by least squares techniques) is based on a regression function of the form

$$(\text{biomass}) = b_1 + b_2 (\text{diameter}) + b_3 (\text{diameter})^2$$

We shall start with the presently used procedure of calculating the error of $\bar{v}$, the average biomass per plot when only the first source of error is taken into account. Because v_a and v_t are nothing but $\bar{v}$ multiplied by appropriately selected factors, it suffices to consider only $\bar{v}$. As v_i represents the biomass of the i-th plot, i= 1,2, ..., n, and the values v_i are statistically independent of each other, (recall that we have assumed a random selection of plots) the variance of $\bar{v}$ can be estimated by the formula (where the sample size n is assumed relatively small with respect to the population size and thus the finite population correction factor was made equal to 1)

$$\text{Var}(\bar{v}) = \text{Var}(v)/n$$

where $\text{Var}(v) = \sum (v_i - \bar{v})^2/(n-1)$ = sample variance of values v_i

It can be shown that the above formula considers only the source of error represented by the sample plots of the first phase, including the random part of the measurement error of the trees of the first phase; all other sources are being ignored.

To take into account the second source of error, one must be able first to express the error of the biomass tables in a form that makes it suitable to its later combination with the error of the first source. This may be a relatively easy task when (1) the biomass tables are constructed from least squares linear regression functions and (2) the basic assumptions of the least squares model are satisfied sufficiently well by the sample of the trees of the second phase. It may be extremely difficult, if at all possible, when non-linear regression functions are used or when the basic assumptions are critically violated. We shall consider only the case where the basic assumptions are satisfied; what happens when they are not satisfied will be briefly discussed in a later section.

Assume that the linear regression function of the form

$$y = b_1 + b_2 d + b_3 d^2$$

is an adequate representation of the relationship between the tree biomass y and diameter d. Past experience has shown that this would be the case with most of the biomass components, as for example, when y is the volume or the oven-dry weight of the merchantable bole. It is then known that the variation of the tree biomass about the regression function is not homogeneous; it increases with the increase of the diameter. In many cases, the conditional standard deviation of the biomass y for given diameter d is approximately proportional to the basal area, or what is the same thing, to the squared diameter value d^2. For example, empirical evidence shows that the standard deviation of the volume (or oven-dry weight of the merchantable bole) of all trees of diameter equal to 4 inches is about one quarter of the standard deviation of the volume of all trees of diameter equal to 8 inches and about one sixteenth of the standard deviation of the volume of all trees of diameter equal to 16 inches.

Under the above assumptions, and some additional basic assumptions not mentioned here, the statistical procedure for the calculation of b_1, b_2 and b_3 is that of weighted least squares. To express the error of the estimates, the usual procedure is to calculate the standard error of the regression coefficients b_1, b_2 and b_3, or the standard error of the various regression estimates $\hat{y} = (b_1 + b_2 d + b_3 d^2)$ or the associated confidence limits. As we shall soon see, the proper way to express the error of the biomass tables (based on the regression function of the above form) is by the <u>covariance matrix</u> of the regression coefficients whose elements are the variance Var(b_i) and covariance Cov(b_i, b_j) terms, i, j = 1,2,3. Let us now show that this is the right expression, and then see how the error of the biomass table can be combined with the error of the first source.

Recall that the values v_i were calculated from the summation of the estimated biomass values y_{ij} of the individual trees. For a given plot i, this can be expressed symbolically as follows

$$y_{i1} = b_1 + b_2 d_{i1} + b_3 d_{i1}^2$$

= estimated biomass of the first tree of plot i

$$y_{i2} = b_1 + b_2 d_{i2} + b_3 d_{i2}^2$$

= estimated biomass of the second tree of plot i

$$y_{imi} = b_1 + b_2 d_{imi} + b_3 d_{imi}^2$$

= estimated biomass of the last, m_i-th tree of plot i

Then, the estimated biomass v_i of the i-th plot is

$$v_i = y_{ij} = b_1 m_i + b_2 \, d_{ij} + b_3 \, d_{ij}^2$$
$$= b_1 z_{1i} + b_2 z_{2i} + b_3 z_{3i}$$

where

$z_{1i} = m_i$ = number of trees in plot i,

z_{2i} = sum of tree diameters d_{ij} in plot i, and

z_{3i} = sum of squared tree diameters d_{ij} in plot i

Using this new notation we can finally write

$$\bar{v} = \sum v_i/n = b_1 \bar{z}_1 + b_2 \bar{z}_2 + b_3 \bar{z}_3$$

where

$$\bar{z}_1 = \sum z_{1i}/n, \; \bar{z}_2 = \sum z_{2i}/n, \text{ and } \bar{z}_3 = \sum z_{3i}/n$$

To estimate the variance of $\bar{v}$ (and by taking the square root calculate the standard error of $\bar{v}$) we shall use a statistical formula which in matrix notation can be expressed in a much more compact form, but which for our purposes can also be expressed as

$$\text{Var}(\bar{v}) = \sum\sum \left[b_h b_k \text{Cov}(z_h, z_k) + z_h z_k \text{Cov}(b_h, b_k) - \text{Cov}(z_h, z_k) \text{Cov}(b_h, b_k) \right]$$

where, for h = k, we have

$$\text{Cov}(b_h, b_k) = \text{Var}(b_k) \text{ and Cov}(\bar{z}_h, \bar{z}_k) = \text{Var}(\bar{z}_k)$$

FOREST INVENTORIES FOR MONITORING FOREST BIOMASS

Because in forest inventories with large samples the last term is negligibly small compared to the first two, the formula can be written as

$$Var(\bar{v}) = \sum\sum \left[b_h b_k \, Cov(\bar{z}_h, \bar{z}_k) + \bar{z}_h \bar{z}_k \, Cov(b_h, b_k) \right]$$

When the true <u>biomass tables are known without error</u> (an impossible case), or when the tables have a small error which can be ignored, or simply when one does not wish to take the error of the biomass tables into account, the variance and covariance terms of b_1, b_2 and b_3 are made equal to zero and the expression of the variance of $\bar{v}$ becomes

$$Var(\bar{v}) = \sum\sum b_h b_k \, Cov(\bar{z}_h, \bar{z}_k)$$

If in addition, the sample plots are selected by simple random sampling as assumed here, and we still ignore the effect of the finite population correction factor, we have

$$Cov(\bar{z}_h, \bar{z}_k) = Cov(z_h, z_k)/n = \sum (z_{hi} - \bar{z}_h)(z_{ki} - \bar{z}_k)/n(n-1)$$

and it can be shown (by simple although relatively lengthy algebraic manipulations) that we can write the identity

$$Var(\bar{v}) = \sum\sum b_h b_k Cov(\bar{z}_h, \bar{z}_k) = \sum (v_i - \bar{v})^2 / n(n-1)$$

As the reader can verify this is the same formula we have obtained before, when only the error due to the first source (main sample) was taken into account.

On the other hand assume that the trees from the entire forest area of interest were measured for their diameters. This is equivalent to saying that we have the <u>complete enumeration</u> and all population values z_1, z_2 and z_3 were recorded. If we are willing to ignore the existence of the measurement error, and because the finite population correction factor is equal to zero, all variance and covariance terms of $\bar{z}_1$, $\bar{z}_2$ and $\bar{z}_3$ become equal to zero. Then, it is easily seen that the variance of v becomes equal to

$$Var(\bar{v}) = \sum\sum \bar{z}_h \bar{z}_k \, Cov(b_h, b_k)$$

This expression represents the error of the biomass table as applied to the given forest area of interest.

This procedure can be easily extended to the case where the methods of selecting sample trees in the first and second phase are such that the variances and covariances of the plot averages $\bar{z}_1$, $\bar{z}_2$ and $\bar{z}_3$, and those of the regression coefficients b_1, b_2 and b_3 can be estimated. For example, the sample plots of the first phase can be selected by stratified, cluster, multistage or multiphase sampling. Then, the estimates of the average values per plot, more generally denoted here as $\bar{z}_1$, $\bar{z}_2$ and $\bar{z}_3$, can be calculated, together with their variance and covariance terms by formulae given in standard texts on statistical sampling techniques.

The problem is much more difficult to handle when the sample trees of the second phase are not selected by simple random sampling but by other techniques such as stratified or cluster sampling. Then, the method to use for the calculation of b_1, b_2 and b_3, as well as their variances and covariances are not generally available in the standard texts on regression analysis. If one wishes to derive reliable and valid estimates of b_1, b_2 and b_3, and estimates of their variance and covariance terms, one must adequately modify the classical least squares techniques to fit the specific sampling method used in the selection of the trees. Some of these modifications can be found in the statistical literature, although not in a form readily applicable to the construction of biomass tables.

Another extension is also possible to the case where the biomass table constructed from the sample of the second phase is based on a regression function having a different set of independent variables. Then, the number of regression coefficients b will change from three to more than three and the definition of the corresponding plot variables z of the first phase sample will also change. Assume, for example, that the biomass table is based on a regression function of the form

(biomass)=b_1+b_2 (diameter)+b_3 (height)+b_4 (diameter)^2 (height)

Then, there are four instead of three regression coefficients and the plot variables z_1, z_2, z_3 and z_4 are defined as

z_1 = number of trees in the plot,

z_2 = sum of tree diameters in the plot,

z_3 = sum of tree heights in the plot, and

z_4 = sum of the tree values (diameters)^2(height) in the plot.

Of course, the formula for the variance of $\bar{v}$ will change in the sense that the summation will be taken over four rather than three values b and $\bar{z}$.

A third extension occurs when instead of one, we have several species and each species has its own volume table. Because several regression functions can be expressed, by means of dummy variables into one, giant size regression no additional problem arises. For example, assume that we have two species and their regression functions are

$$y = b_1 + b_2 d + b_3 d^2 \text{ if species 1}$$
$$= b_4 + b_5 d + b_6 d^2 \text{ if species 2}$$

Then, using dummy variables techniques, we define new independent variables

$x_1 = 1$ if species 1 or 0 otherwise

$x_2 = d$ if species 1 or 0 otherwise

$x_3 = d^2$ if species 1 or 0 otherwise

$x_4 = 1$ if species 2 or 0 otherwise

$x_5 = d$ if species 2 or 0 otherwise

$x_6 = d^2$ if species 2 or 0 otherwise

The giant size regression function becomes now

$$y = b_1 x_1 + b_2 x_2 + \ldots + b_6 x_6$$

and it is easy to verify that for species 1 this function is identical to

$$y = b_1 + b_2 d + b_3 d^2$$

and for species 2

$$y = b_4 + b_5 d + b_6 d^2$$

Of course, the six plot variables z_1, z_2, ..., z_6 are now defined as

$$z_k = \sum x_{ki} \text{ for } k = 1, 2, \ldots, 6$$

For example,

z_1 = number of trees for species 1 in the plot

z_2 = sum of diameters of the trees of species 1 in the plot etc.

Another extension is also possible to the case where (1) the second phase sample leads to regional two-way biomass tables by diameter and height, (2) a subsample of trees of the first phase sample of the forest area of interest is measured for diameter and height and (3) the two-way biomass table and the tree data of the subsample are used to construct the one-way biomass table by diameter alone. The error of this last biomass table is dependent not only on the error of the first phase sample and the error of the subsample of the second phase, but also on the procedure used to construct it. If this procedure is conveniently defined, some statistical theory exists and can be used to calculate the error of the one-way biomass table and express it in its form of covariance matrix of regression coefficients.

No extension seems possible, however, to the case where non-linear regression functions are used; a new statistical approach may be necessary. Unless he is compelled by special reasons, the reader is advised against using non-linear regression functions of the allometric type (as for example $\log y = \log b_1 + b_2 \log d$) that are so popular with many researchers in the tree biomass field.

BASIC CONCEPTS USED IN THE SAMPLING OF THE FIRST PHASE

It may be useful now to classify the forest biomass inventory systems into two basic types, those concerned with the measurement of the forest on one occasion only, and those requiring the measurement on successive occasions. The sampling problem is relatively simple when only estimates of the current biomass values at one given point in time are needed. Then, the inventory of the forest areas of interest taken at the given point in time would normally provide the required answers. There is an entire body of statistical sampling techniques that can be used to design efficient sampling systems for forest inventories on one occasion only. These sampling techniques can be found in standard texts on statistical sampling methodology.

The problem becomes more complex when together with estimates of current biomass values, estimates of rates of change are also required. Then, one can still use systems of forest inventories on one occasion to estimate current values and use indirect methods to estimate rates of change. Some of these indirect methods are based on yield tables, that is, tables that estimate the expected value of biomass per acre or hectare as a function of stand age (by species, forest type, site quality, etc.). But these methods seem unsatisfactory since (1) there are statistical objections to using yield tables that are by definition biased when applied to populations other than the populations they were derived from, (2) most of the present-day forests are unique in the sense that their development has been uniquely affected by human factors such as harvesting, thinnings or fertilization and (3) yield tables are often unavailable for many forest types and site conditions or if available, their reliability has never been calculated and expressed in statistical terms. Other indirect methods make use of increment core measurements of past diameter growth on living trees and analysis of dead trees and stumps for the estimation of the mortality and harvested biomass respectively. But these methods are also unsatisfactory from a statistical point of view since the selection of the sample trees from which increment cores are obtained is often non-random and the estimation of the mortality or harvested biomass lacks enough precision and is often biased.

It is much better to make the measurement of both, current biomass values and rates of change as parts of the same system of successive forest inventories. For these systems, one could generally recognize a space and a time dimension. The sampling design should include basic rules stating how to select sample units at a given point in time. These rules refer to the geographical distribution of the sample plots, the space dimension. These may be the same rules occasion after occasion, or the rules may change from one to the next occasion. The design should also include basic rules stating how to select sample units over successive points in time, more specifically rules stating how to chain the sampling designs of the various occasions. This is the time dimension.

For example, a particular system of forest inventories on successive occasions is the well known CFI (Continuous Forest Inventory) system whereby the same set of systematically selected plots is measured and remeasured at regular time intervals. Then the space dimension of the system is represented by the systematic sampling design by which the sample plots are selected over the forest area and the time dimension by the rule that the same plots must be measured on every occasion.

The basic sampling concepts of the designs used on a given occasion (the space dimension of the systems of inventories on successive occasions) are similar, although not identical to those of the designs used by the inventory systems on one occasion. The linking over time of various sampling designs used on various occasions requires, however, additional concepts. We shall now discuss briefly some of these concepts and show the contrasting aspects of the sampling designs used by forest inventories on one or more than one successive occasions.

(1) Although not generally true, it is much more efficient to work with sample units defined as clusters of trees rather than individual trees. The main advantage is that the traveling costs to and from sample trees are much smaller (on a per tree basis) with cluster sampling. The disadvantage is that the average amount of information per tree becomes also smaller as the cluster increases in size; neighbor trees are more similar to each other than trees farther apart. This brings the problem of finding an optimum size for the cluster of trees by appropriately balancing the costs of measuring trees with the loss of information when the sample trees are selected close to each other. There is not one optimum size for the cluster; the optimum may vary according to stand age, forest type, site quality, etc.

Ordinarily, the clusters are defined as groups of trees growing together within sample plots of given shape and fixed area. When the shape of the plots is rectangular, one may be able to define non-overlapping plots covering the entire forest area. Many times, however, the sample plots have a circular shape and one must work with overlapping sets of sample plots, even though in actual practice, when sample plots are selected, the plots in the sample itself are not allowed to overlap.

In inventories of large forest areas of poor accessibility, where the time to travel to and from sample plots is relatively large compared to the time to measure the trees, it is convenient to work with clusters of plots. For efficiency purposes, a cluster of plots is ordinarily made to represent, on the average, one day's work; it is intuitively more efficient to go to a new point in the forest on every new day of work, than go to an old point and finish the sample cluster started the day before.

The plots may be of fixed size (equal area), with all of the trees growing within the plots measured. Then the trees are selected with equal probability, provided of course that the plots themselves are selected with equal probability. For example this would be the case when a set of points are selected completely at random from the forest area of interest and the sample plots are defined as the circular areas, around the points, having the same fixed radius and with all their trees measured for the attributes of interest.

The biomass of large trees is much larger than the biomass of small trees; and, thus, the values of the biomass estimates are affected much more by the large than by the small trees. Because there are relatively many small and only few large trees, selecting trees with equal probability leads to too many small and not enough large sample trees. To obtain relatively more large and less small trees, and thus, to improve the efficiency of the sampling design, one may consider defining sample plots of different areas for trees of different size.

For example, one may select points at random from the given forest area and define three circular plots at each selected point, with the same center, the point, but different radia, say 5, 10, and 20 meters respectively. The small trees of diameter less than 15 cm are measured only in the first plot of 5 meters radius, the medium size trees of diameter between 15 and 25 cm are measured in the second plot of radius equal to 10 meters while the large trees above 25 cm are measured in the third plot of 20 meters radius. Although the plots are selected with equal probability, the trees are not; the probability of selection of the large trees is four times higher than that of selecting the medium size trees and sixteen times higher than that of selecting the small trees.

By making the probability of tree selection proportional to the square of their diameters one obtains the so called Bitterlich sample points. The points are still selected with equal probability. The plot size is unique for every diameter value and this procedure leads generally to efficient sampling designs. There are optical instruments to facilitate the selection of trees with probability proportional to squared diameter. The procedure is not described here, it is very ingenious in theory but very simple to apply in practice with the appropriate instruments.

(2) For <u>operational inventories</u> where the forest area is measured on one occasion only, the plot and tree <u>measurements should be made according to the currently accepted standards of merchantability</u>. This is because, with operational inventories, the main objectives are to obtain estimates of the current biomass values, estimates which are then applied towards solving specific problems of limited scope. On the other hand with <u>management or national inventories</u> that require the measurement of the forest on several successive occasions, <u>additional measurements</u> of the plot and tree attributes must be made that are <u>independent of the current concepts of merchantability and accessibility</u>. For example, trees of all species should be measured, whether of commercial or non-commercial value, whether growing on presently accessible or non-accessible areas. Or, the total height should be measured instead of, or in addition to the merchantable height of the tree.

Criteria of merchantability and land accessibility change fast. Trees of lower quality or non-merchantable species today may become merchantable tomorrow, and areas which are not economically accessible today may soon become accessible as scarcity of biomass resources may make them more valuable. To compare biomass estimates from successive inventories of the same forest area and, thus, obtain estimates of the changes occurring in the forest between successive measurements, same measurement standards must be applied.

On the other hand, when the estimates are presented to the management, they must be presented in meaningful units conforming to the current standards, even though the original measurements as taken in the field or as computed in the office are in different units. For example one should not present the management with estimates of the oven-dry weight of the total tree biomass, when estimates of construction timber of specific species are required. Consequently, the inventory system should have a built-in facility to transform the field measurements taken in fixed units, to units that become meaningful in terms of current applications.

(3) The assessment of forest biomass resources of large forest areas should not be considered separately from the assessment of the entire forest environmental system and the biomass inventory should be <u>broadly viewed as an integrated, multiple resource inventory</u>.

The traditional merchantable and unmerchantable biomass resource data should be collected together with other environmental data, such as, for example, data on water, air, soil, wildlife and wildlife habitat, fish and insects, grass, shrubs, recreational potential, etc. Present informational needs should be considered together with needs as perceived in the future. The problem of integrating inventories is of great concern today, at least in the United States. For example, a recent, 1978 Symposium held in Tucson Arizona, attended by over 350 foresters had as a main theme of discussion the problem of integrating inventories of renewable forest natural resources.

Because the forest land data on biomass (and all other environmental factors) will be used by managers and policy makers together with data for other types of land or water, it may be better to think of forest inventory as part of an overall system of inventorying all other land and water resources including range, agricultural or open (with no apparent use) lands, rivers, lakes and oceans. The decision maker at the national or regional level should consider alternative uses of all land and water resources in view of the overall needs, present and future. In this sense, it may be highly desirable to devise national and possibly international systems of land and water classification, and then use this classification system as a basis on which to build a comprehensive system of biomass (forest or other) resources.

Note that this basic concept does not necessarily apply to operational inventories since they are generally concerned with small areas and very specific problems. They do ordinarily apply to management or national forest inventory used by policy makers and managers of large forest areas.

(4) The sample plots may be selected by a variety of methods. The most convenient, and in many cases the most efficient procedure is to select the plots in a systematic fashion by randomly overlaying a dot grid over the forest area. The distance between dots is chosen so as to yield a desired sample size (or a desired precision for the biomass estimates) since each dot defines the location of a sample plot, sample point or cluster of plots or points.

When the forest area is heterogeneous (as the case may be with forests that contain a variety of stands of different density, species composition, age class, site quality, etc.) it is advantageous to (1) subdivide the forest into homogeneous areas called strata (2) sample each stratum separately and (3) combine the biomass estimates of the individual strata to obtain estimates for the overall forest area. Described in this way the concept is known as prestratification. When (1) the forest area is independently subdivided into strata after the selection of the plots is completed, and (2) the biomass estimates for the entire forest area are obtained by combining the estimates from individual strata, the concept is know as poststratification.

The efficiency of sampling is much higher with prestratification since the distribution of the sample plots among the various strata can be optimized so as to maximize the precision of the estimates for given sampling costs or minimize the costs for required precision. For this reason it is extensively used with operational inventories of heterogeneous areas where only estimates of current biomass values are required. On the other hand, it is not efficient to use prestratification with inventory systems that require forest measurement on successive occasions. Due to accidents such as fires, windstorms, insect or disease epidemics, harvesting operations, etc. the boundaries of the various strata change with time. For this reason, it is customary to apply post stratification but not prestratification in management and national inventories where the forest is measured on successive occasions.

The subdivision of forest areas into strata and their location on the map can be made by using past historical data about the forest stands. Most of the time, however, the borders of the various strata are delineated from detailed photo-interpretation ofthe aerial photographs of the forest area. Sometimes the forest is very heterogeneous and the strata are small, are numerous and have very irregular shapes. Then, the stratification work may become extremely laborious, costly and time consuming. If, in addition, biomass estimates are not required by individual strata, one can preserve the advantages of stratified sampling with optimum allocation by making use of a double sampling for stratification (or a two-phase sampling) procedure of the following type.

In the first phase, a very large number of points (hereby called photo-points) are randomly or systematically located on the aerial photographs of the forest area of interest, and through photo-interpretation, each point is classified as belonging to one, and only one of previously defined strata. The proportion of points falling within a given stratum provides an estimate of the true proportion, and, through multiplication by the total forest area, an estimate of the stratum area. In the second phase, a stratified random subsample (with optimum allocation) of these photo-points is selected and for each photo-point so selected, a field plot is established at the corresponding location on the ground. The trees from these plots are then measured for the usual attributes such as diameter at breast height, species, total height, etc. and their biomass estimated through the use of appropriately constructed biomass tables. Finally, the stratified sampling theory is applied to obtain biomass estimates per unit area, by stratum or for the entire forest.

There are recent attempts to make a more extensive use of small and large scale aerial photography. The present use of aerial photography is mostly limited to stratification of forest areas, either directly by delineation of stratum boundaries on the photograph (and then by their transfer on the map) or indirectly through the use of double sampling for stratification techniques. Old procedures trying to make use of aerial volume tables, that is, tables giving average biomass per unit area by forest type, species composition, tree height, density class, site quality, etc. as interpreted or measured on aerial photographs, have proved unsuccessful. New inventory systems have now evolved that make use of multistage or double sampling with regression techniques.

For example, large scale photographs of randomly or systematically selected sample plots can be used as the first phase of a double sampling with regression design. The biomass of these sample plots is estimated subjectively by photo-interpretation or objectively by photogrammetric measurements of their trees. A small random subsample of these plots is then selected and through ground measurements, their actual biomass is determined as accurately as possible. Finally, the two sets of data, the photo and the ground measurements are suitably combined through regression techniques and estimates of the forest area biomass calculated. Note that with this procedure the bias which exists in the photo-plot biomass is greatly reduced if not completely eliminated by the regression functions.

It is almost as easy to take large scale photographs of simple plots as it is to take photographs of long and narrow strips. Therefore, a two-stage cluster sampling design can be defined as follows. The primary units consist of narrow strips of forest land of various lengths, and the strips can be divided into rectangular sample plots of equal size, the secondary units. In the first stage a random sample of strips is selected and covered by large scale photographs. From each sample strip so photographed, several sample plots are randomly selected. This constitutes the second stage sample. The trees from these plots are then measured photogrammetrically for their size, usually height and crown width. Note that with this design the sample trees of the first phase as selected by the two stage design are measured for height and crown diameter not diameter at breast height. This implies that the second phase sample (not being discussed here) must consider sample trees with measurements leading to construction of two-way biomass tables by tree height and crown diameter.

For inventories of large forest areas, the two stage design above can be improved by using one or more additional stages and, thus, obtaining a multi-stage cluster sampling design. For example, from small scale photographs of the area, more or less homogeneous stands are identified, delineated and grouped by strata. A stratified random selection of stands (using equal probability or preferably probability proportional to size) constitute the first stage sample. The sample units so selected in this first stage are then subsampled by the two-stage cluster sampling design described above, or any other, appropriately defined multi-stage or multi-phase sampling technique.

(5) The successive inventories of the same forest area must be interrelated if the inventory system is to be cost-efficient. It is relatively easy to show that to estimate rates of change efficiently, one should use permanent sample units. But it is not easy to see, however, that to estimate current biomass values on successive occasions, it is also efficient to use permanent sample units combined with temporary ones.

It can be shown that changes occurring in a given forest area over the time interval separating two successive measurements, cannot be estimated with any reasonable precision by the difference between the estimates of the current biomass values as calculated from two, statistically independent inventories. It is difficult for the non-statistically trained scientists to see why this would be true, even when the current biomass estimates on each of the two measurement occasions are highly precise. For the statistically trained readers we shall now show how this can happen by considering a specific illustrating example.

Let B_1 and B_2 be the average biomass per hectare values as estimated from the inventory data of the first and second occasion respectively and let $D = (B_2 - B_1)$ be their difference showing the average change in biomass between the two measurement times. For the sake of argument we shall assume that (1) D is equal to 5 percent of the value B_1, that is $D = .05B_1$, and (2) the standard errors of B_1 and B_2 are both equal to 3 percent of B_1, (not B_1 and B_2 respectively), that is, $SE(B_1) = SE(B_2) = .03B_1$. Using a well known statistical formula, and remembering that the variance is the square of the standard error we can write

$Var(D) = Var(B_1) + Var(B_2) = .0009B_1^2 + .0009B_1^2 = .0018B_1^2$ and
$SE(D) = .0424B_1$

This implies that the standard error of D (that is the value the confidence limits of the true but unknown value of the change in biomass between the first and second measurement times are approximately equal to $D + 2SE(D) = .05B_1 + .0848B_1 = D + 1.7D$ that is, the lower and upper 95 percent confidence limits are equal to -7D and 2.7D respectively. Because the value of zero change is included in the confidence interval, it is also said that the change D is not significantly different than zero. Note that although B_1 and B_2 are estimates with high precision, the precision of the estimate D is very low.

Assume now that the two successive forest inventories are based on the same set of permanent sample plots. In the absence of harvesting between measurements it is not uncommon to find that the correlation coefficient between the two successive measurement values of the same sample plot is .98. Assuming this value, and using a well known statistical formula the covariance of B_1 and B_2 is equal to 98 percent of the product of the standard errors of B_1 and B_2, that is

$$Cov(B_1,B_2) = .98 SE(B_1) SE(B_2) = (.98)(.03 B_1)(.03 B_1) = .000882 B_1^2$$

Using now another well known statistical formula, we can write

$$Var(D) = Var(B_1) + Var(B_2) - 2\, Cov(B_1,B_2)$$
$$= .0009 B_1^2 + .0009 B_1^2 - (2)(.000882) B_1^2 = .000036 B_1^2$$

and $SE(D) = .006 B_1$

Because the 95 percent confidence limits are approximately equal to

$$D + 2 SE(D) = .050 B_1 + .012 B_1 = D + 25 \text{ percent of } D$$

the lower limit is $.038 B_1$, or about 75 percent of the value D, the upper limit is $.062 B_1$, or approximately 125 percent of D, and it is common to say that D is significantly different than zero. Note that, (1) in relative terms, the precision of D is still low compared to the precision of either B_1 or B_2, that is, a standard error of 12.5 percent for D compared to the standard error of 3 percent of B_1 and B_2, and (2) because D is significantly different than zero, we can say that the change in biomass between the two measurement times can be estimated.

To estimate current biomass values it does not really matter whether the sample plots are permanent or temporary as long as only the current measurement values of the plots are used in the estimation process. To see how a mixture of permanent and temporary plots on each of the two measurement occasions can be used to obtain more efficient estimates of the current biomass values (for the same number of plots measured on each occasion if not for the same sampling costs) we shall use intuitive arguments applied to the following illustrative example whereby (1) the sample size on occasion 1 is n, (2) the sample size on occasion 2 is also n, and (3) m n of the sample plots are permanent and common to both occasions, that is, m plots are measured and remeasured on occasion 1 and 2 respectively. This implies that (4) u=n-m plots of occasion 1 are temporary and not remeasured on occasion 2 and finally (5) u=n-m plots of occasion 2 are temporary and not measured on occasion 1.

To estimate the current biomass values at the second measurement time (occasion 2), one can use the values of the n=u+m second measurement plots as measured on occasion 2. Then, it would not have mattered whether all n plots were permanent, temporary or any combination of temporary or permanent plots. But one can ordinarily derive better estimates of the current biomass values by combining the values of all plots from both occasions. Using the two measurement values of the permanent plots, one can determine a relationship such that, through its use, one is able to <u>estimate</u> the current value (as it exists at the second measurement time) of any temporary plot that was measured only on occasion 1 and not remeasured on occasion 2. This means that on occasion 2 one has m <u>measured</u> plot values (of the m permanent and u temporary plots of occasion 2) and u <u>estimated</u> plot values (of the u temporary plots of occasion 1). By suitably combining the n=u+m measured and u estimated values one can derive current biomass estimates for occasion 2. And if the relationship between the two values of the permanent plots is strong, and in the absence of harvesting or catastrophical accidents between measurements (such as fires, insect epidemics or windstorms) the relationship is very strong, then the estimates based on all n+u permanent and temporary plots of both occasions are much better than the estimates based on the n measured values of occasion 2 alone.

Thus, to estimate rates of change, it is extremely advantageous to use permanent sample units. They provide efficient estimates of the growth, more specifically efficient estimates of growth components such as growth on live trees, periodic mortality, ingrowth and drain. On the other hand, to estimate current values it is more efficient to use a combination of permanent and temporary plots. Because the costs of establishing and maintaining permanent plots is much higher than those of establishing temporary plots, the problem is now that of finding that combination of permanent and temporary plots that yield at least required precision for estimates of both current values and rates of change for minimum sampling costs.

The statistical theory of combining permanent and temporary plots on two or more occasions, including the theory of optimization of sample size has been well developed in the last 25 years. For more on this theory, the interested reader is referred to an old work by Ware and Cunia (1962) and some additional 15 to 20 papers that appeared since then in the forestry literature. Known under the name of Sampling with Partial Replacement (SPR), this theory is currently being applied to several large size management and national forest inventory systems.

BASIC CONCEPTS USED IN THE SECOND PHASE SAMPLING

Some of the basic sampling concepts used in the second phase are similar to those used in the first phase, but others are specific to the problems of tree measurement and biomass tables construction. Because the biomass tables are ordinarily constructed by least squares techniques and the basic underlying assumptions of these techniques must be satisfied, at least sufficiently well, if valid results are to be obtained, the sampling concepts discussed below are intimately related to four main areas of concern, namely (1) representativeness of the sample data, (2) measurement of the sample tree biomass, (3) statistical models used in the data analysis and construction of the biomass tables, and (4) combination of the error of biomass tables of the second phase with the error from the sample plots of the first phase.

(1) The sample trees must be selected by a <u>sampling procedure that makes them representative</u> of the population of interest. If any of the sampling techniques from any standard textbook on statistical sampling methodology are properly used, and if the sample trees are selected from the forest area to which the biomass tables will be later applied, the condition of representativeness would normally be satisfied. But if the phase two sampling of the current inventory is deleted in favor of using previously constructed biomass tables, the problem becomes more complex. One must then be concerned with (1) whether the sample trees from which the biomass tables were constructed were indeed representative of the tree populations from which they were selected and (2) whether the two tree populations, that is, the population from which the biomass tables were constructed and the population to which the biomass tables are applied, are sufficiently similar to each other so that the error due to their diference, if any, could be ignored.

We shall not discuss here the problem of whether the population of trees represented by a given biomass table and the population of trees to which the biomass table is applied are similar to each other. This is a problem which is ordinarily solved by subjective means. We shall only consider the problem of whether a sample of trees is representative of the population of trees from which it was drawn. And from the outset it is useful to distinguish between the <u>target</u> population defined as the population for which estimates are required and the <u>sampled</u> population defined as the population actually represented by the sample.

An important step to any sampling technique is the definition of the target population. If a tree is a part of the population, the tree must be given a chance of being included in the sample. It is not necessary that this chance, or probability of selection in the sample be equal for all trees, but it is necessary that this chance be non-zero and known at least in relative terms. For example, if the forest inventory is concerned with trees of all crown classes, one should not construct biomass tables from sample trees selected from dominant and codominant trees only. If the forest area is large, the sample trees must be selected by some random process from its entirety and not from easily accessible areas, or from ordinary clearcutting logging operations or highly selective thinning operations. Trees of special form, trees with unusual defects or partially defoliated trees should not be automatically excluded from sampling simply because they do not conform with the image of the average tree that one may have.

It is interesting to note that a survey of the biomass literature of the past 10 to 15 years shows that most authors did not feel important to state <u>how</u> the sample trees were selected or <u>what</u> population of trees they were supposed to represent. It was somehow assumed implicitly that the sample of trees so selected constituted a representative sample of some undefined population. Sometimes, the authors stated that the trees were selected "at random" but a closer look showed that the selection procedure was not random in the strict statistical sense. For example trees were only selected if they were dominant or codominant only, or had only one stem, or were healthy with no signs of defoliation, or belonged to stands having a closed cover, etc. Seldom, if ever the sample of trees was representative of the target population and then, most of the time the sampling procedure was such that no valid statistical procedure existed for a proper analysis of sample data and construction of biomass tables.

This brings us to another point. Not all sampling methods that yield representative samples are acceptable. Two additional conditions must be satisfied. It is necessary first that the sampling method be such that valid statistical procedures exist for the analysis of the sample data and the construction of biomass tables. And then, the sampling method must be cost-efficient, that is provide biomass tables of sufficient precision at reasonable costs. This implies that one should not use a sampling method which yields representative samples of trees but for which there is no valid statistical method of data analysis; nor should one use a sampling method which although sound from a statistical point of view is too expensive to apply for all practical purposes.

(2) The sample trees of the second phase should be measured first for as many of the variables measured on the trees of the first phase as possible, and second, for additional variables not measured in the first phase, as for example, total height, diameters at various stem heights and biomass of various tree components. There are three <u>major components</u> of the tree, the main stem, the crown and the stump-root system and minor components or <u>subcomponents</u> of the major components such as wood and bark, dead and live branches by various size classes, leaves or needles, twigs,

sampling method may be sound and the subsample may be representative of the tree component it represents, the estimation of the oven-dry weight of this component would be biased. This is because the estimates are of the type known as ratio or regression estimates. For example, if F and D denote the known fresh and the unknown oven-dry weight of the entire tree component respectively, and f and d the corresponding known weights as measured on the subsample, then the unknown value D is estimated by the formula $D=F(d/f)$. It is well known by statisticians (although not generally known by biomass researchers) that this estimate is biased. It is also known that if the subsample units are selected by random sampling and if the number of these units is sufficiently large, the bias is relatively small and its effect can be easily ignored. This implies that to reduce the bias in the estimation of the oven-dry weight of a tree component one is much better to have a relatively large subsample of small units randomly selected than a relatively small subsample of large units. And as stated before, if one wants to evaluate the error of the estimate of the oven-dry biomass, it is then necessary to measure the biomass of each subsample unit separately. The measurement of composite subsamples may be cost-efficient but does not provide one with an estimate of the error.

(3) The statistical procedure for analyzing the sample tree data of the second phase must be consistent with the sampling method if valid results are to be obtained. For example, one would never use the ordinary sample arithmetic mean to estimate the population mean when the sample elements are selected by stratified sampling; he would normally use a more complex type of mean known as stratified mean. But a review of the biomass literature shows that many researchers have used the least squares estimates (that are equivalent to simple arithmetic means) to construct biomass tables even though the sample trees were selected by stratified or cluster sampling.

The regression least squares techniques yield valid results only when its underlying basic assumptions are strictly satisfied. With finite populations these assumptions are almost never satisfied and questions arise as to how critical these assumptions are. Based on theoretical arguments, experimental data and simulation studies it can be stated that with sufficiently large sets of biomass data that are representative of some sampled population of interest, only a few of the assumptions are critical. For a detailed discussion of this problem the reader is referred to two papers by Cunia (1979a, 1979b); we shall content ourselves here with a few summary comments.

The assumption that the conditional distribution of the tree biomass is normal in the statistical sense, is not critical for the validity of the inventory results. The central limit theorem applies and the effect of non-normality is practically limited to a distortion of the calculated prediction limits of the regression function. But as these limits are not needed when the error of the inventory estimates is calculated the normality assumption is not important. Similarly, the effect of the assumption that the independent variables (in our case the tree diameter, height etc. but not biomass) must be fixed, not random variables, or that they be measured without error is negligibly small. The assumption that the conditional variance of the biomass is homogeneous may become quite critical if the classical, unweighted least squares regression method is applied. But if proper transformations are used or the weighted least squares regression method is applied, the effect of this assumption is minimal. Finally, the assumption that the true regression function is of the assumed form may become critical only when the assumed form is poorly selected. If good care is taken to work only with regression functions of the type commonly used today, and there is enough empirical evidence to suggest functions which are good, one does not encounter critical problems.

One important assumption that may have a critical effect on the validity of the biomass tables and the calculation of their error, is that of statistical independence of sample units. The sample trees are selected by a variety of techniques, many of which yield representative samples of the population of trees. But unless the trees are selected by the so-called method of simple random sampling with replacement, the sample trees are not statistically independent of each other and the least squares regression technique yields results which, strictly speaking are not statistically valid. To yield valid results, the least squares method must be appropriately modified.

For example, the sample trees of the second phase may be selected by a sampling method that uses the concepts of clustering and stratification. The average sampling costs per tree are reduced if groups of trees (clusters) rather than individual trees are randomly selected. But this results in loss of information; the average amount of information per tree decreases with measurement of any additional tree selected from a given group. On the other hand, the average amount of information per tree increases when stratification is used. By forcing trees to come from different tree sizes or different forest areas, one is able to obtain trees from a much wider spectrum of tree conditions than the spectrum obtained by simple random sampling.

The statistical literature is scarce on the problem of (1) what happens when the ordinary least squares regression techniques are applied to samples of trees selected by methods other than simple random sampling and (2) how to modify the least squares techniques to take into account the sampling procedure. The reader interested in a detailed discussion on what may happen and some suggested modifications is referred to two earlier papers by Cunia (1979a, 1979b).

As indicative of what may happen one may refer to some empirical results reported by Cunia (1981a). Working with two samples of trees, each sample consisting of some 280 trees selected in clusters of five trees, Cunia has shown that the biomass tables as constructed by the ordinary least squares method and an appropriately modified least squares procedure are practically the same. The estimates of the error of these tables, however, were quite different; using the ordinary least squares method leads to confidence limits that are only about 60 percent as large as those obtained by the modified procedure. Of course, this may be true only for the empirical data used in the study, that is, the two species and forest areas considered and the cluster size of five trees. With other species and forest areas and different cluster sizes the results may be different.

(4) The error of the biomass tables must be expressed in a suitable form and means should be found to take this error into account when the error of biomass estimates is calculated. This is a complex problem and no general solution exists. If linear regression models are being used, and the selection procedure for the sample trees of the second phase is such that a modified least squares method can be devised to estimate the covariance matrix of the linear regression coefficients, then the problem can be solved in an adequate way. Some of these concepts applied to an oversimplified sampling design were discussed in an earlier section and are not repeated here. One thing is, however, important and should be stressed out here in strong terms. Because (1) both, linear and allometric regression functions can be used to construct equally good biomass tables and (2) the theory to express the error of the biomass tables in a form convenient for its later application to the calculation of the error of biomass estimates exists only for the linear models, there is no advantage to use the allometric functions so much in favor with researchers in forest biomass; the use of linear models is strongly recommended.

SUMMARY COMMENTS

We started with a short assessment of the forest biomass resources in the world and stated that (1) these resources have reached a relatively stable state in the temperate regions and (2) the resources of tropical regions are being depleted or degraded at a relatively fast pace. A global monitoring system maintained by some international body is critically needed to keep watch over what happens to biomass resources and make periodical assessments. Assessment of biomass resources refers in general to estimates of current values and rates of change, of quantity of biomass by quality classes and its distribution by geographical regions. Because the forest area is too large for a complete measurement, the biomass estimates must be derived from forest inventories by sampling.

If specific biomass estimates are required for specific points in time, the inventory is of the one-shot type. But if estimates are required of both, current values at given points in time and rates of change in these values, one should use systems of forest inventories on successive occasions. In any case the structure of the basic sampling design used by most, if not all inventory systems consists of two phases. In the first phase a sample of trees yields an estimate of the frequency of trees per unit area by quality classes and the sample of the second phase is used in the establishment of the relationship between biomass and tree size (by quality class). Applied to the frequency function of the first phase, the relationship of the second phase yields biomass estimates by unit area and tree quality classes.

We have then continued with an outline of the basic sampling concepts used in the sampling designs of the first and second phase. We did not attempt to describe specific inventory systems using these concepts. The interested reader can refer to several earlier papers by Cunia (1974, 1975, 1976, 1978a, 1981b). But it may be interesting to say a few words of how we see future developments in the inventory systems.

Except for use of more sophisticated and efficient statistical sampling designs we see little future change in the basic inventory systems on one occasion (operational inventory) or inventory on successive occasions applied to relatively small areas (management inventory). The remote sensing technology, combined with advanced statistical methodology will find a much wider application to inventory systems on successive occasions applied to large areas at the national or global level.

With the advent of high altitude aircraft and spacecraft, and the development of infrared sensors, multispectral scanners, or sidelooking airborne radar (SLAR), the forest area of the entire globe can be covered at regular, short-time intervals at relatively low cost. The quantity of data generated by satellite or aircraft multispectral sensors is staggering and means must be found to reduce it to a manageable size. Some substantial reduction of the data can be obtained by multivariate statistical techniques such as principal components analysis or canonical analysis.

It is outside the scope of this paper to discuss in detail any of these techniques. But a few words may be appropriate. For example, the principal components techniques transforms a set of original p variables to a new set of p variables such that most of the information contained in the old set is now concentrated into the first two or three components. Eav (1979) has stated that it is possible to pack into three components about 99 percent of the information contained in four spectral bands of the Landsat; that about 90 percent of the information from eight Landsat bands (two successive scans of the same area) may be concentrated into three components; that for the 16-channel Landsat data the first three components may contain as much as 80 percent of the information. This means that the many variables generated by multispectral bands can be replaced by three or four new variables containing most of the information.

Similarly, the canonical analysis, also known as multiple discriminant analysis derives linear transformations of the original variables that emphasize the differences among land and water classes. This is accomplished by (1) moving the origin of the axes to the point defined by the means of the old variables and (2) rotating the original axes so the the first new axis is placed according to the maximum possible separability among the land class mean vector, the second axis is made orthogonal to the first and according to the maximum remaining class separability and so on. The axes are also scaled so that the ellipsoidal dispersion patterns are transformed into spherical patterns and finally the space is reduced to a lower dimension for ease of classification. A test made with eight variables obtained by merging Landsat data for two dates showed that 99.6 percent of the separability between four land types (river, railroad yard, suburban area, and areas of vegetation) was accounted for by the first two canonical axes.

Independent of what reduction technique is being applied, one needs ground measurements to relate multispectral variables to the variables of interest as measured on sample plots. Presently, the correlation between remote sensing and ground measurement is relatively weak and has a somewhat limited application to systems of forest inventories on successive occasions. It is expected, however, that the strength of this correlation will increase sufficiently well in the near future so as to dramatically improve the methodology of forest inventory systems of large areas.

REFERENCES

Cunia, T. 1974. Forest inventory sampling designs used in countries other than the United States and Canada. In: Inventory design and analysis - Proceedings of a workshop (W.E. Frayer, G.B. Hartman and D.R. Bower, Ed.), pp. 76-110. Colorado State University, Fort Collins, Colorado, USA.

Cunia, T. 1975. Current inventory designs in Europe and the United States. In: Canadian forest inventory methods - Proceedings of a workshop (V.G. Smith and P.L. Aird, Ed.), pp. 43-61. University of Toronto Press, Toronto, Canada.

Cunia, T. 1976. Statistical advances in the methodology of forest inventory. In: Proceedings, XVI IUFRO World Congress, Div. VI, Oslo, Norway.

Cunia, T. 1978a. A short survey of the worldwide forest inventory methodology. In: Integrated inventories of renewable natural resources - Proceedings of a workshop (H.G. Lund, V.J. LaBau, P.F. Folliott and D.W. Robinson, Ed.), pp. 114-120. USDA, Forest Service, Rocky Mountain Forest and Range Experiment Station, General Report RM-55, Fort Collins, Colorado, USA.

Cunia,T. 1978b. On the objectives and methodology of National Forest Inventories. In: National Forest Inventory-Proceedings of a workshop, pp. 11-28. Institutul de Cercetari si Amenajeri Silvice, Bucharest, Romania.

Cunia,T. 1979a. On tree biomass tables and regression: some statistical comments. In: Forest resource inventories - workshop proceedings (W.E. Frayer, Ed.), pp. 629-642. Colorado State University, Fort Collins, Colorado, USA.

Cunia, T. 1979b. On sampling trees for biomass tables construction: some statistical comments. In: Forest resource inventories - workshop proceedings (W.E. Frayer, Ed.), pp. 643-664. Colorado State University, Fort Collins, Colorado, USA.

Cunia,T. 1981a. Cluster sampling and tree biomass tables construction. In: XVII IUFRO World Congress, Interdivisional Proceedings, pp. 151-163. Kyoto - Japan.

Cunia, T. 1981b. A review of the methods of forestinventory. In: WOODPOWER - new perspectives on forest usage (J.J. Talbot and W. Swanson, Ed.), pp. 9-20. Pergamon Press, New York, NY, USA.

Eav, B.B. 1979. Present and potential uses of multivariate techniques in the analysis of remote sensing data. Unpub. MS. Lockheed Electronics Co., 1830 NASA Road 1, Houston,TX, USA.

Lanly, J.P. 1982. Tropical forest resources, Food and Agriculture Organization of the United Nations, Rome, Italy.

Milescu, J. and A. Alexe. 1969. Padurile pe glob. (in Romanian) Editura Agrosilvica, Bucharest, Romania.

Ware, K.D. and T. Cunia. 1962. Continuous forest inventory with partial replacement of samples. Forest Science Monograph 3. Society of American Foresters, Washington, D.C., USA.

Young, H.E., L. Hoar and T.C. Tryon. 1976. A forest biomass inventory of some public lands in Maine. In: Oslo Biomass Studies (H.E. Young, Ed.), pp. 283-302. Univ. of Maine, Orono, ME, USA.

AERIAL PHOTO BIOMASS EQUATION

Joseph Kasile

Division of Forestry
The Ohio State University
Columbus, Ohio 43210
U.S.A.

INTRODUCTION

Aerial photo volume tables have been used by foresters and others to estimate board feet and cubic feet of standing timber gross volume in the United States for a number of years (Paine 1981). These tables have been very useful to assist in the stratfication of timber stands, thereby making ground sampling more efficient.

The introduction of whole-tree chippers into the timber harvesting of the eastern hardwood region of the United States rendered aerial photo volume tables obsolete. Many forest stands that have little board foot volume can have a considerable tonnage of whole-tree chips. Even cubic foot aerial photo volume tables are difficult to accurately convert to chip tonnage because utilization standards used to develop the cubic foot tables are no longer applicable to the whole tree harvesting process.

Study Sites

Southeast Ohio, of the United States, containing fully stocked stands of second growth hardwoods, was the study area. This area is part of the non-glaciated region of southeastern Ohio. Figure 1 gives the approximate location of the sample locations.

At each location, ground samples of site index were obtained. Carmean's curves (1978) were used to determine site quality for 13 tree species. Since each tree species has its own individual growth characteristics, site indexes of these various tree species were

Figure 1. Sampled whole tree chipper sites, Ohio, U.S.A.

converted to those of black oak (<u>Quercus</u> <u>velutina</u> Lam.) for a location to location comparison (Carmean 1979).

A one-way analysis of variance and using Tukey's pairwise comparison test (Guenther 1964) showed significant differences in the site index of black oak for these locations. The range of site indexes indicates that most of the accepted level of site productivity found for this region was included in the sample locations.

Eighteen study sites were used in the study. The harvested area for each study site ranged from 16 to 47 acres (6.5 to 19 hectares).

METHODS

Black and white pancromatic aerial photographs of scale 1:15840 were used to make measurements of stand height, crown closure percent, and average crown width of dominant and codominant trees. These photographs were taken in the spring of 1979.

Each of the eighteen study sites was then harvested with a whole tree chipper operation. The total weight of all harvested material was obtained from the sale records of the operators. A field survey to determine the actual area cut was done after the harvest was complete.

Ten measurements of stand height, crown width, and crown closure percent were taken from aerial photographs at each harvest location. Table 1 gives the results of these measurements. Most of

Table 1. Aerial photo measurements and yield for eighteen forest stands in southeast Ohio, U.S.A.

Stand	Tree Height[1]	Crown Width	Crown Closure Percent	Harvested Tons per Acre
1	68 ± 13	38 ± 6	100 ± 0	82
2	74 ± 8	28 ± 4	100 ± 0	88
3	84 ± 11	35 ± 5	96 ± 1	128
4	59 ± 7	36 ± 6	98 ± 1	62
5	40 ± 10	22 ± 5	100 ± 0	38
6	58 ± 4	28 ± 5	100 ± 0	68
7	71 ± 9	31 ± 7	100 ± 0	92
8	88 ± 6	34 ± 4	100 ± 0	121
9	93 ± 8	41 ± 7	97 ± 0	130
10	75 ± 6	43 ± 4	100 ± 0	98
11	77 ± 7	41 ± 2	100 ± 0	94
12	67 ± 11	37 ± 2	100 ± 0	77
13	70 ± 13	37 ± 4	100 ± 0	88
14	89 ± 5	35 ± 2	93 ± 2	116
15	98 ± 7	38 ± 3	85 ± 2	124
16	89 ± 14	44 ± 4	77 ± 3	122
17	82 ± 9	40 ± 6	100 ± 0	109
18	83 ± 7	40 ± 4	100 ± 0	113

[1] Measurements are in feet ± 95 percent confidence bound for the mean of ten measurements per stand.

the forest stands had 100 percent crown closure, and because the harvested stands tended to be mature, no average stand tree height was less than 40 feet.

The following variables were tested using maximum R^2 stepwise regression analysis (SAS 1980): stand height, stand height squared, crown diameter, crown diameter squared, crown closure percent, crown closure percent squared, stand height times crown closure percent, stand height squared times crown closure percent, stand height squared times crown closure percent squared times crown diameter, and stand height times crown diameter squared. Aerial photo volume tables have indicated that one or more of these variables should be good predictors of harvested forest biomass.

Stand height was the only significant ($P = .01$) variable. The coefficient of determination ($r^2 \times 100$) equaled 94.5 percent and the equation was:

$$\text{Biomass (tons per acre harvested)} = -32.62 + 1.7255 \text{ Average Stand Height (feet)}.$$

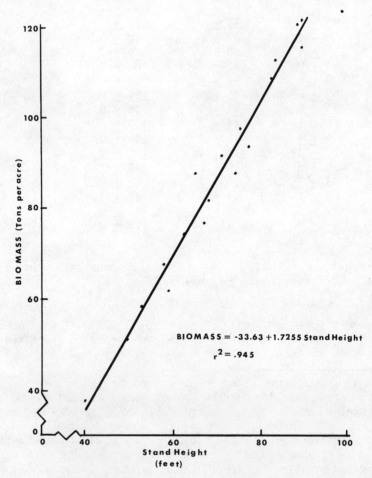

Figure 2. Regression of harvested biomass on tree height for 18 forested areas in Ohio, U.S.A.

DISCUSSION

Average crown diameter, crown closure percent, and stand height squared or combinations of these variables were not significant. Since these stands were typical of the areas harvested with whole tree chipper operations in southeast Ohio, the crown closure was 100 percent or close to that for all the data. It is not surprising that crown closure percent was not significant, because of this lack of variability in crown closure percent. A wider range of percentages may have made crown closure percent a good predictor as it has been important in most aerial photo volume tables. Average crown diameter was not significant, and for fully stocked stands smaller crown diameters are usually indicative of more stems per acre; thus crown diameter had no effect on a stand basis for these fully stocked stands.

For fully stocked hardwood stands of Ohio that are being considered for harvest in a whole tree chipper operation, the predictor equation developed by this study should provide a reasonable estimate of potential tonnage per acre that may be produced. Forest managers of Ohio now have a quick, inexpensive tool for planning their harvest operations by first using aerial photographs to predict harvest tonnage of various stands and then plan their harvest operations to insure a continuous uniform supply to the pulp mill.

REFERENCES

Carmean, Willard H. 1978. Site index curves for northern hardwoods in northern Wisconsin and upper Michigan. U.S.D.A. Forest Service Research Paper NC-160. 16 pages.
Carmean, Willard H. 1979. A comparison of site index curves for northern hardwood species. U.S.D.A. Forest Service Research Paper NC-167. 12 pages.
Guenther, William C. 1964. Analysis of Variance. Prentice-Hall, Inc. Englewood Cliffs, N.J. 199 pages.
Paine, David P. 1981. Aerial Photography and Image Interpretation for Resource Management. John Wiley and Sons, Inc., New York, N.Y. 571 pages.
Statistical Analysis System. 1978. SAS Institute Inc., Box 8000, Cary, N.C. 27511.

FOREST BIOMASS UTILIZATION IN GREECE

George Tsoumis

School of Forestry and Natural Environment
Aristotelian University
Thessaloniki, Greece

INTRODUCTION

The concept of "forest biomass", as it pertains to the present discussion will include only wood of small dimensions and from the following sources: coppice forests, thinnings, logging residues, shrubs and wood processing residues.

Wood of large dimensions (sawlogs, veneer logs, etc.) is not included, as well as other types of forest biomass -- namely, foliage, pine resin, Christmas trees, fruits and seeds, etc.

The utilization of wood of small dimensions is the main problem of forest biomass utilization in Greece. Such material constitutes the majority of production from the forests due to the forest situation in the country (Tsoumis 1964, 1978, 1980).

Forest Situation in Greece -- Production of Wood

Greece is a country with many mountains but few forests. The forests that once existed have been largely destroyed over thousands of years of inhabitation and dependence on wood for heating, cooking, construction, boat building, etc., by grazing and other misuse (Tsoumis 1964).

---

[1]In this paper, wood of small dimensions includes, at the upper scale, logs up to about 20 cm in diameter. Some are used as saw logs, according to length and quality, but larger diameter logs of short length (1-2 m) are also included.

Although about three-fourths of the country is mountains, the forested area is only 18.6 percent and forests producing wood of large dimensions cover about one-third of that area. About half of all forests are coppice, which are harvested every twenty to thirty years, and shrubs, both of which produce wood of small dimensions[2]. In addition, the wood of large dimensions is no more than fifty to sixty percent of the total volume harvested from forests producing such wood; the rest is wood of small dimensions, tops, branches and wood of low quality.

In 1980, the total wood production was 2,824,298 m^3; of this quantity 448,027m^3 (15.9%) was sawlogs; 163,882 m^3 (5.9%) small dimension wood utilized for particleboard and fiberboard; and 2,038,695 m^3 (72.1%) small dimension fuelwood. The rest, 173,694 m^3 (6.1%) was partially wood of small dimensions utilized for boxes, mine timbers, parquet flooring, etc.(Ministry of Agriculture 1981) Therefore, small dimension wood was about 80 percent of the total[3]. Fuelwood production has been declining lately due mainly to expanded use of other sources of energy such as oil, gas and electricity, but it is indicative of the potential production of such wood that in former years the production was more than double. For example, in 1959 it amounted to 4,529,994 m^3 (CEPE 1976).

In addition, wood of small dimensions is derived from other sources such as some logging residues remaining in the forests including bark, since debarking is mainly done in the forests), small dimension wood from private, fast-growing hybrid poplar plantations and from burnt forests, and residues from wood-using industries and workshops.

The quantity of logging residues is small and bark is of no practical value and remains scattered in the forests. However, private poplar plantations produce every year an estimated 300,000m^3 of wood which is not included in the above figures for wood production (Tsoumis 1980) and 40 to 50 percent of such production is wood of small dimensions. Forest fires are frequent in Greece, especially in the pine forests of the dry costal regions, and small dimension wood that becomes available yearly from this source may reach 200,000m^3, according to estimates. Woodworking residues should be

---

[2]The forested area is covered as follows: high elevation conifers 19.7%, Mediterranean conifers (mostly utilized for pine resin) 18.7%, deciduous hardwoods (mostly coppice) 43.0%, and shrubs (evergreen hardwoods, "maquis") 18.6% (Tsoumis 1980).

[3]The above figures do not include the wood of such items as posts and short supports (vineyards, etc.) which are given in numbers, not volume.

estimated to be 30 to 50 percent of the volume brought to wood-using industries such as sawmills, veneer and plywood operations and workshops. This volume is about 1,000,000 m^3 of local production and imported tropical round timber.

In total, the forest biomass under consideration should amount to about 3,000,000m^3 based on 1980 data. The greater part of this, two-thirds, is used as fuelwood. However as previously mentioned, the production capacity is much higher and fuelwood consumption is declining. If this declining trend continues as expected, it has been estimated statistically that by 1990 the consumption will come down from the present value of about 2,000,000m^3 to less than 1,000,000m^3; a high estimate is 980,000m^3 while a low estimate is 445,000 m^3 (Sakkas 1979).

In summary, a very considerable volume of small dimension wood is now produced in Greece, actually the majority of local production and this will increase in the future. The species composition of such wood is as follows: the majority is hardwood, mainly oak which constitutes about one-third (30.2%) of the forests of Greece, but all oak forests are coppice (Tsoumis 1964, 1980); other species are beech, chestnut, poplar, and evergreen hardwoods which are shrubs; softwoods including pine, fir and spruce contribute less.[4] Manufacturing residues include tropical woods.

PRESENT UTILIZATION[5]

In addition to fuelwood for heating and cooking (mainly in the rural regions), wood of small dimensions finds outlets in the following uses:

Charcoal. In 1980, 11,589 tons were produced in traditional charcoal kilns and consumed about 58,000m^3 of fuelwood. The main species utilized is oak while other species include beech, ahorn, shrubs, etc.

Particleboard. Total yearly production from 15 units is about 380,000m^3 which corresponds to about 700,000m^3 of wood (1980). This is made up as follows: 151,585m^3 is small dimension wood or "Industrieholz" from state forests and is mainly softwoods such as pine

---

[4]In 1978, the composition of small dimension wood (round form) was as follows: oak 1,056,000m^3 beech 412,500m^3, other hardwoods 133,500 m^3, shrubs 310,500m^3 and softwoods 506,000m^3 (Ministry of Agriculture 1981).

[5]Data are based on cited references (CEPE 1976, Ministry of Agriculture 1981) and other information supplied by industries.

and fir; 200,000m³ from private sources such as poplar plantations; 350,000m³ are industrial residues including tropical woods.

Fiberboard. There is one factory producing hardboard which amounted to about 10,000 tons in 1980. Small dimension wood (fuelwood) of beech and some chestnut is utilized amounting to a total of about 20,000m³.

Pulp. One factory is producing thermomechanical pulp and consumes about 40,000m³ of small dimension wood, mainly pine and some spruce. Two other factories produce pulp from straw. Most of the pulp and paper utilized in Greece is imported.

Boxes and Crates. Most are fruit boxes usually made by peeling small poplar logs; other boxes and crates utilize beech and softwoods. A total of 119,777m³ of wood was used in 1980, according to statistics. The actual quantity should be greater, however, because wood from private poplar plantations is extensively used but not included in the statistics.

Parquet Flooring. This is produced mainly from small dimension wood of oak, beech, chestnut and pine as well as some imported tropical woods. Short logs of large diameter, selected from old trees of low quality, and small diameter wood are thus transformed into a high value product. In 1980, 21,624m³ of local production was used in this manner.

Posts and Short Supports. Fence posts, short and small diameter supports for rural construction, supports for vineyards, frames for air-drying tobacco leaves, etc. are included here. These products, as mentioned earlier, are given in numbers, not in volume of wood. In 1980, a total of 613,920 such items, including 200,300 posts, were produced from the forests. The main species for posts is chestnut due to its high natural durability; other items are made of pine, oak, poplar, Mediterranean cypress, etc.

Mine Timbers. This is local wood of small dimensions consisting of oak, pine and chestnut used in round form. Production in 1980 amounted to 6,950m³.

Hewnwood. This is a commodity preferred in rural regions for beams and supports in building and is produced by hand (axe) squaring of small diameter logs such as thinnings. Production is declining because workers with the required dexterity are lacking and sawn products are a suitable replacement. In 1980 only 400m³ were used in this manner compared to 63,233m³ in 1972. Both softwoods such as pine and fir, and hardwoods such as chestnut and oak are used.

Other Products. This category includes pallets (various wood species), barrel staves (pine, oak, etc.), baskets (weaving willow, etc.), toys (beech and other species), carvings (juniper, pine, olive, boxwood, cypress), smoking pipes (outgrowths of (<u>Erica arborea</u>) and others. The production of outgrowths for pipes was 1,630 tons or about 2,000m^3 in 1978.

<u>Manufacturing Residues</u>. Slabs, edgings, sawdust, etc. in addition to their utilization in particleboard manufacture, are utilized to produce steam which is used for heating, drying and steaming (beech, veneer logs), or as a source of energy to operate woodworking or other machines.

<u>Bark</u>. This material is not utilized as such except for some inclusion in particleboard. In the past, bark from certain species such as Aleppo pine was used as a source of tanning in processing leather.

FUTURE PROSPECTS

Prospects for the future include increased use of the forest biomass under consideration for products and energy, as well as silvicultural measures to lower production of small dimension wood in the forests. In the products and energy category the main items for future consideration are the following:

<u>Particleboard</u>. Wood of small dimensions may be used to increase the production of particleboard. Experimental particleboard, made from small dimension oak wood with bark ("fuelwood"), was shown to have good quality according to DIN standards but was heavier than usual at about 850 Kg/m^3 (Tsoumis et al. 1977). The mixture of other, lighter species such as poplar and fir will correct this property and such research is under way.

<u>Fiberboard</u>. Interest has been expressed in producing medium density fiberboard (MDF). This product can be made with low quality wood (small dimension, residues with bark) and hardwoods, and has certain advantageous properties in comparison to hardwood and particleboard.

<u>Pulp</u>. Increased production of pulp has been investigated and is a subject for future work. The realization of such production will be favored by application of methods which do not require great quantities of wood.

<u>Other Products</u>. Increased production of other products previously mentioned could be expected, but there are problems. For example, exploratory research in making parquet flooring from small dimension oak wood, 10-20 cm in diameter and from trees 20 to 30 years of age, was not successful. Its quality was mostly low due to the heartwood and sapwood color distribution on the surfaces and to warping due to presence of juvenile wood (Theodorou 1979).

Energy. Improved methods of producing energy from forest biomass need to be applied. Pyrolysis with more efficient carbonization and small unit gasification present possibilities. In relation to such usage, as well as for products, chipping of the tree in the forests could be helpful in better utilization.

Mention should be made that the above proposed possibilities for products and energy are subjects of the following research studies, in addition to the previously mentioned ones:

A project will investigate, from a technical and an economic point of view, the possibilities of increased utilization of small dimension wood from coppice forests and from shrubs. This includes inventory studies, harvesting methods and economics, evaluation of the material as a source of wood and fiber, and technical and feasibility studies for production of products and energy.

Another project involves a proposal to harvest shrubs every ten years and utilize their biomass for energy and products. The idea is that such "ecosystems" are prone to burn, for climatic reasons. Therefore, their harvesting will reduce the danger of fires while at the same time making considerable quantities of biomass available (Margaris 1979).

As indicated earlier, improvement of future prospects is also sought by silvicultural and related measures such as (1) transformation of coppice to high forests; (2) increase of the rotation or harvesting period of coppice forests from the present 20-30 years up to 100 years or more; (3) reforestation in coppice forests and shrub formations by introduction of conifers; and (4) cultivation of high and other forests by thinning, pruning, and removal of low quality trees to decrease the proportion of small dimension wood. However, these measures, especially 1, 2 and 3, are at a disadvantage because of the many years needed to accomplish the goals sought.[6] Related to the above are the efforts to increase biomass production by increased cultivation of fast-growing forest trees such as hybrid poplars, eucalypts, pines such as _Pinus radiata_, and other species (Panetsos 1981).

CONCLUSION

Forest biomass utilization in Greece is a problem mainly with regard to the large production of wood of small dimensions. In addition to its use as fuelwood, such wood is currently processed into

---

[6]The same measures have also contributed to some extent to the reduction of fuelwood production, as mentioned earlier.

various products. Efforts are being made to broaden the technical utilization and reduce the production of small dimension wood.

REFERENCES

CEPE (Center of Planning and Economic Research). 1976. Plan of Development 1976-1980. Forestry Sector. Athens (Greek).
Damalas, G. 1976. An econometric investigation of wood products consumption in Greece. Inst. For. Res., Athens. (Greek with English summary).
Margaris, N. 1979. Can we harvest Mediterranean type ecosystems to obtain energy and organics? In: Biological and Sociological Basis for a Rational Use of Forest Resources for Energy and Organics. S.G. Boyce, Ed. pp. 121-128. USDA Forest Service, S.E. For. Exp. Sta. Ashville, N.C., U.S.A.
Ministry of Agriculture. 1981. Activities of the Forest Service, 1980. Athens (Greek).
Panetsos, K. 1981. Biomass yield of short-rotation Platanus species in Greece. In: Components of Productivity of Mediterranean-climate Regions. N. Margaris and H. Money, Eds., pp. 235-238. Dr. W. Junk Publ., The Hague/Boston/London.
Philippou, I. 1981. Die Holzindustrie Griechenlands. Proc. IX Interna. Symposium: Mechanisierung bei der Forstnutzung. Thessaloniki.
Sakkas, G. 1979. An economic appraisal of forestry and forest products industries in Greece, 1950-1990. Ph.D. Dissertation, Dept. Forestry and Wood Science, Univ. College N. Wales, U.K.
Theodorou, Ph. 1979. Utilization of the wood of oak coppice forests (small dimension wood) in the parquet industry. Thesis, Lab. Forest Utilization, Univ. Thessaloniki. (Greek).
Tsoumis, G. 1964. Forestry in Greece. Yale Forest School News 52 (3) (3):35-37.
Tsoumis, G. 1978. L'économie forestière grecque -- situation actuelle et perspectives. Publications CEA (Conféderation Européenne de l'Agriculture) 59:172-178. [Die griechische Forstwirtschaft -- Aktuelle Lage and Ausblick. Veroffentlichungen CEA 59:185-190.]
Tsoumis, G. 1980. Forest Utilization in Greece. Lab. Forest Utilization, Univ. Thessaloniki.
Tsoumis, G., Passialis, C. and Ph. Siamidis. 1977. Experimental particleboard from oak fuelwood. Technica Chronica 3:12-18. (Greek, English summary).

BIOMASS--THE AGRICULTURAL PERSPECTIVE

William J. Sheppard and Brian Young

Battelle Columbus Laboratories
505 King Avenue
Columbus, Ohio 43201
U.S.A.

WHY USE BIOMASS?

In considering the interest that the agricultural industry might have in biomass production for energy, it is important to consider the motivation for this activity. It is also appropriate to ask why an energy user would want to use biomass and why a government would want to encourage the production and consumption of biomass energy. Without trying to exhaust all the possible reasons, three come to mind quickly:

o to make or save money
o to solve a problem
o to decrease risk.

The farmer or forester obviously wants to have a large income. If a new crop or product will allow him to earn more, he will consider new crops or selling to new markets. He may also have a problem, for example, in cold weather regions such as Minnesota and the Canadian prairie provinces, maize residues left in the field keep the soil cold in the spring, delaying planting. Collection and sale of these residues could solve this problem. Obviously, the feed lot operator can have a problem if manure accumulates. Finally, the farmer can reduce risks. He may plant several crops to reduce the impact of bad weather on one crop that is economically more rewarding but is sensitive to drought. Planting a crop with energy as well as food use reduces risk of a market oversupply. For the farmer, producing his own fuel is one way to reduce risk. When it is the time to sow or the time to reap, he needs fuel immediately and cannot wait for a governmental solution of the latest oil problems.

The user of energy or chemical feed stocks also wishes to make as much money as possible and is thus willing to use biomass if it is cheaper. He may also have a problem to solve such as the need for improving the octane number of his gasoline without adding octane improving equipment at the refinery. He would be willing to buy ethanol for blending, particularly if the problem were short term. The user also wishes to reduce risk and may, therefore, install a boiler that can burn biomass residues as well as coal and oil.

Governments want to "make money," that is, to see the Gross National Product increase as well as to increase tax income. Domestic biomass production, with local value added, is more desirable than imported finished fuels. Biomass use can solve problems such as preserving farming as a way of life without flooding the market with an oversupply of food. The reduction of risk by replacing foreign energy supplies with domestic supplies is a way of reducing the risk of international supply problems. Reducing the risks to the economy of a strike in one industrial segment is another possible goal of government.

WHAT IS BIOMASS WORTH?

Looking at the money side of biomass use for energy or feedstock, one needs to consider the other uses that can be made of the biomass. There is no point looking at markets for biomass as a fuel or chemical feedstock even at a low cost if its use as food or animal feed commands a better price. In this comparison, the value of the material to the user must be considered. As a start, let us compare food and fuel values. Sugar at $0.66 per kg in the supermarket sells for $40 per gigajoule. Gasoline at $0.31 per liter (including taxes) is equivalent to $10 per gigajoule. The higher value for the sugar can be attributed to the fact that it has extra value, namely, nutritional, preservative, and flavor values. Thus, until all food markets for sugar are satisfied at all prices down to the fuel value, no sugar will be offered to produce fuel at the fuel valued price.

Of course, once the food market at the current price is satisfied and prices drop, other markets may be able to use sugar before the price reaches the fuel price. Another large market that can afford to pay a nutritional premium is the animal feed industry, which is now a large user of molasses. Thus, there is a hierarchy of needs, and willingness to pay, in the order food, feed, then fuels. Further consideration of other markets leads to the extended hierarchy shown in Table 1 (with apologies for the forced alliteration). Intermediate between feed and fuel uses are those that utilize the one-, two-, and three-dimensional properties of cane fiber or sucrose derivatives, for example, rayon or polyethylene, fiber, paper, and Cellotex board. The chemical and bulk physical properties are reflected in use as a chemical feedstock, fertilizer, and aid to improve the quality of the soil, prevent erosion, and hold moisture in the soil. These all have a greater value per kg than just using the chemical bonding energy released on combustion. Also, there are values higher than the nutritive value of food. These are related to higher physiological activity of the material such as use as medicine, intoxicant, flavor, or fragrance. The beverage use of alcohol at retail, tax paid, commands at least $10 per liter, which is equivalent to $15 per liter for gasoline. Thus, sucrose as a sweetener is worth more per kg than corn starch, which provides calories but no flavor. Functional aids, that is, ingredients which improve the properties of food, shelf life, or preparation ease have a premium. Sucrose as a preservative in marmalade is worth more than a food that provides calories only. The hierarchy of values is summarized in Table 1.

It should be noted that the high value products are usually sold in smaller volumes than the lower valued products. For example, the annual consumption of sugar is about 40 kg per capita in the United States versus about 1100 kg of gasoline per capita. Economic incentives lead the producer to fill the highest value market even at low volumes then to "spillover" to the next lower price, larger volume market. Frequently, product differentiation is used to avoid loss of the premium price in the first market while gaining sales volume in the second. For example, alcohol is sold in beverage (taxable and tax free), medicinal, and denatured solvent and fuel grades.

Within each category there is also a heirarchy. Meat is worth more than fats and oils, which in turn are worth more than starchy foods, which merely provide calories. In the fuels category, transportation fuel and home heating oil have higher value than fuel for running factories or generating electricity.

Occasionally, there is an inversion of the order, but only in unusual circumstances, for example, the snowbound cabin dweller's burning his furniture for warmth. For chemical uses of sucrose to rank higher than food and feed would require quite a revolution in human values. On the other hand, at the food and feed price for sucrose, economic incentives would be more likely to lead to the use of sucrose as a chemical intermediate than as a source of fuel.

WHERE WILL IT COME FROM?

Just as the values for biomass in various uses forms a hierarchy of values, the cost of obtaining biomass forms a hierarchy of costs, shown in Table 2. Some sources must be grown and have an associated cost, for example, growing grain for ethanol production

Table 1. Hierarchy of biomass uses.

Farmaceutical
Firewater
Flavorant, fragrance
Functional aid for food
Food
    Flesh, fowl, fish
    Fats and oils
Fillers
Feed
Fiber
Film and sheet
Framing and lumber
Feedstock for chemicals
Filler for plastics
Fertilizer
Friability aid, erosion control, etc.
Fuel
    For family Ford
    Flying
    Family furnace
    Factory and electricity generation

and planting trees for forest crops. Even crops and forests that grow wild must be managed and the production harvested and hauled to market.

Field crop and forest residues can be obtained at the cost of collection and hauling to market. There may be additional hidden costs such as replacement of soil nutrients and erosion problems or there may be benefits such as removal of insect eggs. Still cheaper is the use of factory residues since the cost of collection and transportation is borne by the principal product. Examples are sugarcane bagasse at a sugar mill, molasses at a sugar mill or refinery, food processing wastes at a cannery and slab wood, wood scraps, bark, and sawdust at a sawmill, pulp mill or furniture factory. There are other wastes that accumulate at a processing location that have a large negative value and must be disposed of at a cost, for example, manure at a feed lot and whey at a cheese factory.

Table 2. Sources of biomass.

Field crops
Forest crops
Field crop residues
Forest crop residues
Factory residues
Feed lot residues

BIOMASS-THE AGRICULTURAL PERSPECTIVE

HOW MUCH IS AVAILABLE?

In the sections that follow, data are given for crop and residue availability, mostly from the work of Brian Young. Primarily the data are from the United States and Canada and are given in U.S. units (1 ton U.S. = 0.909 tonne; 1 pound = 0.454 kg, 1 mile = 1.61 km.)

Field crops can be divided into:

o grain crops--maize, wheat, rice
o sugar crops--sugar cane, beets, sweet sorghum, Jerusalem artichokes
o starchy roots--potatoes, cassava
o oil seed crops--soybean, sunflower, rape (cottonseed, maize, germ)
o hydrocarbon crops--<u>Euphorbia lathyris</u>, guayule.

Data from grain, oil seed and hydrocarbon crops are presented in the following section. With the current price structure in the United States, sugar crops are too expensive for biomass use today. However, with the world price at $0.15 per kg for raw sugar, opportunities are great. Among the root crops, cassava (maniac) has potential in dry areas with small farms and low labor costs.

GRAIN CROPS

All of the major grains are sold as commodities and are readily available in all parts of the country. The primary regions of production are listed below:

o corn--U.S. Midwestern states
o wheat--U.S. plains states and Saskatchewan
o barley--U.S. northern plains and Pacific northwest states and Alberta
o sorghum--U.S. plains states.

Table 3 displays present and future production and prices.

Oilseed Crops

There are seven principal oilseed crops grown in the United States and Canada at the present time. Several other oilseed crops are grown in limited quantities but near-term exploitation of these crops for final use is not likely due to uncertainties concerning commercial cultivation. Among these crops are Chinese tallow tree and crambe. While castorbean is grown in significant quantities in other parts of the world, imported oil is committed to industrial or

Table 3. Estimated U.S. and Canadian production and prices for the major grain crops--1982, 2000, and 2020.

Crop	U.S. and Canadian production and prices			
	1982		2000	2020
	(M tons)[a]	($/ton)[b]	(M tons)[c]	(M tons)[d]
Corn	191,592	88.57	268,851	302,042
Wheat	91,909	121.67	107,994	119,084
Barley	22,129	105.00	27,429	31,204
Sorghum	17,842	83.60	26,397	29,825

(a) Calculated by projecting actual 1980 production forward at rates equal to those projected for each crop (1975-1985) by Ernest et al. (1979) (with the exception of wheat, which was considered constant).

(b) Season average price, 1981.

(c) Calculated by projecting 1982 data forward at rates equal to those projected for each crop (1985-2000) by Ernest et al. (1979).

(d) Calculated by projecting 2000 data forward at rates equal to those projected for each crop (2000-2020) by Ernest et al. (1979).

Sources: USDA 1961, USDA 1981, USDA 1981a, Ernest et al. 1979.

medicinal use and is moderately expensive. Pertinent information on the principal oilseed crops is presented in Tables 4 and 5.

Latex-Bearing Crops

There are only two latex-bearing crops for which reliable data have been generated on yields and husbandry practices for reasonably large-scale production in North America. These are euphorbia (<u>Euphorbia</u> <u>lathyris</u> L.) and guayule (<u>Parthenium</u> <u>argentatum</u>).

Although the extractable oil of euphorbia is chemically similar to crude oil and yields of up to 18.5 bbl crude oil per acre have been achieved, recent studies at the University of Arizona indicate that the cost of fuel produced from euphorbia is approximately twice that of its fossil fuel equivalent. Since significant increases in yield of euphorbia oil per acre are not likely to be achieved in the near future, this crop should not be considered for fuel production at the present time.

Table 4. Present production of selected oilseed crops in the United States and Canada.

Oilseed	1982 Production U.S. & Canada (M tons)	Raw Oil Prices (U.S. dollar) Avg. '78-'80 ¢ per lb.	Aug. 1, 1982 ¢ per lb.
Soybean	65,345	25.6	18.5
Cottonseed	4,008(a)	28.0	21.0
Sunflower (oil varieties)	1,952(b)	29.2	24.0
Peanut	1,780	36.7	24.3
Flaxseed	816(b)	28.6	27.0
Safflower	215(b)	46.0	70.0
Rapeseed	1,967(b)	41.5(c)	56.0(c)

(a) 1.67 lb seed/lb lint
(b) '79-'80 for U.S. data
(c) Refined oil

Source: Battelle estimates from Agricultural Outlook (August 1982). USDA and Canadian Grain Statistics Handbook (1981).

Table 5. Projected availability of oilseed crops, United States and Canada.

Oilseed	2000 Seed Production M/tons	2020 Seed Production M/tons
Soybean	88,970(a)	98,879(b)
Cottonseed	4,000(b)	4,800(b)
Peanut	3,200(b)	3,800(b)
Sunflower	2,839(c)	3,336(c)
Flaxseed	1,186(c)	1,394(c)
Safflower	312(c)	367(c)
Rapeseed	2,859(c)	3,359(c)

(a) Battelle estimates from Ernest et al. (1979), projections for U.S. soybean production and historic Canadian production 1971-1981.

(b) Ernest et al. (1979) projections.

(c) Battelle estimates based on average of Ernest et al. (1979) projected increases of soybean and peanut.

Table 6. Productivity of Guayule.

Potential Yield of Rubber (lbs/acre)	Composite of Guayule Plant (%)	
	Moisture	Rubber
1500-2000	45-60	8-26

Source: Battelle, National Academy of Science, 1977.

Guayule produces a latex compound of polyisoprene of the cis-1, 4 isomer (Campos-Lopez 1976) in relatively large amounts (Table 6). Recent investigations by Firestone Rubber Company indicate that the cost of rubber produced on large plantations would be about $1100/ton without by-product credits. Commercial size plantations have not existed in the United States since 1946.

Crop Residues

The amounts of various crop residues in the United States and Canada are displayed in Table 7. Harvest and transport costs for corn, small grains, and sorghum are displayed in Table 8. These values, compiled in 1979, represent harvest and transport by farmers and assume an acquisition cost of zero.

In a study conducted by Purdue (1979), corn, small grains, sorghum, and rice residues were identified as attractive for recovery for energy uses. Rice hulls are a special problem because of the high ash content. The total availability of these residues is greatest in the corn belt states. The top six states were identified as Minnesota, Illinois, Iowa, Indiana, Ohio, and Wisconsin.

U.S. Logging Residues

The logging residues considered here are defined as consisting of the above ground unused portion of trees cut or killed by logging. These residues may be from growing stock, non-growing stock, or non-commercial trees growing on commercial forest lands. Growing stock trees are defined by the U.S. Forest Service as "live trees of commercial species qualifying as desirable or acceptable trees." Non-growing stock trees include "rough, rotten and dead trees" of commercial species.

Table 9 displays projected logging residue generation through the year 2020. Complete data on current or even relatively recent logging activity is not available. Because of this, the 1980 data displayed in Table 9 are from projections made in 1977 (Howlett and Gamache 1977).

Table 7. Generation of crop residues in the United States and Canada.

	Residue (M Dry Tons)		
	1982	2000	2020
Corn[a]	111,698	156,740	176,090
Wheat[a]	148,892	155,511	111,481
Barley[a]	26,179	26,405	29,855
Sorghum (grain)[a]	11,205	16,577	18,730
Cotton[b]	2,898	2,746	3,062
Potatoes, Irish[c]	1,942	2,323	2,602
Sugar Beets[c]	2,525	3,129	3,826
Sugarcane[b]	2,482	2,820	3,592
Peanuts[d]	2,022	3,456	4,104
Soybean[d]	118,826	161,836	179,861
Sunflower (oil varieties)[c]	10,076	14,648	17,212

[a] Calculated from residue and dry weight conversion factors (Ernest et al. 1979) applied to production data (Grain Crops section).

[b] Data from Ernest et al. 1979.

[c] 1982 production assumed to be the same at 1980. Future residue general increase at same rate as that given by Ernest et al. 1979.

[d] Calculated from residue and dry weight conversion factors (Ernest et al. 1979) applied to reduction data (Oilseed Crops).

Table 8. Harvest and transport costs of crop residues in selected states (1979)[a].

Crop	State	Cost ($/dry ton)
Corn	Illinois	13.51
	Indiana	13.69
	Iowa	13.93
	Minnesota	18.09
	Nebraska	20.21
	Ohio	15.63
Small Grains	California	13.47
	Illinois	15.76
	Minnesota	15.07
	South Dakota	16.84
	Washington	15.40
	Wisconsin	21.51
Sorghum	Colorado	18.50
	Kansas	33.87
	Missouri	19.38

[a] From Purdue 1979; transportation distance of 15 miles.

Table 9. Projected logging residue generation by geographical region in selected years.

	Projected Residue Generation (MMDTE)[b]					
	1980		2000		2020	
	Low	High	Low	High	Low	High
North	20	25	22	42	33	56
South	38	52	38	71	52	94
Rocky Mountain	4	7	5	9	5	11
Pacific Coast	21	26	16	26	17	34
TOTAL	83	110	81	148	107	195

(a) Regions are defined as follows:

North: Maine, New Hampshire, Vermont, Massachusetts, Connecticut, Rhode Island, Delaware, Maryland, New Jersey, New York, Pennsylvania, West Virginia, Michigan, Minnesota, North Dakota, South Dakota (east), Wisconsin, Illinois, Indiana, Iowa, Kansas, Kentucky, Missouri, Nebraska, Ohio.

South: North Carolina, Virginia, Florida, Georgia, Alabama, Mississippi, Tennessee, Arkansas, Louisiana, Oklahoma, Texas.

Rocky Mountain: Idaho, Montana, South Dakota (west), Wyoming, Arizona, Colorado, New Mexico, Nevada, Utah.

Pacific Coast: Alaska (coastal), Oregon, Washington, California, Hawaii.

(b) Low and high values for each year correspond to two demand-price situations:

o Low projections assume a low level of demand and rising relative prices for wood products.
o High projections assume a high level of demand and stable relative prices for wood products.

Source: Howlett and Gamache 1977.

Logging residues generally have no positive value to forest landowners and in many cases expenses must be incurred to remove part of the residues in order to replant harvested forest land. Table 10 provides two examples of the cost of delivery of logging residues to a utilization point (assuming zero acquisition cost). Future delivery costs are likely to be higher as acquisition costs become positive in response to demand. However, this may be moderated by the development of more efficient residue chippers and machinery which can separate high quality chips from bark and other low quality components of chipped residue. The presence of bark in chipped logging residue impedes the utilization of these residues in a manner similar to mill residues.

Table 10. Estimates of collection, reduction, and transportation costs for logging residues in two regions, 1982[a].

Region	Cost ($/DTE)
Pacific Northwest	
Collection	42.50-69.80
Chipping	9.00-23.00
Transportation (50 mi)	13.00-93.60
Total delivered cost	66.90-93.60
South	
Collection	14.27
Chipping	8.59
Transportation (20 mi)	8.59
Labor	3.84
Total delivered cost	35.29

(a) After Howlett and Gamache (1977), original cost data in 1976 dollars projected to 1982 dollars using 6.0% annual inflation and a rate of increase for wood products above inflation of 1.4% (after Spiewak et al. 1982). DTE = dry ton equivalent.

U.S. Mill Residues

Mill residues are defined as those residues produced by the primary wood producs industry. The industry is considered to composed of three major sectors--the lumber, plywood, and miscellaneous wood products industries. Estimates of residue generation by industry sector and residue type are presented in Table 2. However, it should be noted that much of this residue is utilized as raw material for a number of secondary products including woodpulp and particleboard. In addition many individual manufacturers utilize residue for their own energy needs. Table 12 displays the amount and percentage of total residue left unused by each industry sector and residue type. These values were calculated by applying 1970 land management and technology to 1980 actual production.

The major sources of competition for the various types of residues are as follows:

o   Coarse residues - conversion into chips or flakes for the manufacture of woodpulp and particleboard; and further processing of chips for the manufacture of fiberboard.

o   Fine residues - planer shavings are in strong demand for the manufacture of particleboard. Smaller amounts are utilized as a packing medium, as bedding in kennels and stables and for manufacture of woodpulp. Sawdust is mostly

Table 11. Estimated residue generation in the lumber plywood and miscellaneous wood products industries in 1980[a].

Industry Sector	Coarse[b] Residues (MMDTE)	Fine[c] Residues (MMDTE)	Bark[d] Residues (MMDTE)	Total Residues (MMDTE)
Lumber	40.7	15.1	15.8	71.6
Plywood	12.0	0.7	3.4	16.1
Miscellaneous	3.2	1.2	1.7	6.1

(a) Miscellaneous wood products industry data is from 1979.
(b) Size range for hogged fuel 1/32-4 in.
(c) Size range 1/32 in. or less.
(d) Size range for hogged fuel 1/32-4 in.

Source: Battelle estimates based on Howlett and Gamache, 1977. Agricultural Statistics 1981, and Forest Products Review Winter/Spring 1982.

o   fuel or is discarded. Minor amounts are used as a mulch, for kennel and stable bedding, as a packing medium, and as an absorbent material for grease and oil. Manufacturing end products include composite flooring, woodpulp, wood flour, and artificial stone products.

o   Bark - there is little competition for bark, and it is mostly used for fuel or discarded. Minor amounts are used as mulch and in the manufacture of charcoal.

Table 12. Estimated amounts and percentages of mill residues left unused in 1980[a,b].

Residue Type	Lumber MMDTE		Plywood MMDTE		Miscellaneous MMDTE		Total MMDTE	
Coarse	6.10	15%	0.84	7%	0.32	10%	7.26	13%
Fine	9.36	62%	0.14	20%	0.66	55%	10.16	60%
Bark	6.32	40%	1.36	40%	0.68	40%	8.36	40%

(a) Miscellaneous wood products industry data from 1979.
(b) Assumes 1970 level of management and technology.

Source: Battelle estimates based on Howlett and Gamache, 1977. Agricultural Statistics 1981, and Forest Products Review Winter/Spring 1982.

Table 13 identifies the relative production and utilization of residues by region in the U.S. It should be noted that utilization of residues by individual plants within a given region may vary widely.

Table 13. Percent of total wood and bark mill residues produced by region and percent of residues left unused in each region(a).

Region(b)	Percent of Total Residue Generated	Percent of Residue Unused in Each Region
Northeast	7.7	35
North Central	7.4	33
Southeast	13.2	39
South Central	19.4	28
Pacific Northwest	32.3	15
Pacific Southwest	10.2	37
Northern Rocky Mountain	7.7	32
Southern Rocky Mountain	2.1	56
TOTAL	100.0	

(a) From 1970 data.
(b) Regions are defined as follows:

Northeast - Connecticut, Maine, Massachusetts, New Hampshire, Rhode Island, Vermont, Delaware, Maryland, New Jersey, New York, Pennsylvania, West Virgina.

North Central - Michigan, Minnesota, North Dakota, South Dakota, (East), Wisconsin, Illinois, Indiana, Iowa, Kansas, Kentucky, Missouri, Nebraska, Ohio.

Southeast - North Carolina, South Carolina, Virginia, Florida, Georgia.

South Central - Alabama, Mississippi, Tennessee, Arkansas, Louisana, Oklahoma, Texas

Pacific Northwest - Oregon, Washington, Coastal Alaska

Pacific Southwest - California, Hawaii

Northern Rocky Mountain - Idaho, Montana, South Dakota (West), Wyoming

Southern Rocky Mountain - Arizona, Colorado, Nevada, New Mexico, Utah

Source: Howlett and Gamache 1977.

Table 14 gives estimated representative prices for wood and bark residues in selected regions in the U.S. These values should not be interpreted as all-inclusive; residue prices at particular sites are dictated by local demand and should be determined individually.

Future residue prices are difficult to project as they will be dictated by demand which in turn is subject to: 1) changes in technology and management practices by residue producer and consumer industries, and 2) fossil fuel prices. However, wood products prices are expected to increase 1.4% per year in real terms through the end of the century (Spiewak et al. 1982).

Canadian Logging and Mill Residues

Table 16 and 17 display estimated logging and mill residues in Canada for the years 1979 and 2000. These estimates were made by assuming residue generation rates determined for U.S. industries at 1970 levels of management and technology. Estimates for miscellaneous mill residues are not available. Logging residues were not estimated by region due to a lack of region-specific residue generation coefficients. However, total logging residues generation by region is determined primarily by the level of logging activity.

Food Processing Residues

Of the variety of food processing residues, data have been analyzed for whey and molasses, Tables 18 and 19. Sugarcane bagasse has been covered in the section on crop residues. Others not covered here are fruit and vegetable canning residues and nut hulls. Nut hull use for energy generation is extensively practiced already in California.

Table 14. Some estimated representative prices for wood and bark residues in selected regions, 1982 ($/DTE).

	West Coast	South	Maine
Chips			
Softwood	47-71	39-44	63
Hardwood	NA	28-32	NA
Shavings	7-10	12	NA
Sawdust	5	2-3	8
Bark	4-9	1-2	NA

DTE = dry ton equivalent.
Source: Battelle estimates based on survey conducted by Howlett and Gamache 1977 and producer price indexes for lumber and plywood.

Table 15. Estimated residue generation in the lumber, plywood and miscellaneous wood products industries for the year 2000 and 2020.

Industry and Year[a]	Coarse Residues (MMDTE)	Fine Residues (MMDTE)	Bark Residues (MMDTE)	Total Residues (MMDTE)
Lumber				
2000 - Low	36.3	11.8	14.0	62.1
- High	55.1	17.9	21.3	94.3
2020 - Low	29.7	8.2	11.4	49.3
- High	61.7	16.9	23.8	102.4
Plywood				
2000 - Low	13.9	0.8	4.0	18.7
- High	19.6	1.1	5.6	26.3
2020 - Low	14.1	0.8	4.0	18.9
- High	25.0	1.4	7.2	33.6
Miscellaneous Wood Products				
2000 and 2020	3.9	1.4	2.1	7.4
Total				
2000 - Low	54.1	14.0	20.1	88.2
- High	78.6	20.4	29.0	128.0
2020 - Low	47.7	10.4	17.5	75.6
- High	90.6	19.7	33.1	143.4

(a) Low and high values for each year correspond to two demand-price situations:

o Low projections assume a low level of demand and rising relative prices for wood products.
o High projections assume a high level of demand and stable relative prices for wood products.

Source: Howlett and Gamache 1977.

Whey

Available whey includes all whey which is not further processed for food, feed, or industrial use. This whey is mostly fed to farm animals as-is, or is spread on or knifed into crop land. Some is dumped into sewer systems. Excess whey is basically a negative value by-product which is supplied without charge to farmers. Arrangements vary from plant to plant, but generally small users pick up excess whey at the plant while in some cases excess whey is delivered to large users.

Table 16. Estimated Canadian logging residue generation -- 1979, 2000[a].

Year	Logging Residues[b] (MMDTC)[c]
1979	28.6
2000	49.5

[a] Residue generation estimated using a residue coefficient of 5.0 DTE residue 110^3 ft. timber cut developed by Howlett and Gamache (1979) for average softwood logging in the U.S.

[b] 1979 value derived from actual production, 2000 value derived from logging level projected from 1979 levels at the same rate of increase used in USDA-FS (1973) projection for Canada.

[c] MMDTE = Million dry ton equivalents.

Sources: Howlett and Gamache 1979, USDA-FS 1973, Statistics Canada 1981.

Table 17. Estimated mill residue generation in the Canadian lumber and plywood industries--1979, 2000[a].

Industry and Year	Coarse Residues (MMDTE)	Fine Residues (MMDTE)	Bark Residues (MMDTE)	Total Residues (MMDTE)
Lumber[b]				
1979	19.9	7.4	7.7	35.0
2000	27.4	8.9	10.6	46.9
Plywood				
1979	3.6	0.2	1.0	4.8
2000	7.6	0.4	2.2	10.2

[a] Residue generation calculated using residue generation coefficients (units of residue per unit of primary product) developed for U.S. industry by Howlett and Gamache (1979). Data for miscellaneous wood products industry not available.

[b] 1979 values derived from actual production data, 2000 values derived from production levels projected from 1979 levels at the same rate of increase used in USDA-FS (1973) projection for Canada.

Sources: Howlett and Gamache 1979, USDA-FS 1973, Statistics Canada, 1981.

Table 18. Production of whey in top four states and all of U.S. (millions of tons)[a].

State	No. Plants Producing Non-Cottage Cheese	Fluid Whey[b]	Available Fluid[c] Whey
Iowa	13	0.99	0.48
Minnesota	32	2.39	1.15
New York	45	2.19	1.06
Wisconsin	334	7.04	3.41
U.S. Total	737	22.90	11.08

[a] Calculated from cheese production: 9 lb/1 lb non-cottage cheese
6 lb/1 cottage cheese
[b] Average solids contents of whey: 6.5 percent.
[c] In 1980, 48.4% of total whey production was not further processed.

Molasses

The price of molasses has increased at about 4 percent in real terms during the period from 1966-1978. However, the price of molasses has dropped dramatically in recent months. Animal feedstuffs and alcohol production are the major end-uses of molasses.

Animal Manure

Table 20 displays the manure production in the United States by beef cattle on feed, broilers, and hens and pullets of laying age. These groups offer the most attractive possibilities for obtaining manure for energy production for three reasons: (1) as groups, these are large producers of manure, (2) these animals are highly concentrated, and (3) the manure from these groups must be hauled away. Dairy cattle, sheep and hogs are generally raised in smaller and/or less confined groups and the manure may be difficult to collect or has a positive value to the farmer as a fertilizer.

A survey conducted in 1977 estimated delivered prices of cattle and chicken manure to be $3 - $4.50 and $19.50 - $26 per fresh ton, respectively.

Table 19. Projected world molasses production and prices.

1982		1990	2000
million tons	$/ton	million tons	million tons
43.82	50	49.98	59.12

Source: Battelle estimate. 1979 price of New Orleans black-strap molasses: $81.70/ton.

Table 20. Manure generation by cattle on feed, broilers and hens and pullets of laying age in the U.S.--1980, 2000 and 2020.

Group	Manure Generaton (M tons/day)		
	1980	2000	2020
Cattle on feed[a]	43.49	49.02	55.25
Broilers[b]	14.25	26.24	48.32
Hens and pullets	12.36	13.90	14.89

[a] Based on historic increase of 0.60% per year, 1966-1981.
[b] Based on historic increase of 3.10% per year, 1966-1980.
[c] From Ernest et al. 1979.

Source: USDA 1981, Ernest et al. 1979.

CONCLUSIONS

In thinking about biomass from agriculture, the following should be kept in mind:

o Growing crops has an opportunity cost since food crop sales are not received. In addition, marketing may be more expensive if marketing channels are not as well developed as for food and fiber.

o Use of residues involves collection and transportation cost even if they are "free."

o Some factory and feedlot residues are truly free or even provide the user with a credit. Others that have alternative uses may have an opportunity cost.

o In principle, one can get into and out of biomass in a season or two, whereas building a coal or oil shale processing plant can take a decade and represents a huge cost if shut down due to lack of markets. Thus, the small scale of most biomass processing relative to synfuels is an advantage as well as a disadvantage.

o Some biomass is highly seasonal and some is hard to store from season to season.

o Some residues cannot be obtained in larger amounts without increasing the price to the point that they are really co-products.

o Finally, biomass for energy economics are highly species and region specific. Each case must be investigated in detail.

REFERENCES

Agricultural Outlook. 1982. August. p. 33-34.
Compos-Lopez, E. and Palacios, J. 1976. J. Polymer Sci. Polymer Sci. Polymer Chem. 14:1561. See also: Compos-Lopez, E. et al. 1978. Guayule: State of knowledge. In: Guayule: Rencuentro en al Desierto, Consejo Nacional de Ciencia y Tecnologia, Mexico City. pp. 375-410.
Canada Grains Council. 1981. Canadian Grain Statistics Handbook. Winnipeg, Manitoba.
Ernest, R.K., Hamilton, R.H., Borgeson, N.S., Schooley, F.A. and Dickenson, R.L. 1979. Mission Analyses for the Federal Fuels from Biomass Program. Vol. III: Feedstock Availability. U.S. Dept. of Energy, Washington, D.C.
Howlett, K. and Gamache, A. 1977. Silvicultural Biomass Farms Vol. VI:Forest and Mill Residues as Potential Sources of Biomass. Prepared by Mitre Corp. for Energy Research and Development Administration, Washington, D.C.
National Academy of Science. 1977. Guayule: An Alternative Source of Natural Rubber. Washington, D.C.
Purdue University. 1979. The Potential of Producing Energy from Agriculture. Prepared for OTA by Purdue University, West Lafayette, Indiana.
Scott, Tony. 1982. Personal communication. Whey Products Institute, Chicago, Illinois. Sept. 1982.
Spiewak S., Nichols, J.P., Alvic, D., Delene, J.G. Fitzgerald, B.H., Hightower, J.R., Klepper, O.H., Krummel, J.R. and Mills, J.B. 1982. Technical Analysis of the Use of Biomass for Energy Production. Oak Ridge National Laboratory, Oak Ridge, Tennessee.
Statistics Canada. 1981. Canadian Forestry Statistics 1979. Statistics Canada, Ottawa, Catalogue 25-202.
USDA. 1961. Agricultural Statistics 1961. U.S. Department of Agriculture, Washington, D.C.
USDA. 1973. The Outlook for Timber in the United States. U.S. Department of Agriculture, Forest Service, Forest Resource Report No. 20, Washington, D.C.
USDA. 1980. Dairy Products Annual Summary, 1980. U.S. Department of Agriculture Crop Reporting Board; Economics and Statistics Service, Washington, D.C.
USDA. 1981. Agricultural Statistics 1981. U.S. Department of Agriculture, Washington, D.C.
USDA. 1981a. Agricultural Prices Annual Summary 1981. U.S. Department of Agriculture Crop Reporting Board, Statistical Reporting Service, Washington, D.C.
U.S. Dept. Commerce. 1982. Forest Products Review, Winter/Spring. Bureau of Industrial Economics, Washington, D.C.

THE AQUATIC RESOURCE

Mentz Indergaard

Institute of Marine Biochemistry
University of Trondheim
The Norwegian Institute of Technology
N-7034 Trondheim - NTH, Norway

I. THE WILD MARINE PLANTS: A GLOBAL BIORESOURCE

INTRODUCTION

The surface of the earth is about 508×10^6 km^2, of which 71 percent is marine waters and about 5 percent is freshwater lakes and rivers. Much of the freshwater area is found in the temperate and cold regions where it is covered with ice in winter. This limits the growing season for the freshwater plants severely. The freshwater macrophytes are most noticeable when they clog inland waterways. The periodic removal of this biomass has in recent years been connected to its possible use by conversion to biogas. Conversion studies are in progress with reference to its local importance as a renewable source of energy, especially in developing countries (Choudhury et al. 1982).

The marine macrophytes -- the seagrasses -- hold an important position in the marine ecology in local areas by their production of detritus (Phillips 1980). Their chemical composition and occurrence is such that they have only found minor, local use. Most of the 50 species occur in the tropical and subtropical regions where they inhabit shallow nearshore coastal waters. The marine phytoplankton provide as much organic matter as all terrestrial vegetation, in the order of 10^{10} tons carbon per year. This wildgrowing marine biomass is not harvested directly. The production value is best expressed by the yield of the higher trophic levels of zooplankton, shellfish and fish. The direct harvesting of phytoplankton would require considerable energy input in order to get rid of excess water. Even a spectacular natural phytoplankton bloom is only a thin soup with a few milligrams of dry weight per liter.

SEAWEEDS. BIOLOGICAL AND BIOCHEMICAL FEATURES

The seaweeds are not weeds. They are the large red, brown and green algae. Their main characteristics are summarized in Table I. Seaweeds reproduce by means of spores which swim to a new habitat or are carried there by water currents. They have root-like structures which attach them to the subtrates but do not absorb water and nutrients like the roots of other plants; there is no need for this as they are surrounded by all the sustenance they need. Their root-like structures are aptly called hold-fasts; they have stems which support leaf-like blades. Here the similarity to other plants ends as the stem is only for support and generally does not transport foodstuffs and water. The majority of the seaweeds are attached to rocky substratum, but there are a few species living free-floating, like in the Sargasso Sea. The life cycles of the seaweeds are much more complex and varied than those of the flowering plants.

SEAWEED PRODUCTIVITY

The photosynthesizing capacity of the seaweeds ranks among the foremost in the plant world. Their productivity is huge because they do not have to waste energy in internal transport of water and nutrients, and they live in an environment that is rather stable in temperature, humidity and salinity the whole year. This productivity is illustrated in Table II. Their net primary production is in the order of 1500 g C M $^{-2}$yr^{-1}, equal to intensively cultivated alfalfa in the U.S.A. (Mann and Chapman 1975). Remember that almost all the seaweeds investigated are wild plants. They are not genetically improved over decades as the agricultural terrestrial vegetation, and they are not tended as carefully as the monocultures in agriculture.

SEAWEED RESOURCES AND CHEMICAL COMPOSITION

Some estimates of resources and harvesting of wild seaweeds are shown in Tables II and III.

Of all the seaweed species only a few have been thoroughly analysed regarding chemical composition and a few examples are given in Tables IV and V.

SEAWEED UTILIZATION

Man's utilization of the seaweeds goes back to ancient times. The extent and purpose of the uses have varied with time and geographic location. Whereas long and extensive utilization of this resource characterizes the situation in a few Oriental countries

Table I. Characteristics of seaweeds.

Class	RHODOPHYCEAE (Red algae)	PHAEOPHYCEAE (Brown algae)	CHLOROPHYCEAE (Green algae)
No. of genera/species	500-600/3500-4500	250/1500-2000	450/ca.7000
Spore flagella	No flagella	2 flagellae unequal length lateral insertion	2, 4 or 8 flagellae equal length apical insertion
Pigments	Chlorophyll a c- and r-phycocyanin r-phycoerythrin α- and β-carotene xanthophylls	Chlorophylls a, c_1, c_2 β-carotene xanthophylls	Chlorophylls a, b α-, β- and γ-carotene xanthophylls
Habitat	Mostly SW Benthic	SW Mostly benthic	FW (90%) SW (10%)
Morphology	OCT Unicellulars rare	OCT	Unicellulars Colonies OCT
Max. wet weight of individual plant	50 kg (Eucheuma)	200 kg (Macrocystis)	2 kg (Ulva)

OCT = organised cellular tissue
SW = salt or brackish water, FW = fresh water

Table II. Potential and actual harvests of seaweeds ($\times 10^3$ tons wet wt.). From Michanek 1975).

	Red algae		Brown algae	
Area	Recent harvests[a]	Potential output[b]	Recent harvests[a]	Potential output[b]
18 Arctic Sea	−	−	−	−
21 NW Atlantic	35	100	6	500
27 NE Atlantic	72	150	208	2000
31 WC Atlantic	−	(10)	1	1000
34 EC Atlantic	10	50	1	150
37 Mediterranean/ Black Sea	50	1000	1	50
41 SW Atlantic	23	100	75	2000
47 SE Atlantic	7	100	13	100
51 W Indian Ocean	4	120	5	150 (1000-Kerguelen)
57 E. Indian Ocean	3	100	10	500
61 NW Pacific	545	650	825	1500
67 NE Pacific	−	10	−	1500
71 WC Pacific	20	50	1	50
77 EC Pacific	7	50	153	3500
81 SW Pacific	1	20	1	100
87 SE Pacific	30	100	1	1500

[a] Recent levels of harvests based upon estimates for 1971-73.

[b] Broad indications of possible annual output.

such as Japan, Korea and China, the general picture is one of variable use in relatively short periods of very small to fair fractions of the total seaweed resource.

Most of the species listed in Table VI are only collected locally in the Far East for human consumption.

The utilization of seaweeds may roughly be divided into industrial use of phycocolloids in the Western world, and the human consumption of seaweeds as vegetables in the Far East.

Table III. European harvest of wild seaweeds.

Country	Tons wet weight/yr	Main end products
Norway[a]	160,000	Alginate, Seaweed meal
Ireland[b]	65,000	Alginate, Seaweed meal
Portugal[c]	50,000	Agar, Carrageenan
France[d]	40,000	Alginate, Seaweed meal, Carrageenan, Agar
Baltic U.S.S.R.[c]	40,000	Furcellaran, Agar, Carrageenan
U.K.[c]	20,000	Alginate
Iceland[f]	20,000	Seaweed meal, Alginate
Denmark[c]	10,000	Furcellaran
Spain[c]	10,000	Agar, Carrageenan
Italy[e]	750	Agar, Carrageenan

a) Jensen 1982 (pers. comm.), b) Guiry and Blunden 1980, c) Michanek 1975, d) Gayral 1982 (pers. comm.), e) Bombelli 1982 (pers. comm.), f) Ludviksson 1982 (pers. comm.).

PHYCOCOLLOIDS

The most prominent chemical component in both flowering plants and the algae are the carbohydrates. The carbohydrates in the algae are mostly polysaccharides quite unlike those in the flowering plants (see Table VII).

Phycocolloids are polysaccharides in seaweeds that have the ability to give viscosity, gel strength and stability to aqueous mixtures, solutions and emulsions. The most important are alginic acid from brown algae and various sulphated galactans from red algae such as agar, carrageenan and furcellaran.

A few introductory remarks are needed before taking a closer look at the phycocolloids. There are such variations in seaweed species, their habitats, accessibility and topography of the substratum, in chemical properties, availability and utilization and in the methods and costs of harvesting and processing that it is quite impossible to consider the seaweed industry as a homogeneous entity. It is also found operating under all economic systems in the world which makes direct comparisons difficult. The relatively few major commercial enterprises are keeping their production figures rather secret as the seaweed colloids market is highly competitive.

Anon. (1981) estimated the world market value of phycocolloids to be U.S. $350 million. In 1972 these algal gel extractives were

Table IV. The chemical composition of a few selected, commercially exploited seaweeds. All figures as percent of dry matter. (From Jensen 1966; Jackson and North 1973; Chapman 1980).

	Ascophyllum	Laminaria	Macrocystis	Chondrus
Water[a]	70-85	75-90	86-87	71-82
Ash	15-25	20-45	36-38	25-28
Alginic acid	15-30	20-45	13-24	0
Carrageenan	0	0	0	52-64
Laminaran	0-10	0-20	6-7	0
Mannitol	5-10	3-15	n.d.	n.d.
Fucoidan	4-10	2-4	n.d.	0
Other carboh.	ca.10	traces	n.d.	n.d.
Protein	5-10	7-15	ca.6	7-17
Fiber	5	5-10	5-8	n.d.
Fat	2-7	1-2	0-4	n.d.
Tannoids	2-10	ca.1	n.d.	n.d.
K	2-3	4-12	14	7
Na	3-4	2-5	5	7
Mg	.5-.9	.5-.8	2	7
I	.01-.1	.1-1.1	.1-.4	.08

a) As percentage of fresh weight.
The variations in the brown algae genera Asc., Lam. and Macr. are due to seasonal variations. Chondrus is a red algae genus.
n.d. = no data.

estimated to be essential in the production of goods worth some 22 billion U.S. dollars of the U.S. gross national product, when, in comparison, the U.S. plastics industry neared the 50 billion dollar level (Doty 1979).

Alginates, the salts of alginic acid, find a very wide use in food, textile printing and the pharmaceutical industry. About half the U.S. output of alginates is used as stabilizer in ice cream and as a suspending agent in milk-shakes. Alginate is applied in bakery icings, in canned food, as a binder for pharmaceuticals such as pills, and in the production of cosmetic creams. Further it is used widely in paints, insulating and sealing compounds, paper products, oil drilling lubricants and coolants, etc. (Naylor 1976). (Table VIII).

From red algae there is an array of more or less sulphated galactans that serve principally similar industrial purposes as the alginates. They might be separated into two important groups: The agar which is a superb gelling agent with water, and the carrageenans which form no gel with pure water but need the addition of salts to do so. The industry's ability to produce carrageenans to standardized specifications, allied with competitive and relatively stable prices, has led to a rapid expansion in their usage over the last two decades. Of the phycocolloids it has by far the widest food industry applications. It is particularly reactive with casein and thus finds use in a range of milk products. Beverages and bakery products, dietetic foods, dressings, sauces and frozen food are also among the products into which carrageenan finds its way.

Agar is especially characterized by its high gel strength at very low concentrations: 1% gels melt at 80-100°C and settle at 35-50°C. In solution it has low viscosity and is transparent. This makes it highly valuable as a medium thickener in microbiology. In addition it is non-toxic and resists liquification, it can be sterilized repeatedly without losing its other properties and it remains liquid when cooled to 42°C, thus permitting the thorough distribution of organisms at a temperature which will not harm them, and once set can be kept at incubation temperature without melting. In foodstuffs it is used in ice-cream, cream cheese, yoghourts, etc., and has industrial application in dentistry, plastic surgery and cosmetics. A non-sulphated fraction of agar called agarose has found wide use in biochemical and medical research such as in gel filtration and electrophoresis for work with proteins. Because of a shortage of red algae for high quality agar extraction, the pure agarose now fetches prices in the range U.S. $ 200-400 per kg.

The agar sales potential is estimated to be as great as that of algin and carrageenan. However, unless someone can work out the methods for growing agarophytes (particularly species of Gelidium) under controlled conditions, Moss (1977) predicts that production and sales will permanently decline. It is possible that other colloids will replace all but the bacteriological and biochemical applications of agar, which worldwide consume probably less than 1000 tons annually.

A less known phycocolloid is extracted from Furcellaria in a quantity of about 1200 tons/year (Naylor 1976). It is used in food products, introduced as a substitute for agar. A notable feature is the successful combination of furcellaran with other gums or thickening agents, such as carrageenan, locust bean gum, guar gum and sucrose to produce gelling products with special properties.

A few other galactans from red algae are also utilized to a minor degree along the same lines as agar and carrageenan (Naylor 1976, Chapman 1980).

Through their superior quality, the algal gelling agents are successful in their competition with such synthetic substances as carboxymethylcellulose and other synthetic gums. But in many cases

Table V. Minerals and vitamins in seaweed meal from Ascophyllum nodosum (Phaeophyceae). (From Jensen 1966).

Minerals	mg/kg dry matter	Vitamins	mg/kg dry matter
Potassium	20,000	Carotene	30-60
Phosphorus	1,000	Ascorbic acid	5-20
Magnesium	5,000	Thiamine	1-5
Calcium	20,000	Riboflavin	5-10
Sodium and		Nicotinic acid	10-30
chlorine	25,000	Pyridoxine	.3
Sulphur	30,000	Pantothenic acid	1.4
Iron	300	Folic acid	.1-.5
Manganese	30	Biotin	.1-.4
Zinc	50	Menadione	10
Copper	5	B_{12}-vitamin	.003-.06
Molybdenum	.5	Tocopherol	150-300
Iodine	500		
Cobalt	1		
Selenium	.1		

low cost is considered before quality, especially in times of recession. Neish (1980), however, states: "Within the next five years, then, the marine gum market will go in one of two vastly different directions. It may grow at roughly the rate of population growth and maintain its role in traditional applications, or it may show vast growth at the expense of other gums and stabilizers."

Tables IX and X give a fair indication of the processed amounts of agar and carrageenan. Ninety-three percent of the raw material for carrageenan is processed in Denmark, U.S.A. and France.

SEAWEEDS FOR HUMAN CONSUMPTION

Over 80 percent of the estimated first hand value of the world's total seaweed output in 1973 was in fact attributable to the

Table VI. Number of multicellular algae of commercial use. (Tabulated from Bonotto 1979).

Class	No. of genera	No. of species
Rhodophyceae	59	220
Phaeophyceae	43	88
Chlorophyceae	5	27

sale of semiprocessed edible products in Japan, Korea and China. In the West no such tastes and traditions have been acquired, although some coastal communities still uphold use of a few species. In Western Europe and Eastern North American dried seaweeds such as Palmaria, Porphyra and Chondrus are still in use. Until the beginning of this century species like Laminaria saccharina, Alaria esculenta, Gigartina stellata and Ulva lactuca also found use as vegetables in Europe.

On the Pacific Coast of North America, the stipes and bladders of the giant brown seaweed Nereocystis luetkeana are marketed locally in desalted, flavoured and candied pieces used as a confectionery and specialty food product (Levring et al. 1969). In Chile they also use Ulva, but consumed even more widely is the stipe of the giant bull kelp Durvillea antarctica, which is washed and dried and sold for use mainly as an admixture in thick soups. On the continent of Africa, seaweeds appear to be largely neglected as a direct source of food. An interesting occurrence is the use of the unicellular blue-green alga Spirulina as food. It is particularly rich in proteins and vitamins. It grows on the surface of pools and annual yields of up to 50 tons per hectare may be obtained (Becker and Venkataraman 1982).

In many parts of the Indo-Pacific region are several species being utilized as food, among them Caulerpa, Sargassum, Turbinaria, Porphyra, Gracilaria, Laurencia and Rhodymenia. Hawaii is a typical example of the decline in using algae for food with the advent of abundant western food. A hundred years ago there were about 75 different kinds of seaweeds regularly eaten there. Now seaweed eating is mainly reserved for special occasions and much of the knowledge of the qualities of the different species are lost with the older generation.

In Japan no such decline is apparent in the traditionally high and diverse consumption of edible algae. Annual consumption of the three main seaweed food products eaten in Japan -- "nori", "wakame" and "kombu" (called "haidai" in China) -- reached an average of some 3.5 kg per household in 1973, a 20 percent increase in ten years. Taking into account various other minor seaweed foods, annual consumption of edible seaweeds in Japan averaged approximately one kg per person, a remarkably high figure for such paper-light food items used primarily as an additive or garnish to more bulky dishes, providing relish, taste and "feel" to the meal.

Together the seaweed foods represent a very important industry in Japan, with a value now amounting to some 700 million U.S. dollars per annum, far in excess of the total value of the world production of phycocolloids, rated at some 200 million U.S. dollars by Moss (1977). In 1978 the global harvest of seaweeds was estimated at 1,422,000 tons wet weight. Of this, Japan landed 638,000 tons,

Table VII. Some major polysaccharides in seaweeds.

Type	RHODOPHYCEAE	PHAEOPHYCEAE	CHLOROPHYCEAE
Energy	Floridean glycogen (amylopectin, no amylose)	Laminaran (β-1,3-glucane with β-1,6 branches and mannitol-ends)	Starch, Inulin
Structure	Xylan, Mannan Galactans with sulphate half-esters: AGAR CARRAGEENAN FURCELLARAN and others	ALGINIC ACID (β-1,4-D-mannuronic acid and α-1,4-L-guluronic acid) Fucoidan (α-1,2-L-fucose-4-sulphate and xylose) Ascophyllan (uronic acids and sulphated monomers)	True cellulose, modified cellulose, Mannan, Glucomannan, Xylan, Pectic acid, Complex hemicelluloses and sulphated mucilages

Cellulose occurs in some species of all three classes, but when found only constitutes max. 2-7% of dry matter. Note for fermentation: Seaweeds contain no lignin.

of which 90 percent was used for subsidiary foods or table delicacies and the remainder in industrial production of alginic acid and the like (Anon. 1980).

Supplies from natural sources have been increasingly supplemented by cultivated seaweeds in Japan (Figure 1). "Nori", made from Porphyra, is marketed in thin, uniformly sized sheets of dried and pressed algae and used in soups and with rice and fish balls. The richest Porphyra cultivation sites have a product value of U.S. $ 1000/ha, making Porphyra cultivation the most profitable of all fisheries in Japan. Some 60,000-70,000 Japanese fishermen partake. In the U.S. market it sells for U.S.$ 50 per pound.

"Kombu" prepared from compressed Laminaria is sold in many forms: shredded, powdered, seasoned, roasted and sugared. It is widely used as a soup stock, a vegetable and, after soaking in soy sauce, with rice dishes.

"Wakame" is obtained fairly simply by washing and drying Undaria and is eaten chiefly with soybean soup. We find more or less the same traditional pattern of algae use in Korea and China, where the use of Laminaria japonica dominates. Almost all the production of this species in the People's Republic of China, 1,650,000 tons wet weight in 1979 (Michanek 1981), was from cultivation.

In her "Sea Vegetable Book," J.C. Madlener (1977) says: "Seaweeds ... can be cooked with the same ease and comfort with which we prepare our everyday foods. They can be boiled, baked, steamed, sauteed, fried, blanched, dried, salted, brined or prepared in any way which their terrestrial counterparts, the land vegetables, are

Table VIII. Estimated production of alginates in the latter part of the 1970's (From Anon. 1981).

Country	Tons alginates
United Kingdom	8350
United States	6000
Norway	3250
Japan	1500
France	1300
Canada	700
India	450
Chile	100
Total	21200

The raw material originates from ca.14 countries.
In addition to the amounts given it is estimated that China produces about 10,000 tons/yr of a rather low-quality product.

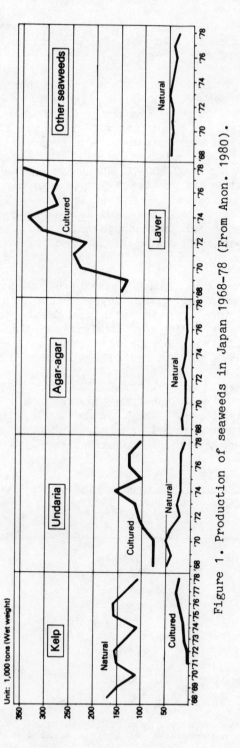

Figure 1. Production of seaweeds in Japan 1968-78 (From Anon. 1980).

Kelp = <u>Laminaria</u>, Laver = <u>Porphyra</u>

Table IX. Estimated average annual harvest of carrageenan-containing seaweeds in the 1970's (From Naylor 1976 and Anon. 1981).

Country	Tons wet weight
Philippines	65,500
Canada	35,000
Chile	20,000
Spain	10,000
Singapore	5,000
Japan	5,000
France	4,000
Ireland	3,600
Mexico	2,500
Portugal	2,000
Madagascar	1,500
Argentina	1,200
Tanzania	400
Others	2,000
Total	157,700

giving about 10,000 tons carrageenan.

93% of the raw material is processed to carrageenan in Denmark, U.S.A. and France.

Table X. Estimated average annual production of agar in the 1970's (From Naylor 1976 and Anon. 1981).

Country	Tons agar
Japan	2,000
Rep. of Korea	1,350
Spain	1,000
U.S.S.R.	1,000 (?)
Portugal	800
Chile	500
Philippines	300
Argentina	280
Mexico	240
Morocco	200
U.S.A.	160
Rep. of China (Taiwan)	150
S.Afr.,France,India,Peru, New Zealand,Brazil	300
Total	8,280

prepared. Most can be eaten raw and many can be successfully frozen. Many species posses unique characteristics of taste and texture which elevate them to the level of fine foods, and as such are appreciated the world over."

HARVESTING OF SEAWEEDS

Harvesting methods for wild seaweeds vary widely in sophistication. They depend primarily upon the physical properties and natural environment of the individual seaweed species. In general, the red algae are much smaller in size and occur in deeper waters than the brown algae; consequently their harvesting is slower, more difficult and more costly. Almost all red algae are gathered by hand with primitive tools, like longhanded grapnels or rakes. Harvesting by free-diving can be rather efficient: In Japan women divers are said to harvest up to 300 kg fresh weight of the agarophyte Gelidium per day. The handpicking of such red algal species as Chondrus in East Canada and Eucheuma, Hypnea and Iridaea in the Pacific is in the range 100-400 kg fresh weight per low tide per person.

The abundance, habitat and size of the larger brown algal species Ascophyllum, Laminaria and Macrocystis have permitted a high degree of mechanization in harvesting practices: 50-300 tons fresh weight per vessel per day with an operating crew of 1-4 persons.

Collection of seaweeds drifted ashore after heavy weather is on the decline. The quality is affected by the casual drying it is committed to when washed ashore. This especially holds true for the seaweeds collected and milled for fodder supplement.

In the production of phycocolloids in the developed countires the collection by hand of seaweeds accounts for roughly half the total expenses of production. This presents an interesting socio-economic aspect for developing countries with under-exploited seaweed resources and a surplus of labour. Under such circumstances, the hand collecting and natural drying of seaweeds can provide supplementary employment and income for all members of coastal families, and a potential source of foreign currency to the country itself.

The drying is essential to prevent rapid degrading of the raw material. For the production of agar it is essential for processing as extracts of wet seaweeds do not gel properly. Once satisfactorily dried, the seaweeds can be stored for several years without appreciable loss of gel content. The average climate in each country in question decides whether sun- and wind-drying is feasible or if one has to rely on the more efficient, but extremely costly artificial drying in heated drums.

INDUSTRIAL PROCESSING OF SEAWEEDS FOR PHYCOCOLLOIDS

Essential considerations for establishing seaweed extraction factories for the production of phycocolloids are discussed by Moss (1977). In the world there are now 30-35 agar factories, 8-10 algin factories and 9-10 carrageenan factories. The profile of uses for each extractive is very broad and will always be able to compete in quality with synthetics. In factory set-ups all other problems are strictly secondary to that of obtaining an adequate supply of the appropriate algae. To this should be added: enough local fresh water supply for its processing. A plant processing 2000 tons of algin per year (equal to 15 percent of the world output today) would need about 7.6 million gallons of water per week (Guiry 1979).

The questions raised in the following apply equally well to seaweeds growing wild along the coasts as to those produced under controlled conditions by cultivation:

i) Algal resources and harvesting: mapping of distribution, density, species composition, life cycles of the interesting species, harvesting approach, harvesting potential for adequate regrowth, etc.

ii) Factory siting: communications, fresh water supply, labour.

iii) Processing and marketing: cost of drying, analysis of phycocolloid content variation, raw material price fluctuation.

Recently an extraction factory operating in Brazil, employing up to 2400 people for harvesting and processing 100,000 tons per year, closed down for reasons that might suggest inadequacy in defining and controlling such fundamentals as observing optimal harvesting times, continuous raw material supply and control of product quality (Michanek 1981).

To the developing nations of the world which hithero have not fully utilized their resources of seaweeds, the raw material cost in the developed nations offers a number of economically and socially attractive opportunities. Even if there are inherent disadvantages in being tied too closely to an individual manufacturer, there might be benefits of initiating joint-venture or technical co-operation arrangements with foreign enterprises already well experienced in the techniques of growing, collecting and processing seaweeds. The successful establishment of one or more of the varied aspects of seaweed commercialization in, for example, Morocco, Chile, Argentina and the Philippines indicates the overall socio-economic potentials in such developments.

SEAWEEDS FOR MANURE AND FODDER SUPPLEMENT

Wherever proximity to the coast has made access to the resource possible, seaweeds have been applied to the land for many centuries as a direct and simple manure. Since 1950 liquid seaweed extracts have enabled this practice to be extended. Around 1980 the total production was ca. 1,000,000 gallons (Anon. 1981).

The seaweeds' content of minerals, vitamins and, especially trace elements, promotes various positive effects such as resistance to plant diseases and pests, inducing fruit setting, increasing germination rates and prolonging shelf life of vegetable products (Senn and Kingman 1978). The consistent use of seaweed manure on the land is shown to obviate any need for crop rotation. A direct and simple way to restore earth depleted of unknown trace elements is to fertilize it with seaweeds or seaweed extracts. This sector of the seaweed industry appears to have potential for further expansion, especially in areas with meager soil, and to avoid expensive, imported fertilizers.

Dried and milled brown algae constitute seaweed meal. _Ascophyllum nodosum_ is the most used species. The nutritional value of this species is lower than that of, for instance, _Alaria esculenta_, but the latter is inaccessible for commercial harvesting. The main value of the seaweed meal is in the content of minerals and vitamins (Table V). The meal is intended to be used as a fodder supplement with ordinary food stuffs. The amount used is 1 to 4 percent.

A recent development has been the use of seaweed meal as a ration for mink being bred for their fur. It is claimed that it gives better litters, greater survival rate and more docile animals. The use of seaweed meal in hen fodder improves the yolk colour because of the fucoxanthin in the meal. Seaweed meal added to the fodder of a set of identical twin cows proved more efficient than the equivalent standard mineral mixture. It increased milk production by around 6 percent in lactating cows (Chapman 1980). Other experiments with the addition of seaweed meal to animal fodder are still in progress.

SEAWEEDS FOR PHARMACEUTICAL USE

The use of algae has a proven record in traditional folk medicine. Seaweeds from all the classes are mentioned in the scriptures as remedies against kidney diseases, stomach pains, fever, parasites, etc.

No drugs from seaweeds are presently commercially utilized, but a few examples from recent research might be mentioned. Circulatory disturbance may be cured by phycocolloids that lower plasma cholesterol and thus work as an antiarteriosclerotic agent.

Insufficient supply of vitamins and minerals might be treated by seaweed consumption as almost all the elements are found in seaweeds. Some seaweeds contain vermifugal substances to fight the exhausting effect that populations sucked by parasites experience. The distribution of seaweed products to inland populations should be supported in particular, as they could most easily form the basis for lasting improvements.

The case of fighting endemic goiter with seaweed meal tablets is noteworthy. Thyroid enlargement caused by low iodine content in food will often be accompanied by mental retardation if acquired early in life, and even by physical retardation, or, in most severe cases, cretinism. Public information, social welfare and iodized salt or seaweed food can theoretically eradicate this unnecessary disease. One might thus look upon the large harvest of _Laminaria_ in China as a gigantic source of high iodine drugs, as they are distributing it to the inland population as an additive to the generally low-iodine diet (Michanek 1979, 1981).

The search for pharmaceuticals in seaweeds is continuing. All the new, complex organic substances discovered in this work, even if they are irrelevant for any possible drug production, serve as inspiration and model-substances to further synthesizing work.

THE FUTURE OF SEAWEED UTILIZATION

Table XI gives an outline of the major uses of seaweeds today. While seaweeds still remain of importance as a food supplement in the Far East, the most notable trend over the last three or four decades has been the very significant growth in the output of seaweed colloids; agar, carrageenan and algin, of remarkably varied commercial application. Moss (1977) is of the opinion that so far this growth shows no drastic signs of abating. Most observers agree that the diverse demand for vegetable gums will be the major factor in influencing further exploitation of the world's resources of marine algae. There is, however, an awareness of the possibilities of new processes to make new products.

In the West the possibilities for expanding the traditional exploitation of marine algae seem limited to a growth in the phycocolloid production following the population growth. There is no reason to expect that greater quantities of seaweeds will be used for food here in the short term. Thus any future large-scale exploitation of marine algae requires the creation of new areas of application. These can be built either on the basis of the biomass production of existing algal material or from algae which have been genetically improved, just like the development of plants in agriculture.

Table XI. Present industrial utilization of seaweeds. (From Jensen 1978 and Anon. 1981).

Algae used	Product (tons/yr)	Quantity of seaweed consumed (tons wet weight/yr)
Seaweed meal (10 x 10^6 US$/yr)		
Ascophyllum nodosum	30,000	ca. 100,000
Alginates (ca. 120 x 10^6 US$/yr)		
Macrocystis, Laminaria, A. nodosum	20,000	ca. 400,000
Agar (ca. 110 x 10^6 US$/yr)		
Gelidium, Gracilaria	8,000	ca. 150,000
Carrageenan (ca. 60 x 10^6 US$/yr)		
Eucheuma, Chondrus, Gigartina	10,000	ca. 150,000
Furcellaran (ca. 5 x 10^6 US$/yr)		
Furcellaria	1,000	ca. 20,000
Nori (ca. 350 x 10^6 US$/yr)		
Porphyra spp.	ca. 25,000	ca. 350,000
Wakame (ca. 300 x 10^6 US$/yr)		
Undaria	ca. 14,000	ca. 130,000
Kombu/Haidai (ca. 350 x 10^6 US$/yr)		
Laminaria	ca. 220,000	ca. 1,600,000

Figure 2 shows the areas of application which could be regarded as having a future. Among the new possibilities for using existing algae on a larger scale is the microbial conversion of wet seaweeds into methane, alcohols, fatty acids, esters and resynthesis into carbohydrates. Research in methane production is chiefly conducted out of need for energy. The total U.S. energy supply converted into the growth area of bioconvertible kelp for methane would require some 3.9×10^6 km^2. This is as much as the global marine waters with a depth down to 10 meters (Doty 1979). The area may be greatly extended by cultivating seaweeds fastened to floating structures. If all the sewage discharged in the U.S.A. was used as fertilizer for seaweed energy farms, the algae producible could be converted to yield somewhat less than 1 percent of the U.S. energy needs. Thus, algal cleaning of the U.S. sewage water is a smaller problem phycologically than the problem of producing seaweed power. In the present context fermentation is defined as microbial degradation and transformation of organic plant material into various substances that are suited for industrial and other purposes. Figure 3 names a few of these.

Methane is an obvious energy carrier produced from primary marine biomass. Seaweeds do not contain lignin and their polysaccharides are much easier hydrolysed than is the cellulose of higher plants. Rough estimates for the production of electric power from methane both in the U.S.A. and in Norway indicate about 5 U.S. cents /kWh which is about 50 percent higher than the present Norwegian price of hydroelectric energy. There is evidence for a very favourable comparison in favour of the Laminaria saccharina compared to the American seaweed Macrocystis as raw material for fermentation purposes. The L. saccharina has a higher content of carbohydrates and lower content of water and ash. All fresh seaweeds have a relatively high water content. Fermentation can handle fresh algal material avoiding the prohibiting expense of drying the seaweeds first. Applicaton of selected marine bacteria should allow fermentation at 24°C instead of the usual 35°C. Estimates for Laminaria indicate 91 kg biogas from 1 ton fresh weight, equal to 650 kWh energy. The lowering of temperature would improve the economy of process.

Studies of the conversion process have so far shown that:

- Degradation can take place in the presence of salt.
- Addition of extra nutrients is not necessary for the bacteria to work efficiently.
- Methane recovery as high as 65 percent may be obtained
- The yield as methane from seaweeds exceeds that of any other tested biomass, including the widely used cattle manure (Ghosh et al 1981).

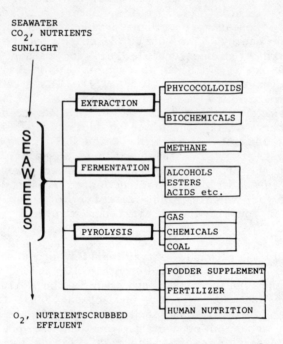

Figure 2. Present and potential future uses of seaweeds.

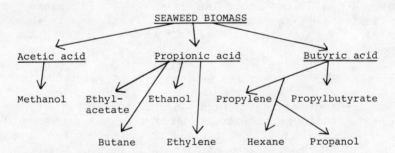

Figure 3. Possible products from the fermentation of seaweeds.

Table XII. Possible future uses of seaweeds (Modified from Jensen 1978).

Item	Production (tons/year)	Algae consumed (tons wet weight/yr)
Alginates	50,000	1.3×10^6
Carrageenan	30,000	400,000
Agar	20,000	500,000
Nori	35,000	400,000
Wakame	30,000	200,000
Kombu	250,000	2×10^6
Seaweed meal	several million	several million
Crude chemicals	several million	several million
Energy	several million	several million
Fine chemicals	several thousands	several thousands

Pyrolysis is the oldest method for chemical conversion of biomass. Through heating without supplying oxygen the organic complex compounds are decomposed to simpler components:

Biomass + Heat → Char + Oil + Gas + Carbon + Ash

The relative quantities of the components formed depend on the temperature; higher temperature gives more gaseous components. This process was run on a large scale in the U.S.A. in the 1920's but was unable to complete with the petrochemical industry (Chapman 1980). Table XII is an effort to summarize the more or less optimistic consideration of the future for the wild seaweed resources as we know them today.

II. LARGE SCALE CULTIVATION OF SEAWEEDS: THE SKY'S THE LIMIT?

INTRODUCTION

In the previous chapter it was documented that the wild marine plants represented by the seaweeds are a weighty bioresource. Micanek (1975) estimated that recent global harvesting of wild seaweeds amounted to only about 31 percent of world red algal supply and 8 percent of world brown algal supply. That should indicate that the potential of the wildgrowing stock is not exhausted. The untapped seaweed resources are, however, growing in faraway, rough waters (see Table III). This makes the harvesting of them marginally profitable. In addition, the species that we have a lot of are not the ones that we want or need. These arguments explain partly why the cultivation of seaweeds have been the object of research in the Western world in the last decade. In the Far East seaweed cultivation has traditions several hundred years back. They do, however, need to enlarge the scientific basis for their continued cultivation efforts. One of the basic criteria for the cultivation work is product need. Plant cultivation is put to use to increase the production and to satisfy our specialized requirements for plant biomass. This can only be achieved by having the right species growing at favourable locations and providing raw material of adequate quality. The development of agriculture may be a suitable analogue in trying to clarify the reasons for and direction of seaweed cultivation.

A FEW EXAMPLES OF SEAWEED CULTIVATION

The product need set off the first successful seaweed cultivation efforts in Japan some 300 years ago. The goal was to increase the production of the red alga Porphyra which "nori" is made from. The whole life cycle of Porphyra was not worked out in detail until 1950. Since then the cultivated production of this seaweed has increased significantly (see Figure 1). All the production is from a cultivation covering at least 600 km^2 of nearshore waters. With the development of various cultivation methods, countries in the Far East have expanded their cultivation of seaweeds for food to now include the genera Laminaria, Porphyra, Undaria, Monostroma and Gracilaria plus a few minor ones. The cultivation of seaweeds for phycocolloid extraction is of a very recent origin.

The Macrocystis beds off California are the source of the U.S. algin industry. When the kelp beds started to deteriorate in the late 1950's, extensive research was initiated. The restoration of the kelp beds included the outplanting of mature plants, the seeding of offshore areas with spores released from mature plants in the laboratory and the control of grazers such as the sea urchins by poisoning them by releasing quick lime (CaO) into the water. The beds are now harvested for some 200,000 tons wet weight each year.

When the carrageenan industry started to experience boom-and-bust-cycles -- the overexploitation of raw material one year leading to bad crops and price rises the next -- one realized that the raw material supply had to be stabilized. With unpredictable supply and widely fluctuating prices the carrageenan buyers were looking elsewhere for synthetic colloids to use. Eucheuma is a tropical red alga and a fine source of carrageenan. When a political collapse in Indonesia led to a cut-off in Eucheuma supply to the U.S., American phycologists started research on Eucheuma in co-operation with U.S. carrageenan manufacturers and Philippine authorities. After about five years, in 1972, it was concluded that Eucheuma mariculture could be profitable for Filipino farmers. By 1974 Eucheuma farming had proliferated rapidly and become an accepted livelihood for local Filipinos, harvest increasing from 500 tons in 1973 to 10,000 tons dry algae in 1974. The farming methods have evolved rapidly. This has included fertilizer use, increased planting density, improved cultivation techniques, pest and disease control by selecting out disease-resistant strains, and containment of fragmented plants which might increase the harvest by 30 percent of this vegetatively growing plant. The goal was to supply half the raw material needed in the world for carrageenan production. Such successful mariculture could certainly not have been introduced in North America or Europe. Farming of 1.5 - 3.0 kg of algae/manhour with prices of U.S. $ 0.1 - 0.4 per kg dry in the Philippines is economically rewarding, for it interferes little with the normal subsistence efforts of the farmers. This is especially true in countries where the family income of the local people is about U.S. $50 per year (Hansen et al. 1981).

ARGUMENTS IN FAVOUR OF SEAWEED CULTIVATION

This chapter was introduced by some explanation of why seaweeds are produced by cultivaton. Addition of factors can be listed:

- The productivity of the wild seaweeds compete favourably on a weight basis with intensively cultivated agricultural plant species. This seaweed productivity potential is promising for the coming genetical improvement of selected seaweed species.

- Controlled growth by cultivation is a methodical prerequisite for the genetical development of superior plant varieties.

- The climate of the sea is very stable; the temperature fluctuations are moderate and the humidity always sufficient. The chemical composition and the buffer capacity of sea water is also favourable. In many areas there is a natural, seasonal renewal of fertilizer substances such as nitrate and phosphate.

- The sea offers much new acreage in a time when cultivation land is hard to find for increased agricultural production.

- Aquatic plant cultivation offers a possibility of combining biomass production with uptake and removal of excessive nutrient salts in coastal waters. These nutrients often stem from agricultural runoff and municipal sewage. They might easily cause unwanted organic production (eutrophication) in already loaded waters. Eutrophication occurs near densely population areas and not on the outer coastline where most natural seaweed beds are. This nutrientscrubbing of the recipient, often called tertiary treatment of effluent, might be one aspect of the polyculture approach (see Figure 4).

THE POLYCULTURE CONCEPT

Monoculturing of phytoplankton is in operation in Japan, Republic of China and Mexico with Spirulina. It is an especially good producer of protein. Under well managed conditons it can produce 100 times as much protein as wheat (Priestly 1978). Species within the genera Scenedesmus, Chlorella, Chlamydomonas and Porphyridium are also being cultivated for various purposes. There are no reliable statistics of world output of cultivated phytoplankton. A problem in the production of phytoplankton is the so-called "crashing" of the population. This occurs when a massive population maintained by heavy fertilization dies or shifts to less productive or less useful species; we get an ecological succession. Thus simple textbook subjects become a major research area in context with the phytoplankton pond production management.

The best way of cultivating and harvesting the thin phytoplankton soup is in establishing food chains and integrated aquaculture (polyculture) to let the experts -- the filter feeders -- harvest and convert the phytoplankton into desired food products (Figure 4).

This is both ecologically and economically desirable. In Taiwan they have successfully integrated seaweeds (Gracilaria) for food and phycocolloids with shellfish and fish for human consumption into a complex polyculture. Chuang et al. (1977) set the production of Gracilaria at 1200 tons dry weight from 300 ha. This is about half the theoretical maximum of the plant conversion of sunlight figured by Ryther (1959) of 98.6 tons dry weight/ha/year.

Animal productivity in the presence of Gracilaria has increased 1.5 to 2 times, with the Gracilaria itself providing a harvest of 5-10 tons fresh weight/ha/year, a figure which seems more reasonable (Doty 1980).

CULTIVATION METHODS IN PRESENT USE

The simplest form of phycoculture is the managed harvests of in situ algal populations. It is here rather naturally seen as part of

THE AQUATIC RESOURCE

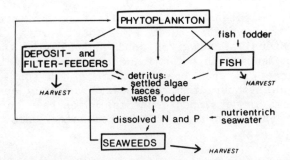

Figure 4. A schematic representation of the material pathways in marine polyculture.

the harvesting of wild algal stocks, as the management only includes regulation of harvesting season and area, species identification and a specification of the tools allowed for harvesting to prevent crop and environment damage.

The first efforts to increase wild seaweed harvests involved manipulation of natural habitats to favour the desired species. In this way existing algal beds could be expanded and new ones created. This approach has been applied in Japan since the 1700s for both red and brown algae. The next step was to put large stones out in shallow coastal waters with muddy bottoms to give the plants something to adhere to. This method was developed further when specially made concrete slabs were put to sea beginning in 1951 to act as substratum for Laminaria (Hasegawa 1976). The remaining work is only to harvest when the plants have become large enough. They now also use twined PVC pipes as substratum (Torii and Kawashima 1978). In China they put out stones weighing 20 kg, 2-3 per m^2 where young sporophytes of Laminaria already have settled (Cheng 1969). Such stone-planted areas may also be sown by placing fertile, half-dried thalli to the sea to liberate numerous zoospores which swim to the stones and settle.

A more sophisticated cultivation technique is the use of artificial substratums such as ropes and floaters. This method is the basis of the nori (Porphyra) industry in Japan and Korea and the Laminaria cultivation in China. It is the most highly evolved example of habitat manipulation in seaweed culture and certainly exceeds most forms of terrestrial agriculture in its complexity as a result of the plants' intricate life cycles. The Porphyra life cycle involves a microscopic stage developing inside a shell. This "conchocelis" stage is cultivated in tank nurseries where fertile Porphyra thalli are introduced. The nets of rope are seeded in tanks incubated with conchospores and then transferred to the sea for development of the macroscopic plants. Fertilization may be carried out during the growing season by using ceramic cylinders containing slowrelease fertilizer pellets or by spraying the fertilizer onto the net structures with the plants. This method is very labour-intensive and gives good control with the outplanting. The population density is optimal and grazers are avoided. The fertilization aspect in such open systems is not too efficient. The amount of manual labour involved, however, hardly seems to fit the economic systems in developed countries which rely on capital-intensive, high technology production processes, unless the product prices are very high.

In the handling of dulse (Palmaria palmata) in Canada, the live storage of harvested wild stock has been introduced to both increase quantity and quality of the product. Artificial drying of the fresh algae insures the highest quality but is very expensive. The dulse harvest is limited to very low tides in the spring and summer, resulting in intermittent and heavy loads on the dehydration and processing facilities, making the operation far from cost-efficient. Impounding the crop in net bags or tanks where it will remain healthy for a few days and up to several months allows increased efficiency in processing. If fertilizer is added to the storage area the seaweeds may reach a high growth rate while still waiting to be dried and processed.

Experiments with cultivating seaweeds in enclosed systems, as in tanks on land or in nets in the sea, have mostly been done with the genera Chondrus and Gracilaria, both carrageenophytes. The method is developed for much the same purposes as previously discussed, that is, for tertiary treatment of municipal and agricultural runoff, for the production of colloids, food and fodder supplement and as a way of promoting polyculture for these and other ends. It must be said that much of the work in this field is very recent and the results are often confined to the investigator, which often is sponsored by industry.

The commercial future of enclosed seaweed culture systems at the nursery level is assured. It gives a high degree of control of a high unit value crop, and the scale of production is rather small.

Enclosed systems for crop growout have so far only been commercially successful in Taiwan in a polyculture system. Experimental work now involves the carrageenophytes Chondrus, Eucheuma, Neoagardhiella, Iridaea, Gigartina, Gymnogongrum and Rhodoglossum, and the agarophytes Gelidium and Gracilaria. It is taking place in Northern France, Nova Scotia, British Columbia, Massachusetts, Florida, Washington, Texas, Hawaii, California and Ireland. Seaweed nutrition is still so poorly understood that cultivators tend to protect against malnutrition by ensuring high waterflushing rates.

Rather than using herbicides it is generally preferred to strive toward culture conditions which give maximum competitive advantage to the desired crop (Hansen et al. 1981). Before commercial seaweed culture can be profitably carried out in the West, enclosed systems and more technical refinement will be required. True commercial success will depend on the development of polycultures in which seaweeds and other crops interact synergetically.

Due to the prospective shortage of energy the cultivation of algal biomass in large quantities for fermentation to methane and other fuels is under way in several countries including the United Kingdom, France, the Federal Republic of Germany, Sweden, Norway and others. The most spectacular effort is the U.S. Food and Energy Farm Project. An ocean-moored structure with a diameter of 100 feet covering some 800 m^2 was set up as substratum for kelp transplants off California. The crucial point in making the operation cost-efficient is the pumping up of deep nutrient-rich water for fertilization of the canopy. The energy to pump up the water was planned to come from the ocean-waves. The retention time of the cold, heavy, nutrient-rich water is so short that the waterflow has to be kept at a high level. A peripheral curtain hanging from the surface to 10 meter depth should provide protection from currents and waves and thus try to keep the nutrients from dispersing too much. Unfortunately, the protective curtain failed during a storm soon after the adult Macrocystis plants had been transplanted. Currents also caused entanglement by the transplants around the various solid structural components and led to abrasion and eventual destruction of all plants. About 30 weeks later it became apparent that the adult plants had released spores that developed into a new generation covering most of the structure, a development which is being followed (North 1980). The Macrocystis material has shown a biodegradability of 80 percent and it gives 281 cm^3 methane /g volatile solids (Ghosh et al. 1981).

PLANT IMPROVEMENT WORK WITH SEAWEEDS

Selection and mutant inducement have to some extent been carried out with Laminaria japonica by the Chinese since the later 1950's. They claim to have selected out varieties that tolerate increased sea temperature, that contain less water and have more iodine than the wild plant average (Anon. 1976, Tsung-ci et al. 1978). The improved varieties are being put into their cultivated production of Laminaria. Work with the red algae is in progress, especially with the cultivated Porphyra. Attempts to obtain high quality Porphyra stock have centered around the two approaches taken by traditional plant-breeders, cross-breeding and selection. Overall improvement have raised yields three to five times. A fast-growing strain of Chondrus, called the T-4, has now been around for almost ten years (Hansen et al. 1981). Clearly the time has come to really exploit the possibilities of strain selection to improve yields of all seaweeds we utilize, and open up prospects for cultivation of improved versions of many other interesting species.

THE FUTURE OF CULTIVATED SEAWEEDS

The potential of the genetic improvement of the seaweeds is theoretically sound but has to be tested more thoroughly on a larger scale, with more species. In coupling this plant material development with the developing biotechnology we may create new processes to make new products. No clear statement valid globally can be made as we are dealing with different traditions. In the Far East seaweeds are mainly an attractive condiment, and in the West we are so far only interested in the phycocolloids and the seaweed meal. The seaweeds are also utilized within totally different economic systems from labour-intensive to capital-intensive. A few highlights can be noted:

Intensive monoculture of red algae in enclosed systems is near a commercial breakthrough. The progress awaits technological and biological improvements in cultivation efficiency and is expected in the mid-80s. As stated as one of the basic criteria for cultivation, it will evolve out of product need.

Of the seaweed uses described in the previous chapter, there is a shortage in the supply of a stable, high-quality agarose colloid. Although it requires a relatively small percentage of red seaweeds provided today, the need must be met, for no substitutes are known. As a result, cultivation of agarophytes and appropriate carrageenophytes under the intensive conditions described will be most promising.

Another highly promising development is the combination of fish, mussels and seaweeds in the described polyculture. This is a way to have the seaweeds fertilized, to divide overhead expenses and to have the seaweeds clean the system of excessive nutrients. It has already proven its case in Taiwan.

To get on with these projects it is often not enough to have the arguments and the experimental evidence. One will also need the confidence and enthusiasm of industrialists and politicians. This is a problem that maybe must be approached in different ways according to each country's regulations, bureaucracy and political machinery.

Hunt (1980) has sketched a few common problems in the transfer of know-how from the phycologists to the general public involving the categories mentioned.

The transfer of algal research insight involves three components:

    i) the information to be transferred (the basic know-how),

ii) the medium used to convey the information, and
iii) the audience.

The barriers are many, among them:

- lack of widespread public awareness of seaweeds
- inadequate or uncreative presentation of research findings
- lack of explicit interpretation of research results and explanation of their potential application
- the use of the technical scientific jargon
- widespread awe or suspicion of scientists.

To reach the envisaged goals, one must train research workers, gain taxonomic knowledge of the seaweeds, make estimates of the wild resources and their harvesting potential, gain knowledge of the growth physiology and fertilization demands of the desired species, carry out pilot-scale cultivation and genetic experiments for ensuring the success of the cultivation, and finally put it all into commercial operation.

At present there is only a handful of applied phycologists outside of Japan and China, and there are even fewer solving the basic scientific problems that are the principal technical barriers. Even the first step, the taxonomy and life cycle of many common and interesting species is not satisfactorily known.

Seaweeds have, however, many arguments in favour of their continued and expanded utilization. They have caught the eye of international scientific bodies, and are now subject to a proposed European co-operation with the title "Marine primary biomass - production and conversion". The International Energy Association (IEA) has also signalled its interest in this renewable bioresource and its potential. This will hopefully help push both developed and developing nations closer to their production potentials regarding this biomass resource.

ACKNOWLEDGMENTS

I gratefully acknowledge valuable comments, criticism and suggestions from Dr. Arne Jensen, and financial support by the Bioenergy Program of the Norwegian Agricultural Research Council.

REFERENCES

Anon. 1976. The breeding of new varieties of haidai (_Laminaria japonica_ Aresch.) with high production and high content of iodine. Sci. Sin. 19(2): 243-252.

Anon. 1980. Fishery Journal no. 12. Yamaha Motor Co., Ltd., AD et PR Div., 2500 Shingai, Iwata-shi, Shizuoka-ken, Japan. 12 pp.

Anon. 1981. Pilot survey of the world seaweed industry and trade. International Trade Center UNCTAD/GATT, Palais des Nations, 1211 Geneva 10, Switzerland. 111 pp.

Becker, E.W. and L.V. Venkatamaran. 1982. Biotechnology and Exploitation of Algae, The Indian Approach. Deutsche Gesellschaft fur Technische Zusammenarbeit, GmbH, Postfach 5180, Dag-Hammarskjold-Weg 1, D-6236 Eschborn 1, Fed. Rep. of Germany. 216 pp.

Bonotto, S. 1979. List of multicellular algae of commercial use. In: Marine algae in pharmaceutical science (H.A. Hoppe, T. Levring and Y. Tanaka, Eds.), pp. 121-139. Walter de Gruyter. Berlin/New York.

Chapman, V.J. and D.J. Chapman. 1980. Seaweeds and their uses. 3rd ed. Chapman and Hall. London/New York. 334 pp.

Cheng, T. 1969. Production of kelp - a major aspect of China's exploitation of the sea. J. Econ. Bot. 23(3): 215-236.

Choudhury, R.P., Paul, P. and S.C. Santra. 1982. Energy and biomass production from freshwater macrophytes. Proc. 2nd EC Conf. on Energy from Biomass. (In press).

Chuang, J.-L., Pan, B.S. and G.-C. Chen. 1977. Fishery products of Taiwan. JCRR Fisheries Series 25B, Taipei, Taiwan, Rep. of China. 90 pp.

Doty, M.S. 1979. The present and future for algal materials. Actas I Symp. Algas Mar. Chilenas: 35-49.

Doty, M.S. 1980. Outplanting Eucheuma species and Gracilaria species in the tropics. In: Pacific seaweed aquaculture (I.A. Abbott, M.S. Foster and L.F. Eklund, Eds.), pp.19-23. California Sea Grant College Program, Univ. of Calif., A-032, La Jolla, CA 92093.

Ghosh, S., Klass, D.L. and D.P. Chynoweth. 1981. Bioconversion of Macrocystis pyrifera to methane. J. Chem. Tech. Biotechnol. 31: 791-807.

Guiry, M.D. 1979. Commercial exploitation of marine red algal polysaccharides - progress and prospects. Trop. Sci. 21(3): 183-195.

Guiry, M.D. and G. Blunden. 1980. What hope for Irish seaweed? Seaweed around our coast. Technology Ireland Sept. :38-43.

Hansen, J.E., Packard, J.E. and W.T. Doyle. 1981. Mariculture of red seaweeds. California Sea Grant College Program, A-032 University of California, La Jolla, CA 92093. Report No. T-CSGCP-002. 42 pp.

Hasegawa, Y. 1976. Progress of Laminaria cultivation in Japan. J. Fish. Res. Board Can. 33: 1002-1006.

Hunt, J. 1980. Transfer of technology in Pacific seaweed research. In: Pacific seaweed aquaculture (I.A. Abbott, M.S. Foster and L.F. Eklund, Eds.), pp. 164-177. California Sea Grant College Program, Univ. of Calif., La Jolla, CA 92093.

Jackson, G.A. and W.J. North. 1973. Concerning the selection of seaweeds suitable for mass cultivation in a number of large, open-ocean, solar energy facilities in order to provide a source of organic matter for conversion to food, synthetic fuels and electrical energy. Final report under contract no. N 605030-73-MN-176, U.S. Naval Weapons Center, China Lake, California.

Jensen, A. 1966. Norsk Institutt for Tang- og Tareforskning. Svensk Kemisk Tidsskrift 78(8): 393-403. (In Norwegian).

Jensen. A. 1978. Industrial utilization of seaweeds in the past, present and future. Proc. Int. Seaweed Symp. 9: 17-34.

Levring. T., Hoppe, H. and O.J. Schmid. 1969. Marine algae; a survey of research and utilization. Cram, de Gruyter et Co. Hamburg.

Mann, K.H. and A.R.O. Chapman. 1975. Primary production of marine macrophytes. In: Photosynthesis and productivity in different environments (I. Cooper, Ed.), pp. 207-223. Intern. Bio. Programme, Vol. 3. Cambridge U. Press, Cambridge.

Madlener, J.C. 1977. The sea vegetable book. C.N. Potter. New York. 288 pp.

Michanek, G. 1975. Seaweed resources of the ocean. FAO Fish. Techn. Paper 138. 127 pp.

Michanek, G. 1979. Seaweed resources for pharmaceutical uses. In: Marine algae in pharmaceutical science (H.A. Hoppe, T. Levring and Y. Tanaka, Eds.), pp. 203-237. Walter de Gruyter. Berlin/New York.

Michanek, G. 1981. Getting seaweed to where it's needed. Ceres Jan.-Feb.: 41-44.

Moss, J.R. 1977. Essential consideration for establishing seaweed extraction factories. In: The marine plant biomass of the Pacific Northwest Coast (R.W. Krauss, Ed.), pp. 301-315. Oregon State University Press.

Naylor, J. 1976. Production, trade and utilization of seaweeds and seaweed products. FAO Fish. Techn. paper 159. 73 pp.

Neish, I.C. 1980. Innovative trends in the marine colloid industry. In: Pacific seaweeds aquaculture (I.A. Abbott, M.S. Foster and L.F. Eklund, Eds.), pp. 6-10. California Sea Grant College Program, Univ. of Calif., A-032, LaJolla, CA 92093.

North, W.J. 1980. Review paper. Biomass from marine macroscopic plants. Solar Energy 25: 387-395.

Phillips, R.C. 1980. Overview of sea grass studies with special reference to tropical species. In: Pacific seaweed aquaculture. (I.A. Abbott, M.S. Foster and L.F. Eklund, Eds.), pp. 54-62. California Sea Grant College Program, Univ. of California, A-032, La Jolla, CA 92093.

Priestly, G. 1978. Algal proteins. In: Food from waste (G.G. Birch et al., Eds.), pp. 114-139. Applied Science Publ. London.

Ryther, J.H. 1959. Potential productivity of the sea. Science 130: 602-608.

Senn, T.L. and A.R. Kingman. 1978. Seaweed research in crop production. Dept. of Horticulture, Clemson University, Clemson, SC 29631. 135 pp. + app.

Torii, S. and S. Kawashima. 1978. Experiments on propagation of Laminaria (Phaeophyceae). I. Propagation of Laminaria japonica by plastic pipe (netron) in sandy substrate. Proc. Int. Seaweed Symp. 9: 473-479.

Tsung-ci, F., Chi-sun, T., Yu-lin, O., Chin-chin, T. and C. Tenchin. 1978. Some genetic observations on the monoploid breeding of Laminaria japonica. Sci. Sin. 21(3): 401-408.

MUNICIPAL SOLID WASTE — A RAW MATERIAL

Robert F. Vokes

Consultant, Parsons & Whittemore, Inc.
New York, New York
U.S.A.

INTRODUCTION

Herman Kahn (American Express 1982) founder of the Hudson Institute, advises us that:

"The world is not running out of resources and energy!"
"The world is not going to be choked by growing pollution!"
"Almost every expert in the field agrees!"

The basis for these optimistic statements is a recognition of a high degree of consciousness on these issues and enormous improvements in technology as well as increased knowledge and experience. Certainly, this NATO-ASI BIOMASS UTILIZATION conference may serve to undergird Mr. Kahn's prognosis.

The following treatise will include information on all of the known components of Municipal Solid Waste (MSW), although emphasis is limited to the organic fraction as a biomass form of raw material for further utilization.

Background and History

The situation, from the beginning of man's existence, is generally known by documentation of world history whereby entire civilizations have been consumed and lost as a result of natural changes in the ecological and environmental balances. For example, the ice age resulted from global changes in the temperature of the atmosphere and created lakes and mountain ranges. Entire forest lands and animal kingdoms were buried and now provide our fossil fuels. Natural regrowth of forests provides biomass fuel.

Man's contribution to ecological changes has also been well documented as he denudes the flora, fauna and fossil fuels from the earth and thereafter moves on to denude additional areas. This practice is still in existence, particularly in many underdeveloped countries (Barnes 1980) and exemplifies a major problem of mismanagement of natural assets, resulting in the loss of many critical raw materials. This practice is also common in many industrially developed nations whereby biomass is lost via disposal of MSW in landfills.

The mismanagement of raw materials as well as land, lakes, rivers and the atmosphere, has come to a head on many occasions in our history. Not only have entire cities been rebuilt, layer upon layer and in new locations, but also the disposal of man's waste products has caused major epidemics of illness such as cholera in Germany in 1914-1917, thus giving rise to the mandate of the burning of all MSW and pathological wastes. In Europe, mass burning of MSW has been practiced for many years and energy recovery has partially replaced the use of other sources of fuels.

The increased awareness of the finite nature of fossil fuels, the establishment of the Organization of Petroleum Exporting Countries (OPEC), the existence of technology for production of alternative fuels from biomass and the new technology for production of organic chemicals from biomass has therefore established the circumstances for an opportunity to employ MSW-biomass as a raw material.

Terminology

The term "Municipal Solid Waste" (MSW) evolved during the 1970's for the purpose of identification of a specific class and kind of waste material. It refers primarily to the waste materials discarded by people, living in the industrially developed nations, pursuant to the use of natural and manufactured products. Major sources of Municipal Solid Waste are generated in the larger cities with relatively high population densities.

Previously, the terms garbage, trash, litter, junk, refuse, waste papers, etc. were in common use and may continue to identify some of the MSW raw materials in certain areas of the world. However, as the problem of disposal has evolved as an opportunity for recycling, it is anticipated that additional terminology will develop, including many of the prior terms. For example, the literature includes frequent reference to current acronyms such as:

RDF - refuse derived fuel
d-RDF - densified refuse derived fuel

The term "Solid Waste" (Eberhard 1981) has been adopted by the National Research Council (NRC) which was established by the National Academy of Sciences (NAS). An advisory board on the "Built Environment" serves the commission on "Sociotechnical systems" for the NRC.

Such terminology, partially standardized and collected in a "Glossary" by the National Center for Resource Recovery (NCRR Bulletin 1978) applies to the whole field of Resource Recovery from both natural and man-made sources of waste materials. This is a rapidly developing field of applied and research technology for the protection of the environment and the management of consumption, recovery and disposal of raw materials from nature. Additional typical new terminology is as follows:

Co-disposal: Simultaneous disposal of solid waste and waste water treatment sludges from municipal and industrial sources.

Co-generation: Simultaneous generation of steam and electricity.

Solid Waste Stream: All solid waste materials produced by man or nature and discarded. Includes demolition waste, papers collected and recycled.

Built Environment: That which is man-made.

Resource Recovery: The practice of separating and using the components of the Solid Waste Stream and MSW.

EPA: Environmental Protection Agency, U.S.A.

DOE: Department of Energy-U.S.A.

DOC: Department of Commerce-U.S.A.

GAO: General Accounting Office-U.S.A.

Fibreclaim: Trade name for wet process separation of cellulose fibers from MSW Biomass and fiber fractionation by length. (Black Clawson Co.).

Hydrasposal: Trade name for wet shredding, pulping and grinding. (Black Clawson Co.).

Megawatt: 1000 Kilowatts

Quad: 1×10^{15} BTU (British Thermal Units).

Sociotechnical Systems: The technical processes and procedures created by man to effect changes or controls of his environment.

Solid Waste: That solid waste material from all sources, municipal, industrial, natural, etc.

Analyses of MSW Composition

To effect "Resource Recovery" from MSW, the basic approach to the application of technology and equipment requires a data base including specific components of the MSW as a raw material. Such data has been accumulated for projects proposed and in operation for many areas of the world. These data are not considered to be fully representative of the total range of components at any particular time or place. However, the best representation of average distribution is indicated based upon the use of standard methods of sampling and statistical averaging of analytical data.

A typical average presentation of data on MSW-Biomass components as a raw material is shown for the U.S.A. vs. European situations as it relates to major municipalities. Table I is a representative sampling of the U.S.A. source of information and Table II a sampling of the Western European scene.

Solid Waste Streams

A solid waste stream includes all solids employed by man, some of which are subsequently discarded for disposal or recycling. This "stream" is the source of MSW.

Table I. Typical MSW composition-U.S.A.

	% OD*		% OD
Paper	40	Combustibles	77 (organic)
Yard waste	15	Non-Combustible	23 (inorganic)
Food waste	12		100
Glass	10		
Metals	(10)		
Fe	8		
Al	1	Average Moisture:	26.25% (wet
Other	1		basis)
Plastics	4	See Table IV Col. 7	
Wood	2		
Rubber & leather	2		
Textiles	2		
Misc. Inorganics	3		
	100		

* OD-oven dry basis

Table II. Typical MSW composition-western Europe.

	% OD*		% OD
Paper	34	Combustibles	73 (organic)
Yard Waste	30	Non-Combustible	27 (inorganic)
Food waste			100
Glass	11		
Metals	(6)		
Fe	4		
Al			
Other	2	Average Moisture 28.5% (wet basis)	
Plastics	6		
Wood	1		
Rubber & leather			
Textiles	2	See Table III, Col. 5	
Misc. Inorganic	10*		
	100		

* Average less UK=4

A portion of the solids used becomes a so-called permanent fixture in the environment, i.e. wood, metals, ceramics, glass, roofing papers, and materials of construction for housing and factories.

Another portion is recycled for reuse as raw materials in the paper industry, steel industry, or glass industry; for wood burning energy production, as animal feed stock, or as compost, etc. The balance is identified as MSW and it flows as continually as the waters of the Nile and the Amazon.

The nature of MSW is characteristic of the geographical area of source and the composition reflects the specific life style and raw materials available and used by the population. Variables in composition are also controlled by state laws, the status of the art of conservation and recycling, by the economy, available technology as well as world politics, i.e. war and peace. Hence, the specific types and amounts of biomass from MSW are constantly in a state of flux although sufficiently uniform and in adequate quantity to qualify as a very reliable source of supply for industrial and institutional purposes.

WESTERN EUROPE

The "solid waste stream" of Western Europe yields an "MSW Stream" which is described via details of specific analyses in Table III.

Table III. MSW composition specific-(percentage, oven-dry).

Material	UK[1]	Holland[2]	France[3]	Italy[4]	Statistical Average
Paper	31	27	34	44	34
Yard waste, Food waste	16	40	35	31	30
Glass	10	14	13	6	11
Metals	9	5	5	5	6
Plastics	3	7	7	7	6
Wood		1	7		1
Rubber, leather and textiles	3	3	2	1	2
Misc. Inorganics	28*	3	3	6	10
	100	100	100	100	100
Moisture Percentage (wet basis)	25	27	27	35	28.5

*UK MSW includes "dust bin" ash and cinders from coal fires in homes

[1] Kenworthy 1975.
[2] Marsh 1972.
[3] Marsh 1973.
[4] Marsh 1977.

A wide variation in the inorganic composition of solid waste exists between the United Kingdom and the Continent of Europe as a result of common use of coal for hearthside heating of homes in the United Kingdom.

Certain other variations are noted as a result of the methods of collection and processing of MSW for resource recovery. For example, in Rome, Italy, the collection of MSW is controlled as to size and composition by mandate and the use of plastic bags only at the homesite. This is a form of source separation whereby certain vegetation, metallic and demolition wastes are routed to landfill or major metal recovery facilities.

Generation Rates

MSW is generated in Western Europe at a rate of about 1.1 to 1.5 lbs/capita/day. This data (Marsh 1973) is limited in scope although it is believed to be reasonably accurate for projection of

MSW as a raw material for biomass utilization, subject to further research.

MSW-Biomass Energy Recovery

Mass burning of unprocessed MSW was originated in Europe in the late 1800's. The oldest plant on record, built in 1896, is in Hamburg, Germany. Approximately 180 such plants are in operation today and the practice has been adopted by Japan and the U.S.A. Installations, built since 1946, are larger and more efficient and are used for producing steam for hot water heating as well as electricity.

It is estimated that nearly 50 percent of the MSW-biomass generated in Western Europe cities is incinerated and that in the order of 75 percent of the incinerator capacity is operated for energy recovery.

UNITED STATES OF AMERICA

Data on the subject of "Solid Waste" in the U.S.A. is rather limited. However, the EPA has collected some information initiated in 1960 through 1978 (NCRR 1980) which indicates that the current MSW production may be in the order of 154-160 million tons per year (Goldstein 1981) and 200 million tons per year in 1990 (EPA 1981).

Data on the specific composition of MSW biomass is shown for certain municipalities and averages from several sites as reported by authorities. Column 7 of Table IV shows a current statistical average estimate.

Generation Rates

The generation of MSW is documented at a growing rate of about 3 percent per year from 1960 to 1978 and a rate of about 4.4 lbs/capita/day in 1981 (EPA 1981). This value will vary according to the development of resource recycling technology which, if expanding in capacity, would tend to decrease the availability of raw materials for recovery from MSW in the future. The principle involved here is, as components of MSW are recycled, the residue is classified as MSW subject to further resource recovery and finally, as MSW for disposal in sanitary landfilling. Since the optimal feasibilities for resource recovery are currently limited to the production of energy and the recovery of metals, glass and ash (from energy production), it is anticipated that projections for the immediate future of resource recovery operations will be based upon these criteria. The MSW-biomass component is forecast to be in the order of 75 percent (Table I) of the expanding growth of MSW generation as indicated above.

Table III. MSW composition specific-U.S.A. (percentage, oven-dry).

Material	Franklin Ohio[1]	Hempstead N.Y.[2]	Dade County Fla.[3]	Woodbury N.J.[4]	EPA[5]	NCRR[6]	Statistical Average[7]
Paper	40	43	47	35	43	33.5	40
Yard waste	15	13	8	15	16	17.5	15
Food	10	8	16	3	15	17.0	12
Glass	12	13	8	11	10	9.9	10
Metals	13	(12)	(8)	4	(10)	(9.2)	(10)
Fe		10	7		9	8.0	8
Al		1	1		1	0.9	1
Other		1				0.3	1
Plastics	4	3	3	6	4	3.6	4
Wood	2	1	3			3.2	2
Rubber & leather	1	2			3	2.6	2
Textiles	2	3	3		2	2.0	2
Misc. Inorganics	1	2	4		5	1.5	3
Total	100	100	100	100	100	100	100
Moisture Percentage	25	25	30		25		26.25

1 Marsh 1973.
2 Marsh 1974.
3 Marsh 1977.
4 Goldberg 1982.
5 EPA 1977.
6 NCRR 1980.

Biomass Composition

For consideration of MSW biomass as a raw material, certain assumptions must be made:

1. That the biomass can be isolated from the MSW raw material;
2. That the biomass can be fractionated to isolate certain major components; and
3. That the isolated biomass can be used as a raw material for economically viable products.

Assumption No. 2 above is thereby proposed as an approach to categorize the biomass components into more specific types of organic raw materials so that ultimate utilization may be facilitated.

Table V presents approximate values of the MSW biomass fractions. Note that these data are subject to wide variations as a result of seasonal and geographical factors.

Uniformity

Although considerable variability of the composition of MSW is indicated, it is possible to apply some controls at the source of supply as well as through the design and operation of the processing technology and equipment.

Evidence of the uniformity of the composition of processed MSW for the production of RDF has been demonstrated by commercial plants in the past several years (City Currents 1982). Mass burning of MSW for energy recovery has also shown that the organic component of MSW

Table V. MSW — biomass composition.
(Biomass = 75% MSW, OD basis)

Biomass % OD	Components	Paper Type (Kenworthy 1975)	% OD	Total
43	Paper	News	40	
		Magazine	12	
		Packaging	29	
		Board	6	
		Tissue	8	
		Mixed	5	100
12	Plastics		31	
	Rubber & Leather		23	
	Textiles		17	
	Wood		29	100
42	Vegetation		50	
	Food		50	100
3	Other Organics		100	100
100				

is sufficiently uniform in quantity and composition so that the use of MSW as a raw material for other than energy production may be considered feasible.

Energy Production

Energy production from the burning of MSW biomass in America is a recent phenomenon, although the rate of growth is increasing as a result of oil embargoes and waste disposal problems in landfills. The present day disposition of the MSW streams in the U.S.A. is approximately as indicated in Table VI.

The present day proposals and programs for energy recovery planned by the Federal Government (DOE-EPA-DOC) in GAO report 1979 reveal that by 1990 there will be:

- 131 MSW-to-energy projects
- 36 million tons MSW per day processed
- utilization of 18% of MSW generated
- 48 million bbl. oil equivalent energy produced
- saving of 3.5 billion U.S. dollars

Note that the current support of such energy recovery programs by the Federal Government has been curtailed, although it is anticipated that private investment in cooperation with state and city governments will increase.

MSW-Biomass -- Problems and Limitations

Availability. "Garbage" strikes are common in large metropolitan areas where unionized manpower is employed to collect source separated MSW at the curbside, physically handle the waste in all kinds of weather and compete with dogs, thieves, traffic, equipment breakdowns, contamination, fires, spoilage, etc. Generation of the MSW does not cease as long as human habitation exists, but controlled availability and material handling presents a problem to municipal governments and cooperating industrial users of the MSW as a raw material.

Essentially 100% of MSW biomass from the urban areas of the world should become available for use as a raw material. Since a

Table VI. Current disposition of MSW streams in the U.S.A.

Stream	%	Elements
Landfill	88	Land recovery
Source Separation	6	Paper and minerals
Energy Recovery	2	Steam and electricity
Incineration	4	No energy recovery

majority of citizens in developed countries reside in urban areas the problems of assured supplies, assured technology, and availability of markets for the converted products represent a highly desirable goal for millions of people. The leadership and the incentives will certainly be obtained in the advanced industrially oriented countries.

Contamination

Contamination refers to the health hazards which arise from human exposure to the MSW as a raw material as well as a converted product or by-product. MSW contains both explosives and poisons in varying quantities neither of which can be readily avoided by known techniques.

Shredding and incineration, in either sequence, effectively controls the contamination during processing to the extent that minimum health safety standards are readily met. Sanitary landfill procedures and standards are also commonplace for municipalities.

Contamination, identified as an "undesirable element" in the composition of MSW and the biomass fraction can be coped with by multiple means and methods already in practice. However, for each "undesirable element" we face a control and possible disposal of residual "undesirable by-products".

Studies of contamination in MSW biomass fiber recovery projects have centered on chemical, i.e. toxic heavy metals, carcinogenic organic compounds, etc., as well as bacteriological and pathogenic organisms.

Essentially, the degree of contamination in MSW-biomass, isolated as a raw material for paper and board manufacture, seems to be in the same order of magnitude as that which exists in the mixed waste papers which are source separated from the MSW stream for commercial paper industry furnishes (Kenworthy 1975).

SUMMARY

MSW biomass is identified as a readily available uniform commodity in perpetual supply. The quantity is in direct proportion to the population, is generated by the population and must be disposed of by the population.

Disposal is relatively simple via sanitary landfilling technology and will be practiced according to the circumstances of existing life styles of various elements of the population. For the majority of the urban population of the industrially developed nations, the MSW biomass will be processed as a raw material for corresponding end products followed by disposal of unwanted residuals.

The methods for processing MSW as a raw material, processing the material to produce and market end products and disposing of the residuals, are being borrowed from existing knowledge and created from new knowledge as required. The work began in Europe in the late 1800's and continues to evolve at an increasing intensity throughout the urban environments of the globe. However, for the immediate future, most of the biomass isolated from MSW will be a raw material in the form of RDF for the production of energy.

REFERENCES

American Express Co. 1982. Newsletter, June. P.O. Box 770, New York, N.Y.
Barnes, G.O. 1980. The Global 2000 Report to the President, Vol. I. U.S. Govt. Printing Office. Washington, D.C.
City Currents. 1982. Bulletin of U.S. Conference of Mayors, Resource Recovery Activities, Washington, D.C.
Everhard, John P. 1981. ABBE. The recovery of energy and materials from solid waste. National Academy Press, Washington, D.C.
E.P.A. 1977. 4th Report to Congress. NCRR, Washington, D.C.
E.P.A. 1981. Unpublished Contract 68 01 6000. Solid Waste Data. Washington, D.C.
Goldberg, E. 1982. Woodbury Times, New Jersey, U.S.A.
Goldstein, I.S. 1981. Organic Chemicals from Biomass. CRC Press, Inc. Boca Raton, Florida.
Kenworthy, I.C. 1975. Utilization of fibre reclaimed from garbage. PIRA Report TS-125, Paper Industry Research Association, Leatherhead, Surrey, England.
Marsh, P.G. 1972. Unpublished report 115, The Black Clawson Fibreclaim Division, The Black Clawson Co., New York, N.Y.
Marsh, P.G. 1973. Unpublished report 116. The Black Clawson Fibreclaim Division, The Black Clawson Co., New York, N.Y.
Marsh, P.G. 1973. Raw materials from refuse. Reclamation Industries International.
Marsh, P.G. 1974. Unpublished report, Table 2.1.4. The Black Clawson Fibreclaim Division, The Black Clawson Co. New York, N.Y.
Marsh, P.G. 1977. Fibreclaim - A progress update. Unpublished paper 158, Black Clawson Fibreclaim Division, The Black Clawson Co., New York, N.Y.
Marsh, P.G. 1977. Unpublished report, Pg. 1/11 Rev. 2, Black Clawson Fibreclaim Division, The Black Clawson Co., New York, N.Y.
NCRR. 1978. Bulletin. Resource Recovery Briefs. Washington, D.C.
NCRR. 1980. Bulletin. 10:4. p. 86. Washington, D.C.
NCRR. 1980. Resource Recovery Update, Vol. 9, No. 5, Washington, D.C.

THE POTENTIAL ROLE OF DENSIFICATION IN BIOMASS UTILIZATION

John J. Balatinecz

Professor of Forestry
University of Toronto
Toronto, Canada

INTRODUCTION

Increases in the price of conventional forms of energy and in many cases reductions in their availability are now generally understood to be inevitable. This is due to a number of factors:

o the finite nature of most conventional forms of energy
o the inordinate power over oil price of the OPEC cartel
o the regional and international discrepancies that exist between energy supply and demand
o the potentially serious health and environmental questions regarding the use of nuclear fuels and coal
o the increasing demands being made from Third World countries for fossil fuels as they continue to industrialize
o the increasing demand for all forms of energy in the developed world.

Energy development to meet future demand will be a costly and complex process. No single alternative is foreseen to emerge as an "only solution" to the demand problem. The utilizaton of biomass for energy and chemical feedstocks is part of the solution.

In Canada, various forms of biomass currently account for approximately 3.5 percent of the total energy consumption (Love and Overend 1978). This biomass is primarily recovered from wood waste and spent pulping liquors at pulp mills with a much smaller amount used for residential space heating. Table 1 summarizes the estimates that are available on the quantity of the various different types of biomass generated in Canada. As can be seen, even

Table 1. Estimates supply of biomass in Canada by type and source.

TYPE/SOURCE OF BIOMASS	QUANTITY PRODUCED (10^6 Oven-Dry Tons)	GROSS ENERGY CONTENT (10^{18} Joules)
Forest Waste		
- Mill Residues	7.5	0.14
- Residues from Forest Operations	31.0	0.58
Dedicated Forest Biomass		
- Unutilized Trees in Currently Logged Areas	20.0	0.37
- Wood Available in Areas not Currently Logged	52.0	0.97
- Energy Farms	(*)	(*)
Animal Waste	12.6	0.28
Crop Waste	16.9	0.25
Dedicated Crops	(*)	(*)
Aquatic Biomass	(**)	(**)
Solid Waste	10.6	0.17
Sewage Sludge	0.4	0.01
Total	151.0	2.77

Source: Love and Overend (1978)

(*) No estimate of the potential contribution of agro-forestry crops specially designed for energy production could be made, although it has been hypothesized that the energy available from forest resources may be underestimated by a factor of 3-5.

(**) No estimates available.

excluding the huge potential offered by dedicated forest or crop energy farms, these sources have a gross energy content of 2.77 x10^{18}J (2.62 x 10^{15}BTU), or almost 25 percent of Canada's total energy demand if a 70 percent conversion efficiency is assumed.

In the less developed countries, 72 percent of the wood that is cut each year is used for fuel (FAO 1978); by comparison only 5 percent of the wood cut in North America is used for fuel. Much of the wood that is currently used in the less developed countries is burned inefficiently.

Thus in Canada, as will be the case in most other countries with a sizable forest products industry, a large energy potential exists in the form of harvesting and mill residues and currently unutilized species. In the less developed countries where wood is already used to a large extent for fuel, a large potential exists for upgrading of biomass for use in more efficient stoves or boilers.

In its natural form biomass is often an inefficient fuel because it is bulky, wet and dispersed. Densification can reduce these problems. The purpose of this report is to present an overview of the potential role of densification in biomass utilization. This report is based on a recent survey by the author of the technical and patent literature on densification, and a questionnaire-survey of several North-American equipment manufacturers.

ADVANTAGES AND DISADVANTAGES OF BIOMASS AS AN ENERGY SOURCE

Biomass has a number of desirable attributes as an energy source. These include:

- ready availability
- renewability
- clean, nearly pollution-free combustion
- low energy and capital requirements for production
- no requirement for special storage facilities.

On the other hand, raw biomass fuel has several disadvantages. These are:

- relatively low heat value per unit volume
- variability of quality and heat value
- difficulty in controlling the rate of burning
- rapid burning, necessitating frequent refuelling
- difficulty in mechanizing continuous feeding
- large area requirement for storage
- economic problems in transportation and distribution
- subject to bio-deterioration.

Several of these undesirable attributes result from the low mass and volume energy density of biomass. The mass energy density (MED) and volume energy density (VED) of biomass are 3 - 4 times less than that for coal or gasoline (Table 2). In bulk form (i.e. pile of wood chips, sawdust or shavings), the biomass densities are even lower due to the "fluff" factor. This results in higher transportation and storage costs as well as lower efficiency of combustion. As with refined petroleum fuels, raw biomass can be upgraded in value for various burning situations. Upgrading may involve one or more of size reduction, drying and densification. Since densification pre-supposes size reduction and drying (or at least moisture control) subsequent discussion will deal with densification.

BIOMASS DENSIFICATION PROCESSES

Biomass densification may be defined as compression or compaction to remove inter- and intra-particle voids. Five densification

Table 2. Energy densities of different fuels by mass and volume.

Fuel	Moisture Content	Density	Heat of Combustion MED* (kJ/g)	Heat of Combustion VED** (kJ.cm^{-3})
1. Biomass	50	1.0	9.2	9.2
	10	0.6	18.6	11.2
2. Densified-biomass	10	1.0	18.6	20.9
	10	1.25	18.6	26.1
3. Charcoal	0	0.25	31.8	8.0
4. Coal (Bituminous)	0	1.3	28.0	36.4
5. Methanol	0	0.79	20.1	15.9
6. Gasoline	0	0.70	44.3	30.9

Source: Reed and Bryant (1978).
* MED: Mass Energy Density
** VED: Volume Energy Density

processes are now practised commercially (Reed and Bryant 1978) including: baling, pelleting, cubing, briquetting and extrusion.

The concept of biomass densification is not new. Briquettes have been manufactured from saw-dust, shavings, bark and other wood residues for the past 75 years. The first U.S. Patent for densification was issued in 1880 (Smith, U.S. Patent No. 233,887). This patent described a process where sawdust and other wood residues were heated to 150°F and subsequently compacted to the density of coal with a steam hammer. Since then several patents were issued on the subject (Edison, S.O. 1902; Mashek, G.J. 1907; Griffin, W.T. 1908; Mulligan, P.C. et al. 1925; Levelton, G.H. 1966; Skendrovic, L. 1970; Beningson, R.M. et al. 1977; Bremer, A.R. 1975; Williams, T.A. 1976; Gunnerman, R.W. 1977; Livingston, A.D. 1977; Johnson, A.O. 1978).

While not precisely explained in the literature, densification is believed to involve compression, deformation and self bonding between adjacent particles of biomass. When heated above the plastic temperature range of 165°C (325°F) wood loses its elasticity and is relatively easily compressed. Particle surfaces come into intimate contact and the thermally softened lignin and other phenolics allow the creation of adhesion between adjacent particles. Moisture plays an important role in densification. It may help in heat transfer and in enhancing the plasticity of the material. If the feedstock is either too dry or too wet pressures required for densification increase dramatically. Normally moisture content should be in the range of 10 percent - 25 percent. Gunnerman (1977) in his patent recommends moisture contents between 16 percent and 28 percent. The pre-heating of feedstock between 50°C and 100°C is usually adequate to soften the lignin. Additional heat is generated from the mechanical work of densification which requires 32 to 80 J/g. With a heat capacity of about 1.7 J/g·C, this would raise the temperature of the densified material by 20°C to 50°C. Thus in commercial densification systems the densified biomass product emerges with a temperature of about 150°C. The hot biomass is fragile and should be handled carefully until cooled (Reed and Bryant 1978).

<u>Compression baling</u> and <u>roll-compressing</u> can reduce biomass volume to one-fifth its loose bulk. These processes are useful for agricultural biomass and certain types of forest biomass (e.g. crops from energy plantations, logging residues).

<u>Pelleting</u> employs a hard steel die which is perforated with frequently placed holes 0.3 to 1.3 cm in diameter. The die rotates against inner pressure rollers, forcing the comminuted biomass through the holes with pressures of 7.0 kg/mm^2. As the biomass is extruded through the die, small dense pellets are broken off at a specified length. <u>Cubing</u> is a modification of pelleting which produces larger cylinders or cubes, 2.5 to 5.0 cm in diameter.

<u>Briquetting</u> compacts a biomass feedstock between rollers with cavities, producing forms like charcoal briquettes. <u>Extrusion</u> uses a screw extruder to force a feedstock under high pressure into a die thereby forming large cylinders (2.5 to 10 cm) of densified biomass. Binding agents such as paraffin or resin are often added to modify properties of the fuel logs.

A commercial application of the extrusion principle is the "Press-to-log" machine which had been designed and made by Wood Briquettes Inc., Lewiston, Idaho. This machine manufactures densified fuel from wood waste (sawdust, shavings, etc.) by compressing the material in a primary compression chamber with a feed screw at a pressure of approximately 2.2 kg/mm^2 (3,000 psi). The pressure and friction generate sufficient heat to plasticize wood and create the conditions for self-bonding. The compressed wood material is extruded through cylindrical holes 10 cm (4 inches) in diameter spaced at regular intervals in and extending through the rim of a larger wheel. The 10 x 30 cm (4- by 12-inch) briquettes produced by this machine are suitable for hand firing in stoves and fireplaces, but not for mechanical stoking. For stoker briquettes, a different

Table 3. Some suppliers of densification equipment.

Company	Machine Type
1. Agnew Environmental Products Inc., Grants Pass, Oregon	Extruder
2. The Bonnot Co., Kent, Ohio	Extruder
3. Hawker Siddeley Canada Ltd., Vancouver, British Columbia	Extruder
4. California Pellet Mill Co., San Francisco, California, and Crawfordsville, Indiana	Pelletizer
5. Reydco Machinery Co., Redding, California	Extruder
6. Sprout-Waldron Div. Koppers Company, Inc., Muncy, Pennsylvania	Pelletizer
7. Taiga Industries, Inc., San Diego, California	Extruder
8. Columbia Fuel Densification Corp. Phoenix, Arizona	Extruder

type of machine extrudes the material through a cluster of holes in 2.5 cm (one-inch) lengths. These 2.5 cm long pellets are suitable for mechanical stoking. The Taiga extruder is a more recent commercial development.

Several manufacturers and suppliers of equipment are listed in Table 3. Others offer total plant "packages" using equipment from one or more suppliers. One such system is the "Woodex" process (Gunnerman 1977) of Bio-Solar Corporation. The system is based on hammer mill, dryer and pellet mill. A 120-ton-per-day plant has been operating since 1976 in Brownsville, Oregon. Several other plants have been installed recently in the U.S. and Canada. Purchasers of fuel pellets include not only forest products plants but also utilities.

PROPERTIES AND USES OF DENSIFIED BIOMASS

The specific gravity (or "relative density") of densified biomass is in the 0.8 - 1.3 range, and moisture content is usually 5 percent to 10 percent. The bulk density of a pile of pellets is 0.5 - 0.8 g/cm^3. The heat of combustion is around 18.0 - 19.0 kJ/g.

Depending on the degree of compaction and the chemical properties of the material, densified biomass usually has improved water-repellency and high resistance to biological deterioration.

The principal advantages of densified biomass are:

o rate of combustion is comparable to coal
o permits burning in grate-fired boilers
o provides uniform combustion
o reduces particulate emission
o makes transportation, storage and feeding more efficient
o reduces (eliminates?) the possibility of spontaneous combustion in storage
o significant reduction in biodegradability.

The burning of wood pellets to supplement or replace coal in utility or industrial boilers is well developed. Suspension and spreader stoker coal firing systems can burn densified biomass fuel with little or no modification. Boilers specifically designed to burn wood wastes (small firetube boilers, bark burning boilers, vortex combusters and fluidized bed combusters) can, of course, also burn wood pellets. Such boilers are commonly used to produce steam, process heat or, when coupled to a turbogenerator, electricity.

An advantage in the substitution of pellets for coal is the reduced emissions of sulphur dioxide due to the low sulphur content of the pellets.

Other potential uses of densified biomass fuel include (Reed and Bryant 1978):

o fueling residential, commercial or industrial central heating systems
o fueling wood stoves
o fueling fireplaces and outdoor grills
o firing external combustion engineers
o raw material for pyrolysis and gasification

ENERGY BALANCE AND ECONOMICS

The relevance and future role of densification in biomass utilization, in spite of its technical merits, will be determined by the energy balance and production economics of densification systems. The energy required for densification is influenced by the physical attributes of the feedstock (e.g. moisture content, size, type, density, etc.), pellet size, equipment used and other parameters. Approximate energy requirements and feed rates for several kinds of biomass in a 300 h.p. pellet mill of the California Pellet Mill Company are presented in Table 4. Generally 1 - 3 per-

Table 4. Approximate energy requirements for pelleting by a 300 h.p. pellet mill.

Feedstock	Production Rate (tons/hr.)	Electrical Energy Used (kWh/ton)	Fraction of Product Energy Consumed (%)
Sawdust	6.1	36.8	2.3
Aspen wood	8.2	27.2	1.7
D-fir bark	4.5	49.2	3.1
Municipal Solid Waste	9.1	16.4	1.0

Source: California Pellet Mill Company and Reed and Bryant (1978)

cent of the energy contained in the feedstock is required for densification, whereas 2 - 3 percent is needed for size reduction. Feedstock drying requires most of the process energy, ranging from 7 - 15 percent depending on the initial moisture content of raw biomass. Thus, the energy efficiency of biomass densification is in the range of 82 - 93 percent.

Reed and Bryant (1978) presented a cost analysis of pelletizing. Cost data was based on the Woodex process of the Bio-Solar Corp.; financial data was based on lumber and paper industry averages. The 300 ton-per-day Woodex plant in Brownsville, Oregon, cost about $1.25 million, including the cost of classifying, feeding, and cooling equipment; hammer mills; dryers; pellet mills; and storage. The estimated breakeven selling price was $19.70/ton (1.20/MBtu). Bio-Solar was selling pellets to Western State Hospital near Tacoma, Washington, for $22/ton f.o.b. plant (May 1978). A sensitivity analysis revealed that feedstock cost has a strong influence on the breakeven selling price.

An economic analysis supplied by Sprout-Waldron (1979) and based on a three-pellet-mill installation (300 h.p. each) revealed that over 50 percent of production costs are accounted for by feedstock and electrical energy costs (Table 5).

Whether or not it is practical and economical to pellet biomass when users have the option of directly combusting green biomass with only milling as a required preparation is a complex issue. Drying, pelleting, investment and other operating costs add about $10 per ton of densified biomass fuel. But certain savings accrue from this additional processing because densified biomass has a higher density

Table 5. Estimated production costs for a three-pellet mill biomass densification facility, capacity 18 tons per hour.

Item	Cost per ton (U.S.$-1979)	Percent
1. Feedstock (with 45% MC)	$ 7.72	35.9
2. Electrical (111 KW/ton)	4.44	20.6
3. Labour (four men)	1.60	7.4
4. Dryer fuel	1.07	5.0
5. Die and roll shells	0.89	4.1
6. Maintenance	0.80	3.7
7. Overhead and supervision	1.65	7.7
8. Depreciation* (11%)	3.35	15.6
TOTAL	$21.52	100.0

Source: Sprout-Waldron Div. of Koppers Co., Inc. (1979)
* Based on $1.3 million for equipment (10 yr. am.)
  $2.0 million for building and installation (20 yr. am.)
  Annual capacity: 123,000 tons

and lower moisture content than green biomass. Converting green biomass into pellets, cubes, briquettes, or rolls increases the number of BTU's that can be transported, stored, or handled at a constant cost per unit of weight or volume. Furthermore, moisture content is a key economic factor in residue combustion systems. The lower the moisture content of the fuel, the higher the combustion efficiency of the boiler, a correlation which translates directly into fuel savings, capital investment savings because of reduced capacity requirements, and emissions control savings. It is in this area of economic analysis where more reliable information is needed about the costs and benefits of biomass densification.

REFERENCES

Beningson, R.M., Beningson, H.E., Narayananrutty, V.P., Nadkarni, R.M., Lamb, T.J., Fox, L.K. and W.V. Keary. 1975. Apparatus for disposal of solid wastes and recovery of fuel product therefrom. U.S. Patent 4,063,903.

Bremer, A.R. 1975. Die-head. U.S. Patent 3,904,340.

Edison, S.O. 1902. Process of treating fibres of annual growth for industrial purposes. U.S. Patent 704,698.

Food and Agriculture Organization, Rome. 1978. Yearbook of Forest Products.

Griffin, W.T. 1908. Method of briquetting carbonaceous materials. U.S. Patent 905,693.

Gunnerman, R.W. 1977. Fuel pellets and method for making them from organic fibrous materials. U.S. Patent 4,015,951.

Johnson, A.O., Holiman, W.E. and S.D. Buchanan. 1978. Methods of handling waste. U.S. Patent 4,081,143.

Levelton, B.H. 1966. Process of producing fuel logs. U.S. Patent 3,227,530.

Livingston, A.D. 1977. Process for treating municipal wastes to produce a fuel. U.S. Patent 4,026,678.

Love, P. and R. Overend. 1978. Tree power: an assessment of the energy potential of forest biomass in Canada. Report of Department of Energy, Mines and Resources, Canada.

Mashek, G.J. 1907. Process of preparing pulverulent materials for molding or briquetting. U.S. Patent 852,025.

Mulligan, P.C. and H.G. Swalwell. 1925. Process of preparing materials for briquetting and the product thereof. U.S. Patent 1,551,966.

Reed, T. and B. Bryant. 1978. Densified biomass: a new form of solid fuel. Solar Energy Research Institute No. 35.

Skendrovic, L. 1970. Domestic refuse and garbage disposal system. U.S. Patent 3,506,414.

Sprout-Waldron Division of Koppers Company, Inc. 1979. Personal communication.

Williams, T.A. 1976. Artificial logs and log-making method and apparatus. U.S. Patent 3,973,922.

MASS PROPAGATION OF SELECTED TREES

FOR BIOMASS BY TISSUE CULTURE

S. Venketeswaran*
V. Gandhi, E.J. Romano and R. Nagmani

Department of Biology
University of Houston
Houston, Texas 77004
U.S.A.

INTRODUCTION

The enormous potential of trees as future sources of biomass yield has been envisaged as early as 1935 by Professor Bergius (Gleisinger 1942). Biomass resources, a source of energy, are conversions of solar energy deposited by photosynthesis in woody plants as large quantities of cellulose, hemicellulose and lignin. Biomass resources which exist in forests, have been relatively unexploited both in the U.S.A. and in the world. The commercial implications of this immense reproductive biomass power is limited to a large extent by the absence of application of modern integrated technological developments. One method to obtain additional energy is to increase the yield of biomass by genetic improvement and then the mass propagation of the genetically improved strains. It is now possible through tissue culture techniques to produce "test tube trees", i.e., plants obtained from only a few cells of a tree by manipulations under laboratory conditions. Thus several thousand young plantlets can be produced from a single tree and these can be transferred to the field and grown as seedlings within a short time. It is also possible to use protoplast isolation, culture, fusion of protoplasts and genetic manipulations to obtain new "hybrid" trees in plants that normally do not cross to produce superior genotypes.

---

*Supported by The BioEnergy Council, Washington, D.C., Energy Laboratory, University of Houston and the University of Houston Coastal Center.

By "cloning" and mass propagation of superior genotypes possessing novel and desirable characteristics, a large number of "superior trees" can be obtained for planting in fields and screening of genotypes at considerably lower costs, space and time. This method of high-frequency cloning may also be useful for forest trees with poor seed set, absence of uniform seed production and seeds susceptable to genetic damage or loss of viability during storage (Karnosky 1981).

The production of plantlets in test tubes and growing them in large numbers as a method of vegetative propagation has been very successfully used in various horticultural and related industries. In recent years, these methods are employed in tree genera like Eucalyptus, Sandalwood, Teak, etc. and several thousand plantlets of these trees have been obtained in one year. Rao and Lee (1982) report successful results on organogenesis and plantlet development on several tropical timber and other useful trees. Somatic embryogenesis, plantlet formation and vegetative propagation of Sandalwood plants through tissue culture has been achieved (Rao and Bapat 1978; Bapat and Rao 1979). Another paper describes the clonal multiplication of a 100-year old "elite" Teak tree (Tectona grandis L.) by tissue culture through regeneration from excised seedlings and estimates indicate that about 3,000 plants can be obtained from a seedling in a year (Gupta et al. 1980).

This paper describes the results of investigations employing cell, tissue and protoplasm culture techniques on selected tree genera, potentially useful as biomass fuel producers and for in vitro mass propagation of large numbers of plantlets/plants of desired genetic stability.

MATERIALS & METHODS

Materials

Emphasis of work has been on selected tree species which show the greatest potential for growth and biomass. These are:

1. Sapium sebiferum Roxb.   -- Chinese Tallow -- Euphorbiaceae
2. Leucaena leucocephala
   (Lam.) de Wit.   -- Ipil-ipil   -- Leguminosae
3. Copaifera multijuga Hayne -- Copaiba Tree   -- Leguminosae

These three species were selected on the basis of background information, growth characteristics in specific geographical areas and initial successful results on experiments on tissue culture in this laboratory (Venketeswaran and Gandhi 1982). A brief description of the three genera is as follows:

Sapium sebiferum

Sapium sebiferum Roxb.: A relatively small deciduous tree (seldom reaching heights above 40 ft.), native of China and distributed throughout the tropics, chiefly in America with tremendous growth characteristics on ravine land, in the foothills, etc., it has become naturalized along the Gulf and southern Atlantic coasts (Texas to North Carolina) where there is a long growing season and ample rainfall. Yields in excess of 5.0 dry tons/acre have been obtained in the second of a coppicing study in the Texas Gulf Coast (Scheld and Cowles 1981). They are generally used as ornamental and shade trees in houses and produce multiple fall colors (yellows, reds and purples). They are tolerant to saline water and are resistant to pathogens. Coppicing characteristics and growth is fast with mean annual girth increment of 2.6 - 5.2 cm. Wood is white, even-grained, moderately hard and light (wt. 513 kg/cm^3) and can be used as fuel. Bark on extraction yields useful organic compounds. The large crop of seeds provide fats from the waxy mass covering the seeds and oil from the kernel, both of which are useful in various ways. Seed meal after extraction of oil is possesses high content of protein and is valued as feed and fertilizer (Wealth of India 1975; Scheld and Cowles 1981).

Leucaena leucocephala

Leucaena leucocephala (Lam.) de Wit.: Familiarly called the Ipil-ipil tree, this leguminous tree is a tropical or subtropical species which grows in marginal lands, hill slopes, etc. It has a tremendous growth rate as a single-trunked tree and can grow as high as 65 ft. (20 m.) in 4-5 years. Originating from Central America, some varieties are spread widely throughout the region. The "Hawaiian type" (Giant Ipil-ipil) is widely distributed and can grow and thrive in Texas. These trees are capable of nitrogen fixation if inoculated with Rhizobium. The wood has a specific gravity of 0.52 and is a hard wood, useful as lumber or pulp and provides many wood products. The wood is also useful as fuelwood with heating values of 8,300 BTU/lb. dry weight (19.3 MJ/kg.). Various other uses of the tree include forage, fuel, soil improvement and reforestation (National Acad. Sci. 1977; Forestry 1978; Anon. 1978).

Copaifera multijuga

Copaifera multijuga Hayne: The Copaiba tree of Brazil belongs to Leguminosae and the woods of all species of Copaifera contain gum or oil canals which produce the copaiba balsam. In the U.S., it is probable that this genus would grow in Southern Florida. Heartwood is reddish-brown with variable density, sp. gr. 0.70 - 0.90 and weight 44-56 lbs./ft^3). According to Calvin (1979), when holes (1½") are drilled into the trunk, a yield of 10-20 liters of hydrocarbon material is produced in about 3 hours and the holes can be plugged. The operation can be repeated every six months. According to the anatomy of the trunk, there is very little lateral communication and another hole drilled would deliver a similar amount of oil at the same time. This tree shows great potential as a wood and oil source (Record and Hess 1972; Chem. and Eng. News 1979).

Methods

Seeds were germinated after various treatments, viz., scarification acid treatment (H_2SO_4), hot water treatment, etc. and standard tissue culture techniques were employed to establish callus tissue. A modified Murashige and Skoog (1962) medium or the B_5 mineral salts (Wetler and Constable, 1982) with the addition of appropriate growth factors was used for callus initiation from embryo explants, other plant parts, etc. and for subsequent maintenance of callus in both solid and liquid media. In vitro organogenesis (roots, shoots or buds) from callus tissue and other explants (plant parts) were achieved by manipulation of the nutrient and hormonal constituents in the culture medium. Protoplast isolation was carried out with certain modifications from standard techniques (Bajaj 1977). Cultures were maintained in a culture room at $25 \pm 1°$ C with no specific control of light conditions.

RESULTS

Tissue cultures of Chinese Tallow were established from germinating embryos and other plant parts, viz., elongating hypocotyl segments, pieces of endosperm, tissues from plants grown in the greenhouse (Fig. 1), etc. by growing them in an agar solid medium containing 1 mg/l of both 2,4-D and kinetin. Callus appeared within 2-3 weeks after inoculation (Fig. 2) and further growth resulted in extensive proliferation of the tissue masses. By frequent subcultures, at intervals of 3-4 weeks, large quantities of tissues were obtained routinely (Fig. 3) which were used for various experiments. Recently, several of the subcultures grow well in 1 mg/l of 2,4-D and 0.5 mg/l of kinetin. By growing solid callus tissues in a liquid medium on a mechanical shaker, it was possible to obtain liquid suspension cultures. The tissues dispersed into a fine suspension of cell (s) and cell aggregates (Figure 4). After 2-3 subcultures, inoculums of the suspensions containing cells and cell aggregates could initiate new cultures. Using appropriate filtration methods of the liquid cell suspension, cultures containing approximately 60-70% single cells were obtained. They were plated onto agar plates (like bacterial cultures). By this method, "strains" of tissue were obtained and recultured in different nutrient media to obtain callus. All these tissues are at present maintained in the laboratory and have been used to obtain differentiation and regeneration of plantlets. Extensive root production

PROPAGATION OF SELECTED TREES FOR BIOMASS 195

Figures 1-4. 1) Chinese Tallow plant. 2) Callus initation from hypocotyle segment. 3) Callus (6-8 weeks old). 4) Cellaggregate from callus; X 600. (ca.).

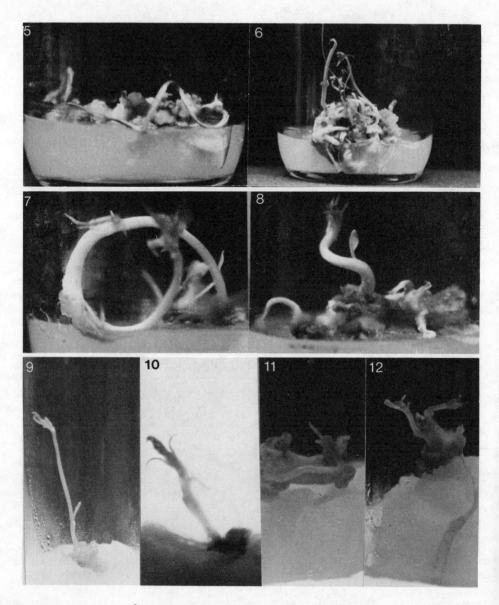

Figures 5-12. 5-8) Development of plantlets from callus.
9-12) Development of shoots, buds from 1 cm. segments grown in test tubes on roller drum.

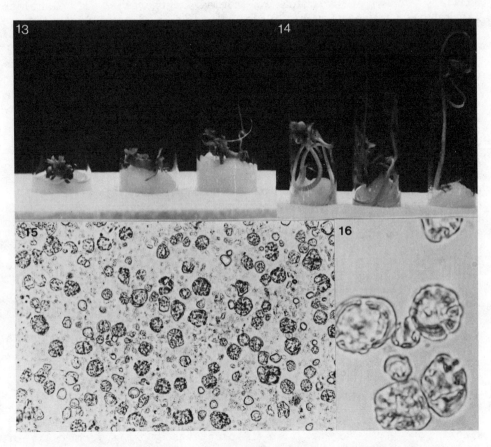

Figures 13-16. 13-14) Regeneration of buds and plantlets.
15-16) Protoplasts isolated from young leaves; X 150 and X 600 (ca.).

production appeared when hypocotyl-derived callus tissues were transferred to media containing 1.0 mg/l α-naphthalene acetic acid (NAA) with 2,4-D and kinetin (both at 0.1 mg/l) or without 2,4-D and/or kinetin. Tissues which have been maintained in a medium with 0.1 mg/l 6-Benzylaminopurine (BAP) for 2-3 subcultures, when transferred to medium with 0.1 mg/l or 1.0 mg/l NAA produced shoots within 3 weeks. The shoots elongated and produced nodes, leaves, axillary buds, etc. indicating normal regeneration of plantlets (Venketeswaran and Gandhi 1981a). Figures 8-11 illustrate the development of plantlets derived from callus tissues.

Regeneration of plantlets from shoot apices and 1 cm. segments of hypocotyls of newly-germinated seedlings were very effective in Chinese Tallow. These explants were grown in test tubes containing cheesecloth and 15-20 ml. of liquid medium supplemented with 0.1, 0.5 or 1.0 mg/l BAP, or 6-γ,γ,-dimethylallylamino purine (2iP). The tubes were maintained on a slanting tissue culture roller drum rotating at 1 rpm. Five to six weeks after inoculation, explants growing in BAP medium responded well whereas those growing in 2iP medium produced inconsistent results. At 0.1 mg/l BAP, the segments produced shoots and buds (about 5-6 per explant). Occassionally, callus appeared at the basal region of the explant in contact with the culture medium. Transfer of the regenerated shoots and buds to a medium containing 0.1 mg/l BAP and 1.0 mg/l NAA induced roots and elongation of the shoots producing "test tube trees" (Figures 9-14). This process can be repeated every 6-8 weeks and it may be possible to obtain as many as 3,000 plants from one seedling in one year, thereby achieving mass propagation in large numbers (Venketeswaran and Gandhi 1981).

Tissue cultures of Leucaena were initiated from plants grown in the greenhouse (Figure 17) or from embryos, hypocotyl segments, cotyledonary pieces, etc. Callus production, proliferation, subsequent growth behavior, and organ differentiation (Figures 18-22) appeared similar to those observations made in Chinese Tallow, except for some minor differences. Certain varieties showed slight differences in response to growth hormone concentrations of 2,4-D, kinetin or NAA. Regeneration of plantlets growing from calli of explants occurred frequently (Figure 23). In recent experiments using the roller-tube method, plantlet regeneration from cut ends of hypocotyl segments or cotyledonary pieces have been excellent. Further experiments are in progress to obtain optimum concentrations of growth substances for routine regeneration.

Vegetatively-produced plantlets from tissue culture cells or tissue explants of both Chinese Tallow and Leucaena are maintained in growth chambers for subsequent transfer to the greenhouse and field conditions at the University of Houston Coastal Center.

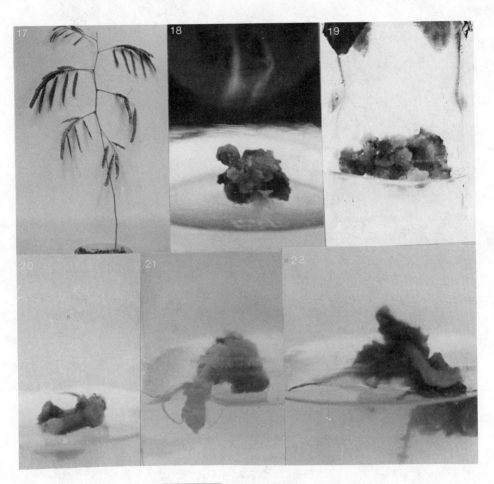

Figures 17-22. 17) <u>Leucaena</u> young plant. 18-19) Callus development and growth. 20-22) Development of shoots, roots and callus.

Figures 23-24. 23) Regeneration of plantlets growing from callus of explants. 24) Protoplasts from young leaves; X 250. (ca.).

Protoplast isolation was carried out from both Chinese Tallow and Leucaena sp. (Figures 15,16 and 24). Leaves from germinated seedlings in the greenhouse were cut into small pieces and treated with an osmotic buffer solution containing 0.2 M $MgCl_2$, 0.1 M KH_2PO_4 and 9% sorbitol for 30-60 minutes. They were incubated in an enzyme solution containing 1.0% each of Drieselase, Pectinase and Hemicellulase with 5% sorbitol for varying periods of time. Other times and concentration of enzymes were also tried. A good yield of protoplasts was obtained from young leaves of seedlings whereas mature leaves from plants or from trees in the field yielded few or no viable protoplasts. Protoplasts were also obtained from callus tissues (Venketeswaran and Gandhi 1980).

After several attempts, it was possible to germinate seeds of Copaiba (Venketeswaran and Romano 1982). Callus establishment, organogenesis, regeneration from callus and in vitro production of plantlets in large numbers using procedures above are being attempted.

DISCUSSION

Biomass production, as an alternate source of energy, by genetic improvement of forest trees has captured world-wide attention over the last 5-6 years (Anon. 1981). Although nearly half of the world population relies mainly on wood for their cooking and heating, only approximately one-sixth of the world's annual fuel comes from biomass, wood being the largest biomass resource (Anon. 1980). On a world-wide basis, bio-energy consumption is about 15% which is equivalent to approximately 40 quads of BTU per year (1 quad = 10^{18} joules) or 20 million barrels of oil per day (Anon. 1981). Therefore, the potential use of wood and its bioconversion into renewable energy has been growing rapidly because of higher prices of oil and natural gas. In several countries, extensive work on silviculture and forest harvesting for biomass and energy is expected to provide the largest component of energy source as an alternative. A forecast of U.S. energy use and supply alternatives toward the year 2,000 includes wood as probably the largest fraction of biomass, although the U.S. Forest Service has predicted a wood shortage by the year 2,000 unless new advances in forest technology are made soon (USDA Forest Resources Report, #20, 1973). The Office of Technology Assessment (OTA) indicates that wood, plants and other biomass fuels could supply as much as 20% of the U.S. annual energy requirements by the year 2,000 representing 10 quads BTU from wood, (BioScience 1980). Currently, the U.S.A. uses about 1.5 quads of biomass energy each year of which about 85% comes from wood.

Production of plantlets in large numbers and mass propagation of trees such as Chinese Tallow and other tropical legumes is very significant. The Chinese Tallow grows naturally in the Gulf Coast region and is widely used as an ornamental or shade tree in houses. It also shows considerable potential for short-rotation intensive silviculture (Scheld and Cowles 1981), and is an excellent choice for biomass production. Similarly, <u>Leucaena leucocephala</u> (Lam.) de Wit. (Ipil-ipil), considered as a potential alternate source of biomass and energy, is a tropical legume and a nitrogen fixing tree capable of growing in Texas. It is also a promising soil-erosion control plant. <u>Leucaena</u> charcoal has a combustion value of 12,980 BTU/lb. dry weight = 30.2 MJ/kg. which is approximately two-thirds of the BTU value of natural gas or fuel oil (Dijkman 1950). The Copaiba tree (oil tree) of Brazil shows promise as an alternate source of energy, although at present the non-availability of seeds and plants puts severe limitations of its use (Chem. and Eng. News 1979).

The State of Texas has large areas of land which are already forested or which can be turned into forests. In particular, the Gulf Coast region has a long growing season and ample rainfall. Therefore, "selected" woody tree crops which show promise as potential sources of energy can be cultivated in large numbers in experimental plots, since many of these trees are adaptable to this environment. Introduction of genera such as <u>Leucaena, Copaifera</u> (tropical legumes) and others like honeylocusts, mesquite and persimmon which are examined as potential tree crops (Siebert et al. 1982) appear appropriate and valuable. Another tree, <u>Paulownia</u>, a crop tree for wood products and reclamation of surface-mined land (Tang et al. 1980), is a promising candidate. <u>Paulownia</u> has been planted as a rapid-growing economic tree crop for reforestation in Brazil during the last decade (Howlett 1975).

Tissue culture methods and clonal mass propagation can be widely applied to several tree genera, since most trees have a long sexual cycle (several years) and some have limited seed availability. Besides, Day (1977) points out that protoplast fusion and cell hybridization offers some promise of tapping sources of genetic variation now unavailable because of sterility barriers. Therefore, the ability to manipulate genetically the genes of the tree, by future experiments employing protoplast culture and genetic manipulations, can lead to the production of trees adaptable to environments previously unable to support them. The "cloned" trees or "superior-quality" trees derived through genetic improvement in large numbers will allow a rapid evaluation of genotypes in experiments of growth rate, cold-hardiness, salt resistance, etc. within short periods of time and with limited space compared to conventional breeding programs which require many hectares of land and longer periods of time (Karnosky 1981). Forest genetics and genetic improvement of trees via tissue culture will play an increasing role in the forests of the future for conversion of biomass into alternate sources of energy. Transfer of tissue culture technology from the laboratory to field conditions will be of prime importance for implementation of genetic improvement and mass propagation in forest tree breeding programs (Wochock and Abo El-Nil 1977).

ACKNOWLEDGMENTS

I express my gratitude to Dr. Joe R. Cowles and Dr. Glenn D. Aumann for their encouragement and support of the research.

REFERENCES

Anon. 1978. International Consultation on Ipil-ipil Research. Papers and Proceedings. pp. 1-172.
Anon. 1980. Proc. Bio-Energy '80 World Congress & Exposition. The Georgia World Congress Center, Atlanta, Georgia, U.S.A. pp. 30-133.
Anon. 1981. The International Bio-Energy Directory. The Bio-Energy Council, Washington, D.C., U.S.A. p. xi.
Bajaj, Y.P.S. 1977. Protoplast Isolation, Culture and Somatic Hybridization. In: Applied and Fundamental Aspects of Plant Cell, Tissue and Organ Culture, pp. 467. Reinert, J. & Y.P.S. Bajaj (Eds.), Springer-Verlag.
Bapat, V.A. and P.S. Rao. 1979. Somatic Embryogenesis and Plantlet Formation in Tissue Cultures of Sandalwood (Santanum album L.). Ann. Bot. 44: 629-630.
Calvin, M. 1979. Personal Communication & Newsletter.
Chem. and Engg. News. 1979. Brazil's biomass program is one of most extensive. p. 35.
Day, P.R. 1977. Plant Genetics: Increasing Crop Yield. Science 197: 1334-1339.
Dijkman, M.J. 1950. Leucaena - a promising soil-erosion-control plant. Econ. Bot. 4: 337-349.
Forestry. 1978. Sector Policy paper, World Bank, Washington, D.C., U.S.A.. pp. 1-65.
Gleisinger, E. 1942. Nazis in the Woodpile. The Bobbs-Merril Co., New York. pp. 262. (Not consulted in original).
Gupta, P.K., A.L. Nadgir, A.F. Mascarenhas and V. Jagannathan. 1980. Tissue Culture of Forest Trees: Clonal Multiplication of Tectona grandis L. (Teak) by Tissue Culture. Plant Sci. Letters 17: 259-268.
Howlett, D. 1975. Forestry in the future of Brazil. Am. For. 81: pp. 14-17 and 44-45.
Karnosky, D.F. 1981. Potential for Forest Tree Improvement via Tissue Culture. BioScience 31: 114-120.
Natl. Acad. Sci. 1977. Leucaena - Promising Forage and Tree Crop for the Tropics. Report, U.S.A., pp. 1-115.
Rao, P.S. and V.A. Bapat. 1978. Vegetative Propagation of Sandlewood Plants through Tissue Culture. Can. J. Bot. 56: 1153-1156.

Rao, A.N. and S.K. Lee. 1982. Proc. Symp. of Tissue Culture of Economically Important Plants. Singapore. (In Press).

Scheld, H.W. and J.R. Cowles. 1981. Woody Biomass Potential of the Chinese Tallow Tree. Econ. Bot. 35: 391-397.

Siebert, M., G. Folger and T. Milne. 1982. Alcohol Co-production from Tree Crops. Annual ASES Meeting, Houston, Texas, U.S.A. 1-7.

Tang. R.C., S.B. Carpenter, R.F. Wittwer and D.H. Graves. 1980. Paulownia - A Crop Tree for Wood Products and Reclamation of Surface-mined Land. Southern J. App. For. 4: 19-24.

U.S.D.A. Forest Resources Report. 1973. #20.

Venketeswaran, S. and V. Gandhi. 1980a. Protoplast Isolation and Culture of Tree Genera for Biomass Production. Abstr. 2nd Internatl. Cong. on Cell Biol. p. 641.

Venketeswaran, S. and V. Gandhi. 1981b. Tissue Culture and Plantlet Regeneration of of Chinese Tallow, Sapium sebiferum Roxb. In Vitro 17: p. 219.

Venketeswaran, S. and V. Gandhi. 1981. Mass Propagation of Two Selected Tree Genera for Biomass Production. In Vitro 17: p. 229.

Venketeswaran, S. and V. Gandhi. 1982. Mass Propagation and Genetic Improvement of Forest Trees for Biomass Production by Tissue Culture. Biomass 2: 5-15.

Venketeswaran, S. and E.J. Romano. 1982. Tissue Culture of Forest Trees for Biomass Energy Production through In Vitro Propagation. Proc. Internatl. Plant Tissue & Cell Cong., Tokyo, Japan. (In Press).

Wealth of India. 1975. C.S.I.R. Publications, Govt. of India. pp. 230-231.

Wetler, L.R. and F. Constabel, (Eds.) 1982. Plant Tissue Culture Methods. 2nd Edition. Natl. Res. Council of Canada, Canada.

Wochock, S.S. & M.A. Abo El-Nil. 1977. Transferring Tissue Culture Technology. TAPPI: 85-87.

THE PRODUCTION OF MICROALGAE

AS A SOURCE OF BIOMASS

      E. W. Becker

           Institut für Chemische Pflanzenphysiologie
           Universität Tubingen
           7400 Tübingen
           Federal Republic of Germany

INTRODUCTION

    Whenever biomass is discussed today, agricultural crop residues as well as forest waste products are the first to be mentioned as the major sources of supply. Only in rare instances is the cultivation of other unconventional sources such as microorganisms (bacteria, yeasts, algae) given consideration as a further possibility for the production of biomass. Yet, this "Single Cell Protein" (SCP) source, to use the collective term, may have significant potential. Thus, it is the aim of this article to give a general introduction and explore the possibilities and characteristics of microalgae biomass production and its utilization.

    The term microalgae used in this context is a pragmatic one without taxonomic significance and includes procaryotic blue-green algae (Cyanobacteria) as well as eucaryotic green algae.

    Mass production of algae can be divided into two major categories:

a) A "clean system", using preferably fresh water fortified with defined amounts of substrates (mineral salts) for photoautotrophic growth. In this system, monocultures of selected algal strains are cultivated in artificial tanks. An exception is the cultivation of the blue-green alga <u>Spirulina</u> in Mexico. Here this alga grows in a natural soda lake (Lake Texcoco) from which it is harvested.

b) A "waste system", using sewage or other waste products as the main substrates. This system, often called "high-rate algal ponds", has the dual advantage of producing algal biomass by simultaneously treating wastes. Municipal, domestic and agricultural waste waters as well as wastes of certain industries contain the essential nutrients for maintaining algal growth. In this scheme, the algae photosynthetically produce oxygen for bacterial break-down of organic matter and utilize the carbon dioxide produced by the bacteria in the sewage or waste.

Interest in algal biomass production originated about 40 years ago with increasing intensity since World War II. This early work was published 1953 in the classic report of Burlew (1953). The fundamentals of algal cultivation were worked out in several countries including the U.S.A., Japan, Germany, Israel, France, and Czechoslovakia. This new technology has been investigated initially for the production of protein-rich food and subsequently also for other purposes like animal feed, utilization in aquaculture, application as biofertilizer, for water renovation and nutrient recycling, extraction of natural compounds or bioconversion of solar energy.

Today, several algal waste water treatment systems are in operation in different parts of the world. Few commercial ventures exist that involve the production of algae for food and feed purposes like the flourishing Chlorella and Spirulina industry in Japan and Taiwan. Here the algae are used for the manufacture of tablets, extracts and other health food items, for which a well established market exists.

Although all these products and applications are distinctly different, they are all based on the photosynthetic conversion of solar energy into bond chemical energy and the transformation of inorganic and organic substrates into biomass. For these purposes, different methods have been developed at various places using sophisticated biotechnologies with a high input of equipment and energy or very simple systems as for rural applications in developing countries.

Algae

In the early work of algal mass cultivation, mainly Chlorella and Scenedesmus strains were investigated for their growth characteristics, particularly for high multiplication rates and high protein content. The genera that now have been considered for these purposes are: Chlorella, Scenedesmus, Uronema, Coelastrum, Dunaliella, and Spirulina. In sewage or waste water cultures also: Micractinium, Oocystis, Oscillatoria, Chlamydomonas, Euglena, and

Ankistrodesmus. However, in clean systems, most of the projects still prefer the cultivation of Chlorella, Scenedesmus, and Spirulina. A diagrammatic view of these three algae is given in Figure 1.

The major differences among these algae are their size and the fact that the blue-green alga Spirulina does not have a cellulosic cell wall, a characteristic which facilitates processing and utilization.

Algal Growth

The growth of an algal culture under the typical regime of batch cultivation passes through different phases:

1) Adaptation phase (lag phase)
2) Exponential growth phase (log phase)
3) Linear growth phase
4) Decreasing growth phase
5) Stationary phase
6) Death phase

The ideal shape of the respective phases is shown in Figure 2. The phases represent the reaction of the algal population to changes of the environmental conditions and depend on the inoculum size, cultivation method, nutrient content, light intensity, etc. During the adaptation phase the algal culture adapts itself to the environmental conditions, the specific growth rate is lower than in the subsequent one and increases with cultivation time.

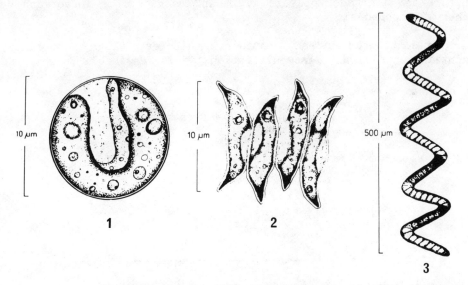

Figure 1. Cell structure of 1: Chlorella sp., 2: Scenedesmus obliquus, 3: Spirulina plantensis.

In the second phase, the algal culture has adapted itself to the given conditions, the light intensity is not limiting, and changes in the nutrient concentration caused by the uptake of the algae are so small that they can be neglected. In a non-limited, non-synchronous culture the increment in algal biomass per time is proportional to the biomass in the population at any given moment.

The algal growth in a dense culture proceeds through a third phase, the linear growth phase, where the cells multiply to such an extent that they begin to screen each other so that gradually a high absorption of incident light may occur. This effect reduces the specific growth rate and the increase in algal biomass becomes linear. This phase continues until a limitation due to the exhaustion of nutrients occurs or where respiration begins to interfere. In well maintained cultures the linear phase remains for several days.

During the fourth phase with a decreasing growth rate, the light supply per algal cell becomes limited and oxidative breakdown of synthesized substances begins to reduce the constant increment of biomass. The growth curve approaches asymptotically a limiting value, which may be described as the maximum attainable concentration of algal biomass.

In the fifth phase, an equilibrium is obtained between the maximum concentration of biomass and loss due to degradation processes. During the final phase the algal cells begin to die, releasing organic, often growth inhibiting material into the medium. This phase is caused by unfavourable environmental conditions, overage of the culture and limited supply of light and nutrients or infections by other microorganisms.

In continuous or semi-continuous cultures, which are normally maintained in algal mass cultivation, it is desirable to keep the

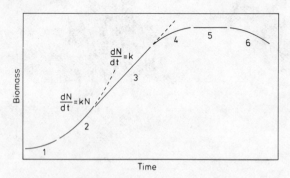

Figure 2. Growth phases of algal cultures (for details see text).

culture in the linear growth phase. This is possible by maintaining the algal density within certain concentration ranges, where the chances of limitations of any kind are minimal.

ALGAL CULTIVATION

Independent of the substrates involved in the cultivation or the final utilization of the algal biomass, large scale mass cultivation in general requires the following basic steps:

a) a pond or artificial tank, wherein the algae are grown with the addition of specific nutrients; b) an agitation system, which prevents the algae from settling and ensures uniform distribution of minerals and equal exposition of the algal cells to light; c) a harvesting system, by which the algae are separated from the culture medium; and finally (in most cases) d) a drying step, primarily for dewatering of the concentrated algal biomass and also, as in the case of green algae, for improving digestibility and nutritional value of the algae by rupturing of the algal cell wall.

A flow diagram of the cultivation, processing and utilization of microalgae is given in Figure 3. The individual factors involved are discussed in the following sections.

Tank Designs

One of the important requisites for successful algal cultivation is the construction of suitable basins which should be cheap, efficient, easy to operate and durable. Therefore, size and shape of the tanks, the material used for construction and the type of system employed for agitation vary depending on the local conditions, the raw material available and the final utilization of the algal biomass. In practice, the construction is always a compromise between a high-performance system with optimal hydrodynamic properties and the economical need to reduce costs.

The size of the tanks used at the various places ranges from $1\ m^2$ to more than $1000\ m^2$ of surface area. In most of the cases the construction material is either concrete (sometimes coated), plastic lining or just compressed gravels and sand.

Regarding shape and agitation, three different designs are commonly in use nowadays (Figure 4):

1) Circular ponds with agitation provided by a rotating arm.
2) Oblong forms ("raceways"), which are constructed either as one unit (2a) or in a joint form of several units ("meander"), with agitation by means of paddle wheel(s) (2b).

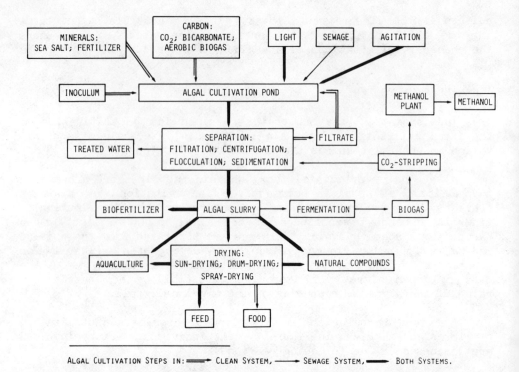

Figure 3. Flow diagram of algal mass cultivation systems.

3) Sloped meander-like constructions, where mixing of the algal suspension is achieved by pumping and gradient flow.

Besides the above mentioned methods of mixing, others have been tried at some places. Some of these methods are shown schematically in Figure 5.

Circular ponds have the disadvantage that high amounts of energy are required for continuous stirring and that the turbulence is low in the centre part of the pond. Horizontal raceways are preferred in most of the algal plants, with channel widths between 1 - 2 m for clean systems and up to 5 m in high-rate ponds (Oswald 1972). The velocity of the flow can be a few cm/s in sewage systems and up to 0.5 m/s in high-performance clean autotrophic units. The regime of mixing varies from place to place and depends on the algal species, algal density, substrate supplied, etc. While some plants operate with continuous agitation, others mix only at intervals and sometimes stop agitation completely during the night.

The principle of sloped algal culture units was developed in Czechoslovakia (Setlik et al. 1970) and later was adopted in

PRODUCTION OF MICROALGAE AS A SOURCE OF BIOMASS

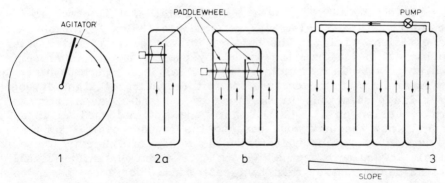

Figure 4. Common designs of algal tanks. 1: Round tank, 2 (a): Horizontal "raceway", (b): Horizontal "meander", 3: Sloped "meander".

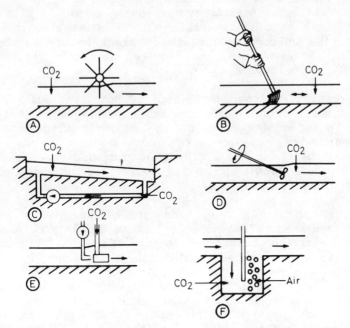

Figure 5. Different methods of agitation in algal culture. A: Paddle wheel, B: Manual stirring, C: Pump and gradient flow, D: Propeller, E: Injection, F: Air lift. (After Markl and Mather).

Bulgaria (Dilov 1980). These units consist of tilted surfaces made out of glass (Czechoslovakia) or concrete (Bulgaria), with baffles for efficient mixing. The culture is continuously recirculated to the highest point from the lowest by pumping and the culture flows down by gravity in a thin layer (3 - 5 cm), which permits high algal concentrations in the culture without reduction of light intensity. A similar system, but in the form of a meander was constructed in Peru (Heussler et al. 1978), where the suspension flows in a long channel from the uppermost part to the lowest point, from where it is pumped back. The culture depth is kept at 20 cm. These relatively low culture depths (<30 cm) are a common feature of all the various designs, except in high-rate ponds, where depths up to 1 m are found. The reason is that at greater depths the light penetration is reduced due to self-shading of the algal cells.

Nutrients

In general, all algae require the same common nutrients for growth; however, each species has its own specific needs which have to be worked out to obtain maximum gowth rates. For algal cultures in laboratory scale, nutrient mixtures with analytical grade chemicals have been developed. These recipes are too expensive and elab-orate for outdoor cultivation and commercial grade chemicals and locally available fertilizers are being used. In this case care has to be taken that no impurities (heavy metals, etc.) are introduced into the culture medium, which might be accumulated by the algae. In sewage systems, the required nutrients are mostly supplied by the waste water.

An exception is the cultivation of Spirulina which requires a high concentration of bicarbonate and high pH-values (9-11) of the medium. To substitute some of the minerals with still cheaper sources, industrial by-products (molasses) or waste materials (blood, bone meal, urine, effluent of biogas digesters) have been tested for their growth promoting properties in algal plants designed for rural applications.

Although there has been considerable success in eliminating costly refined chemicals by simpler raw materials, the algal growth is limited mainly by unsufficient supply of carbon. Like other photoautotrophic plants, algae require inorganic carbon in the form of CO_2 for photosynthesis. Natural oligotrophic waters contain some amounts of CO_2 absorbed from the atmosphere, but these concentrations are too low to achieve optimal growth. Therefore, additional carbon has to be supplied to the cultures. The following reflection shall illustrate the importance of sufficient carbon provision. The carbon content of dry algal material is approximately 50%. To produce 1 kg of dry algae, 0.5 kg of carbon (about 1.8 kg or 900 l of CO_2) have to be added to the culture, assuming 100% efficiency in CO_2-uptake, which is not possible in practice. It is realistic to calculate a consumption of 2.5 kg of CO_2 per kg of algae.

On the other hand, it is not necessary to supply all this carbon in the form of CO_2. A part of this amount can be replaced by organic carbon (sugar, molasses, acetate) since algae can utilize these substrates as well. Experiments have demonstrated that green algae (Chlorella, Scenedesmus) can be produced in mixotrophic cultures with alternate supply of inorganic and organic carbon (Figure 6). Chlorella producing companies in Japan and Taiwan are using acetate as the only carbon source; according to their reports 4 kg of acetate are required to produce 1 kg of algae.

In sewage systems, the carbon is mostly provided in two forms: a) partly from the air by intensive aeration and partly by microbial break-down of organic substances and b) in the organic form, dissolved in the waste water.

Harvesting

Harvesting is one of the Achilles heels in algal production systems because low concentrations of solids ($<0.2\%$) have to be separated from large quantities of liquids to slurries of more than 10% solids. Since the economy of algal production depends to a large extent on the costs for harvesting, different methods have been tested (Mohn 1980; Benemann et al. 1980):

- Centrifugation
- Direct Filtration
- Ultrafiltration
- Microstraining
- Chemoflocculation-Flotation
- Chemoflocculation-Sedimentation
- Biosedimentation

The selection of the most suitable harvesting system is determined by several factors like algal species and concentration, concentration factor, energy costs, investment, reliability, quality and further utilization of the algal biomass, etc.

While it is relatively easy and economic to harvest the fairly large microalgae Coelastrum and Spirulina by simple filtration, it is much more expensive to harvest the minute Chlorella or Scenedemus cells for instance. No doubt, centrifugation is the most reliable harvesting process, suitable for all algal species, but it cannot be recommended, especially for low level technologies, due to high investment and running cost. Microstraining, as described by Dodd (1979) offers a possibility of concentrating most of the algae at lower costs and is widely used in removal of algae from water supplies (Berry, 1961).

For large-scale operation, particularly in waste water treatment, the most cost-effective way to preconcentrate the algae is by chemical flocculation and subsequent removal by sedimentation, flotation or decantation. For this purpose, several chemicals such as lime, alum, iron salts or polymers have been tested successfully. Since residues of these flocculants remain in the harvested material, only very few non-toxic ones are suitable, if the algae are intended as feed and particularly as human food. One of these compounds is chitosan, a natural polysaccharide produced by deacetylation of chitin, obtained from marine arthropods.

Under certain conditions (darkness, high pH, low oxygen content, stagnant water, etc.) most algae tend to settle in the medium without addition of any chemicals. In principle, this method would be the easiest possibility for algal concentration; unfortunately, this effect is difficult to manipulate. The exact factors and processes of algal flocculation and agglutination are not known yet.

Drying

Depending on the final destination of the algal product, the material can be utilized either as wet slurry, obtained directly from the harvesting process, or in the form of dry powder. For most of the proposed applications drying is required, especially if the algal cells will be used as food or feed.

The drying process combines three different effects i.e. dewatering, sterilization, and breakage of the algal cell wall. The latter effect is of particular importance for green algae (Chlorella, Scenedesmus), which possess rigid cellulosic cell walls. These

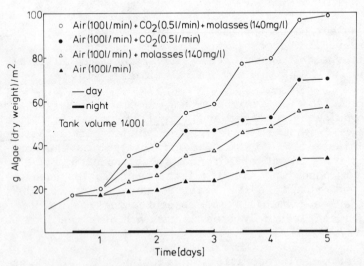

Figure 6. Growth of Scenedesmus at different autotrophic and mixotrophic culture conditions.

cell walls have to be ruptured by either physical or chemical treatments before the algal protein becomes accessible for proteolytic enzymes in monogastric animals. Only ruminants are capable of digesting unprocessed algae. Cell wall breakage is also not required for Spirulina. Blue-green algae do not have cellulosic cell walls and they are digestible without special treatment.

Drying of algae is an expensive process in view of the high moisture content of the slurry and involves about the same costs as harvesting. Among the various drying methods which have been tested for green algae, only drum-drying gave satisfactory results, especially with regard to digestibility (Becker et al. 1976 a, b).

Figure 7 shows the effect of different drying methods on the digestibility of Scenedesmus and Spirulina. As can be seen, simple sun-drying is sufficient for Spirulina to obtain an acceptable product, while for Scenedesmus drum-drying and possibly spray-drying is required.

FACTORS AFFECTING GROWTH

The large scale outdoor cultivation of algae in a clean system involves a very complex system, depending on the interactions of various external and internal factors, as demonstrated in a simplified form in Figure 8. Only a limited number of these interrelated operational variables such as CO_2-supply, pH, medium or mixing velocity can be adjusted during pond operation. Since light and temperature cannot be controlled on large scale cultivation, it must be tried to compensate for by modifying other parameters, which is possible only to a certain extent. In general, algal cultivation therefore is practicable only in lower geographical altitudes (below 35°) where diurnal variations are minimized and light intensities are sufficient to provide optimal growth over a major portion of the year.

In some publications algae have been mentioned as being more efficient than higher plants in their inherent capacity to utilize the energy of the visible spectrum of the light. However, recent data from the literature show that algae and higher plants appear to be about equal in this respect. According to Coombs and Hall (1982) the photosynthetic efficiency (% of total radiation) of algae (California) is 1.5, while it is 3.0 for Sudan grass (California) and 2.8 for sugar cane (Texas) in subtropical countries. For tropical areas, values of 3.8 for sugar cane (Hawaii) and up to 4.3 for bullrush millet (Australia) have been reported.

The major advantages of algae, compared to the conventional crop plants, is their high protein content and the possibility for almost continuous cultivation.

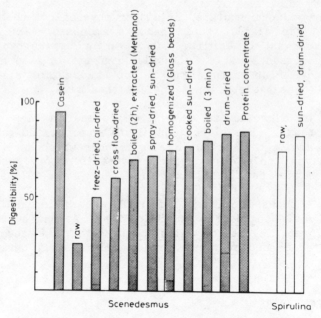

Figure 7. Effect of different processing methods on the digestibility of Scenedesmus and Spirulina.

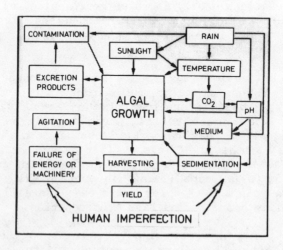

Figure 8. Interaction of main factors, influencing outdoor algal cultivation.

Besides temperature and light, other parameters such as evaporation (10 l/d/m^2 in tropical zones), infection with other microorganisms, oxygen transfer, optimal inoculum concentration and pollution have to be monitored for successful algal biomass production.

YIELD

Generally, outdoor mass cultivation of algae represents a special form of agriculture, being subjected to the usual variations in yield, quality and composition of the product, which are due to climatic fluctuations and other factors. The yield of algal biomass is expressed normally as g (dry matter)/m^2 (cultivation area)/day or metric tons/hectare/year. Reported data on growth rates of algae vary over a wide range and depend on algal species, culture conditions and climatic parameters. For the most common algal species (*Chlorella*, *Scenedesmus*, and *Spirulina*) the literature gives values in the range of 5-50 g/m^2/d (Goldman 1979, Soeder 1981).

However, one has to distinguish between results obtained for a very short period under optimal conditions and values determined over long cultivation periods of one year or more. All recent reliable figures on growth rates of various algae are between 10 and 35 g/m^2/d with an average of 15-25 for green algae and 8-12 g/m^2/d for *Spirulina*. These data indicate that with the algal production technology of today, optimum yields probably have been reached. Assuming a mean yield of 20 g/m^2/d, a yield of 70 tons of algal dry matter per year and hectare could be obtained. This value looks very promising compared with the productivity of conventional crops, i.e. wheat 3.0 tons (0.4 tons of protein), rice 5.0 tons (0.6 tons of protein) or potato 40 tons (0.8 tons of protein). However, the data on algal productivity are extrapolated from short-term experiments because most of the algal production plants are considerably smaller than one hectare and in only a few cases has continuous operation been performed for one year or more. For sewage systems, comparable data are reported, although the biomass harvested is not specified as algal biomass, in many cases bacterial biomass also is included. Even if it is possible to increase the average yields, it seems that an upper limit of 30-40 g/m^2/d is a given characteristic of the photosynthetic process of algal growth, which cannot be improved very much.

CHEMICAL COMPOSITION

As indicated earlier, the research on algal mass cultivation was started primarily in view of food production. Therefore, detailed studies were performed on the chemical composition and nutritional value of several algae. Table 1 summarizes analytical data of some algal species. It can be seen that besides the high

Table 1. Chemical Composition of Different Algae.

a) Overall composition (% of drum-dried matter)

	Scenedesmus	Chlorella	Spirulina
Crude protein (N x 6.25)	45 - 55	45 - 55	55 - 65
Lipids	8 - 12	5 - 9	2 - 6
Carbohydrates	10 - 15	10 - 18	10 - 15
Crude fibre	5 - 12	2 - 9	1 - 4
Ash	6 - 12	6 - 12	6 - 15
Moisture	5 - 10	5 - 10	5 - 10
RNA	4.40	3.90	2.90
DNA	1.60	1.90	1.00
Calcium	0.85	0.20	0.75
Phosphorus	1.90	1.20	1.42
Iron	0.60		0.12

b) Fatty acid composition of algal lipids (% of total lipids)

	Scenedesmus	Chlorella	Spirulina
C 14 : 0	1.3	-	0.9
C 16 : 0	12.1	20.5	27.5
C 16 : 1	9.5	10.9	2.5
C 16 : 2	-	-	2.9
C 16 : 4	-	-	13.1
C 18 : 0	0.9	4.7	-
C 18 : 1	15.3	10.2	3.3
C 18 : 2	9.0	14.9	18.2
C 18 : 3	30.7	29.5	14.6
Others	4.3	0.5	12.0

c) Vitamins (mg/kg algae)

	Scenedesmus	Chlorella	Spirulina
Thiamine	8.2	9.9	55.0
Riboflavin	36.6	36.1	40.0
Pyridoxyl-HCl	2.5	23.0	3.0
Cobalamine	0.44	0.02	2.0
Biotine	0.2	0.15	0.4
β-Carotene	230.0	482.0	500.0
Folic Acid	0.7		0.5
Nicotinate	120.0		118.0
d-Ca-Pantothenate	16.5	20.1	110.0

protein content the algae are rich in minerals and several vitamins, which will enhance the potential of algae as a valuable ingredient for feed and food.

The toxicological testing program of algal biomass (according to international recommendations on testing of novel protein sources) is not completed yet. However, it can be stated that in Chlorella, Scenedesmus, and Spirulina no biogenic toxic compounds (phycotoxins) have been detected so far and it is quite unlikely that any are present at all.

The situation is different, if the possible accumulation of undesired compounds from the surroundings is considered. It is a fact that algae, like many other organisms, are accumulating pollutants (mostly anthropogenic) such as heavy metals or pesticides to such a concentration that further utilization becomes limited or even impossible. Since the presence of these compounds is not an intrinsic characteristic of the algal metabolism per se but the consequence of a polluted environment, it can be avoided by selecting proper places for algal cultivation.

The nutritive value of a protein is determined by its amino acid composition. The amino acid pattern of algae, soya and a suggested FAO pattern of requirements are given in Table 2. The sulphur containing amino acids are marginally deficient; this deficiency can be compensated for by supplementing the algal protein with proteins from other sources rich in sulphur amino acids like cereals.

ALGAL PRODUCTION COSTS

Economical considerations on the production costs of algal biomass are the crucial point which determines the entire viability of the algal technology. Information on cost/benefit analyses of algal mass cultivation plants is only fragmentary, due to the small number of projects connected with the technical production of microalgae. In most of the cases the predictions are either mere extrapolations based on limited studies on pilot plant scale with small cultivation areas or calculations from undertakings with energy-intensive, sophisticated installations, where the production costs are offset by high selling prices of the algae. However, to point out the major technical problems and limitations of the algal cultivation process, some cost computations have been made. Although several price assumptions may be not valid anymore due to soaring prices, especially in the energy sector, the data listed in Table 3 give a general overview on the economical aspects of algal production. A recent computation of the expected production costs of a 5 hectare algal plant in India with an assumed yield of 280 tons per/year predicts prices of U.S. $2.34/kg. An analysis of these

Table 2. Amino acid pattern of algal and soya protein (g/100 g protein)

	FAO*	Chlorella	Scenedesmus	Spirulina	Soya
Isoleucine	4.0	3.5	3.6	6.7	5.3
Leucine	7.0	6.1	7.3	9.8	7.7
Valine	5.0	5.5	6.0	7.1	5.3
Phenylalanine	6.0	2.8	4.8	5.3	5.0
Tyrosine		2.8	3.2	5.3	3.7
Methionine	3.5	1.4	1.5	2.5	1.3
Cystine		0.2	0.6	0.9	1.9
Lysine	5.5	10.2	5.6	4.8	6.4
Tryptophan	1.0	2.1	0.3	0.3	1.4
Threonine	4.0	2.9	5.1	6.2	4.0
Arginine		15.8	7.1	7.3	7.4
Aspartic acid		6.4	8.4	11.8	1.3
Glutamic acid		7.8	10.7	10.3	19.0
Glycine		6.2	7.1	5.7	4.5
Histidine		3.3	2.1	2.2	2.6
Proline		5.8	3.9	4.2	5.3
Serine		3.3	3.8	5.1	5.8
Alanine		7.7	9.0	9.5	5.0

*) Energy and protein requirement report of a "Joint FAO/WHO ad hoc Expert Committee", No. 52 (1973).

Table 3. Algal Production Costs (U.S. $/ton).

Alga	Plant capacity (t/y)	Average production (t/ha/y)	Costs	Reference
a) Clean system.				
Chlorella	3,200	32	375-550	Fisher,1956
Chlorella	?	18	570	Tamiya,1956
Spirulina	10,000	50	370	Clement,1970
Coelastrum	500	45	1,560	Soeder,1978
Coelastrum	500	80	1,140	Soeder,1978
Scenedesmus	500	?	700	Soeder,1971
Scenedesmus	300	50	1,160	Kugler,1974/75
Scenedesmus	40	80	3,300	Schulz,1980
Scenedesmus	400	80	1,360	Schulz,1980
Scenedesmus	800	80	1,200	Schulz,1980
Scenedesmus	280	60	1.900	Becker,1980
Chlorella	?	?	11,000	Kawaguchi,1980
Chlorella	?	?	14,000-21,500	Soong,1980
b) Sewage system.				
Mixed Population	740	74	1,150	Berend,1980
Mixed Population	7,400	74	933	Berend,1980

costs is given in Table 4, demonstrating what percentage of the final price is contributed by the several major steps involved in the production.

The most expensive parts of the proposed algal plant are the production of CO_2 from fuel (utilization of commercially available CO_2 is even more expensive than the construction of an individual CO_2 production plant), the harvesting and the drying step, which together attribute already two thirds of the overall costs.

This example demonstrates that the major aim of research in the field of applied algal technology will be to develop possibilities to utilize cheaper carbon sources like industrial or agricultural wastes and to work out simpler harvesting and drying techniques. With the technology of today, costs of about U.S. $2.00/kg for algae produced in clean systems and below U.S. $ 1/kg for sewage-grown algae can be assumed.

APPLICATIONS OF ALGAL BIOMASS

The possible applications of algal biomass are manifold and the important ones shall only be mentioned here briefly.

Food

Certain algae have served as human food for centuries. Besides the fact that several marine macroalgae are being used as vegetables in the Far East, it was discovered that Spirulina was eaten by the Aztec's in Mexico. They called it "Tecuitlatl". North of Lake Chad in Central Africa, the same alga forms part of the diet of the Kanembou tribe who prepare Spirulina into a kind of sauce called "Dihe".

Table 4. Analysis of the main factors of algal production costs (U.S. $).

	Capital Costs	Fixed Expenses	Variable Costs	Total Price	% of Total Price
Complete plant	0.79	0.40	1.15	2.34	100
Centrifugation*	0.48	0.29	1.04	1.81	23
Drying*	0.71	0.36	0.82	1.89	20
CO_2 Plant*	0.61	0.33	0.89	1.83	22
Tanks*	0.62	0.36	1.15	2.13	9.5
Fertilizer	0.79	0.40	0.96	2.15	8
Labour*	0.79	0.19	1.15	2.13	9.5

*) Production costs without respective parameter.

Although algal mass cultivation was initiated primarily in view of food production, the great expectations in the use of algal SCP, produced in clean systems, could not be met. Algal products are still not competetive with conventional foodstuffs and acceptance by the consumer is poor.

On the other hand, it has been demonstrated clearly that the selected algae are of high nutritional value, which is about 80% compared with reference proteins like egg or casein. Toxicological studies confirmed that properly produced and processed algae, consumed in appropriate amounts as protein supplement, do not cause any form of malnutrition or toxicity. The production of proteinaceous food from algae is still the long-term goal of several projects.

Feed

Both, algae from clean systems or sewage-grown, can be used as animal feed, provided that no toxic substances from the medium (sewage, wastes) are accumulated in the algae. For instance, several projects are testing the possibility of growing algae on pig manure and of feeding these algae, after suitable processing, directly back to the pigs (Taiganides 1979). Algae have been successfully introduced into feedstuff for poultry (broiler and laying hen), cattle, sheep and in aquaculture (fish, shrimps, shellfish), whereby part of the conventional protein sources (fishmeal, soya) could be substituted by algal protein (Mokady et al. 1980, Hintz et al. 1966, Sandbank et al. 1978).

Natural Compounds

It has been demonstrated that it is possible to manipulate the chemical composition of algae by modifying the culture conditions. Under certain circumstances, algae produce high amounts of compounds, which are of commercial value. It was found for instance that under hypersaline conditions the green alga Dunaliella synthesizes large quantities of glycerol and β-carotene (Ben-Amotz et al. 1980). Special enzymes for biochemical research are already extracted from Spirulina and sold commercially. Furthermore, it was proposed to produce labeled compounds with algae by adding nutrients with isotopes (^{14}C, ^{15}N, ^{32}P) to the culture medium. However, these aspects can only be realized when the production costs are compensated by the high commercial value of the final product.

Algae as a Source of Energy

Microalgae convert solar energy into biomass, which could be converted into energy sources such as methane, ethanol or hydrocarbon. Concepts have been developed for the exploitation of algal energy farms. Although some results have been published about the successful conversion of Spirulina biomass into methane (Samson et al. 1982), there are still many problems involved in "energy farming" with algae on a large scale. The future realization of this idea seems to be possible, when, especially in developing countries, small communities develop integrated systems with treatment of sewage in high-rate algal ponds and digestion of the biomass in small biogas fermenters.

Another theoretical possibility to utilize algae as an energy source is the production of hydrogen by biophotolysis. Several approaches are possible using either isolated cellular components or algae cultures. The only algal system demonstrated to meet the basic requirements of biophotolysis uses nitrogen starved cultures of nitrogen-fixing heterocystous blue-green algae. However, development of practical systems is limited by the low efficiency of photosynthesis and severe economic constraints (Benemann et al. 1976).

Algal Biofertilizer

One of the most important problems in rice production is the availability of nitrogen fertilizer. It is in this context that algal biofertilizers have a vital role to play. Since De (1939) suggested that the sustained fertility of tropical rice fields is largely due to the activity of nitrogen-fixing blue-green algae, considerable interest has been generated over the years in this field. In recent years a simple rural-oriented technology has been developed for the use of algal fertilizer in rice cultivation. The nitrogen fixed by the algae (Aulosira, Tolypothrix, Scytonema, Nostoc, Anabaena, Plectonema) and subsequently released on decomposition is taken up by the rice plants.

Besides nitrogen fixation, the algae also synthesize and liberate growth promoting substances (auxins, amino acids) which stimulate the growth of the rice plants. Extensive field trials conducted in India have shown the positive effect of these algae on the growth and yield of rice plants. An application of nitrogen-fixing blue-green algae can substitute for the application of 20-30 kg N/ha per cropping season.

Algae for Sewage Treatment

The most succesful application of algal cultivation seems to be in the field of waste water treatment, whereby the processes of biomass production and waste treatment can be combined. With regard to the costs of algal production, it must be emphasized that high-rate algal ponds, using waste water as substrate, have no or only minor expenses for CO_2 and nutrients. This system has the versality of not only providing photosynthetically produced oxygen for BOD removal, but also for incorporating nitrogen and phosphorus from the sewage into biomass, thereby reducing the eutrophication of receiving waters.

The algal biomass produced can be utilized for various purposes, except as food and biofertilizer. The best applications seem to be for animal feed -- especially in an aquatic food chain without harvesting and processing -- or as raw material for natural compounds.

CONCLUSIONS

The production of algal biomass offers a rich potential of substrate utilization and application (Figure 9). While considering the future development of algal technology it should be differentiated between a "rural technology" and an advanced technology such as commercial algal production systems in the form of photoautotrophic "clean systems". The first approach seems to be feasible for smaller communities, especially in developing countries. This system incorporates the production of algal biomass on agricultural or domestic wastes and the subsequent utilization of the biomass as animal feed, in aquaculture or as source of energy. It combines the possibility of solving sanitation problems and of producing valuable raw material. This technology is labour intensive but with low of energy and equipment, so that algal production costs will be significantly low.

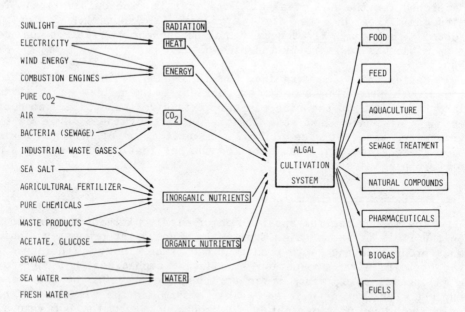

Figure 9. Potentials of algal biomass: Raw material, energy sources, products and applications.

For commercial production, the most promising bases for immediate utilization of microalgae seem to be those cases where the algae can be obtained either as a by-product (sewage treatment) or by utmost utilization of suitable waste products (CO_2). However, algae have found only limited application so far. This seems to be due to biotechnological, economical and administrative problems. The major applications of algae thus produced will be as feed and as raw material for valuable natural compounds.

The recent use of algae as health food seems to be more a fashion than a sound application and the utilization of algae as a supplement in human nutrition is no longer treated as an urgent goal. However, the time might not be far off when a situation arises where the commercial utilization of the more or less "ready to scale up technology" of algal cultivation becomes economic and necessary.

REFERENCES

Becker, E.W., Venkataraman, L.V. and Khanum, P.M. 1976a. Nutr. Rep. Int. 14: 457-466.
Becker, E.W., Venkataraman, L.V. and Khanum, P.M. 1976b. Nutr. Rep. Int. 14: 305-314.
Becker, E.W. and Venkataraman L.V. 1980. In: Algae Biomass. Eds. G. Shelef and C.J. Soeder, Elsevier, North-Holland Biomedical Press 35-50.
Ben-Amotz and Avron M. 1980. In: Algae Biomass. Eds. G. Shelef and C.J. Soeder, Elsevier, North-Holland Biomedical Press 603-610.
Benemann, J.R. and Weissman, J.C. 1976. In: Microbial Energy Conversion. Eds. H.G. Schlegel and J. Barnea, Pergamon Press 413-426.
Benemann, J.R., Koopman, B., Weissman, J.C., Eisenberg, D. and Goebel, R. 1980. In: Algae Biomass. Eds. G. Shelef and C.J. Soeder, Elsevier, North-Holland Biomedical Press 457-495.
Berend, J., Simovitch, E. and Ollian, A. 1980. In: Algae Biomass. Eds. G. Shelef and C.J. Soeder, Elsevier, North-Holland Biomedical Press 799-818.
Berry, A.E. 1961. J. Am. Wat. Works Assn. 53: 1503.
Burlew, J.S. 1953. Publ. No. 600. Washington, D.C., Carnegie Institute of Washington.
Clement, G. and Landeghem, H. 1970. Ber. Dtsch. Bot. Ges. 11: 559-565.
Coombs, J. and Hall, D.O. 1982. Techniques in bioproductivity and photosynthesis, Pergamon Press.
De, P.K. 1939. Proc. R. Soc. London, Series B, 127: 121-139.
Dilov, C. 1980. III. Int. Conf. of Microalgae, Lima, Peru.
Dodd, J.C. 1979. Agricult. Wastes 1: 23-37.
Fisher, A.W. 1956. In: Proc. World Symp. Appl. Solar Energy, Phoenix, Ariz. 243-253.

Goldman, J.C. 1979. Water Res. 13: 1-19.
Heussler, P., Castillo, J., Merino, F. and Vasquez V. 1978. Arch. Hydrobiol. Beih. Ergebn. Limnol. 11: 254-258.
Hintz, H.F., Heitman, H., Weir, W.C., Torell, D.T. and Meyer, J.H. 1966. J. Anim. Sci. 25: 675-681.
Kawaguchi, K. 1980. In: Algae Biomass. Eds. G. Shelef and C.J. Soeder, Elsevier, North-Holland Biomedical Press 25-33.
Kugler, F. 1974/75. In: Fourth Report Algae Project. Kasetsart University, Bangkok, Thailand.
Märkl, H. and Mather, M. 1980. III. Int. Conf. of Microalgae, Lima, Peru.
Mohn, F.H. 1980. In: Algae Biomass. Eds. G. Shelef and C.J. Soeder, Elsevier, North-Holland Biomedical Press 547-571.
Mokady, S., Yannai, S., Binav, P. and Berk, Z. 1980. In: Algae Biomass. Eds. G. Shelef and C.J. Soeder, Elsevier, North Holland Biomedical Press 655-660.
Oswald, W.J. 1972. Proc. 6th Intl. Water Poll. Res. Conf. Pergamon Press.
Samson, R. and LeDuy, A. 1982. Biotechnol. Bioeng. 24: 1919-1924.
Sandbank, E. and Hepher, B. 1978. Arch. Hydrobiol. Beih. Ergebn. Limnol. 11: 108-120.
Schulz, P.A., Heussler, P. Moya, R. and Merino, F. 1980. III. Int. Conf. of Microalgae, Lima, Peru.
Setlik, I., Sust, V. and Malek, I. 1970. Algol. Stud. Trebon, 1: 111-164.
Soeder, C.J. and Pabst. W. 1971. Ber. Dtsch. Bot. Ges. 83: 607-625.
Soeder, C.J. 1978. Arch. Hydrobiol. Beih. Ergebn. Limnol. 11: 259-273.
Soeder, C.J. 1981. U.O.F.S., Publ. Series C, No. 3, 63-72.
Soong, P. 1980. In: Algae Biomass. Eds. G. Shelef and C.J. Soeder, Elsevier, North-Holland Biomedical Press 97-113.
Taiganides, E.P., Chou, K.C. and Lee, B.Y. 1979. Agricult. Wastes 1: 129-141.
Tamiya, H. 1957. Ann. Rev. Plant Physiol. 8: 309-334.

HYDROGEN FROM WATER - THE POTENTIAL ROLE OF GREEN ALGAE
AS SOLAR ENERGY CONVERSION SYSTEM

B. Mahro and L.H. Grimme

Department of Biology and Chemistry
University of Bremen, D-2800 Bremen 33
Federal Republic of Germany

INTRODUCTION

The photobiological production of hydrogen is discussed as a promising biological process for solar energy conversion (Weaver et al. 1980; Hall 1982). The H_2 photoproduction of green algae represents one of those possibilities to convert solar energy in a storable and useful fuel. But H_2 photoproduction by green algae, as a consequence of its extreme sensitivity to oxygen, is hampered for economical and technological applications. Furthermore, the basic mechanism of the algal photosynthetic reaction is not yet clear. H_2 photoproduction may be based on a pure photosynthetic process including photolysis of water as ultimate source of electrons and protons for hydrogen or, on the other hand, H_2 photoproduction may be based on a kind of light supported fermentation of organic substrates. Because of the better solar light conversion efficiency the prospects for a practical application of H_2 photoproduction by green algae would become better if water could get established as the original source of hydrogen produced.

To get an answer to this question we investigated whether three typical features of normal photosynthesis occur also under conditions of H_2 photoproduction. We investigated:

a) Whether the photoproduction of H_2 is a high light saturable reaction

b) Whether the photoproduction of H_2 is inhibited by diuron (DCMU) a specific inhibitor of photosynthetic electron transport, and

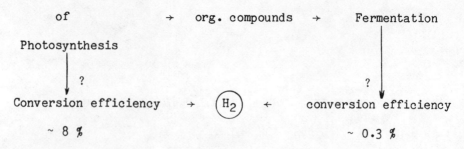

c) Whether the electron transfer under H_2 productive conditions is coupled to photophosphorylation, as it is under conditions of aerobic photosynthesis.

The greater the coincidence of properties of aerobic photosynthesis and of anaerobic H_2 photoproduction, the greater the probability that photosynthesis and H_2-photoproduction may be based on the same mechanism.

METHODS

Experiments were carried out with the green alga <u>Chlorella fusca</u> (strain 211 - 15, algal collection Göttingen, FRG), H_2 was determined gaschromatographically (Mahro and Grimme 1982), the adenine nucleotides with a luminometric technique (DeLuca and McElroy 1981).

RESULTS

a) H_2 photoproduction of <u>Chlorella fusca</u> is saturated by high light intensities, however to intensities up to 20000 lux only. The higher the light intensities used during the initial phase of H_2 production, the sharper the decrease in the rate of H_2 production during the following period. This possibly indicates a light supported depletion of an endogenous and limited donor-pool for the photoproduction of H_2.

b) H_2 photoproduction of <u>Chlorella fusca</u> is inhibited by DCMU to about 80% of a untreated control, 10 minutes after illumination. This shows a participation of a photosystem II reaction in H_2 photoproduction as in aerobic photosynthesis. The remaining extent of H_2 photoproduction, which is not inhibited by DCMU must be based on a mechanism other than photolysis of water and photosynthetic electron transfer from PS II.

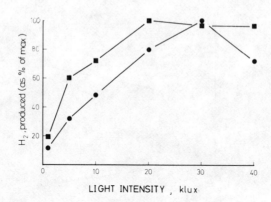

Figure 1. The influence of different light intensities on the H_2-photoproduction of *Chlorella fusca* with (■) and without (●) sodium dithionite.

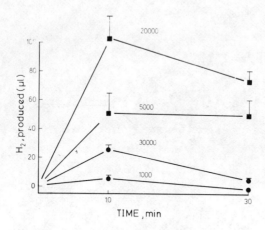

Figure 2. The influence of optimum- and low-light intensities on the production of H_2 by *Chlorella fusca* suspension with (■) and without (●) sodium dithionite during the first 10 and 30 min.

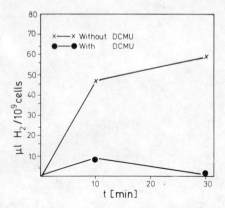

Figure 3. The H_2-photoproduction of <u>Chlorella fusca</u> after 10 and 30 min of illumination of (30000 lux) with and without DCMU.

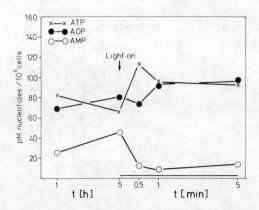

Figure 4. Changes in AMP, ADP and ATP concentration in <u>Chlorella</u> cells during 5 hours of anaerobic dark incubation and after 30, 60 and 300 sec of illumination (30000 lux).

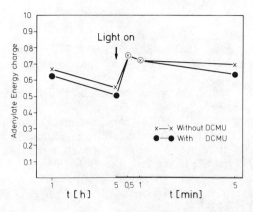

Figure 5. Changes in Adenylate Energy Charge in <u>Chlorella</u> cells during 5 hours of anaerobic dark incubation and after 30, 60 and 300 sec of illumination (30000 lux) with and without DCMU.

c) H_2 photoproduction of <u>Chorella fusca</u> is concomitated by a phosphorylation (6 µM ATP $mg^{-1}chl\ h^{-1}$). This photophosphorylation is not inhibited by DCMU. This is a distinct difference to aerobic photosynthesis where non-cyclic phosphorylation, based on the non-linear electron transport, is fully inhibited by DCMU.

CONCLUSIONS

The experiments show that H_2 - photoproduction and aerobic photosynthesis are similar in some, but not in all features investigated. Possibly H_2 photoproduction is based on a depletion of an endogenous pool during the initial phase of the reaction and is supported stepwise by water decomposition in the course of the reaction as it can be seen from the kinetics of H_2 production under light saturation and from the experiments with DCMU.

It seems to be justified to regard green algae as a potential conversion system of solar energy for the production of hydrogen from water. Further experiments, however, have to be carried out to enhance and to stabilize the rate of H_2 photoproduction (Pow and Krasna 1979; Mahro and Grimme 1982). Whether H_2 photoproduction by green algae really can be brought to an economical and technologically reasonable degree of applicability is not yet established. Perhaps the result of all basic research on biological energy conversion systems will show that they should be used better for the production of more complex molecules rather than hydrogen.

REFERENCES

DeLuca, M.A., McElroy, W.D., Eds. 1981. Bioluminescence and chemiluminescence. Basis chemistry and analytical applications. Academic Press, New York.
Hall, D.O. 1982. Solar energy through biology: Fuels from biomass. Experientia 38: 3-10.
Mahro, B. and Grimme, L.H. 1982. H_2-Photoproduction by green algae: The significance of anaerobic pre-incubation periods and of high light intensities of H_2 photoproductivity of Chlorella fusca. Arch. Microbiol. 132: 82-86.
Pow, T. and Krasna, A.I. 1979. Photoproduction of hydrogen from water in hydrogenase-containing algae. Arch. Biochem. Biophys. 194: 413-421.
Weaver, P.F., Lien, S., and Seibert, M. 1980. Photobiological production of hydrogen. Solar Energy 24: 3-45.

UTILIZATION OF AQUATIC BIOMASS FOR WASTEWATER TREATMENT

G. Lakshman

Saskatchewan Research Council
30 Campus Drive
Saskatoon, Canada S7N 0X1

INTRODUCTION

Municipal effluents from the U.S.A. and Canada generate annually more than 400,000 tonnes of phosphorus and nitrogen, a large portion of which is dumped directly into surface waters with little or no treatment. Rapid deterioration of water quality due to algal blooms and nuisance vegetation stimulated by the flow of these nutrients has reduced the usefulness of surface waters for recreation and human consumption rather alarmingly in recent decades. Eutrophication of freshwater lakes and the increasing need to treat water supplies before consumption have had undesirable socio-economic impacts all over North America. It has been well documented that the removal of phosphorus and nitrogen at the source is the most effective means of assuring that this trend does not continue unabated. While most urban centres can possibly afford advanced physicochemical treatment systems, thousands of towns and small communities in the developed world can support only the conventional lagoon treatment, hardly a solution to the ubiquitous problem.

As a result, the frightening prospects of permanently endangering the aquatic life as well as losing valuable water resources have spurred on a renewed interest in developing cost-effective, appropriate technology applicable at individual point-source (National Academy of Science 1976; Tourbier and Pierson 1976; Wolverton and McDonald 1975; Seidel 1973). The use of aquatic biomass such as algae, emergent macrophytes and floating weeds has been well exemplified as alternatives for removing polluting nutrients and contaminants from wastewaters through recent work (Tourbier and Pierson 1976; Lakshman 1982) in North America and Europe. Artificial and natural wetlands, and greenhouses with algal and water hyacinth

hyacinth systems have been shown to be practical treatment schemes for providing equivalent to tertiary treatment to primary or partly treated sewage.

A systematic approach to the utilization of aquatic biomass such as cattail and bulrush (common aquatic plants in North America and Europe) was conducted by Lakshman (1979; 1981; 1982) from growth chamber experiments to a large scale demonstration system at Humboldt, Saskatchewan which successfully treated primary as well as aerated sewage from the town. This work has demonstrated that the aquatic biomass can be utilized to develop a practical alternative to the conventional sewage treatment systems using precipitation and bacterial degradation mechanisms.

The results from the growth chamber and the large scale tests are summarized below.

GROWTH CHAMBER STUDIES

The objectives of these studies were to:

1. Determine if common aquatic plants such as cattail and bulrush grown in untreated sewage could remove phosphorus and nitrogen from the effluent within reasonably short retention times,

2. Evaluate the rate of nutrient removal as a function of the nutrient concentration, and

3. Develop prediction equations for designing large scale facilities.

Cattail and bulrush were grown in gravel substrate contained in separate trays (large tubs) measuring 91 by 84 by 46 cm high. A tray with gravel but without plants served as a control. Primary (untreated) sewage was pumped into each of the trays and retained until the phosphorus level was significantly reduced, at which time the trays were drained and recharged with a fresh batch of sewage. Composite samples obtained from the trays were analyzed for total organic carbon (TOC), total kjeldahl nitrogen (TKN) and total phosphorus (TP) at regular intervals. Each batch feed and discharge constituted an experimental run, the duration of which varied from 6 to 35 days. Data were obtained for 20 runs lasting a total of 527 days, well over three times as long as the average growing season on the Canadian prairies.

The experiments showed successfully that cattail and bulrush can remove significant quantities of phosphorus and nitrogen from raw sewage within about two weeks. Table 1 shows the data from the experimental runs. While the performance of the control in removing

Table 1. Initial and final concentrations (mg per litre) of total phosphorus (TP) and total kjeldahl nitrogen (TKN) in the treatment and control trays during the experimental runs.

Run #	Duration (Days)	TOTAL PHOSPHORUS				TOTAL KJELDAHL NITROGEN			
		Initial Concn. (mg/l)	Final Concentration (mg/l)			Initial Concn. (mg/l)	Final Concentration (mg/l)		
			Control	Bulrush	Cattail	Cattail		Control	Bulrush	Cattail	Cattail

Run #	Duration (Days)	Initial Concn. (mg/l)	Control	Bulrush	Cattail	Cattail	Initial Concn. (mg/l)	Control	Bulrush	Cattail	Cattail
1	35	16.8	0.2	0.2	0.2	0.6	30.0	2.2	2.6	1.1	1.9
2	33	19.0	0.7	0.9	1.1	1.4	34.4	2.3	2.0	2.7	0.4
3	12	9.7	5.9	0.5	0.5	0.2	26.0	14.0	6.0	4.0	1.6
4	16	13.4	8.3	1.5	0.5	1.6	26.4	18.4	3.2	1.7	1.8
5	28	8.2	4.6	0.8	0.7	0.8	31.2	10.8	2.8	4.0	2.8
6	26	4.6	1.5	1.5	1.1	0.9	31.2	6.0	1.6	11.2	4.4
7	33	15.6	4.0	0.1	3.6	8.2	35.2	24.0	2.4	10.4	4.4
8	21	11.6	8.0	1.2	0.8	3.4	37.6	17.6	2.0	4.0	5.2
9	19	29.0	16.2	0.7	0.5	5.6	44.0	26.4	3.2	6.0	8.4
10	34	17.6	11.6	0.8	1.3	13.6	40.0	28.0	1.7	16.0	22.0
11	35	8.2	1.7	0.6	0.4	0.7	27.6	8.4	1.2	2.7	1.9
12	13	3.9	1.5	1.5	0.6	0.7	10.3	4.0	1.3	1.8	1.8
13	6	9.5	5.8	2.5	1.9	2.1	10.3	8.4	2.7	5.9	5.6
14	14	11.5	15.0	0.8	6.0	18.0	25.8	17.6	2.8	8.4	12.4
15	14	12.0	20.2	6.6	4.0	11.8	21.6	21.2	2.8	14.4	16.4
16	14	9.0	20.8	3.2	12.5	11.9	24.0	20.5	4.0	13.5	11.8
17	14	16.4	24.5	3.7	22.2	11.6	25.8	22.2	9.0	18.3	15.6
18	14	15.6	27.1	13.7	32.1	27.7	31.0	28.0	15.0	26.0	21.0
19	34	17.0	7.4	0.95	14.9	1.8	34	18.4	1.0	11.6	1.6
20	29	6.8	2.9	1.0	0.9	0.7	22.8	8.2	0.9	1.8	1.8

nutrients diminished rapidly with time, that of the experimental trays containing plants continued to be significant throughout the 527 days of continued operation. The rate of nutrient removal increased with the initial concentration in the following manner,

$$Y = a X^b$$

where Y is the rate of phosphorus or nitrogen removal, g/d, and X is the initial concentration in mg/L, a and b are regression constants. Table 2 gives the expected percentages of TP and TKN removal for various retention periods.

Table 2. Expected percentages of TP and TKN removal for various retention periods.

Retention Period (Days)	Percent Removed	
	TP%	TKN%
7	45.0	39.6
14	84.4	74.8
21	95.6	89.5
28	98.7	95.6

DEMONSTRATION EXPERIMENTS

At Humboldt, a town of 5,000, about 100 km east of Saskatoon, two lagoons and three raceways planted with cattail and bulrush were set up to demonstrate the feasibility of the treatment technique on a large scale. The lagoons, about one metre deep, were about 0.5 ha each in area and one of them was lined with a synthetic liner to determine if the lined bottom had any effect on the treatment efficiency. The three raceways were 90 m long, 6 m wide and 1 m deep. One raceway was stocked with plants in soil and in the other sites the plants were grown in gravel. Treatment schedules consisted of batch feeding the lagoons and raceways with anaerobic (primary) and continuously aerated effluents from the town and discharging the treated effluents after various retention periods. This system was designed to treat only a part of the town's total wastewater production. Periodic samples taken from the sites were analyzed for TOC, TKN, TP and BOD as primary parameters. Although the study was started in late August 1979, during the following two years the treatment runs started about the middle of May and continued to the end of November. The study has successfully shown that:

1. Large lagoons and narrow raceways stocked with cattail and bulrush can efficiently remove significant quantities of phosphorus, nitrogen and BOD from untreated and partly treated effluents within two to three weeks. The application rate at Humboldt varied from 13 to 380 kg/ha for phosphates, from 4 to 890 kg/ha for BOD and from 26 to 277 kg/ha for TKN. The corresponding concentrations ranged from 2 to 71 mg/L for phosphates, from less than 10 to 144 mg/L for BOD, and from 10 to 44 mg/L for TKN. The depths of operation ranged from 0.2 m to 0.5 m. Nutrient reduction (Y, in%) as a function of the initial loading (X, kg/ha) was correlated highly in the form,

$$Y = a X^b$$

Table 3 shows the observed nutrient reductions for various retention periods.

Table 3. Measured PO_4 and BOD reductions for various retention times.

Approximate Retention Time (Days)	PO_4 Reduction (%)		BOD Reduction (%)	
	Mean Minimum	Mean Maximum	Mean Minimum	Mean Maximum
7	36	75	60	86
14	48	82	66	90
21	61	92	58	92

AQUATIC BIOMASS FOR WASTEWATER TREATMENT

2. Lagoons which removed roughly 4 times more nutrients than the raceways on a unit area basis performed better.

3. Rate of phosphorus removal (Y, $g/d/m^2$) varied with the plant density (X, plants/m^2) in the form,

$$Y = (0.09) \exp(0.0113X)$$

and the rate of BOD removal (Y, $g/d/m^2$) varied with the plant density (X, plants/m^2) in the form,

<u>1980</u>: Y = 0.134 + (0.015) X
<u>1981</u>: Y = -0.713 + (0.012) X

The data for 1979 were not considered for this analysis as no plant density values were available.

The three-year data from the Humboldt Demonstration Project have shown that a practical and economically viable technique can be developed for treating wastewaters using the common aquatic plants such as cattail and bulrush.

It should be emphasized that the biomass utilization which begins with the treatment of wastewaters continues with the utilization of the harvested biomass for protein, fertilizer and energy as summarized below.

ANIMAL FEED

Cattail and bulrush grown in nutrient rich environment contain high crude protein and DOM (Digestible Organic Matter) values compared to the plants grown in freshwater wetlands. Table 4 shows the data obtained from the growth chamber studies. The harvested biomass has excellent feed quality as a supplemental animal ration.

CALORIFIC VALUE

Aquatic biomass can be air dried in the open and can be used as a source of heat in wood-burning stoves. Table 5 shows the calorific values for cattail and bulrush in comparison with the more common agricultural sources.

ETHANOL PRODUCTION

Cattail which is abundant in the millions of natural wetlands in North America can become a prime candidate as an energy crop. Research programs in Minnesota, Florida and Indiana have indicated

Table 4. Analysis of the harvested biomass for its value as animal feed. All values are on dry matter basis, except percent moisture. DOM = digestible organic matter. N/D = Analysis not done.

DATE OF HARVEST	FEED PARAMETERS (%)						
	CRUDE PROTEIN	CRUDE FIBRE	DOM	ASH	PHOS- PHORUS	CALCIUM	MOISTURE
A. CATTAIL							
Sept. 8, 1976							
Sample 1	17.2	31.8	55.2	10.8	0.6	0.8	86.0
Sample 2	21.8	29.7	56.2	11.5	0.5	0.9	84.3
Sample 3	17.4	N/D	52.9	N/D	0.5	N/D	88.6
Sept. 27, 1976	19.1	31.0	61.4	12.9	0.5	1.1	74.1
July 4, 1976	21.3	28.4	59.3	13.7	0.7	0.9	88.5
B. BULRUSH							
Sept. 8, 1976							
Sample 1	19.2	26.8	61.2	18.3	0.3	0.6	78.7
Sample 2	21.8	27.9	59.4	14.1	0.5	0.6	81.8
Sample 3	21.3	N/D	51.1	N/D	0.4	N/D	83.9
Nov. 15, 1976	18.8	39.7	50.2	8.2	0.3	0.4	63.0
Jan. 4, 1976	20.4	27.0	52.2	9.9	0.2	0.5	75.4
C. ALFALFA	18 - 20		55 - 60				
D. STRAW	2.5 - 5.0		30 - 45				

that cattails may offer better prospects for energy production in terms of annual biomass avilability, energy content and economics than most common agricultural substrates. The total non-structural carbohydrates in cattail vary from 15 to 28 per cent (dry weight) in aerial shoots and from 35 to 54 per cent (dry weight) in rhizomes. The accumulation of reducing sugars can be as high as 30 per cent (dry weight) in the rhizomes. One of the programs at the Saskatchewan Research Council is directed towards obtaining laboratory and field data on the ecomonic availability of fermentable sugars from managed cattail crops.

Added to the above mentioned utilization aspects of aquatic biomass in terms of feed, energy and environment there are pharmacological and industrial aspects of aquatic plants that are beginning to be explored. This may well augur the re-discovery of the enormous potential of aquatic plants.

Table 5. Calorific values for some substrates.

Substrate	Calorific Value (btu/lb)	(KJ/K)
Wheat Straw	7,725	18,000
Corn Cobs	7,768	18,100
Refuse Derived Fuel	7,120	16,590
Manure	6,010	14,000
Corn Stover	6,500	15,145
Alfalfa	4,464	10,400
Coal (Lignite)	6,818	15,886
Biogas	16,363	38,126
SRC Experiments		
Cattail Leaves	8,200	19,106
Cattail Rhizomes	7,700	17,941
Bulrush Leaves	7,800	18,174
Bulrush Rhizomes	7,750	18,058

REFERENCES

Lakshman, G. 1979. An Ecosystem Approach to the Treatment of Wastewaters. J. Environ. Qual. 8: (3): 353-361.

Lakshman, G. 1981. A Demonstration Project at Humboldt to Provide Tertiary Treatment to the Municipal Effluent Using Aquatic Plants, Saskatchewan Research Council, Publication No. E-820-4-E-81.

Lakshman, G. 1982a. Natural and Artificial Ecosystems for the Treatment of Wastewaters, Saskatchewan Research Council, Publication No. E-820-7-E-82.

Lakshman, G. 1982b. A Demonstration Project at Humboldt to Provide Tertiary Treatment to the Municipal Effluent Using Aquatic Plants, Final Report, Saskatchewan Research Council, Publication No. E-820-11-B-82.

National Academy of Sciences. 1976. Making Aquatic Weeds Useful: Some Perspectives for Developing Countries. Advisory Committee on Technology Innovation, Natl. Acad. of Sci., Washington, D.C.

Seidel, K. 1973. Macrophytes and Water Purification. Verh. Int. Verein Limnol. 18: 109-121.

Tourbier, J., and R.W. Pierson, Jr. (Eds.). 1976. Biological Control of Water Pollution. Univ. of Pennsylvania Press, Philadelphia.

Wolverton, B.C., and R.C. McDonald. 1975. Water Hyacinths for Upgrading Sewage lagoons to Meet Advanced Wastewater Treatment Standards. Part I, NASA Technical Memo. TM-X-72729.

THE IMPORTANCE OF SPARTINA MARITIMA IN THE RECYCLING OF

METALS FROM A POLLUTED AREA OF RIVER SADO ESTUARY, PORTUGAL

F. Reboredo

Dept. de Química e Biotecnologia, F.C.T.
Univ. Nova de Lisboa
Lisboa, Portugal

INTRODUCTION

The determination of trace metals in aquatic environments have received particular attention, due to the large "inputs" of heavy metals and other residual elements as a consequence of the high degree of industrialization and urbanization.

The use of biological indicators is one of the most common methods for quantitative and qualitative determination of heavy metals in an area (Melhuus et al. 1978; Agadi et al. 1978).

Sediments also store large amounts of pollutants (Reboredo et al. 1981; Cosma et al. 1982), and their analysis constitutes a key factor in detecting sources of pollution in aquatic systems. On the other hand, the vertical distribution of heavy metals in sediments give us a clear indication of chronological accumulation (Bower et al. 1978).

Halophytes have been studied, especially their ecology (Rozema 1978); however, several authors have emphasized the importance of some halophytic species in recovery and recycling of trace metals accumulated in the sediments of the salt marsh (Windom 1975; Breteler et al. 1981).

Our study deals with trace metal determination -- Cu, Zn, Al and Fe, in Spartina maritima (Curtis) Fernald and sediments of the Sado River estuary, and constitutes the first contribution in our country concerning the importance of this species in the decontamination of the estuarine environment, menaced by man's expansion.

SAMPLING AND ANALYTICAL METHODS

Samples from biological material and sediments were collected in January 1982, during low tide, from three points as indicated in Figure 1. The halophytic pioneer species Spartina maritima colonizes bare mud of coastal areas of the Sado River estuary. This area receives an important input of residual elements due to the high agricultural and industrial activity around and upstream from Setubal, and also the untreated domestic sewage from this populous city.

Biological material was divided into roots, stems and leaves and rinsed carefully with distilled water and dried at 70°C to constant weight. Sediments were placed in polyethylene containers and kept at 0°C until the drying step at 70°C to constant weight.

To obtain an homogenized sample, dry sediments were crushed in a mortar, and dried again for 24 h. Each gram of dry sediment and biological material was submitted to an acid digestion procedure using nitric-perchloric acid (4:1), followed by atomic absorption spectrometry analysis, according the method previously described for sediments (Reboredo 1981).

The results are based on mean values of five replications and are expressed in µg/g dry weight (D.W), or µg/g dry weight x 10^3.

RESULTS

Trace Metal Concentration in Sediments of the Salt Marsh

Cu - The distribution of Cu in sediments, collected at different sampling points, do not present a considerable variation. In Station 1 the highest value was found in the first layer. Mean concentrations in other layers are similar (Table 2).
Mean values in Station 2 are very similar, while in Station 3, mean values in first two layers are alike and the lowest value was found in the third one (Table 2).

Zn - A decrease with depth was observed in all sampling points (Table 2).

Al - In all sampling points, an increase with depth was observed (Table 2).

Fe - In Station 1 and 3 mean values decrease with depth. In Station 2, the highest value was also found in the first layer; however, the mean value of second layer is a litte bit less than the corresponding value of the third one (Table 2).

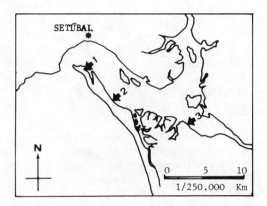

Figure 1. Location of sampling points in Sado River Estuary.

Trace Metal Concentration in Biological Material

The highest mean values in different sampling points were always observed in Spartina maritima roots. For Cu, Zn and Al, the highest root mean values were observed in Station 2, while the lowest were found in Station 3 for Cu and Zn, and Station 1 for Al (Table 1).

For Fe, the highest root mean value was found in Station 1, while the value observed in Station 2 was much lower. The lowest value was observed in Station 3 (Table 1).

DISCUSSION

The distribution of Zn and Fe in sediments of the Sado River estuary seems to be controlled by pollution sources. The first layer presents, in all sampling points, the highest mean values. These values must correspond to a recent accumulation which had it's origin in the high industrial activity of the area (production of pulp and fertilizers, and shipbuilding).

The pattern of Cu distribution in sediment layers do not present a considerable variation, especially in Stations 2 and 3. Mean values observed can probably be considered as background values.

An interesting finding is that the third layer in each sampling point presents the highest value of Al, while the first one the lowest value. This can probably be related with the nature of sediments, emphasizing the importance of their characterization.

Trace metals are preferentially accumulated by roots of Spartina maritima, rather than above-ground tissues. A similar finding was observed by Breteler et al. (1981), for Hg in relation to Spartina alterniflora.

If we assume that root mean values correspond to complete recovery, it appears that this recovery presents a considerable variation according to the element studied and the sediment layer (we do not consider the first layer, because the root system is located especially in the second and third layers).

The maximum recovery value for Al is 7.8% while the minimum is 4.0% For Fe 24.0 - 12.2%. For Cu and Zn a large variation was observed, 95.2 - 17.2% and 103.5 - 25.6% respectively.

For Cu, Zn and Fe, the recovery generally increases when concentrations decrease with depth. The contrary is observed for Al. Although the results do not describe the pathways by which trace metals are recycled between components of the estuarine ecosystem, they at least indicate a direction for further investigation.

First, it is necessary to know the annual input of trace metals in the Sado River estuary.

Second, what is the degree of accumulation in the sediment component and in aquatic organisms.

Third, calculate the area occupied by Spartina maritima and their net annual primary production. This value may provide clear information about the input of trace metals from sediments to the species.

Windom (1975), reports an average annual uptake of Fe and Cu by Spartina alterniflora (leaves and stalks) along the southeastern Atlantic coast of 2100×10^3 kg and 10×10^3 kg, respectively. For both cases, these values correspond to 3% of total input.

Wolfe (1975) estimates an input of Fe and Zn from sediments to Spartina alterniflora in the Newport River estuary of 1230 mg/m^{2}yr and 5.5 mg/m^{2}yr, respectively.

Other factors must also be considered, such as, the transfer of elements accumulated in sediments to the water column and detritus formation in salt-marshes.

Our findings support the importance of Spartina maritima in removing and recycling of trace metals and agree with previous the

Table 1. Trace metal composition (μg/g D.W) of root (R), stem(S) and leaf(L) of _Spartina maritima_.

RSD - Relative Standard Deviation

(*) Means of five replications

Plant Part	Local Sampling	Cu Mean*	RSD(%)	Zn Mean*	RSD(%)	Al Mean*	RSD(%)	Fe Mean*	RSD(%)
R	1	14.7	8.0	139.4	3.0	1751.8	10.7	4727.3	0.2
S		1.8	19.5	137.5	9.5	85.7	15.2	109.1	17.4
L		2.4	18.3	31.0	13.5	229.9	17.6	318.2	17.5
R	2	52.0	12.3	222.9	18.7	4095.2	14.2	3945.8	3.9
S		3.9	16.8	52.1	7.1	108.0	28.6	127.3	16.0
L		4.0	17.9	31.2	19.2	143.9	21.3	481.8	1.8
R	3	9.2	11.9	62.5	4.0	3000.0	11.8	2150.0	11.4
S		3.0	24.8	49.2	3.8	338.9	6.9	77.3	16.1
L		6.2	13.5	29.0	4.7	380.0	7.2	386.7	7.6

Table 2. Trace metal composition as µg/g D.W(Cu and Zn)or µg/gD.W. 10³(Al and Fe) of sediments of the salt marsh, where Spartina maritima was collected.

(*) Means of five replications   RSD - Relative Standard Deviations

Depth (cm)	Local Sampling	Cu Mean*	RSD(%)	Zn Mean*	RSD(%)	Al Mean*	RSD(%)	Fe Mean*	RSD(%)
0-5	1	42.3	9.1	202.9	7.4	32.04	6.2	24.80	12.2
5-10		30.0	9.8	180.2	10.6	38.88	11.6	20.56	6.9
10-15		33.5	8.6	156.1	2.9	44.28	17.4	19.72	3.4
0-5	2	55.0	9.1	285.2	2.4	41.40	13.0	30.56	8.4
5-10		56.6	1.8	255.8	15.7	52.60	6.7	25.86	15.5
10-15		54.6	5.4	215.4	3.1	57.12	4.7	26.74	13.4
0-5	3	55.3	23.4	317.9	13.3	35.98	15.2	20.78	9.2
5-10		53.5	9.3	244.0	13.8	39.60	25.6	17.66	9.1
10-15		36.9	4.0	122.4	2.2	45.62	11.5	16.56	5.3

Table 3. Distribution of trace metals with depth in salt marsh sediments. Copper and Zinc mean values expressed as µg/g D.W, and Aluminum and Iron as µg/g D.W · 10³.

(*) Assuming the highest mean value in the species studied corresponds to complete recovery.

Local Sampling	Depth(cm)	Cu Mean	Cu Recovery(%)*	Zn Mean	Zn Recovery(%)*	Al Mean	Al Recovery(%)*	Fe Mean	Fe Recovery(%)*
1	5-10	30.0	49.0	180.2	77.4	38.88	4.5	20.56	23.0
1	10-15	33.5	43.9	156.1	89.3	44.28	4.0	19.72	24.0
2	5-10	56.6	91.9	255.8	87.1	52.60	7.8	25.86	15.3
2	10-15	54.6	95.2	215.4	103.5	57.12	7.2	26.74	14.8
3	5-10	53.5	17.2	244.0	25.6	39.60	7.6	17.66	12.2
3	10-15	36.9	24.9	122.4	51.1	45.62	6.6	16.56	13.0

(*) Assuming the highest mean value in the species studied corresponds to complete recovery.

work of Banus et al. (1974) and Windom (1975). However, more clarification is needed to determine the best management of salt marshes, preserving their role as a buffer environment and as a potential decontaminating factor.

ACKNOWLEDGEMENTS

This study was supported by a grant from INIC.

REFERENCES

Agadi, V.V., Bhosle, N.B., Untawale, A.G. 1978. Metal concentration in some seaweeds of Goa (India). Bot. Mar., 21, 247-250.
Banus, M, Valiela, I., Teal, J.M. 1974. Export of lead from salt marshes. Mar.Pollut. Bull, 5, 6-9.
Bower, P., Simpson, J., Williams, S., Ly, Y.H. 1978. Heavy metals in the sediments of Foundry Cove, Cold Spring, New York. Environ. Sci. Technol., 12, 683-687.
Breteler, R.J., Valiela, I., Teal, J.M. 1981. Bioavailability of Hg in several Northeastern U. S. Spartina ecosystems. Estuar. Coast. Shelf Sci., 12, 155-166.
Cosma, B., Frache, R., Baffi, F. Dadone, A. 1982. Trace metals in sediments from the Ligurian Coast, Italy. Mar. Pollut. Bull. 13, 127-132.
Melhuus, A., Seip, K. L., Seip, H.M., Myklestad, S. 1978. A preliminary study of the use of benthic algae as biological indicators of heavy metal pollution in Sorfjorden, Norway. Environ. Pollut., 15, 101-107.
Reboredo, F.H.S. 1981. Determinacao de metais pesados em sedimentos do estuario do rio Tejo por espectrofotometria de absorcao atomica, por processo de chama e sem chama. Relatorio Final de Estagio, Fac. Ciencias de Lisboa, 54p.
Reboredo, F.H.S., Carrondo, M., Ganho, R., Oliveira, J.F.S. (1981) Use of a rapid flameless atomic absorption method for the determination of the metallic content of sediments in the Tejo estuary, Portugal. Proc. 3rd Int. Conf. of Heavy Metals in the Environment, p. 587-590, Amsterdam, 15-18 September.
Rozema, J. 1978. On the ecology of some halophytes from a beach plain in the Netherlands. Ph.D. Thesis, Free Univ. of Amsterdam, 191 p.
Windom, H.L. 1975. Heavy metal fluxes through salt-marsh estuaries. In: "Estuarine research" vol. I, Ed. by L.E. Cronin, Academic Press, London, P. 137-152.
Wolfe, D.A. 1975. The estuarine ecosystem(s) at Beaufort, North Carolina. In: "Estuarine Research" vol. I. Ed. by L E. Cronin, Academic Press, London, p. 645-671.

THE ANATOMY, ULTRASTRUCTURE AND

CHEMICAL COMPOSITION OF WOOD

      Wilfred A. Côté

      N.C. Brown Center for Ultrastructure Studies
      State University of New York
      College of Environmental Science and Forestry
      Syracuse, New York 13210, U.S.A.

INTRODUCTION

    At a conference devoted to biomass utilization, it is to be expected that wood, a unique renewable material produced by living trees, be given major consideration. Since it is estimated that about 80 percent of available biomass is derived from the forest, it is essential that this component be characterized thoroughly as to anatomy or structure, ultrastructure and chemical composition. Only with this background is it possible to undertake effective utilization of woody biomass for the production of energy, chemical feedstocks or food.

    Whole tree utilization is a concept that is finding wider acceptance in the context of biomass utilization. This implies that not only is the woody stem used, but that the bark, the branches, and the roots are also removed from the forest and utilized. In the present discussion, only the woody portion of the tree stem will be considered. The structure of bark and other portions of the tree will undoubtedly continue to receive more attention as interest in biomass utilization grows. In recent years many species have been added to the list of bark studies, but wood has been studied more thoroughly since it has been the major component of the tree to be utilized.

    Before discussing the structure and composition of the individual elements of wood, it is advisable to examine the gross structure of the material as it is produced in the living tree. This can be done with the unaided eye or with the use of hand lens. Then the anatomical structure of the material can be examined with a light microscope and a scanning electron microscope. The ultrastructure of wood, actually the sub-light microscopic, requires a transmission electron microscope to provide the necessary resolving power to image the wood cell wall and its components. As each of these levels is considered, the distribution of chemical constituents within the material can be noted.

GROSS STRUCTURE OF WOOD

Wood is a cellular composite. The majority of its cells are oriented longitudinally, that is, parallel to the length of the tree. This accounts for the tendency to split lengthwise along rather than across the grain. The cellular components are arranged within concentric rings which are growth increments representing usually one season's growth. The well-known practice of determining the age of a tree by counting its rings is a valid one, at least in the temperate zones. In tropical regions the growth increments may reflect wet and dry seasons rather than calendar years.

One can describe the gross structure of a tree stem as a series of hollow cones, each having the thickness of one growth increment, placed one over the other. By cutting across the tree stem, a transverse section is produced on which the concentric growth rings can be seen (Figure 1). In the center of the disc in this illustration is the pith which represented the growing point of the living tree at one time. In other words, this was the tip or apex of the cone produced in that particular year. Surrounding the wood or xylem in Figure 1 is the bark or phloem. At the interface between the two is the cambium, the layer of cells that initiated both types of tissue, the wood cells to the inside and the bark cells to the outside. Obviously the cambial zone expands as the tree stem thickens, but it remains a thin band of living cells as shown in Figure 2.

Unlike many man-made materials, wood is anisotropic. Its structure is different in three directions and so, therefore, are its properties. In addition to the transverse or cross-section mentioned above, there is a radial aspect or section which is exposed when a cut is made from the center of the stem, the pith, to the bark, in a longitudinal direction. Figure 3 illustrates this view as well as the tangential section which is a plane tangent to stem.

Besides the gross structural features described so far, one can generally detect sapwood and heartwood differences, especially on a transverse surface. As can be seen in Figure 4, some wood species exhibit a heartwood that is darker than the sapwood. This color change accompanies aging of the tree and is not found in very young trees. In fact, there are some species in which it is very difficult to detect sapwood/heartwood color changes.

Figure 1. Cross-section of the woody stem of a conifer, Douglas-fir, showing growth rings in the wood; pith; and bark. The cambium is located at the interface between the wood (xylem) and the bark (phloem).

Wood rays represent one additional feature which can be detected without the aid of magnifiers. These are thin bands of parenchyma cells which are oriented radially from the center of the tree to, and into, the bark. They can be seen in Figure 2 because of the magnification of this photomicrograph, but they are generally too thin to be seen in cross-section except in some hardwoods such as oak. Reference will be made to them again when wood anatomy is discussed.

ANATOMY OF WOOD

Up to this point the general discussion could be applied to the wood of any tree. However, there is great variation among trees and consideration of wood anatomy cannot proceed without at least a brief review of this topic.

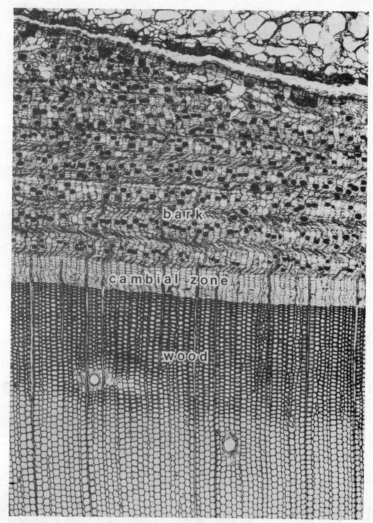

Figure 2. Photomicrograph of a cross-section of eastern white pine showing a small portion of a growing stem. The cambial zone across the center of the photo is the site of cell initiation for both the bark and the wood.

Most people realize that there are two broad categories of trees, the conifers or softwoods, and the broad-leaved trees or hardwoods. The botanical designations are Gymnosperms and Angiosperms, respectively. The wood produced by trees of these two groups is very different because of the cell types and cell distribution patterns characteristic of each. In essence this is what constitutes wood anatomy. The peculiar mix of cell types and tissue

ULTRASTRUCTURE AND CHEMICAL COMPOSITION OF WOOD 253

Figure 3. Wedge-shaped segment of wood illustrates the three aspects of this anisotropic material, the cross-, radial-, and tangential sections.

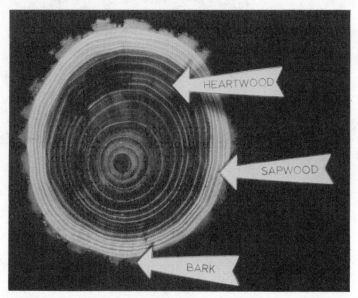

Figure 4. This cross-sectional view of a stem of black locust wood illustrates how some species have a very dark colored heartwood which contrasts with the sapwood. From Core, Côté and Day (1979).

systems is different for every species and wood identification is based on the recognition of these unique features and patterns.

Softwoods, such as the pines, firs and spruces, exhibit a relatively simple structure. Most of the cells are longitudinal tracheids and they are arranged in radial rows. As can be seen in Figure 5, the coniferous tracheid is a long, slender element measuring to about 5 mm in length.

The longitudinal elements found in hardwoods such as the oaks, poplars and maples, are also illustrated in Figure 5 for size comparison. The cell that characterizes the wood of all broadleaved trees is the vessel element or vessel segment. It is a cell with a

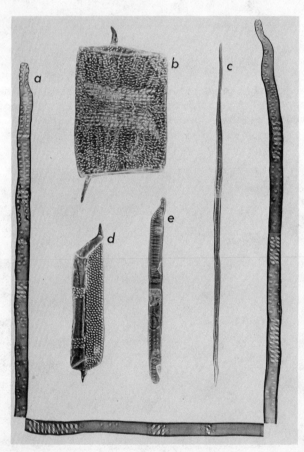

Figure 5. Photomicrographs of individual wood cells: a. coniferous longitudinal tracheid; b. springwood vessel element from a hardwood; c. libriform fiber from a hardwood; d. and e. hardwood vessel elements. From Kollmann and Côté (1967).

large tubular lumen, but it is generally much shorter than coniferous tracheids. In fact, there may be a range of shapes and sizes depending on location within the growth increment.

Other cell types of the hardwoods include libriform fibers, fiber tracheids, vasicentric tracheids and parenchyma cells. There is usually a mix of most of these in a given hardwood.

The cell types mentioned above do not include ray parenchyma which are found in the radial bands of tissue discussed briefly earlier, the wood rays. Both hardwoods and softwoods have such rays made up of short brick-shaped elements oriented horizontally in most cases. The total wood volume devoted to ray tissue is variable. In a softwood such as Pinus strobus the rays constitute less than ten percent of the wood volume. In an average hardwood the rays account for a greater percentage of the total volume than in softwoods. However, some genera such as Tilia may have ray volumes as low as 5 percent while woods of the genus Quercus have ray volumes greater than 30 percent of the total wood.

The nature and location of rays in the two broad classes of wood can be appreciated by studying Figures 6 and 7, scanning electron micrographs of small cubes of softwood and hardwood. These views which make it possible to see wood in all three aspects in a single photograph represent a significant improvement over the former method of examining this material. Individual sections of the cross-, radial and tangential surfaces had to be prepared for mounting on microscope slides and then each could be photographed at equal magnifications. The resulting micrographs could then be glued to a cube after careful matching so that the rays and other features were correctly joined at the corners.

While the study of wood anatomy has been made easier in some respects, there is no substitute for the individual examination and evaluation of large numbers of sections for each wood sample. Statistical methods must be applied to generate valid or meaningful data on cell length, vessel diameter, ray dimensions and make-up, and many other features which are variable from tree to tree within a single species and indeed within a single tree.

In addition to this natural variability in normal wood, abnormal wood is produced under certain conditions. For example, the wood on the underside of a leaning conifer has cells whose structure is very different and even the chemical composition of the wood is abnormal. Fortunately the striking difference in anatomy makes it possible to detect this wood which is weak and brash. Similar variations from the normal are found in the hardwoods. It should be noted here that some of the new concepts proposed for rapid production of biomass through short rotation tree plantations may overlook an important fact. The wood of very small trees does not have the

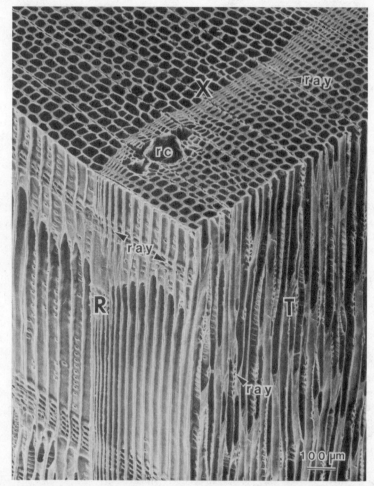

Figure 6. Scanning electron micrograph of a small wood cube of eastern white pine showing the cross- (X), radial (R) and tangential (T) surfaces. Note that the tracheids are arranged in radial rows (X). The rays can be seen on all three faces of the cube. The large opening on the cross-section is a resin canal (rc) which can be found in many coniferous species.

structure nor the chemical composition of larger, mature trees and specific investigations would be required to establish the nature of this biomass.

The subject of wood anatomy can hardly be covered adequately within the framework of a relatively short presentation. There are numerous recent publications emphasizing the woods of a particular region. For North America, Panshin and de Zeeuw (1980) or Core,

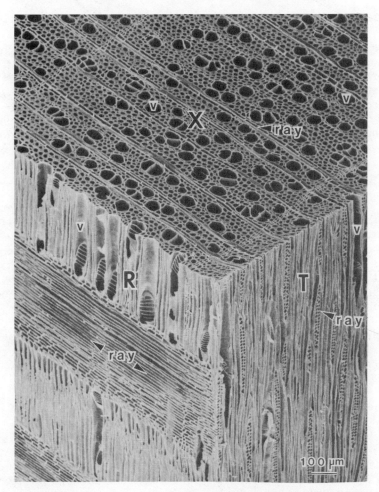

Figure 7. Scanning electron micrograph of a small wood cube of sweetbay. The characteristic structure for hardwoods is the vessel (V) which can be observed all three aspects as can the rays. Note the more random distribution of elements on this cross-section than in the softwood of Figure 6.

Côté and Day (1979) offer both introductory and detailed coverage. Kribs (1968) deals with the structure of woods from tropical Africa and South American as well as from the Pacific region. The woods of Europe as well as other regions of the world have been included in publications of the Forest Products Laboratory (Princes Risborough) in England. Each of the above publications contains extensive references to other sources.

ULTRASTRUCTURE OF WOOD

As suggested earlier, ultrastructure implies sub-light microscopic structure. It refers to structure in wood that is too fine to be resolved with the light microscope, that is, smaller than 0.2 micrometer or 2000 Angstrom units. The transmission electron microscope is needed for imaging the smallest structural units in wood cells since modern instruments can resolve detail separated by less than 5 Angstrom units. Even the scanning electron microscope, though much more limited in its resolving power, can provide considerable ultrastructural evidence since most modern SEM's will resolve well below 100 Angstrom units.

There are many important structures in the ultrastructural domain of wood. Their importance lies in the influence they have on processing of wood as in pulping for paper or penetration with chemicals for wood preservation. The strength properties of wood can be related to the ultrastructural organization revealed through electron microscopy. By various techniques of specimen preparation, it is also possible to demonstrate the location or distribution of the major chemical constituents of wood.

In this brief survey emphasis will be given to the organization of the individual wood cell walls, to the role of the three principal chemical components of wood and to a few structural features that are of broad interest.

The Wood Cell Wall

While wood is used in the manufacture of composite materials such as plywood, particleboard and flakeboard, it is itself a composite material. It has three major chemical components: cellulose, hemicelluloses and lignin. In addition, its cell wall organization can be described as the forerunner of the three-ply panel concept. It is a remarkable example of natural design which accounts for many of the strength properties of this renewable material.

Although the overall morphological characteristics of coniferous tracheids, vessel elements, libriform fibers and other longitudinal elements are quite dissimilar, the walls of each cell type are surprisingly similar. Generally there are three layers in the secondary wall of the wood cell. This is the portion of the wall produced by the cell itself from cytoplasm enclosed by the primary wall at the time of cell division in the cambium. Biosynthesis of the three major chemical constituents proceeds in the living tree and each resulting "wood cell" is an empty, dead cell with only the primary and secondary walls remaining. In most instances this stage has been reached within days or weeks, and certainly within the same growing season.

The framework of the cell wall is cellulose which invariably occurs in the form of long slender strands called microfibrils. The cellulose chains are organized in parallel array which yields a crystalline-like material in terms of x-ray and birefringence phenomena. As much as 50 to 60 percent of this cellulose could be classed as "crystalline" while the remainder is "amorphous" or less highly oriented and therefore more accessible. The strands may be as small as 17 Angstrom units in diameter, representing perhaps the joining of a minimal number of cellulose chains. These have been termed "elementary fibrils". They have been found to be as large as several hundred Angstrom units in diameter in wood and this presumably represents the aggregation of many smaller fibrils into what may be called a "macrofibril".

Closely associated with the cellulose framework are the hemicelluloses which are thought to fill the voids around the microfibrils. For this reason they have been classed as matrix substances. Both the nature and the amount of hemicellulose in the cell wall varies widely between the broad groups of softwoods and hardwoods although the amount of cellulose remains about the same at approximately 42 percent of the total wood substance. Hardwoods are characterized by their high content of a partly acetylated, acidic xylan which amount to 20 to 35 percent of the wood. Glucomannan, a second hemicellulose is present in small amounts in the hardwoods. On the other hand, coniferous woods contain smaller total amounts of hemicellulose represented by a partly acetylated galactoglucomannan (20%) and xylan which accounts for less than 10% of the wood substance.

The third major component, lignin, occurs in larger amounts in softwoods than in hardwoods, thus making up for the difference in hemicelluloses. In hardwoods the lignin content ranges from about 21% to 27% while in softwoods 26% to 34% is representative. Lignification of the cell wall begins before the secondary wall has been completed. It begins at the cell corners near the junctions with other cells and progresses inward toward the lumen. Lignin has been classed as the incrusting substance in wood because it serves to harden the material, giving it resistance to compression, for example. It also provides for rigidity. Cellulose has high tensile strength for its weight, but all three constituents together impart the properties which make it a true composite material.

The distribution of lignin is quite uniform across the secondary wall in normal wood (Figure 8). In the middle lamella or intercellular layer, it has a higher packing density as it serves as a cementing substance joining the cells. In pulping for papermaking, lignin is removed from the wood thus freeing the individual fibers. In wood exposed to biological attack such as decay fungi, lignin provides some resistance to degradation although there are fungi which produce enzymes capable of metabolizing lignin as well.

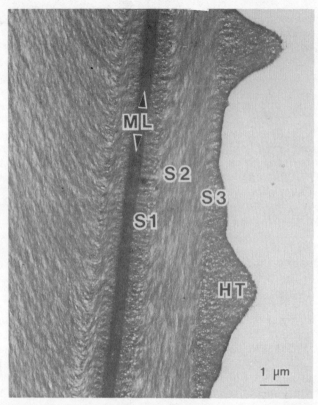

Figure 8. Transmission electron micrograph of a "lignin skeleton" from Douglas-fir wood. The S1, S2 and S3 layers of the tracheid can be identified in this carbohydrate-free specimen. Note the helical thickenings which are part of the S3 and the high electron density of the middle lamella (ML). (From Côté 1967.)

As suggested above, the cell wall is organized in a three-layered structure. The microfibrils produced immediately inside of the primary wall are oriented approximately perpendicular to the long axis of the cell. This outer layer, also designated as the S1, is relatively thin in most cells. There is a shift in orientation then to one that nearly parallels the cell axis. This occurs in the middle layer or S2 which is the thickest layer of the cell wall. In latewood or summerwood cells the S2 becomes much thicker than the S1 or the S3 which is formed as the inner layer lining the lumen. Its orientation parallels that of the S1, thus creating the three-ply design. The S3 is usually quite thin and often exhibits greater decay resistance than the other portions of the wall.

Figure 9 is a diagrammatic representation of a typical wood cell wall. This scheme can be applied to both hardwoods and soft-

ULTRASTRUCTURE AND CHEMICAL COMPOSITION OF WOOD   261

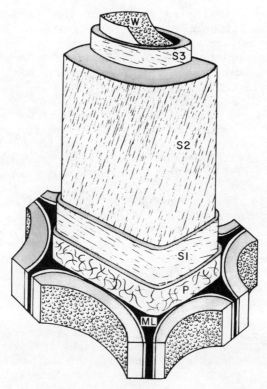

Figure 9. Diagrammatic representation of the organization of typical (normal) wood cell walls. The primary wall (P) has randomly distributed cellulose microfibrils while the S1, S2 and S3 are oriented in a three-ply design. The warty layer (W), when present, lines the cell lumen. Portions of neighboring cells separated by middle lamella (ML) are shown. (From Côté 1967.)

woods, but deviations occur, especially in abnormal woods such as reaction wood which was mentioned earlier. There are other occasional variations in the organization of wood cell walls in normal wood cells as well. Perhaps these occur because of unusual growth conditions or other factors.

Cell Wall Sculpturing

In some species, an additional layer is produced to cover or partially cover the S3. It has been termed the warty layer because of the small protuberances that may be part of a membrane or deposited directly onto the S3 without evidence of any other structures. Because of their very small size, the warts were not known until wood was examined with the aid of the transmission electron microscope. The cytological significance of the warty membrane is not fully understood although studies have uncovered evidence of wart formation or deposition from the cytoplasm. The chemical nature of warts is also incompletely known. The warty membrane is apparently more resistant to decay than the other portion of the wall so it is assumed that it is probably related to lignin.

The warty layer can be found in both softwoods and hardwoods, but there appears to be no explanation for its presence in some species and its absence from others. There is one instance of possible taxonomic significance of this unique structure. In the genus Pinus, the so-called "hard pines" have warts in their tracheids while the "soft pines" do not. Wart shapes and sizes do appear to be characteristic for a given species, but the inconsistency of the occurrence of warts as well as their small size make this an impractical feature for wood identification. The transmission electron micrograph of Figure 10 was selected to illustrate the relative thinness of the warty membrane in a hard pine as well as to show the nature of the warts themselves. The underlying microfibrillar structure was exposed by the fortuitous removal of a portion of the warty membrane.

Another hard pine is represented in Figure 11, a micrograph of a cross-section of portions of three longitudinal tracheids. In this case it is possible to observe the three-layered secondary walls of these elements and the electron-dense middle lamella or inter-cellular layer. Note also that the warts can be seen in sectional view in each of the three tracheid lumens.

What might be termed as more realistic sculpturing of the secondary wall are the helical thickenings that occur in a number of species, both softwoods and hardwoods. These are composed of cellulose microfibrils and form part of the S3 layer itself. (The warty membrane is non-cellulosic and it could therefore be argued that it is not part of the framework of the cell wall.) Figure 12 dramatizes the sculptured look of the lumen lining in four tracheids of Douglas-fir wood. Not only are the helical thickenings present, but also involved are the domes of the bordered pits which protrude into the cell cavity. These will be discussed in greater detail shortly.

The resolution in Figure 12 is limited, due in part to the relatively low magnification of the scanning electron micrograph, but also because of the inherent resolving power limitations of the SEM. To demonstrate that the transmission electron microscope offers improved resolution as well as to show the ultrastructure of helical thickenings, Figure 13 was selected. The microfibrillar organization of the S3 is clearly illustrated and, in addition, one can observe microfibrils of the S2 layer through a gap left in the S3 at left center.

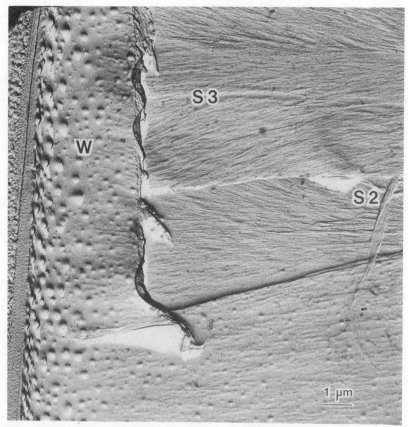

Figure 10. Transmission electron micrograph (replica technique) of a small portion of a Virginia pine tracheid lumen. Some of the warty membrane (W) was torn away revealing the microfibrillar layers of S3 and S2 underneath. (From Côté and Day 1969.)

A number of other types of sculpturing can be found in wood cells. However, space does not permit more complete coverage of these features and they would have only incidental importance in biomass utilization. There is one feature of the cell wall that must not be overlooked because of its role in accessibility of wood to fluids such as processing chemicals: the bordered pit.

Bordered Pit Structure

In the discussion of Figure 12 mention was made of the bordered pit domes in the tracheid lumens. At the center of those structures one can observe an aperture which leads to a chamber formed by a matching bordered pit of the neighboring tracheid. These are actually bordered pit pairs that provide for lateral movement of fluids

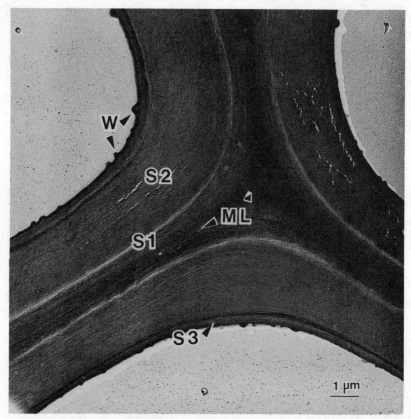

Figure 11. Transmission electron micrograph of an ultra-thin cross-section of portions of three longitudinal tracheids in Table-Mountain pine, a hard pine. The three-layered secondary walls are visible with the electron-dense middle lamella at their junction. Note the view of the warts (W) in profile. (From Côté and Day (1969).)

from one tracheid to the other provided that the membrane and torus in the center is not displaced and aspirated.

The scanning electron micrograph of a cross-section of Douglas-fir (Figure 14) offers a three-dimensional view of such a bordered pit pair that was cut through in the sectioning. Incidentally, one can also observe the helical thickenings lining the lumens of the tracheids. In this case, the torus or valve-like structure was drawn into the aperture at the right, thus sealing it and preventing mass movement of liquid. Since there are many other pits on each tracheid, this does not necessarily prevent penetration of the wood structure. However, during the seasoning of wood most bordered pit pairs do become aspirated.

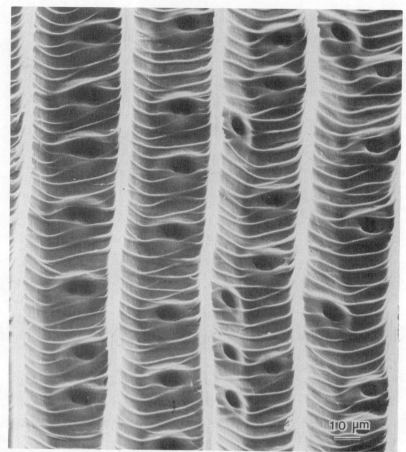

Figure 12. Scanning electron micrograph of portions of four longitudinal tracheids in Douglas-fir wood. The lumens were exposed by making a longitudinal cut through the tracheids. The sculpturing of the cell wall by helical thickenings and bordered pit domes is particularly striking in this view.

A transmission electron micrograph (Figure 15) of the face view of a bordered pit membrane and torus, this one from a hard pine, suggests that the torus is quite flexible since it conforms closely to the aperture it was drawn against. Note the microfibrillar orientation on the torus surface and the cellulosic strands extending outward to the periphery of the pit chamber.

While hardwoods also have bordered pits, they lack the flexible membrane and torus. Instead, the membrane has the appearance of the primary wall; that is, a randomly distributed array of microbrils forming a kind of filter network stretches across the pit chamber.

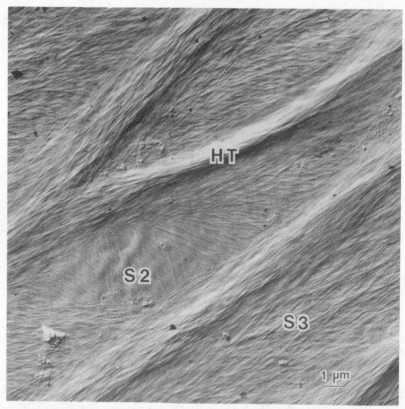

Figure 13. Transmission electron micrograph (replica technique) of the lumen lining of a portion of a Douglas-fir tracheid. The microfibrillar nature of the S3 with its characteristic helical thickenings (HT) is clearly visible because of the absence of a warty membrane. (From Côté (1967.)

Since bordered pit pairs provide for lateral movement of liquids, utilization of woody biomass can be aided by such accessibility. Penetration in softwoods depends largely on these pathways while in hardwoods there is generally good conduction via the vessels in the longitudinal direction. Then lateral movement from the vessels into surrounding cells is by diffusion.

CONCLUSIONS

If woody biomass is to be utilized for energy by burning, its anatomy, ultrastructure and chemical composition are obviously of lesser importance than if it will be used for the production of

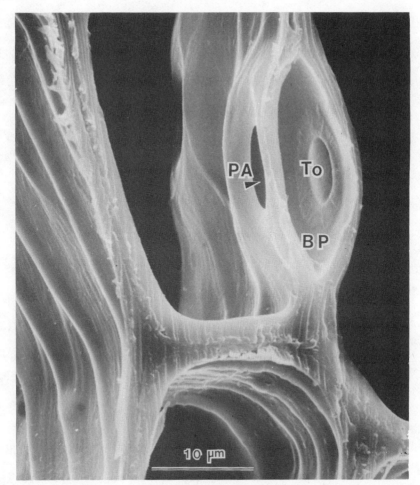

Figure 14. Scanning electron micrograph of a cross-section of Douglas-fir wood affording an unusual view of a bordered pit (BP) chamber as well as the cell cavities of four tracheids. The aperture (PA) to the bordered pit and the torus (To) sealing the other aperture can be seen. (From Core, Côté and Day 1979.)

fiber or chemical feedstocks. The physical nature of the wood and its ease of penetration or accessibility to fluids is of concern as has been pointed out.

The chemical composition of wood is known within reasonable limits even though it varies considerably from species to species and especially between the hardwoods and softwoods. The distribution of the major chemical constituents, cellulose, hemicelluloses and

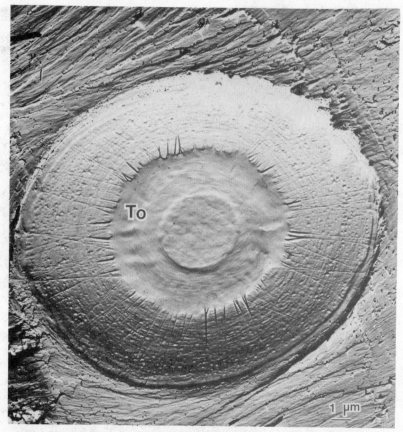

Figure 15. Transmission electron micrograph (replica technique) of the face view of a bordered pit membrane in the wood of pitch pine. The torus (To) is sealed against the aperture in such aspirated pits. Note the warts lining the pit chamber. (From Côté and Day 1969.)

lignin, within the wood cell walls can be determined with some degree of accuracy for cellulose and lignin, but it is more difficult to determine the exact location of hemicellulose with the specimen preparation techniques currently available.

During the past three decades ultrastructural evidence has been developed for some species, but there has been no unified effort to survey the more important commercial species. Nevertheless, enough work has been reported to allow some generalization about the organization of the wood cell wall, its sculpturing and various unique features. Considering that wood is a most important source of biomass, the new emphasis on its effective utilization is likely to generate new research interest in wood chemistry and ultrastructure of wood.

REFERENCES

Anonymous. 1960. Identification of Hardwoods--A Lens Key. Second Edition. Forest Products Research Bulletin No. 25, Dept. of Scientific and Industrial Research, London.

Brazier, J.D. and G.L. Franklin. 1961. Identification of Hardwoods--A Microscopic Key. Forest Products Research Bulletin No. 45, D.S.I.R., Charles House, 5-11 Regent St., London, W.W.1.

Core, H.A., Côté, W.A. and A.C. Day. 1979. Wood Structure and Identification. Second Edition. Syracuse Wood Science Series No. 6, Syracuse University Press, Syracuse, N.Y.

Côté, Wilfred A., Jr. 1967. Wood Ultrastructure--An Atlas of Electron Micrographs. Univ. of Washington Press, Seattle.

Côté, W.A., Jr. and A.C. Day. 1969. Wood Ultrastructure of the Southern Yellow Pines. State University College of Forestry at Syracuse Univ., Syracuse, N.Y. Tech. Pub. 95.

Kollmann, F.F.P. and W.A. Côté, Jr. 1968. Principles of Wood Science and Technology, Vol. 1, Solid Wood, Springer-Verlag, Berlin, New York.

Kribs, D.A. 1968. Commercial Foreign Woods on the American Market. Dover Publications, Inc., New York.

Meylan, B.A. and B.G. Butterfield. 1972. Three-Dimensional Structure of Wood. Syracuse Wood Science Series No. 2, Syracuse University Press, Syracuse, N.Y.

Panshin, A.J. and Carl DeZeeuw. 1980. Textbook of Wood Technology, Fourth Edition. McGraw-Hill Book Co. New York.

Phillips, E.W.J. 1948. Identification of Softwoods by Their Microscopic Structure. Forest Products Research Bulletin No. 22, D.S.I.R., London Ministry of Technology. Reprinted 1966. 56 pp.

CELLULOSE: ELUSIVE COMPONENT OF THE PLANT CELL WALL

Ed J. Soltes

Professor of Wood Chemistry
Forest Science Laboratory
Texas A & M University
College Station, Texas 77843
U.S.A.

INTRODUCTION

It has been repeated often that plant cellulose is the most abundant natural material on this earth. At approximately 50 percent of all biomass, annual production is about 50 billion tons (Goldstein 1981). Because it is natural and thus potentially available in perpetuity assuming that plants are managed properly, and because it is abundant, cellulose should be an increasingly important resource for our material, energy and food needs. However, cellulose for the most part remains a somewhat elusive resource. Why?

One major problem in cellulose utilization is that cellulose is generally found in nature in a plant cell wall substance called lignocellulose. Lignocellulose is an intricate matrix of three components -- cellulose, lignin and hemicellulose. Conditions used to liberate cellulose from this matrix are often responsible for degrading cellulose, and thus it is difficult to obtain "pure" cellulose in high yield. There is still some controversy, for example, about the molecular weight of cellulose as it appears naturally in wood, because it cannot be isolated from wood without some depolymerization. As isolated polymers, celluloses can be hydrolyzed by acids or cellulolytic enzymes to yield glucose, although only acids can readily hydrolyze cellulose in lignocellulose. For conversion of cellulose into glucose by mild acid hydrolysis, additional problems surface. The high activation energies required to hydrolyze the ordered, so-called crystalline, structures of cellulose present in lignocellulose contribute to slow hydrolysis rates, during which time the acid used for hydrolysis can degrade product glucose, severely limiting final glucose yields (Goldstein 1982). Other components in lignocellulose and/or their arrangement in the lignocellulosic matrix, appear to have an inhibitory effect on cellulose accessibility in its conversion into glucose by the celluloytic enzymes of microorganisms. This inhibition is found not only for cellulolytic enzyme isolates but also for parent microorganisms, including the microflora used by the ruminant for digestion. The ruminant, despite its abilities to convert cellulose into glucose, has problems in accessing the cellulose present in the lignocellulosic tissues of many plants. Plant tissues are then composed of additional substances and/or are ordered in such a way that much of the cellulose present is not readily available for utilization.

The objective of this paper is to critically review selected observations on the nature of the cellulose accessibility problem. Developing useful models for plant cell wall accessibilities, identifying new and effective approaches to extract cellulose from lignocellulosics, improving cell wall digestibiity, and making it more accessible to hydrolytic agents available to the ruminant and conversion scientist alike, are prerequisite accomplishments for the expanded utility of wood and agricultural residues in meeting more of our food, feed, fiber and fuel needs from renewable resources. Resolution of this problem has to be one of the most important and challenging biomass research endeavors today.

CELLULOSE AS A CHEMICAL FEEDSTOCK

The conversion of cellulose to glucose is usually considered the first step in the potential large-scale utilization of cellulose. If acceptable yields of gluocse could be obtained through either acidic or enzymatic hydrolysis processes, glucose could be readily fermented to ethanol in high yield using commercially proven processes. This ethanol could be the basis for a chemical industry (Goldstein 1974; 1977; 1978). Dehydration of ethanol to yield ethylene, and conversion to butadiene using processes which were once proven commercially, but made obsolete by formerly inexpensive petroleum, are suggested approaches. Our chemicals dependence on petroleum is largely due to its content of olefins, and the production from ethanol of ethylene (the largest volume organic chemical used as a building block for petrochemicals and plastics) and of butadiene (used in the production of synthetic rubber) are obvious ways to emulate and decrease our dependence on petroleum. Lactic acid from glucose can be converted to acrylic acid and acrylates. Glucose, additionally, as a natural and ubiquitous chemical of nature, is readily fermented into a wide assortment of antibiotics, chemicals and vitamins (Seeley 1976). Glucose fermentation, one of man's oldest processes, is still at the same time in its infancy in terms of its potentials.

Acidic and Enzymatic Hydrolysis of Cellulose

It has been known for over 150 years that cellulose can be converted to glucose by acid hydrolysis. The basic dilute and concentrated acid processes that have been developed over the years are well summarized by Goldstein (Goldstein 1981b). There are yield limitations, generally 50-60 percent, in the use of high temperature dilute acid processes due to slow hydrolysis of highly ordered cellulose with subsequent degradation of product glucose (Goldstein 1982). The use of concentrated acids requires corrosion-resistant materials and acid recovery cycles, although yields of product glucose can approach or exceed 85 percent. There are champions of both approaches, and research continues today towards the development of cost-effective processes. Most activity has been in the use of sulfuric and hydrochloric acids, although there has been work with other acids (Hall et al. 1956). There is currently renewed interest in the use of liquid hydrogen fluoride (Selke et al. 1982), because of the potentials for near quantitative recovery of glucose.

Enzymatic hydrolysis of cellulose, or at least the idea that enzymes can be used industrially for this purpose, is a much newer activity for cellulose conversion scientists. Most of this activity has centered around the use of the cellulose complex isolated from the fungus Trichoderma reesei although in very recent years, a host of fungi, bacteria, yeasts or their enzyme isolates have been receiving attention. Research in the late seventies was spurred by the U.S. Department of Energy's gasohol programs, most of it based on T. reesei cellulase to produce glucose for fermentation with Saccharomyces cerevisiae to produce alcohol. More recently there has been increasing attention paid to different fermenting microorganisms with abilities to convert cellodextrins (Lastick et al. 1982) and xylose (Detroy et al. 1982; Jeffries 1982) to alcohol. Trends are towards complete fermentation of hemicellulose and cellulose in lignocellulosics to ethanol. Just as the acid hydrolysis of cellulose has been shown to be relatively complex, the enzymatic hydrolysis of cellulose is also complex. Many obstacles still remain to its routine application on an industrial scale. Although isolated celluloses can be readily hydrolyzed, albiet with similar problems to acid hydrolysis with respect to cellulose crystallinity, the major problem appears to be some structure in lignocellulosics which inhibits access of the cellulose to cellulase enzymes.

For both acidic and enzymatic hydrolysis, as for ruminant nutrition, susceptibility of lignocellulose to digesting agents must be enhanced: physical barriers must be removed. The facile solution appears to be chemical pulping to remove hemicelluloses and lignin, and although it has been demonstrated that cellulose does become much more accessible to acids, enzymes and rumen microflora in such tissues, the expense of pulping processes is generally thought to be excessive for this purpose.

There are additionally physical methods to enhance accessibility of cellulose in lignocellulosics. Most of these relate to communition, or the reduction of particle size by grinding to increase surface areas and improve the porosity of lignocellulosics to digestive agents. Explosive decompression is another approach receiving attention today (Marchessault and Malhotra 1982; Muzzy et al. 1982). The use of sulfur dioxide, ammonia, steaming, and other physico-chemicl pretreatments for enhancing cellulose digestion have been well documented, both for man's purposes in cellulose hydrolysis (Goldstein 1981b), and for the ruminant's purposes in nutrition (Huber et al. 1982; Han et al. 1982).

Potentially new solutions to the old problem are the use of solvents which can selectively dissolve cellulose out of the lignocellulosic matrix (Turbak 1982), and thus make it available for use in a clean lignin-free state, or the use of appropriate swelling agents to remove cellulose as a soluble derivative (Durso 1982). Alternatively, pulping with organic solvents, such as the organosolv process using ethanol (Sarkanen 1980; Klausmeier 1982; Lipinsky 1982) exhibit promise for the selective removal of lignin and hemicelluloses from lignocellulosics to leave, as a product, reactive cellulose. In all of these latter options, cellulose is recovered instead of glucose, leading to more utilization possibilities.

Cellulose as a Chemical Material

Our imaginations must not be limited to the use of cellulose as a raw material for the production of glucose. Cellulose, itself, is a relatively reactive polymer from which many types of derivatives can be manufactured, and which can be crosslinked and grafted onto other polymers to produce functionally interesting and useful materials. Functions for these polymers are limited only by our desires to pursue them. Cellulose research, unfortunately, is dying in the U.S. The level of research and development on cellulosic materials, even as a percentage of sales, is only a small fraction of that devoted to chemicals and synthetic polymers. Many cellulose prodicts are available today, but are generally produced in an energy- and materials-intensive fashion from dissolving pulps. Overall yields from wood are about 35 percent. More and different polymers could be available if cellulose was more accessible to these interests. More research must be initiated and promoted for the liberation of cellulose from its lignocellulosic matrix, not as glucose but as cellulose itself, in yields and quanlities suited for such use. Use of cellulose as cellulose instead of as a source of glucose represents utilization at a higher level of entropy. Falkehag (1977) states: "We should attempt to use renewable resources at the highest possible systems level. We should not increase entropy and destroy a composite material, a fiber or a monosaccharide when we don't have to in order to meet our needs." This should be a guiding principle for biomass utilization. It is unfortunate that we are often limited in our efforts by the words "...we should attempt to..." and "...when we don't have to...".

Cellulose is used and marketed as cellulose and as cellulose derivatives. Regenerated cellulose is used in the manufacture of rayon and acetate fibers, cellophane, cellulose ester plastics and cellulose ether gums. Further utilization will depend on improving the cost and performance of cellulosics relative to petrochemical polymers. The following four areas require work (Goldstein 1977): (1) improvement of the production and yield of chemical cellulose using more cost-effective and less energy-intensive processes; (2) improvement of the conversion processes used for the preparation of cellulose derivatives; (3) new regeneration methods for cellulose, and methods for shaping of products; and, (4) improvement of the properties of cellulose-derived products.

What type of new cellulose-derived materials should be pursued? An obvious answer might be petrochemical polymer emulation. However, it is perhaps noteworthy that petrochemical polymers got their start emulating cellulose derivatives. Cellulose derivatives were our first films, "synthetic" fibers, and molded articles, but were generally supplanted by petroleum-based polymers on the basis of cost rather than performance. The functional properties of plastics, synthetic rubber, petrochemical and cellulose-derived polymers in general, have very little to do with the monomer entities that are used to make the polymers. Since functional properties are derived from the physical and chemical properties of the polymer itself, it is possible to produce functionally similar polyers from dissimilar monomers. Thus, the useful properties of polyethylene are primarily based on the macromolecular properties of polyethylene and have little to do with ethylene, and cellulose acetate has little in common with glucose or glucose acetate, although both polyethzlone and cellulose acetate are film formers and serve some similar product needs, such as the transparencies we call acetate sheets. If we keep in mind that cellulose is a polymeric material, and not a polymer of glucose or cellobiose, a material which can be derivatized or grafted onto other polymers with differing and useful properties, and/or molded into various shapes or drawn into fibers and films, only then we can begin to explore its potentials.

Before we can use cellulose for whatever purpose, we must first find access to it. As suggested before, however, there are problems in accessing cellulose in lignocellulose. Let's turn to a discussion of the composition and construction of the plant cell wall in an attempt to understand the nature of this limited accessibility.

THE PLANT CELL WALL

A previous paper in this volume (Côté 1983) contains scanning electron micrographs of the cell structure in woody plants. At first examination, there appears to be a wide variety of cell structures in individual plants -- and so it should be. Nature constructs different plant species, and arranges for different properties in plant species through the use of this variation. The result is not only the wood of oak and the wood of pine but also, for example, the substance of wheat straw, corn cob and sorghum stem. Upon more studied examination, however, we can see that nature has used functionally similar cell structures but in varying shape, size and prominence. Nature has also used similar chemical composition in the construction of her plant cell walls, and we know that the chemical composition of plant cell walls is usually cellulose and hemicellulose with varying amounts of lignin, and this chemical composition and its physical matrix is what we refer to when we speak of lignocellulose. There are other chemical components of plants, and we usually refer to these as extractives, reflecting the ease with which these plant components can be "extracted" from plants. These extractives are the contents of non-structural cells such as various parenchyma cells, or the excretions of epithelial cells found in intercellular spaces called resin canals, gum canals, etc.

As might be expected, those plant cells given structural roles in plants are generally thick-walled and relatively stiff, both needed for plant support. The relative thickness of the walls of structural cells and their stiffness requirements are often reflected in examining the cell wall structures of big vs. small plants, for example a tree or woody shrub vs. a blade of grass. How marvelous an engineered structure is a mature tree stem, that can support a hugh crown and usually survive the worst of nature's weather forces, and do so in some tree species for thousands of years! The stem of the same tree when young, or a forage species when juvenile, is more flexible. It can be argued that structural rigidity is not required in very small and/or juvenile plants, and that the flexibiity allows the juvenile plant to survive nature's forces and accidental brushes with animals. Thus, there are structural changes in plants as they mature, and some understanding of these structural changes, as we will see, is important to our understanding of cell wall accessibilities.

Even the novice, in examining scanning electron micrographs of cross sections of mature plant tissues, will see both thin-walled and thick-walled structures, to a large extent reflecting those tissues in the plant whose function is non-structural and those which are at least partially if not wholly designed for structural purposes. Some non-structural tissues are thin-walled even in maturity, while structural tissues are thin-walled when juvenile and develop thickness upon maturity.

What is the nature of this thickness in structural tissues? As is seen here (Figure 1) for a mature structural wood fiber, there

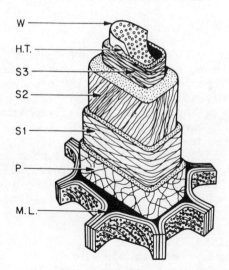

Figure 1. A schematic diagram to illustrate the structure of the plant woody cell wall and the orientation of the cellulose microfibrils within the various layers. ML-middle Lamella, P-Primary wall, S1-outer layer of the secondary wall, S2-middle layer of the secondary wall, S3-innermost layer of the secondary wall, HT-Helical thickening, and W-warty layer (After Côté 1965).

are several secondary layers present inside a primary wall. Of importance to our discussion is the S2 layer in the secondary wall, both in its fine structure and its predominance. The parallel winding lines drawn on the S2 layer represent cellulose microfibrillar orientation, much like the fibers in hemp rope or strands in steel cables. The analogy is not accidental: the fiber, the hemp rope and the steel cable all derive tensile strength from this type of parallel winding and helical orientation. To stretch this analogy further, the thicker the S2 layer, the thicker the hemp rope or steel cable, the stronger the structure. Further examination of thick vs. thin-walled structural tissues indicates that the difference is essentially the thickness of the S2 layer. As most plant cells mature, the general development of wall thickness for both non-structural and structural cells is similar. A thin primary wall is first formed, followed by deposition of secondary layers inside this primary wall. Differentiation between thin-walled and thick-wall tissues occurs here: some secondary walls will be thin, others will be thick. For structural cells, this thickness appears as the highly ordered S2 layer; for parenchyma cells, the organization resembles a thickened primary wall with somewhat less order. A thick-walled structural fiber, because of its predominat S2 layer, contains more highly ordered cellulose.

STRUCTURES OF CELLULOSE

This order in cell wall structure is carried through several stages down to the molecular level. Let's start with cellulose. Cellulose is often represented as a linear polymer of glucose, more precisely a homopolysaccharide composed of β-D-glucopyranose units, which are linked together by (1 → 4)-glycosidic bonds (see Figure 2). Some important features of this molecule are that it is linear, and that all of the hydroxyl groups are equatorial with respect to the sugar rings, and thus available for strong intra- and/or intermolecular bonding. Bundles of cellulose molecules can thus aggregate to form microfibrils. In these microfibrils, cellulose is found in both highly ordered (so-called crystalline) and less ordered (so-called amorphous) states.

Native cellulose, i.e., cellulose as it occurs in nature, is found in a distinctly ordered structure called cellulose I. When the cell wall is disrupted, or when cellulose is isolated, it can exist in several other ordered or non-ordered/disordered forms. The nature of these ordered forms, or the crystalline structures of cellulose, has been the subject of numerous investigations and it is still possible to involve scientists in lengthy arguments about the nature of cellulose crystallinity. The cellulose I structure is destroyed when native cellulose is swelled with alkali or dissolved, and another ordered form of cellulose, called cellulose II or regenerated cellulose, is formed. Other cellulose forms can also be created: upon being subjected to certain chemical treatments or heat, cellulose III and cellulose IV are formed. It is especially in attempting to understand the nature of these latter cellulose forms that much current controversy is based, although recent work even suggests a reinterpretation of cellulose I and cellulose II conformations.

Unit cell structures (crystalline lattices) of cellulose I and cellulose II have been proposed and refined over the years, and some more recent versions of these unit cell structures are those of Kolpak and Blackwell (Kolpak and Blackwell 1976; Blackwell, Kolpak and Gardner 1977; Kolpak, Weih and Balckwell 1978). In the structures proposed, the nature of order is intermolecular bonding between cellulose chains. The cellulose chains in native cellulose microfibrils (cellulose I) are parallel, that is, oriented in the same direction, whereas the cellulose chains in regenerated cellulose (cellulose II) are anti-parallel. Because of these parallel/antiparallel orientations, different intermolecular bonding can exist. Cellulose II is thermodynamically more stable than, and cannot be converted back to, native cellulose. The explanation proposed is that the anti-parallel arrangement in cellulose II results in very strong hydrogen bonding, stronger than the hydrogen bonding available to the cellulose chains in parallel arrangement in cellulose I.

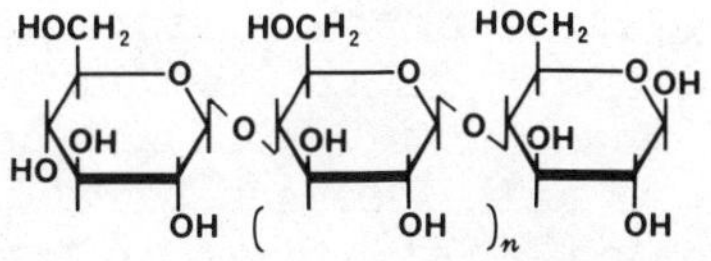

Figure 2. Cellulose -- a polymer of glucose.

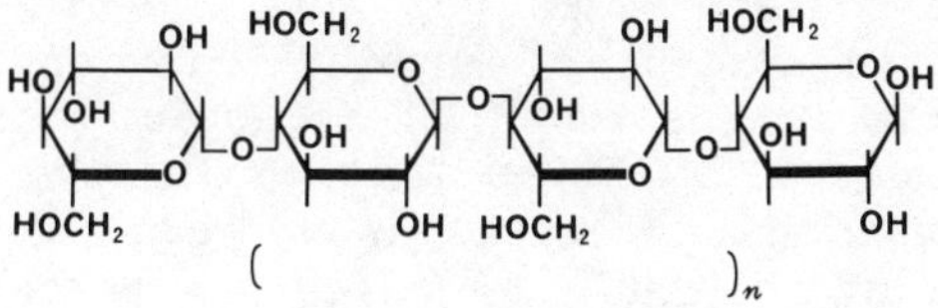

Figure 3. Cellulose -- a polymer of cellobiose.

The primary sources of information concerning the molecular structure of cellulose have been x-ray and electron diffractometric studies, conformational analyses, and vibrational spectroscopy. Recent work by Atalla (1979, 1982), through the examination of celluloses by Raman spectroscopy and solid-state C-13 nuclear magnetic resonance spectroscopy, provides for a reinterpretation of the nature of cellulose conformations. It is confirmed first that there are two different cellulose forms in cellulose I and cellulose II, but there is no support in this work for the parallel/anti-parallel chain hypothesis. Rather, it is argued by Atalla that adjacent anhydroglucose units in cellulose are non-equivalent, that is cellulose is in reality a polymer of cellobiose (or more properly, anhydrocellobiose units), rather than a polymer of glucose (Figure 3). Further, it is proposed that there are three different and coexisting anhydrocellobiose conformations -- two ordered states, K1 and K2, and one disordered state, K0 -- which, in differing proportions account for various cellulose conformations (see Table 1). Thus, cellulose I contains a relative abundance of K1 anhydrocellobiose states, mercerized and regenerated celluloses contain more K2 anhydrocellobiose. Cellulose III is seen as a disordered state, and cellulose IV as a mixture of cellulose I and cellulose II (Atalla et al. 1977).

Table 1. Abundance of anhydrocellobiose states in various celluloses (Atalla 1982).

Native cellulose (I)	K1 > K0 > K2
Mercerized cellulose	K2 > K0 > K1
Regenerated cellulose (II)	K2 > K1 > K0
Cellulose III	K0 > K2 > K1
Cellulose IV	K2 > K1, K0 variable

This new interpretation of the nature of cellulose conformations is still too new to be assimilated and accepted or refuted by cellulose researchers. Concepts of molecular order affect our understanding of the nature of cellulose accessibility problems which are associated with this order. Celluloses differ in susceptibility to acidic and enzymatic agents, and these differences are usually attributed to relative abundances of inaccessible crystalline cellulose and accessible amorphous cellulose. Atalla has alternately suggested (Atalla 1979) that the different conformations of glycosidic linkages in cellulose are responsible. The basic difference between the accepted concept of crystallinity and that proposed by Atalla is respectively one of INTER-molecular order as opposed to INTRA-molecular order. Can it be that the nature of susceptibility is due to the presence of several anhydrocellobiose forms in celluloses which are hydrolyzable at differing rates, rather than due to relative abundances of crystalline and amorphous cellulose? For example, it is generally accepted that in native, highly ordered cellulose I, there is an association between the C6 hydroxyl group of one anhydroglucose unit and C3 hydroxyl group of another anhydroglucose unit. In the inter-molecular hypothesis, the two anhydroglucose units are in different but proximate cellulose chains. In the intra-molecular hypothesis, the anhydroglucose units are adjacent to each other in the same cellulose molecule. There are several similarities in the response of celluloses to acidic and enzymatic hydrolysis, which can be explained by either hypothesis.

For both acidic and enzymatic hydrolysis, a rapid initial conversion to glucose and cellodextrins (amorphous cellulose, or readily hydrolyzed glysocidic linkages ?) is followed by a period of relatively slower conversion, the rate of conversion in this second period depending on the prior history of the cellulosic substrate (relative crystallinity, or content of less readily hydrolyzed glycosidic linkages ?). In general, non-native forms of cellulose are degraded more rapidly in this second period. Further, glycosides in acid hydrolysis must assume a planar half-chair configuration in a carbonium ion intermediate. The resistance of crystalline cellulose to mild acid hydrolysis has been ascribed (Goldstein 1982) to the rigidity of anhydroglucose rings held tightly by intermolecular hydrogen bonds. It can be alternately hypothesized that the higher activation energy required for such hydrolysis is due to intramolecular hydrogen bonds.

Nature appears to add some support to the concept of cellulose as a polymer of anhydrocellobiose units. Cellulolytic enzymes, whether isolated by man for his use, or used by microorganisms for their use, convert celluloses into cellobiose first before using cellobiases to convert cellobiose into glucose. Many cellulolytic microorganisms prefer to grow on cellobiose. Only when cellobiose has been fully utilized will microorganisms generally turn to utilizing glucose (Humphrey 1979). Perhaps this extends to the enzymatic hydrolysis of other sugar polymers as well -- amalyases working on starch produce maltose, and not glucose (Soltes 1980).

Whichever hypothesis is correct, it is nonetheless true that cellulose does change conformation when subjected to various processing operations, and that cellulose in its different conformations exhibits different properties, including susceptibility to hydrolytic agents. Thus, it is important to recognize that there is not one cellulose, but a number of celluloses. To quote Durso (1982): "... cellulose is not a compound; rather it is a material whose chemical and physical usefulness depends on how it has been modified by all steps prior to the 'final' use. In other words, it has a memory which truly never forgets." Further, (Durso 1979): "...each of us trying to understand our investigations upon some aspect of this material must appreciate that we are observing an interaction of our devices upon a particular state of the substrate."

ACCESSIBILITY AND THE PLANT CELL WALL

In lignocellulosics we find, by experiment, regions of accessibility and inaccessibility to various digestive agents. Accessibility and inacccessibility of lignocellulosics are often equated to, and confused with, the relative susceptibility to hydrolysis of cellulose amorphous and crystalline regions, although the two may be different. First, accessibility and inaccessibility are not absolutes, that is, differing digestive agents, be they chemicals, enzymes or microogranisms, acting on a common cellulosic substrate, will usually digest that substrate at differing rates and/or to differing extents. Lignocellulosics can be accessible to acidic hydrolytic agents, but are for the most part inaccessible to enzymatic hydrolytic agents, unless milled to small particle sizes. It can be argued that milling reduces cellulose crystallinity, and thus increases accessibility. Shifts in crystallinity can occur: there is evidence in both acidic hydrolysis (Morehead 1950) and enzymatic hydrolysis (Caufield & Moore 1974) of isolated celluloses that there are shifts in the respective crystalline/amorphous ratios during hydrolysis; and at least during enzymatic hydrolysis with Trichoderma reesei cellulase, that the accessibility of the crystalline component is enhanced, and enhanced to a greater degree than the amorphous component. It may be wrong to assume, however, that the accessibility of lignocelluloses is the same as accessibility of isolated celluloses. Cellulose is not the only component in the plant cell wall. In mature plant cell walls, we find also lignin and hemicelluloses.

Lignin and the Plant Cell Wall

Lignin is often referred to as an "encrusting substance." It may be "encrusting" in the sense that the presence of lignin in lignified tissues imparts rigidity to plants, that its presence in conducting tissues decreases the permeation of water across cell walls and thus plays an important role in the intricate transport of water, nutrients and metabolites in the plant, that lignified tissues in the living plant effectively resist attack by microorganisms by impeding the penetration of destructive enzymes into the cell wall, and that in general, both for the needs of the plant and for any utilization purpose, it appears to play a role in limiting fiber accessibility. Lignin may also be "encrusting" in the sense that it is a "bonding agent", which in wood is part of a composite structure remarkably resistant towards impact, bending and compression forces (Sarkanen and Ludwig 1971).

There, however, appears to be a continung misconception of lignin as an "encrusting substance" in another sense. The cells of woody plants are bound together by a substance referred to as the middle lamella. This middle lamella is composed of highly lignified material, and is usually seen in microscopy as the compound middle lamella (includes the normally highly lignified primary cell wall, from which it is difficult to separate). The compound middle lamella is typically at least 70% lignin. The misconception that appears to persist is that this is where the bulk of the lignin is found. It was shown many years ago (Fergus 1968) that although the lignin concentration of wood secondary cell walls is significantly lower than that of the compound middle lamella, the volume of the secondary wall is such that the bulk of the lignin is actually found in the secondary wall. This observation has been confirmed in more recent work (Saka 1980).

The second and related misconception is that the inaccessibility of fibers is due to the impenetrability of this middle lamella which surrounds fibers. This appears to be true to a large extent in living plants. What is not apparent from examining the usual cross sections of cell walls by microscopy is that many fibrous structures are not open at their ends, that is, their lumens are not continuous. There are exceptions, of course, such as the vessel elements making up the pores of hardwood species. However, for the most part in living plants, the only access to cell walls is via the lumens of cell walls. For example, in the digestion of wood tissues by wood-destroying fungi, scanning electron microscopy provides ample evidence that cell walls are digested from the inside out, with little evidence suggesting disruption of the compound middle lamella by digestive agents. Isolated fibers, whose middle lamellae have been removed through physical or chemical treatment, are indeed more accessible, and this might suggest that the lignin "encrustation" of the middle lamella was responsible for the original inaccessibility of the lignified fibers.

Hemicellulose and the Plant Cell Wall

There is a third component of plant cell walls -- hemicellulose -- that may play a role in cell wall accessibility. Several studies have been reported on the topochemistry of delignification of spruce wood (Procter et al. 1967; Ahlgren and Goring 1971; Wood et al. 1972), and of Douglas-fir wood (Saka 1980). During kraft pulping, in which both lignin and some hemicellulose is removed, secondary wall lignin is preferentially removed in the initial stages of delignification, followed by rapid dissolution of lignin from the middle lamella (Procter et al. 1967). However, acid chlorite pulping, which is highly selective for lignin removal, that is, does not remove hemicellulose, removes lignin from secondary wall and middle lamella at the same relative rate (Ahlgren and Goring 1971). Further work by Saka (1980) on soda, soda-AQ and kraft pulping of Douglas-fir supports a conclusion that the topochemical effects of delignification of wood by various delignification agents are closely related to the abilities of these agents to remove hemicellulose. The removal of hemicellulose governs the pore size in the cell wall and thus the rate of delignification (Kerr and Goring 1975).

There are thus topochemical considerations that must be taken into account in attempts to explain plant cell wall accessibility. Goring has postulated (Goring 1977) the following model for the arrangement of cellulose, hemicellulose and lignin in the wood cell wall (Figure 4).

It is currently accepted that the lignin in the cell wall of wood is a crosslinked three-dimensional polymer which forms an amorphous network in which cellulosic material is embedded. This model by Goring is but one of several recently proposed in which the hemicelluloses in the cell wall are arranged, perhaps together with loosely packed "amorphous" cellulose, around a highly ordered cellulose core. Here, microfibrils, consisting of 2 to 4 protofibrils bonded on their radial surfaces, are arranged in a honeycomb pattern, with interrupted lamellae preferentially coplanar and parallel to the middle lamella, and are surrounded, at least on their tangential surfaces, by a cortex of oriented hemicelluloses, including perhaps amorphous cellulose, and the spaces between these interrupted lamellae are filled with lignin or a lignin-hemicellulose complex. This model was proposed to explain the action of pulping chemicals on wood, but may also serve to explain inaccessibility of lignocellulosics to other digestive agents. It can be argued that cellulose is "protected" by hemicellulose, that is, the "encrustation" is not due to lignin but hemicellulose. There are a number of reported observations which tend to support this hypothesis.

If lignin were responsible for inaccessibility, one might expect that a holocellulose, that is, wood that has been delignified but retains hemicellulose composition, would be accessible to cellulolytic enzymes. This does not appear to be the case. Work reported for the enzymatic hydrolysis of beechwood and sprucewood holocelluloses (Sinner et al. 1979) concludes that: "(1) a partial degradation of the hemicelluloses is imperative before the cellu-

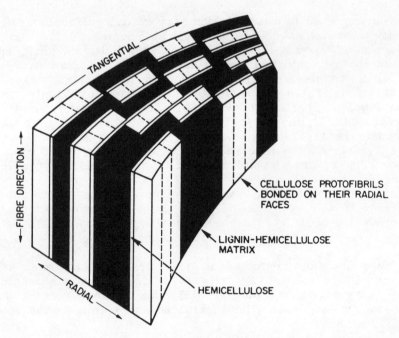

Figure 4. Schematic of the proposed interrupted lamella model for the ultrastructural arrangement of lignin and carbohydrate in the wood cell wall. (Reprinted with permission from Goring 1977; copyright 1977 American Chemical Society).

lose fibrils can be attacked; (2) the hemicelluloses seem to be deposited between the cellulose fibrils or even to be encrusting them; and, (3) the enzymatic hydrolysis of the cellulose is governed by the porosity of the tissue (enzyme diffusion), the impediment of the hemicelluloses, and the properties of the cellulose (e.g., crystallinity)."

In enzymatic hydrolysis of lignocellulosics, it is first necessary to grind such materials before accessibility to enzymatic hydrolytic agents is witnessed. It is commmon to explain that milling results in decrystallization of cellulose in lignocellulose, thus making it more accessible. It has been shown (Caulfield and Moore 1974) that particle size reduction and surface area increase upon ball milling are probably more important than the reduction of crystallinity. Another study on the influence of cellulose physical structure on thermohydrolytic, hydrolytic and enzymatic degradation of cellulose (Philipp et al. 1979) concludes that there is a strong influence of pore structure and fibrillar morphology on enzymatic hydrolysis. The cellulose of red oak sawdust is converted enzymatically to the extent of 6%. However, vibratory milling for 240 minutes resulted in a material in which the carbohydrates of red oak were 93% convertible to sugar (Millett et al. 1979). Similar results were reported for Douglas-fir and newsprint. Lignin is not removed in milling. Milling, however, can disrupt hemicellulose "encrustation" and expose cellulose surfaces for enzymatic hydrolysis.

In acid hydrolysis of lignocellulosics, relative to enzymatic hydrolysis, the cellulose appears to be more readily accessed. It is well known that hemicelluloses in plant cell walls are readily hydrolyzed by acidic reagents, even under relatively mild conditions. In mild acid hydrolysis of wood, a prehydrolysis step removes hemicellulose as its hydrolyzed and soluble monosaccharides, and creates cellulose surfaces for hydrolytic action.

Water and The Plant Cell Wall

There is still yet another component of plant cell walls - water. Water is a key component not only in life processes for the plant, but also in our efforts to use biomass materials for other purposes. As was shown previously (Cote 1982), pits are used by plant cells for communication, and the bordered pits of conifers can aspirate, that is become unavailable for fluid transport when wood is dried, thus affecting water permeability (Thomas 1982). There are additional problems created when wood or wood-derived cellulosics are dried, that are unrelated to the permeability loss associated with aspirated pits. Microorganisms in vivo, such as various rots, generally cannot use, or at least find difficulty in using, dried lignocellulose. We often process biomass materials in ways which make it difficult to gain access to desired components. In the pulping and bleaching of wood, when lignin and hemicellulose are removed, fibers "collapse", and it becomes necessary to "activate" the cellulose in order to prepare cellulose ethers and esters (Durso 1976; Parks 1959; Nelson and Oliver 1971). To a large extent the collapse and subsequent inaccessibility is due to the removal of water. As most cellulose chemists are aware, there is a big difference in accessibility between never-dried and even once-dried cellulosic materials. It has been shown that cellulose ethers can be prepared directly from never-dried defibrated hardwoods and never-dried unbleached pulps (Durso 1976; 1981) and from never-dried sorghum bagasse (Soltes 1982), but only with difficulty when such materials are dried prior to use. There are additionally differences in accessibility of dried vs. never-dried biomass materials to aqueous and enzymatic agents and microorganisms, whether in nature or when applied in vitro for our purposes. See Table 2. Never-dried pulp is much more accessible to both acid and enzymes. Wet beating of the air-dry pulp does not recreate the never-dried pulp state.

The effect of drying is again illustrated in the following work which demonstrates that cellulose can be removed from never-dried lignocellulosics as a soluble derivative. Durso (1978; 1982) has categorized cellulose derivatives on the basis of whether or not reactions retain or destroy the original morphology of cellulose. The two types and representative cellulose products, are given in Table 3.

Table 2. Influence of drying on acid and enzymatic hydrolysis of beech pulp (Philipp et al. 1979).

Sample	Zero-Order Rate Constant (/DP/hr x 1000) 7.5% HCl/100C	% Residue after 68 hr/40C Treatment with T.viride cellulase
Never-dried	0.25	<1
Air dry	15	19
Air dry + wet beating	8	8.6
Air dry + 2hr/140C	22	40
Freeze-dried	22	-

(Reprinted with permission from Philipp et al. 1979. Copyright 1979 American Chemical Society.)

For the preparation of cellulose products of Type I, the lignocellulose matrix must only be sufficiently "expanded" to accomodate the derivative group and/or made "plastic" enough to form the final shape. For the Type II products, the lignocellulose matrix must be in solution form. This suggests that it should be possible, for the preparation of cellulose products of Type I, to use lignocellulosic materials instead of dissolving pulps, swell them to allow the introduction of derivatizing agents which react with cellulose, and "dissolve out" cellulose as a soluble derivative. This is apparently the case: cellulose derivatives can be produced and extracted directly from never-dried wood (Durso, 1981). As an added benefit of this approach, because cellulose is not subjected to harsh dissolving pulp processing, there is a significant improvement in yield and solution properties, while minimizing DP loss. For the purpose of this discussion on the nature of accessibility, it is important to add that use of dried wood results in decreased accessibility of the cellulose to the derivatizing agent leading to poor yields and substitution levels.

Any discussion of plant cell wall accessibility would be incomplete if cellulolytic microorganisms were not covered. As the most important of these is related to ruminant nutrition, we will restrict this discussion to plant cell wall accessibility to rumen microorganisms.

Table 3. Cellulose derivatives (Durso 1982).

TYPE I: Cellulose Derivatives Retaining Original Morphology.

    cellulose nitrate   cellulose sulfate
    methyl cellulose   ethyl cellulose
    carboxymethyl cellulose   cyanoethyl cellulose
    ion-exchange cellulose   hydroxyethyl cellulose
    microcrystalline cellulose   graft copolymers
    cross-linked celluloses   cellulose triacetate

TYPE II: Cellulose Derivatives with Original Morphology Destroyed.

    cellulose xanthate   culprammonium cellulose
    cellulose acetate   vulcanized fiber

RUMEN DIGESTIBILITY OF PLANT CELL WALLS

The nutritive value of forages for ruminants depends on the abilities of rumen microorganisms to degrade plant cell walls and to ferment the available carbohydrates. The metabolic products of this fermentation, and the microbes themselves, provide energy and protein for the ruminant (Hungate 1966). However, data from *in vitro* and *in vivo* investigations indicate that forages vary in the degree to which they are degraded and utilized by the ruminant.

Scores of papers have been written concerning the chemical and physical factors affecting rumen digestibility. Gross chemical compositions of whole plant substance, especially % lignin, % silica, detergent extractions, etc., are still used by animal nutrition scientists as measures of digestible content. However, it is becoming increasingly apparent that different plant parts exhibit varying digestibilities, and more recent work emphasizes the differences in chemical and physical composition of these plant parts. Scanning electron microscopy has assisted in these investigations by visualizing, and through micrographs, recording, the actual digestion of various plant species and their component plant parts.

The general chemical composition of plants, again, reflects a mixture of chemical entities derived from the non-structural (cell content) and structural (cell wall) components of plants. The soluble, non-structural cell contents of plants is highly digestible in the rumen, and represent a uniform nutritional entity which is 95% digestible (Waldo et al. 1972; Mertens 1977). The structural components, and especially cellulose, however, exhibit a much larger variation in potential for digestion. A ruminant, through its rumen microflora, is limited in its ability to degrade and digest certain types of cell wall structures, as is readily apparent in the content of fiber in rumen feces. Figure 5 gives a series of scanning electron micrographs of cross sections of coastal bermudagrass subjected to in vitro digestion by rumen microoganisms (Soltes et al. 1980). It is seen that the vascular bundles containing lignified thick-walled xylem tissues are not digested by the ruminant microflora. The cellulose content of these tissues represents a significant, and for some species, major portion of total cellulose in the plant.

The Role of Lignin in Cellulose Accessibility to Ruminants

All structural carbohydrates, if in their lignocellulose matrix, are then not completely digestible to ruminants. One of the traditional factors associated with indigestibility, that is what scientists usually hold responsible, is the lignin content of cell walls. It is well established that with increasing maturity there is both a decrease in digestibility and an increase in lignin content (Kamstra et al. 1958; Tomlin et al. 1965; Smith et al. 1972). For any given plant species containing lignified tissues, juvenile tissues give lower lignin assays than mature tissues. Lignin is deposited in the later stages of plant wall development, although there appears to be some confusion in the literature as to whether this deposition occurs during or after secondary wall thickening. For some plants, especially non-woody plants, it has been reported that lignin deposition is the last stage of cell wall development, that is, cell walls are more or less fully developed before lignification. This situation may be subject to reinterpretation because recent work proves that, at least for two woody species, lignification of cell walls is coincident with cell wall thickening (Saka 1980).

Disruption of cellulose and hemicellulose digestion by lignin has been reported (Dehority et al. 1962; Wilkins 1972). However, lignin-carbohydrate complexes differ among forage species (Allison and Osborn 1970), and although the increased lignin content with maturity appears to be highly associated with digestibility within species, lignin content offers a poor measure of digestibility in comparisons between species. For example alfalfa, and legumes in general, exhibit lignin contents higher than that for grasses of similar digestibility (Sullivan 1964). Although it had been suggested earlier (Van Soest 1965) that differences in digestibility between forage species could be predicted by relating lignin/cellulose ratios, more recent work (Mertens 1977) has shown a nonsignificant correlation between digestion and lignin ratio for a diverse array of forage species.

It is clear only that lignin content does increase with plant maturity and that digestibility does decrease with plant maturity. The author suggests that also coincident with increasing lignin content in plant maturation is cell wall development, and that some parallels can be drawn between the ruminant's abilities and the bio-

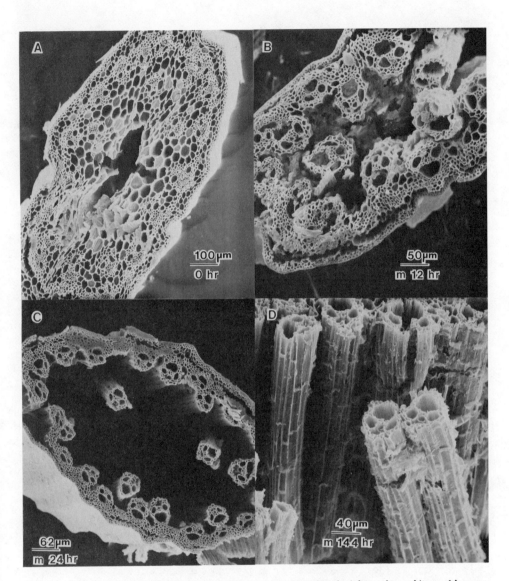

Figure 5. Scanning electron micrographs depicting in vitro digestion of cell wall structures in coastal bermudagrass by rumen microorganisms. A-original structures, B-after 12 hr digestion, C-after 24 hr digestion, D-after 144 hr digestion. (Soltes et al. 1980).

mass conversion scientist's attempts to access cellulose from lignocellulosics for their respective purposes.

ACCESSIBILITY VS. PLANT MATURITY IN LIGNOCELLULOSICS UTILIZATION

When lignocellulosics are utilized for their carbohydrate content by microorganisms, and higher life forms including all monogastric animals, juvenile tissues are found to be more digestible than mature tissues. This is true for humans also: we can eat bamboo shoots but would have problems in digesting bamboo; we can experience some discomfort in eating mature vegetables, such as radishes and turnips, onions and asparagus, to name a few. A horse as a monogastric animal will have some problems with more mature hay. What has this to do with lignocellulosics utilization for feed, fuels and chemicals?

A biomass conversion scientist, in attempting to digest wood, or agricultural products in general, will find that some types of cell walls are relatively inaccessible. The resource which we call biomass normally encompasses mature underutilized plants, residues in the harvest of useful plants, and residues in the processing of the harvested portion of those plants. Invariably, these residues are composed of mature tissues, often stems, of plants. Stems are generally unpalatable and indigestible to animals, and it was known long ago (Johnston-Wallace 1937) that leaves are preferred to stems by grazing animals. It is very tempting to conclude that there is operative the same cause and effect relationship. It is very tempting to suggest also that the plant wants it that way. Nature could have designed certain wall structures in plants, especially those fibrous, structural elements responsible for supporting the plant, to be unpalatable to herbivores and resistant to microorganism attack -- and for obvious reasons in the living plant. Many plants die when their stems are cut or diseased. Could it be that this inherent natural inaccessibility, or relative indigestibility, of residues is responsible for the difficulties experienced when the plant is harvested for some utilization purpose by the ruminant and conversion scientist alike?

There may be strong incentive to develop new, non-traditional uses for biomass residue materials as we concern ourselves with supplying our material needs from renewable resources. However, if there is a strong inverse correlation between accessibility and plant maturity, then we may be looking at the wrong biomass resource. The utility of any plant in normal agricultural activity is normally tied to harvest of a portion of that plant when it reaches maturity. The utility of a plant for cellulosics utilization, however, may be tied to the harvesting of juvenile tissue. If it is sugar we are after, why wait until the plant metabolizes the sugar into less accessible forms? For example, it is known that in many tuberous plants, such as the Jerusalem artichoke, much of the sugar needed to grow the tuber is held in the above ground portion of the plant until it is relocated into the tuber as a polysaccharide just before maturity. How much easier it is to harvest the above ground portion of the plant just before maturity, and extract simple accessible sugars, with the added benefit that the juvenile tuber is ready to reshoot for further growth, for additional harvest. If it is cellulose we are after, why wait until mature tissues are formed? How much easier, for example, it is to harvest juvenile coppicing woody species containing wood with higher accessibility, and leave root stock for regrowth for future harvests.

Additionally, we bemoan the fact that biomass is usually wet when harvested, about 50% water. This water is considered unnecessary in biomass utilization, a burden to transportation, and there is currently research into the preprocessing of biomass at the site of production to remove most of the water to effect savings in transportation costs. As we have discussed, microorganisms _in vivo_, or _in vitro_ for our purposes, generally cannot use, or at least find difficulty in using, dried lignocellulose. Never-dried materials usually exhibit superior accessibility to dried materials. Combining several thoughts, it would appear that the biomass resource we should be looking at is that obtained in the harvest of juvenile tissues, and that which is kept wet or at least held at not too much below the fiber saturation point. Some attention to these factors should result in biomass streams with improved accessibilities.

How do we obtain such biomass streams? It may be easy enough to process green, wet biomass, but where can we find a juvenile biomass resource? Plant breeders have traditionally been concerned only with the breeding of plants which offer high yields of primary agricultural products, and these in physical forms which are readily harvested. As we demand more of agriculture, it is encouraging to note that some plant breeders are now starting to look at the properties of the whole plant, as well as those of the plant's component parts, and considering those properties which affect biomass utilization on a larger front (Soltes 1980; Clark et al. 1981). It is possible to identify varieties of row crops, for example, which offer higher stem digestibilities. It is also possible to identify fast growing annual and perennial plants which can be harvested several times each year, with total biomass production in excess of that obtained if the plant were allowed to mature before harvesting. The product of these harvests would additionally be in a juvenile form assisting conversion efforts by scientists and ruminants alike. Perhaps with such assistance from plant breeders, we may one day harvest agricultural crops which have been designed for maximum utility for food, feed, fiber, chemical and fuel values, rather than food or feed alone.

CONCLUSIONS

Cellulose is a versatile raw material, a raw material which holds promise as a renewable resource for the production of many chemicals and materials needed or desired by man. The elusive lignocellulosic matrix of the plant cell wall, in which most of natural cellulose is found, will continue to tax the prowess of researchers until cost-effective solutions can be found to liberate cellulose from this matrix. Only time and the imaginations of man are limiting factors to cellulose utilization.

ACKNOWLEDGMENTS

The author wishes to acknowledge the assistance of several former graduate students of the Animal Science and Soil and Crop Sciences Departments at Texas A & M University who, through informative conversations over the past three years, provided him with an enhanced appreciation of problems in cellulose utilization. Special thanks are due to J. Clark, W.D. Pitman, and K.R. Pond.

REFERENCES

Ahlgren, P.A. and D.A.I. Goring. 1971. Removal of wood components during chlorite delignification of black spruce. Can.J.Chem. 49:1272-75.

Allison, D.W. and D.F. Osbourn. 1970. The cellulose-lignin complex in forages and its relationship to forage nutritive value. J.Agr.Sci. 74:23.

Atalla, R.H., B.E. Dimick and S.C. Murphy. 1977. Studies on polymorphy in cellulose: Cellulose IV and some effects of temperature. In: Cellulose Chemistry and Technology (J.C. Arthur, Jr., Ed.) Washington, D.C., American Chemical Society Symposium Series No. 48, 30-41.

Atalla, R.H. 1979. Conformational effects in the hydrolysis of cellulose. In: Hydrolysis of Cellulose: Mechanisms of Enzymatic and Acid Catalysis (R.D. Brown, Jr. and L. Jurasek, Eds.) Washington, D.C., American Chemical Society Advances in Chemistry Series No. 181, 55-69.

Atalla, R.H. 1982. The structures of cellulose and their transformations. Paper presented at the symposium of Feed, Fuels and Chemicals from Wood and Agricultural Residues, E.J. Soltes, org., American Chemical Society National Meeting, Kansas City, MO, Sept. 1982, to be published in: Wood and Agricultural Residues: Research on Use for Feed, Fuels and Chemicals (E.J. Soltes, Ed.) Academic Press, New York.

Blackwell, J., F.J. Kolpak and K.H. Gardner. 1977. Structures of native and regenerated celluloses. In: Cellulose Chemistry and Technology (J.C. Arthur, Jr., Ed.) Washington, D.C., American Chemical Society Symposium Series No. 48, 42-55.

Caulfield, D.F. and W.E. Moore. 1974. Effect of varying crystallinity of cellulose on enzymic hydrolysis. Wood Sci. 6:375-379.

Clark, J., E.J. Soltes and F.R. Miller. 1981. Sorghum -- a versatile multi-purpose biomass crop. Biosources Digest 3:35-51.

Côté, W.A., Ed. 1965. Cellular Ultrastructure of Woody Plants. Syracuse University Press, Syracuse, N.Y.

Côté, W.A. 1983. The anatomy, ultrastructure and chemical composition of wood. In: this volume.

Cowling, E.B. and T.K. Kirk. 1976. Properties of cellulose and lignocellulosic materials as substrates for enzymatic conversion processes. Biotechnol. and Bioeng. Symp. No.6, 95-123.

Dehority, B.A., R.R. Johnson and H.R. Conrad. 1962. Digestibility of forage hemicellulose and pectin by rumen bacteria in vitro and the effect of lignification thereon. J. Dairy Sci. 45: 508.

Detroy, R.W., R.L. Cunningham and A.I. Herman. 1982. Fermentation of wheat straw xylans to ethanol by Pachysolen tannophilus. Paper presented at the Fourth Symposium on Biotechnology in Energy Production and Conservation, Gatlinburg, TN, May 1982, to be published in: Biotechnol. and Bioeng. Symp. No. 12.

Durso, D.F. 1976. Cellulose ethers directly from defibrated hardwoods. Svensk Papperstidn. 79:50-1.

Durso, D.F. 1978. Chemical modification of cellulose - a historical review. In: Modified Cellulosics (R.M. Rowell and R.A. Young Eds.) Academic Press, New York, 23-37.

Durso, D.F. 1981. Process for the preparation of cellulose ether derivatives. U.S. Patent 4,254,258 (Mar. 31, 1981).

Durso, D.F. 1982. Cellulose derivatives ... arranging for the future. Paper presented at the symposium on Feed, Fuels and Chemicals from Wood and Agricultural Residues, E.J. Soltes, org., American Chemical Society National Meeting, Kansas City, MO, Sept. 1982, to be published in: Wood and Agricultural Residues: Research on Use for Feed, Fuels and Chemicals (E.J. Soltes, Ed.) Academic Press, New York.

Falkehag, I. 1977. Utility of organic renewable resources. In: Engineering Implications of Chronic Materials Scarcity, proceedings of the Henniker IV Conference, Aug. 1976, Office of Technology Assessment, Washington, D.C., 178-209.

Fergus, B.J. 1968. Lignin Distribution and Delignification of Xylem Tissue. Ph.D. Dissertation, McGill University, Montreal, Canada.

Gharpurapy, M.M., L.T. Fan, and Y.H. Lee. 1982. Caustic pretreatment study for enzymatic hydrolysis of wheat straw. Paper presented at the symposium on Feed, Fuels and Chemicals from Wood and Agricultural Residues, E.J. Soltes, org., American Chemical Society National Meeting Kansas City, MO., Sept. 1982, to be published in: Wood and Agricultural Residues: Research on Use for Feed, Fuels and Chemicals (E.J. Soltes, Ed.) Academic Press, New York.

Goldstein, I.S. 1972. The potential for converting wood into plastics and polymers and into chemicals for the production of these materials. NSF-RANN Report, National Science Foundation, Washington, D.C.

Goldstein, I.S. 1977. The place of cellulose under energy scarcity. In: Cellulose Chemistry and Technology (J.C. Arthur, Jr., Ed.) Washington, D.C., American Chemical Society Symposium Series No. 48, 382-387.

Goldstein, I.S. 1978. Chemicals from wood: outlook for the future. Eight World Forestry Congress, Jakarta, Indonesia.

Goldstein, I.S. 1981a. Composition of biomass. In: Organic Chemicals from Biomass (I.S. Goldstein, Ed.) CRC Press, Boca Raton, Fla., 9-18.

Goldstein, I.S. 1981b. Chemicals from cellulose. In: Organic Chemicals from Biomass (I.S. Goldstein, Ed.) CRC Press, Boca Raton, Fla., 101-124.

Goldstein, I.S. 1982. Acid processes for cellulose hydrolysis and their mechanism. Paper presented at the symposium on Feed, Fuels and Chemicals from Wood and Agricultural Residues, E.J. Soltes, org., American Chemical Society National Meeting, Kansas City, MO, Sept. 1982, to be published in: Wood and Agricultural Residues: Research on Use for Feed, Fuels and Chemicals (E.J. Soltes, Ed.) Academic Press, New York.

Goring, D.A.I. 1977. A speculative picture of the delingification process. In: Cellulose Chemistry and Technology (J. C. Arthur, J.r, Ed.) Washington, D.C., American Chemical Society Symposium Series No. 48, 273-277.

Hall, J.A., J.F. Saeman and J.F. Harris. 1956. Wood saccharification: a summary statement. Unasylva 10:7-32.

Han, Y.W., E.A. Catalano and A. Ceigler. 1982. Treatments to improve the digestibility of crop residues. Paper presented at the symposium on Feed, Fuels and Chemicals from Wood and Agricultural Residues, E.J. Soltes, org., American Chemical Society National Meeting, Kansas City, MO, Sept. 1982, to be published in: Wood and Agricultural Residues: Research on Use for Feed, Fuels and Chemicals (E.J. Soltes, Ed.) Academic Press, New York.

Huber, J.T., A. Hargreaves, C.O.L.E. Johnson and A. Shanan. 1982. Upgrading residues and by-products for ruminants. Paper presented at the symposium on Feed, Fuels and Chemicals from Wood and Agricultural Residues, E.J. Soltes, org., American Chemical Society National Meeting, Kansas City, MO, Sept. 1982, to be published in: Wood and Agricultural Residues: Research on Use for Feed, Fuels and Chemicals (E.J. Soltes, Ed.) Academic Press, New York.

Humphrey, A.E. 1979. The hydrolysis of cellulose materials to useful products. In: Hydrolysis of Cellulose: Mechanisms of Enzymatic and Acid Catalysis (R.D. Brown, Jr. and J. Jurasek, Eds.) Washington, D.C., American Chemical Society Advances in Chemistry Series No. 181, 25-54.

Hungate, R.E. 1966. The Rumen and Its Microbes. Academic Press, New York.
Jeffries, T.W. 1982. A comparison of Candida tropicalis and Pachysolen tannophilus for the conversion of xylose to ethanol and other products. Paper presented at the Fourth Symposium on Biotechnology in Energy Production and Conservation, Gatlinburg TN, May 1982, to be published in Biotechnol. and Bioeng. Symp. No. 12.
Johnson-Wallace, D.B. 1937. The influence of grazing management and plant associations on chemical composition of pasture plants. J. Am. Soc. Agron. 29:441.
Kamstra, L.D., H.L. Moyon and O.G. Bently. 1958. The effect of stage of maturity and lignification on the digestion of cellulose in forage plants by rumen microorganism in vitro. J. Anim. Sci. 17:199.
Kerr, A.J. and D.A.I. Goring. 1975. The role of hemicelluloses in the delignification of wood. Can. J. Chem. 53:952-9.
Klausmeier, W.H. 1982. Configurations for a forest refinery. Paper presented at the conference on Feed, Fuels and Chemicals from Wood and Agricultural Residues, E.J. Soltes, org., American Chemical Society National Meeting, Kansas City, MO, Setp. 1982, to be published in Wood and Agricultural Residues: Research on Use for Feed, Fuels and Chemicals (E.J. Soltes, Ed.) Academic Press, New York.
Kolpak, F.J. and J. Blackwell. 1976. Determination of the structure of cellulose II. Macromolecules 9:273-278.
Kolpak, F.J., M. Weih and J. Blackwell. 1978. Mercerization of cellulose: 1. Determination of the structure of mercerized cotton. Polymer 19:123-131.
Lastick, S.M., D. Spindler and K. Grohmann. 1982. Fermentation of cellulose materials by yeasts. Paper presented at the symposium on Feed, Fuels and Chemicals from Wood and Agricultural Residues, E.J. Soltes, org., American Chemical Society National Meeting, Kansas City, MO, Sept. 1982, to be published in: Wood and Agricultural Residues: Research on Use for Feed, Fuels and Chemicals (E.J. Soltes, Ed.) Academic Press, New York.
Lipinsky, E.S. 1982. Disruption and fractionation of lignocellulose. Paper presented at the symposium on Feed, Fuels and Chemicals from Wood and Agricultural Residues, E.J. Soltes, org. American Chemical Society National Meeting, Kansas City, MO, Sept. 1982, to be published in: Wood and Agricultural Residues: Research on Use for Feed, Fuels and Chemicals (E.J. Soltes, Ed.) Academic Press, New York.
Marchessault, R.H. and S. Malhotra. 1982. The wood explosion process: characterization and uses of lignin and cellulose. Paper presented at the symposium on Feed, Fuels and Chemicals from Wood and Agricultural Residues, E.J. Soltes, org., American Chemical Society National Meeting, Kansas City, MO, Sept. 1982, to be published in: Wood and Agricu-

ltural Residues: Research on Use for Feed, Fuels and Chemicals (E.J. Soltes, Ed.) Academic Press, New York.

Mertens, D.R. 1977. Dietary fiber components: relationship to the rate and extent of ruminal digestion. Fed. Proc. 36:187.

Millett, M.A., A.J. Baker and L.D. Satter. 1976. Physical and chemical pretreatments for enhancing cellulose saccharification. Biotechnol. and Bioeng. Symp. No. 6., 125-53.

Millett, M.A., M.J. Effland and D.F. Caulfield. 1979. Influence of find grinding on the hydrolysis of cellulosic materials -- acid vs. enzymatic. In: Hydrolysis of Cellulose: Mechanisms of Enzymatic and Acid Catalysis (R.D. Brown, J. and L. Jurseak, Eds.) Washington, D.C., American Chemical Society Advances in Chemistry Series No. 181, 71-90.

Morehead, F.F. 1950. Ultrasonic disintegration of cellulose fibers before and after hydrolysis. Text. Res. J. 20:549-553.

Muzzy, J.D., R.S. Roberts, C. Fieber, S. Fass and T. Mann. 1982. Pretreatment of hardwood by continuous steam hydrolysis. Paper presented at the symposium on Feed, Fuels and Chemicals from Wood and Agricultural Residues, E.J. Soltes, org. American Chemical Society National Meeting, Kansas City, MO, Sept. 1982, to be published in: Wood and Agricultural Residues: Research on Use for Feed, Fuels and Chemicals (E.J. Soltes, Ed.) Academic Press, New York.

Nelson, R. and D.W. Oliver. 1971. Study of cellulose structure and its relation to reactivity. J. Polymer. Sci., Part C 36:305.

Parks, L.R. 1959. Classification of pulps according to supermolecular structure of cellulose. Tappi 42:317 .

Philipp, B., V. Jacopian, F. Loth, W. Hirte and G. Schulz. 1979. Influence of cellulose physical structure on thermohydrolytic, hydrolytic and enzymatic degradation of cellulose. In: Hydrolysis of Cellulose: Mechanisms of Enzymatic and Acid Catalysis (R.D. Brown, Jr. and L. Jurasek, Eds.) Washington, D.C., American Chemical Society Advances in Chemistry Series No. 181, 127-144.

Procter, A.R., W.Q. Yean and D.A.I. Goring. 1967. The topochemistry of delignification in kraft and sulfite pulping of spruce wood. Pulp Paper Mag. Can. 68:T445-53.

Saka, S. 1980. Lignin Distribution as Determined by Energy Dispersive x-ray Analysis. Ph.D. Dissertation, North Carolina State University, Raleigh, 138pp.

Sarkanen, K.V. 1980. Acid-catalyzed delignification of lignocellulosics in organic solvents. In: Progress in Biomass Conversion, (K.V. Sarkanen and D.A. Tillman, Eds.) Academic Press, New York, Vo. 2, 127-144.

Sarkanan, K.V. and C.H. Ludwig. 1971. Lignins: Occurrence, Formation, Structure and Reactions. Wiley-Interscience, New York, 916pp.

Seeley, D.B. 1976. Cellulose saccharification for fermentation industry aplications. Biotechnol. and Bioeng. Symp. No. 6, 285-292.

Selke, S., M.C. Hawley and D.T.A. Lamport. 1982. Reaction rates for liquid phase HF saccharification of wood. Paper presented at the symposium on Feed, Fuels and Chemicals from Wood and Agricultural Residues, E.J. Soltes, org., American Chemical Society National Meeting, Kansas City, MO, Sept. 1982, to be published in: Wood and Agricultural Residues: Research on Use for Feed, Fuels and Chemiscals (E.J. Soltes, Ed.) Acdemic Press, New York.

Sinner, M., N. Parameswaran and H.H. Dietrichs. 1979. Degradation of delignified sprucewood by purified mannanase, xylanase and cellulose. In: Hydrolysis of Cellulose: Mechanisms of Enzymatic and Acid Catalysis (R.D. Brown, J.r and L. Jurasek, Eds.) Washington, D.C., American Chemical Society Advances in chemistry Series No. 181. 303-330.

Sjostrom, E. 1981. Wood Chemistry: Fundamentals and Applications. Academic Press, New York. p.4.

Smith, L.K., H.K. Goering, and C.H. Gordon. 1972. Relationships of forage compositions with rates of cell wall digestion and indigestibility of cell walls. J. Dairy. Sci. 55:1140.

Soltes, E.J., S.C.K. Lin and K.R. Pond. 1980. Scanning electron microscopy of the in vitro digestion of coastal bermudagrass by rumen microorganisms and the enzymes of Trichoderma reesei. Unpublished manuscript. Forest Science Laboratory, Texas A&M University, College Station, TX.

Soltes, E.J. 1980. Alternate feedstocks for ethanol production. In: Alcohol Fuels Symposium Proceedings, Center for Energy and Mineral Resources, Texas A&M University, College Station, TX.

Soltes, E.J. 1982. Unpublished results.

Sullivan, J.T. 1964. The chemical composition of forages in relation to digestibility by reminants. USDA, ARS 34-64.

Thomas, R.J. 1982. Wood anatomy and permeability. Paper presented at the symposium on Feed, Fuels and Cehmicals from Wood and Agricultural Residues, E.J. Soltes, org., American Chemical Society National Meeting, Kansas City, MO, Sept. 1982, to be pulished in: Wood and Agricultural Residues: Research on Use for Feed, Fuels and Chemicals (E.J. Soltes, Ed.) Academic Press, New York.

Tomlin, D.C., R.R. Johnson and B.A. Dehority. 1965. Relationship of lignification to in vitro cellulose digestibility of grasses and legumes. J. Anim. Sci. 24:161.

Turbak, A.F. 1982. Newer cellulose solvent systems. Paper presented at the symposium on Feed, Fuels and Chemicals from Wood and Agricultural Residues, E.J. Soltes, org., American Chemical National Meeting, Kansas City, MO, Sept. 1982.

Van Soest, P.J. 1965. Symposium on Factors Influencing the Voluntary Intake of Herbage by Ruminants: voluntary intake in relation to chemical composition and digestibility. J. Anim. Sci. 24:834.

Waldo, D.R., L.W. Smith and E.L. Cox. 1972. Model of cellulose disappearance from the rumen. J. Dairy Sci. 55:125.

Wilkins, R.J. 1972. The potential digestibility of cellulose in grasses and its relationship with chemical and anatomical parameters. J. Agric. Sci. 78:457.

Wood, J.R., P.A. Ahlgren and D.A.I. Goring. 1972. Topochemistry in the delignification of spruce wood. Svensk Papperstidn. 75: 15-9.

SOME STRUCTURAL CHARACTERISTICS

OF ACID HYDROLYSIS LIGNINS

>John Papadopoulos

>Centre Technique de l'Industrie
>des Papiers, Cartons et Celluloses
>Domaine Universitaire
>38020 Grenoble Cedex
>FRANCE

INTRODUCTION

Among the three major wood components lignin is the least utilized by the chemical industry. Most of the proposed wood hydrolysis schemes were optimized for the production of a single product, mainly the production of glucose, underrating any possible benefits gained by the utilization of lignin. The potential from the utilization of lignin to the overall economics of any wood hydrolysis process was recognized only a few years ago (Goldstein 1975, 1979) and since then considerable effort has been devoted towards the development of integrated wood conversion processes.

Lignin can be used either in its macromolecular form or degraded to lower molecular weight aromatic components. In the presence of acids, degradation of lignin occurs, resulting in the formation of low molecular weight phenolic products (Wallis 1971). However, the liberated lignin fractions do not accumulate but react through self condensation with the formation of higher molecular weight adducts (Scheme 1) (Lai and Sarkanen 1971). Condensation reactions of lignin become of particular importance since they, to a considerable extent, govern further utilization of lignin in the degraded or macromolecular form (Falkehag 1975).

The importance of the characterization of the acid hydrolysis lignins was recognized many years ago and several reports have appeared in the literature since then describing structural features and properties of these lignins (Lai and Sarkanen 1971, Winston et al. 1981, Goldstein et al. 1981). However, with few exceptions no

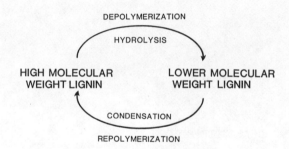

Scheme 1. Schematic representation of the major competing reactions occurring to lignin in presence of acidic media.

attention was given to the systematic and comparative characterization of lignin residues generated from treatment of the same wood species with various acids and under different reaction conditions. This report deals with the characterization of the lignin residues isolated from dilute sulfuric, concentrated hydrochloric acid and anhydrous hydrogen fluoride treatment of birch wood (<u>Betula</u> <u>verrucosa</u> L.) at various hydrolysis conditions as well as the comparison of the lignin residue isolated with different acids under conditions permitting optimum recovery of carbohydrates.

RESULTS AND DISCUSSION

In its native form, lignin is a three dimensional, highly branched, amorphous macromolecule. Because of its insolubility in common organic solvents and appreciable carbon-carbon bonding between phenylpropane units, its detailed structure is not very explicity defined. However, a considerable amount of knowledge about its structure has been accumulated over the years through biosynthetic and degradative investigations. Since the lignin residues generated from the acid hydrolysis processes are also insoluble in common solvents, it seemed proper to adopt characterization techniques similar to these employed in the past, under well defined experimental conditions. Thus, the extent of lignin condensation was estimated by nitrobenzene oxidation while the differences in depolymerization among the various lignin preparations were defined on the basis of the new functional groups generated, and in particular, new phenolic hydroxyl groups.

Preparation of Lignin Residues

It is well known that lignin residues isolated with different acids and under varying reaction conditions exhibit considerable diversity in their structure. Structural differences are likely to affect their properties. Consequently any comparison of their performance on a random basis becomes quite uncertain. Considering that all acid hydrolysis processes aim also at high yields of sugars, reaction conditions for the isolation of the various lignin residues were chosen that enable recovery of the optimum quantity of carbohydrates.

For dilute sulfuric acid, prehydrolysis was carried out with 0.5% sulfuric acid at 140° C for one hour and hydrolysis with the same acid concentration at 180° C for two hours. The yield for xylose and glucose were 49 and 61 percent of the original present in birch wood, respectively.

The HCl-lignin residues were prepared by prehydrolysis of birch wood meal with 30% hydrochloric acid for 90 minutes, followed by hydrolysis with 43% HCl at 50° C for two hours. The yields for xylose and glucose were 93 and 87 percent of the original present in the wood, respectively.

Finally, the HF lignin residue was isolated by treating birch wood with anhydrous hydrogen fluoride at ambient temperature for 30 minutes. The yields of the mixture of xylose and glucose were 91 and 93 percent, respectively, of the original present in the wood.

Alkaline Nitrobenzene Oxidation

Alkaline nitrobenzene oxidation has long been recognized as an effective method for characterizing the extent of carbon-carbon linkage formation of the various lignin preparations. The oxidation is carried out with nitrobenzene, under alkaline conditions (2 N NaOH) in pressure autoclaves at elevated temperatures (170° C). Under these conditions the various phenylpropane units which are joined by linkages other than carbon-to-carbon are mainly oxidized to aromatic aldehydes (vanillin and syringaldehyde). Formation of new carbon-to-carbon linkages is followed by decrease in the yield of total aldehydes and this can be used as an indirect determination of the extent of lignin condensation (Kratzl and Gratzl 1960).

Although condensation of lignin with carbohydrate decomposition products, formed under acid catalyzed conditions, cannot be excluded, it seems likely that internal condensation of the lignin moieties is more profound. Carbon-to-carbon linkages can be formed between aromatic nuclei, an aromatic ring and a side chain, or between two side chains (Scheme 2). It has been previously shown that, in the presence of concentrated HCl, the extent of lignin condensation increases by increasing exposure (hydrolysis) time, acid concentration or temperature (Papadopoulos et al. 1981). From Table 1 it becomes apparent that significant differences exist in the various lignin preparations. Thus, the HCl-lignin residue is the least condensed, while H_2SO_4 and HF lignin residues appear to condense at about equal rates. Considering that hydrolysis with HCl and H_2SO_4 results in the release of small amounts of acid soluble, low molecular weight lignin, in the hydrolyzates (11 and 7 percent,

REPOLYMERISATION-CONDENSATION

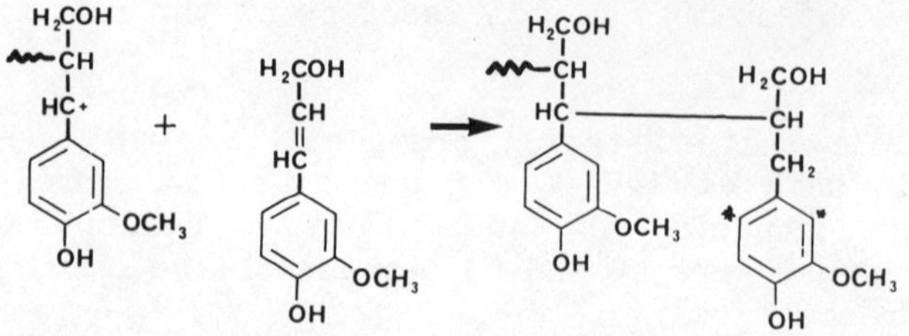

Scheme 2. General schematic representation of the formation of carbon-carbon bonds during acid-promoted condensation reactions in lignin.

respectively, based on original lignin), while HF does not, it is likely that HF lignin residues are the most condensed of the three.

Alkaline Depolymerization of Lignin Residues

Because the isolated acid hydrolysis lignin residues are insoluble in known solvent, their characterization by ordinary methods such as NMR, GPC, UV, .. was impossible. It therefore seemed appropriate to attempt their characterization by degradation to lower molecular weight soluble fractions, by well known methods like the ones employed during alkaline pulping. The structural features of the fractions liberated after alkaline depolymerization of lignin are well defined and information generated in this way could be extrapolated to the original characteristics of the acid hydrolysis lignin residues (Goldstein et al. 1981).

In the presence of alkali and elevated temperatures, lignin depolymerizes either via cleavage of carbon-to carbon bonds or via splitting of alkyl-aryl ether linkages. Alky-aryl ether linkages account for about 60 percent of the possible linkages encountered in

Table 1. Total aldehydes derivable from lignin residues isolated after hydrolysis of birch wood with different acids.

	Total aldehydes[1]	S/V
Birch Wood	42.6	1.1
H_2SO_4 -- Lignin	19.3	1.3
HCl -- Lignin	25.5	1.8
HF -- Lignin	19.8	0.9

[1]Expressed as molar percentage.

hardwood lignins (Nimz 1974) and therefore ether bond cleavage should be of particular importance in the depolymerization of lignin. In general, α-aryl ether linkages are easily hydrolysed by base alone (soda pulping) while significant cleavage of β-O-4 type linkages can be achieved by action of catalytic systems such as sulfide ions (kraft pulping) or anthraquinone/anthrahydroquinone (Soda-AQ pulping).

Methylation with diazomethane of a series of HF lignin residues resulting from hydrolysis of birch wood at different times and temperatures showed no significant increase of the methoxyl group content (DeFaye et al. 1982). This essentially led to the postulate that no detectable cleavage of alkyl-aryl ether linkages in the presence of HF occurs in the range of the reaction conditions studied.

In order to test this hypothesis the lignin residues with lowest carbohydrate content were subjected to alkaline treatment (2N NaOH) at 170° C as well as with alkali and 0.1% AQ under the same reaction conditions. As has been already mentioned under these conditions cleavage of alkyl-aryl ether linkages occurs with subsequent liberation of new phenolic groups. Exploratory experiments on HF lignin residues with alkali alone led to small increases in phenolic hydroxyl groups, while utilization of AQ as catalyst causes liberation of higher amounts of phenolic hydroxyl (Table 3).

Differences were also observed to the syringaldehyde to vanillin (S/V) ratio. In general, under conditions promoting acid catalyzed condensation of lignin, the S/V ratio increases compared to the original ratio observed in the lignin of wood. This can be explained either on the basis of the preferential liberation of syringyl units in the hydrolyzates or the selective condensation of guaiacyl moieties.

Table 1 shows an increase in the S/V ratio for both HCl and H_2SO_4 lignin residues compared to the original ratio in birch wood lignin. In contrast, the S/V ratio became smaller when lignin was exposed to anhydrous hydrogen fluoride. This implies that during treatment of wood with HF, condensation proceeds very fast for both guaiacyl and syringyl units.

Methylation Studies of the Lignin Residues

In the presence of acid catalyzed solvolytic conditions, lignin depolymerises, mainly via cleavage of aryl-alkyl ether type linkages. Hydrolysis of ether linkages leads to formation of new free phenolic hydroxyl structures (Scheme 3). Since phenolic hydroxyl groups are not consumed under the same reaction conditions, investigations of phenolic hydroxyl content can lead to valuable esti-

DEPOLYMERISATION- HYDROLYSIS

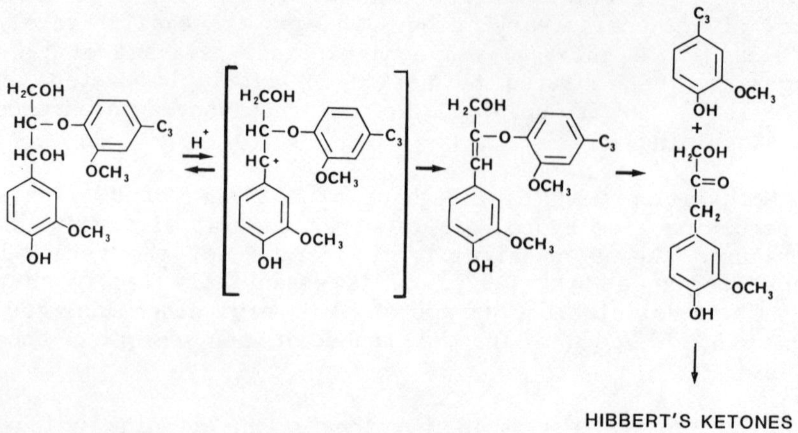

Scheme 3. General schematic representation of the acid-promoted hydrolysis of an aryl-alkyl ether linkage.

mation of the extent of depolymerization for the different lignin preparations.

Phenolic groups can be methylated by diazomethane and the increase in methoxyl group content can be used as an estimation of the increase in phenolic hydroxyl and therefore of the extent of depolymerization. The HF lignin residues showed the smallest increase in methoxyl groups and therefore seem to be the least degraded. As was postulated previously, it is very likely that aryl-alkyl ether linkages are not cleaved in the presence of anhydrous hydrogen fluoride (DeFaye et al. 1982). Assuming a direct relationship between phenolic hydroxyl group increase and increase in methoxyl groups after methylation, the HF lignin residues exhibit the lowest phenolic hydroxyl content (Table 2), which is consistent with the postulate that very little if any hydrolysis of the aryl-alkyl ether linkages occurs in the presence of anhydrous hydrogen fluoride. In contrast, hydrolysis of these linkages proceeds to a larger extent in the case of dilute sulfuric acid hydrolysis (twofold increase compared to HF lignin residue). This suggests that extensive cleavage of these linkages can be achieved even with low acid concentrations, but with the use of high temperatures. However, prolonged exposure to alkali at high temperatures leads to condensation of lignin (Figure 1), and to the formation of higher molecular weight products. It was, therefore, chosen that the three lignin residues be hydrolysed for only 60 minutes with alkali and in the presence of AQ as catalyst.

Molecular weight distributions were determined on Sephadex LH-60 with 0.1M LiCl/DMF as solvent. The three lignin residues were

Table 2. Increase in phenolic hydroxyl as expressed by the increase in methoxyl content after methylation.

	Methoxyl content [1]	Methoxyl content after methylation[1]	Methoxyl content increase
H_2SO_4 -- Lignin	16.8	34.1	18.3
HCl -- Lignin	17.3	32.5	15.2
HF -- Lignin	18.9	28.6	9.7

[1] Corrected for its carbohydrate content, in percent.

depolymerized with alkali under identical reaction conditions. From the elution profiles (Figure 2) it becomes apparent that HF lignin is the most degraded under alkaline conditions, probably because alkyl-aryl ether linkages are cleaved under alkaline conditions giving rise to lower molecular weight lignin entities.

SUMMARY AND CONCLUSIONS

The structural changes occurring in lignin during hydrolysis of wood with mineral acids were investigated. The structure of acid hydrolysis lignin residues depends on the result of competition between acid promoted hydrolysis and condensation reactions. Thus, HCl lignin residues appear to be the least condensed, and HF residues the least hydrolysed.

Table 3. Increase in methoxyl content of the HF-Lignin residue (30 min., 23°C) after alkaline depolymerization.

Reaction time (min)	Methoxyl content after methylation
2N NaOH, 170°C	
0	28.6
60	33.6
120	33.9
2N NaOH, 0.1% AQ, 170°C	
60	38.3
120	40.9

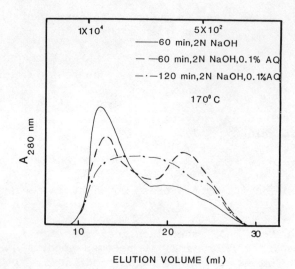

Figure 1. Gel chromatography of alkaline degraded HF-lignin residue on Sephadex LH-60 in 0.1M. LiCl/DMF.

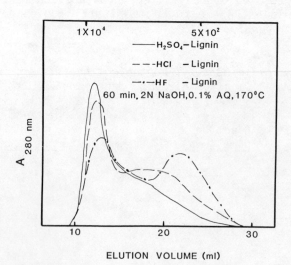

Figure 2. Gel chromatography of various lignin residues degraded under alkaline conditions on Sephadex LH-60 in 0.1M. LiCl/DMF.

REFERENCES

DeFaye, J., A. Gadelle, J. Papadopoulos and C. Pedersen. (1982). Hydrogen Fluoride Saccharification of Cellulose and Lignocellulosic Materials. J. Appl. Pol. Symposium (In Press).

Falkehag, I.S. 1975. Lignin in materials. Applied Polymer Symposium N° 28, 247-257.

Goldstein, I.S. 1975. Potential for Converting Wood into Plastics. Science 189, 847-852.

Goldstein, I.S. 1979. Chemicals from Wood. Unasylva 32 (125):2-9.

Goldstein, I.S., R.J. Preto, J.L. Pittman and T.P. Schultz. 1981. Hydrolytic Depolymerization of HCl Hydrolysis Lignin. International Symposium of Wood and Pulping Chemistry "Ekman Days", Stockholm, IV: 9-13.

Kratzl, K. and J.S. Gratzl. 1960. Zur Theorie der Hydrolyse und Kondensation des Lignins.

Lai, Y.Z. and K.V. Sarkanen. 1971. In "Lignins" (K. V. Sarkanen and C.H. Ludwig, eds.), Wiley-Interscience, N.Y. Chapter 5, 185-186.

Nimz, H. 1974. Beech Lignin-Proposal of a Constitutional Scheme. Angew. Chem. Int. Ed. Engl. 13: 313-327.

Papadopoulos, J., C.L. Chen and I.S. Goldstein. 1981. The Behaviour of Lignin during Hydrolysis of Sweetgum Wood with Concentrated Hydrochloric Acid at Moderate Temperatures, Holzforschung. 35: 283-286.

Wallis, A.F.A. 1971. In "Lignins" (K.V. Sarkanen and C.H. Ludwig, eds.), Wiley-Interscience, N.Y., Chapter 9, 345-372.

Winston, M.H., C.L. Chen, J.S. Gratzl and I.S. Goldstein. 1981. Characterization of the lignin residue from hydrolysis of Sweetgum Wood with Superconcentrated Hydrochloric Acid. In Abstracts of 181st National Meeting of ACS, Atlanta, Ga.

GENERAL CONCEPTS OF WASTE UTILIZATION FOR THE PRODUCTION OF MICROBIAL BIOMASS AND BIO-ENERGY

B. Pekin

Ege University
Bornova
Izmir, Turkey

PART I: PRINCIPLES OF MICROBIAL AND ENZYME ACTIONS

INTRODUCTION

Some agricultural products obtained from plants and animals are nutritionally valuable foods for men and animals. It has been known for a long time that microbial spoilage of food materials cause the formation of some toxic substances as well as the formation of undesirable tastes and odours in foods. On the other hand, some types of microbial spoilage of foodstuff are accepted. For example, the growth of molds in some cheeses ensure the flavour and are accepted. Growth of some microbes in food materials can also increase their nutritional value. For example, some fermented foods such as cheeses, yogurt, etc. can furnish microbial protein during a daily diet of the people. Mass cultivated microbial biomass has been used as a source of nutrition for humans and domestic animals. For example, in 1910 brewer's yeast was used in Germany as a feeding supplement for animals and in Japan, several processes for algae food cultivation are in operation today. After World War II some countries, namely England, the United States, Germany, etc., have been interested in producing microbial biomass for animals and the term, single-cell protein (SCP) has been accepted.

Serious interest in using SCP for nutritional purposes began to reappear after 1950 and increased later, in the early 1970's, because of the increasing world population.

Microbial biomasses which are used as animal and human feed must have certain properties such as high nutritional value (essential amino acids, vitamins and minerals), digestibility, good quality and no toxicity. As an example, comparison of the amino acid distribution in yeasts, egg and other materials are shown in Table 1.

Many microbial biomasses contain high level nucleic acids, particularly RNA. Nucleic acids cause the formation of high levels of uric acid in the blood. Since the solubility of uric acid is low, high levels of uric acid in blood may cause the precipitation of uriates in tissues and give symptoms similar to gout disease. It is estimated that nucleic acid intake for an adult is about 2 g. daily. Several methods of decreasing RNA contents of microbial biomass have been developed. These are heat-shock, osmotic-shock, chemical and enzymatic breakdown, extraction or hydrolysis of cellular RNA, etc.

As an example for the above-mentioned methods, autolytic breakdown of the RNA of bakers' and brewers' yeast has been induced by subjecting a suspension of yeast in NaCl solution at pH 6 to a heat-shock. This was produced by bringing the suspension to 50°C and then adding enough water at 100°C to increase the temperature momentarily to 68°C. It was followed by addition of cold water to reduce it to 50°C after 1 minute and incubation was continued at this temperature. The results are comparable with an efficient chemical for reduction of RNA content. The results for different incubation times of the suspension (60 g. yeast solids and 0.5 mole NaCl/l) at 50°C, are given in Table 2. (S. Pekin and B. Pekin 1982).

Table 1. Comparison of the amino acid distribution in yeasts, egg and other materials. (mg/gN)*

Amino acid	Torula yeast	Brewers' yeast	Egg	Casein	Soybeen	Cotton seed
Tryptophan	86	96	103	84	86	74
Threonine	315	318	311	269	246	221
Isoleucine	444	324	415	412	336	236
Leucine	501	436	550	632	482	369
Lysine	493	446	400	504	395	268
Total sulfur amino acids	153	187	342	218	195	188
Phenylalanine	319	257	361	339	309	327
Valine	392	368	464	465	328	308
Arginine	451	304	410	256	452	702
Histidine	169	169	150	190	149	166

*By permission from R.I. Mateles and S.R. Tanenbaum (eds)., 1968. P.94.

Table 2. The results of heat-shock procedure at 68°C for bakers' and brewers' yeast.

	Incubation Time (hr)	Temperature	Moisture	Total-N %	RNA %	CCP/NA
Bakers' yeast	2	50	70.16	8.45	3.12	15.93
	2.5	50	69.77	7.90	2.26	20.85
	3	50	68.85	7.97	1.86	25.78
	3.5	50	73.54	7.50	1.89	23.77
	4.5	50	73.50	7.60	1.86	23.80
Brewers' yeast	2	50	77.35	8.90	3.04	13.30
	2.5	50	76.34	8.85	1.85	23.90
	3	50	77.80	8.82	1.52	35.27
	3.5	50	77.94	8.80	1.49	35.91

Results were expressed in terms of the content of nucleic acid (NA, total purines in μmol/g x 0.0643), crude protein (CP, Percentage total N x 6.25), and corrected crude protein (CCP = CP-NA), all being stated as percentage of dry matter.

Some other problems have also been observed with microbial biomass. For example, consumption of more than 15 g. of Candida utilis each day causes gastrointestinal disorders. 12-25 g. of Aerobacter aerogenes or Hydrogenomonas eutopha per day causes nausea, vomiting and diarrhea. Some microorganisms produce toxic compounds such as mycotoxins.

Recently, some methods have been improved for extracting protein from microbial biomass and using it as a meat analogue. Microbial biomass is also a very important source of enzymes and many extraction methods are known. Both microbial biomass and enzymes can be used for waste utilization and bio-energy production. Other applications of microbial biomass are vaccine production, waste water treatment and production of primary or secondary metabolites by fermentation techniques.

As we know, at present, the world obtains energy mostly from fossil fuels such as petroleum and coal. Petroleum supplies about 40 percent of the total energy. It is clear that the demand for petroleum will exceed the supply. Therefore energy crises and pollution problems will be more serious. As a result of this situation, we need to find new alternative energy sources. One of the best energy sources is the Sun. The total amount of the solar energy coming to the earth is about 1.3×10^{21} K.Cal/year. The energy use of the world outside the Iron-curtain area in 1972 was about 4.3×10^{16} K.Cal/year, which is about 0.003 percent of the energy of Sunlight (Harima 1978).

The energy of sunlight is converted into chemical energy by photosynthesis and stored in plants. In the world, it is estimated that more than 1.4 billion tons of wood and 1 billion tons of non-wood agricultural materials such as wheat, rice straw, and cane bagasse, etc., are harvested annually. When all of these materials are used as energy sources, 9.6×10^{15} K.Cal/year of energy is obtained. This assumption is made by estimating that the average energy production from them is 4 K.Cal/g.

Since the U.S. energy use in 1972 was 18.2×10^{15} K.Cal, the energy from wood and agricultural materials can only provide about half of the U.S. energy need. Many processes have been developed to produce energy-related materials such as alcohol, methane (biogas), hydrogen, biofuel, etc., from the waste materials by using microbial biomass and enzymes. The most important solid waste materials have three main origins: forestry residues, urban refuse, agricultural and industrial wastes.

It seems to me that anaerobic digestion of crop residues, animal manure and urban refuse to produce biogas will be one of the key processes of the future. This technique is also widely used for stabilization of primary and secondary waste sludge in municipal waste water treatment plants (Moreira 1978).

The cellulosic wastes are also degraded into glucose by the enzymes (cellulases) of some microorganisms, such as _Trichoderma_ species, then the glucose is fermented to give ethanol by _Saccharomyces cerevisiae_ or other specific microorganisms. The difficulty of this process comes from the degradation of cellulose by enzymes. It is necessary to modify the cellulose by chemical methods or to find new microorganisms with higher cellulose activity which can easily attack the native crystalline cellulose more effectively. Recently, a lot of progress has been made in genetic engineering enabling the development of new species by Recombinant DNA techniques.

Strong research efforts also continue to obtain better cost-effective substrate pretreatment processes. In the meantime, mild alkali treatment of cellulosic materials has been accepted for satisfactory digestibility.

To produce SCP from cellulosic solid waste materials by fermentation, two important processes may be considered: submerged culture or solid state fermentation.

Performing the second process is easier with lower energy requirements than submerged culture. The main disadvantages of it

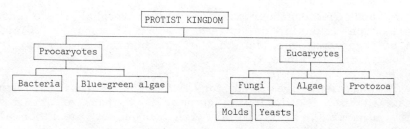

Figure 1. Classifications of microorganisms belonging to the kingdom of protists. (Reprinted by permission from Bailey and Ollis 1977, p.13).

are the high handling requirements and the difficulty of achieving good homogeneity of the fermented materials. I believe that these problems will be solved very soon.

Besides these, some other agricultural, industrial and environmental wastes such as starch, molasses, whey, sulphite-waste liquor, methane, methanol, n-alkanes and natural gas can be also used to produce SCP or energy.

The classifications of microorganisms belonging to the kingdom of protists are shown in Figure 1.

Among the simple organisms shown in Figure 1, protozoa are not employed in industrial processes, but they participate in biological waste water treatment with other microorganisms.

It is a well known fact that substrates containing certain nutrients have to be supplied to the microorganisms to provide the necessary energy for cell growth, maintenance and for the synthesis of metabolites.

The protozoa, fungi and most of the bacteria are heterotrophic; that is, they obtain their cell carbon from organic compounds. The remainder of the protists, consisting of the algae and some of the bacterial groups, obtain their cell carbon from carbon dioxide and are called autotrophic. On the basis of nutrients and energy yielding metabolisms, there are also two main classes of microorganisms: 1) Phototrophs, and 2) chemotrophs.

Phototrophic microorganisms use solar energy, chemtrophs use chemical compounds as their source of energy. Both classes are further subdivided: Photolithotrophs (Photoautrotrophs) where growth is dependent on exogenous inorganic electron donors, and Photoorganotrophs (Photoheterotrophs) which require organic donors. Chemotrophs, on the other hand, are subdivided into Chemolithotrophs (Chemoautrotrophs) which grow by oxidizing exogenous inorganic compounds, and Chemoorganotrophs (Chemoheterotrophs) where growth is dependant on oxidation of exogenous organic compounds. Classification of organisms according to their carbon and energy requirements are summarized in Table 3.

The nutrients required for growth of microbial biomass are classified into the following groups: i- Sources of the major elements such as C, H, O and N. ii- Sources of the minor elements like P, K, S, Mg. iii- Sources of trace elements such as Fe, Zn, Co, Si, etc. iv - Vitamins and hormones. The term "growth factor" is used to refer to essential organic nutrients such as amino acids, which are incorporated whole in the cell structure. Typical source of C, N, S and P are shown in Table 4.

In a suitable culture, besides nutrients to provide energy and essential materials for biosynthesis, growth of microbial biomass during the exponential (or logarithmic) growth phase depend on the following conditions: i- pH. ii- Temperature. iii- water activity. iv- Absence of inhibitors, etc. During the growth of microorganisms in culture media, substrates are converted into cell materials such as protein, fats, lipids, etc. As a result of biosynthesis, biomass growth occurs. As an example, conversion of sugar into cell protein and fat during growth of microbial biomass are shown graphically in Figure 2.

Growth of some microbial biomass in a culture media are illustrated in Figure 3 schematically.

Table 3. Classification of organisms by carbon and energy source.

Type of organisms	Carbon source	Energy source
Photolithotrops-(Photoautotrophs[x]) (Higher plants, eucaryotic algae, blue-green algae, and some bacteria)	CO_2	Light
Photoorganotrophs (Photoheterotrophs[x]) (Some bacteria, some eucaryotic algae)	Organic compounds	Light
Chemolithotrophs (Chemoautotrophs) (Some bacteria)	CO_2	Chemical
Chemoorganotrophs (Chemoheterotrophs) (Higher animals, protozoa, fungi, and most bacteria)	Organic compounds	Chemical

[x]It is clear from the Table that heterotrophs use organic compounds as a carbon source, autotrophs require CO_2 to supply their carbon needs.

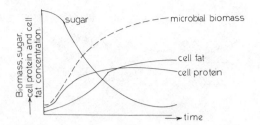

Figure 2. Conversion of sugar into cell protein and cell fat during growth of Rhodotorula glutinis. (Reprinted by permission from S. Aiba, A.E. Humphrey and N.F. Millis 1973, p.112.)

PRINCIPLES OF MICROBIAL GROWTH AND GROWTH RATE

Under suitable culture conditions during infinitely small time intervals (dt), the increase of biomass concentration (dx) is proportional to the amount of biomass concentration (x):

$$dx = \mu x \, dt \qquad (1)$$

or

$$dx/dt = \mu x \qquad (2)$$

Where, dx/dt = Growth rate and μ = Specific growth rate (That is the growth per unit amount of biomass; $\mu = \frac{1}{x} \frac{dx}{dt} \quad \frac{g \cdot biomass}{g \cdot biomass\text{-}hr}$

When μ is constant, integration of Eqn.2 gives

$$Ln\,x = Ln\,x_0 + \mu t \qquad (3)$$

or

$$Log\,x = Log\,x_0 + (\mu/2.303)\,t \qquad (4)$$

Table 4. Typical sources of some elements for growth of microbial biomass.*

Element	Source
Carbon	CO_2, sugars, proteins, fats
Nitrogen	Proteins, NH_3, NO_3^-
Sulfur	Proteins, SO_4^{2-}
Phosphorus	PO_4^{3-}

*From J.E. Bailey and D.F. Ollis, 1977, P.223.

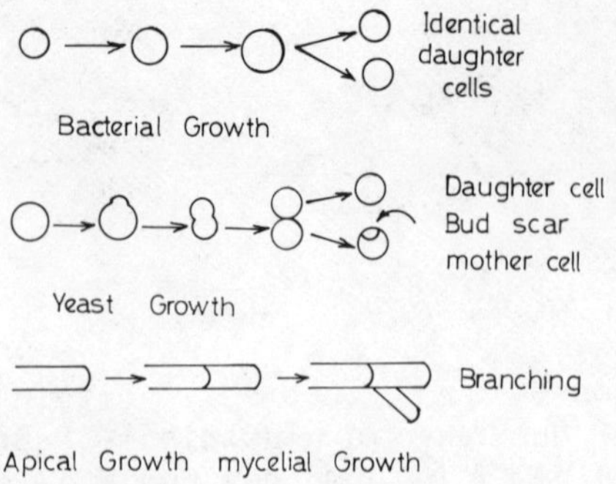

Figure 3. Schematic illustration of typical forms of growth by bacteria, yeast, and molds. (Reprinted by permission from D.I.C. Wang et al. 1979, p. 59).

Eqn.3 can be also written as follows:

$$\ln(x/x_0) = \mu t \tag{5}$$

or

$$x = x_0 e^{\mu t} \tag{6}$$

According to the eqns. 3,4 or 5 plots of $\ln x$, $\log x$ or $\ln(x/x_0)$ against t give straight lines and specific growth rate (μ) can be calculated.

When $x = 2x_0$, equation 5 may be written as follows:

$$t_d = \frac{\ln 2}{\mu} = \frac{0.693}{\mu} \tag{7}$$

Where, t_d is termed as doubling time of microbial biomass.

Problem 1. Under suitable culture conditions specific growth rate for a microbial biomass population is $\mu = 0.0077$ min.$^{-1}$ What is the doubling time of the biomass?

During most of a batch cultivation, the relation between growth rate and essential nutrient (or substrates) was observed in 1949 by Monod. The Monod equation is given as follows:

$$\mu = \frac{\mu_m S}{K_s + S} \tag{8}$$

Where,

μ_m = Maximum specific growth rate
S = Substrate concentration
K_s = Monod or saturation constant

Table 5. K_s values for growth of some microorganisms on diverse substrates.*

Microorganisms	Substrate	K_s (mg/liter)
Saccharomyces cerevisiae	Glucose	25.0
Escherichia	Glucose	4.0
Escherichia	Lactose	20.0
Candida	Glycerol	4.5
Candida	Oxygen	4.5×10^{-1}
Pseudomonas	Methanol	0.7
Pseudomonas	Methane	0.4
Klebsiella	CO_2	0.4
Klebsiella	Mg^{2+}	5.6×10^{-1}
Klebsiella	K^+	3.9×10^{-1}
Klebsiella	SO_4^{2-}	2.7

*By permission from S.J. Pirt, 1975, P.12. (Some part of Table 2.1).

Rearranging Eqn.8, we obtain the following equation

$$\frac{1}{\mu} = \frac{K_s}{\mu_m}\frac{1}{S} + \frac{1}{\mu_m} \qquad (9)$$

A plot of $1/\mu$ against $1/S$ gives a straight line with an intercept on the ordinate at $1/\mu_m$, and an intercept on the abscissa at $-1/K_s$. The slope of the straight line is equal to K_s/μ_m. Therefore, μ_m and K_s constant can be calculated from the plot, easily. As an example K_s values of some micoorganisms are given in Table 5.

Growth and product formation by microorganisms are bioconversion processes in which the substrates in culture media are converted into microbial biomass and metabolite products. These conversions may be expressed as yield coefficients. Therefore, the following equations can be given below:

$$Y_G = Y_{x/s} \equiv -\frac{\Delta x}{\Delta S} \qquad (10)$$

$$Y_{p/s} \equiv -\frac{\Delta P}{\Delta S} \qquad (11)$$

and

$$Y_{p/x} \equiv -\frac{\Delta P}{\Delta x} \qquad (12)$$

Where, $Y_{x/s} = q$ is the yield coefficient between the formation of biomass and consumption of substrate in per unit culture. This is also known as growth yield (Y_G).

$Y_{p/s}$ is the yield coefficient which is product produced and substrate consumed over some period.

$Y_{p/x}$ is the yield coefficient between product and biomass.

Problem 2. Since some quantities of glycerol and higher alcohols are produced depending on the strain of yeast and fermentation condition, the overall conversion of glucose to ethanol by yeast can be represented stoichiometrically as

$$C_6H_{12}O_6 \longrightarrow 2C_2H_5OH + 2CO_2$$

180 g.   92 g.   88 g.

From the above equation, calculate the theoretical (Gay-Lussac) yield of ethanol which can be obtained from the fermentation of glucose.

Dividing of both sides of the Monod equation (eqn. 8) by Y_G, the following equation is obtained:

$$\frac{\mu}{Y_G} = \frac{\frac{\mu_m}{Y_G} S}{K_s + S} \tag{13}$$

or

$$q = \frac{q_m S}{K_s + S} \tag{14}$$

where,

$$\frac{\mu}{Y_G} = q \quad \text{(Specific metabolic rate)}$$

$$q_m = \frac{\mu_m}{Y_G} \quad \text{(Maximum specific metabolic rate)}$$

Rearranging Eqn.14 we obtain

$$\frac{1}{q} = \frac{K_s}{q_m} \frac{1}{S} + \frac{1}{q_m} \tag{15}$$

q is used to estimate the demands for substrate (especially oxygen) at different growth rates related to substrate concentrations. Similar to Eqn.9, a plot of $1/q$ against $1/S$ gives a straight line: By using this plot K_s and q_m can be calculated.

Problem 3. Data on Bakers' yeast in phosphate buffer medium at 23.4°C and various oxygen concentration are given in the following Table.* Calculate K_s, (K_{O_2}) and $q(O_2)m$ values of Bakers' yeast.

P_{O_2} (mm-Hg)	q_{O_2} ($\mu l.O_2$/mg.biomass-hr)
0.0	0.0
0.5	23.5
1.0	33.0
1.5	37.5
2.5	42.0
3.5	43.0
5.0	43.0

* Dr. A.E. Humphrey: Ch.E. 503 Biochemical Engineering Lectures during the period 1975-76 at the University of Pennsylvania.

Note: P_{O_2} = Oxygen partial pressure (mm-Hg) in fermentation media

qO_2 = Oxygen uptake rate (μlO_2/hr-mg. biomass) measured by Warburg manometer.

From the equations (1 and 10) we obtain

$$-\frac{ds}{dt} = \frac{1}{Y_G} \frac{dx}{dt} \qquad (16)$$

Where,

$-\frac{dS}{dt}$ = The rate of substrate consumption during microbial growth.

It must be mentioned here that certain substrates, primarily energy sources such as carbohydrates, are also used for microbial cell maintenance. Therefore, microorganisms can use some or part of these substrates for physiological activities such as endogenous respiration as well as biosynthesis.

For this reason, the yield expression must include a factor to account for utilization of the substrate for maintenance. Therefore, especially at low substrate concentration, Eqn.16 must be written as follows

$$-\frac{ds}{dt} = \frac{1}{Y_G} \frac{dx}{dt} + mx \qquad (17)$$

or, using Eqn.1

$$-\frac{1}{x}\frac{dS}{dt} = \frac{1}{Y_G}\mu + m \qquad (18)$$

Problem 4. The following data were obtained by A.E. Humphrey and colleagues* at suitable cultivation conditions for certain thermophilic microorganisms. Accepting that cellulose is an energy and growth limiting substrate, calculate μ, Y_G and m for this microorganism under these conditions.

Time (Hour)	Cellulose (g/l)	Biomass (g/l)
0	10.00	0.500
2	9.125	0.688
4	8.313	1.063
6	7.500	1.513
8	6.813	2.000
10	6.250	2.313
12	5.688	2.550
14	5.219	2.750
16	4.781	2.906
18	4.344	3.063
20	4.031	3.156
22	3.688	3.281
24	3.375	3.375
26	3.125	3.406
28	2.938	3.469
30	2.688	3.500
32	2.500	3.563
34	2.313	3.594
36	2.188	3.594

*Data are taken from Biochemical Engineering Lectures of Dr. A.E. Humphrey in the University of Pennsylvania during 1975-76 term.

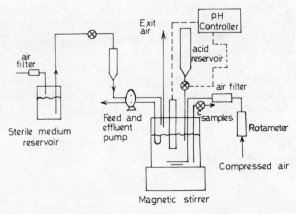

Figure 4. Continuous-culture laboratory set-up for a Chemostat. (Reprinted by permission from D.I.C. Wang et al. 1979 p. 99.)

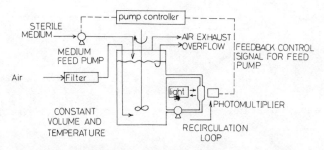

Figure 5. Typical laboratory set-up for a turbidostat. (Reprinted by permission from D.I.C. Wang et al. 1979, p. 100).

CONTINUOUS CULTURE

We have described growth and product formation kinetics in batch culture in the previous section. During growth of micoorganisms in batch culture, the medium becomes depleted of nutrients and there is an accumulation of waste products of metabolism. For this reason continuous culture processes have been developed. In this technique, fresh nutrient medium is added to the well-stirred culture vessel (which is also called a reactor or fermentor) at a rate that is equal to the removal of broth containing biomass and products from the vessel, thereby maintaining a constant volume of culture. This type of continuous-flow culture is referred to as a Chemostat. A variant of the Chemostat, in which the density of microbial biomass in the culture vessel is maintained constant by addition of fresh medium and monitoring the culture's optical density, is known as a Turbidostat. In Figure 4 and Figure 5 typical laboratory set-ups for a Chemostat and Turbidostat are shown.

Today, a single-stage, multistage and cell recycle chemostats are being used for both laboratory and industrial purposes in continuous culture processes.

The specific growth rate in a chemostat is determined by the flow rate (F) of the medium divided by the culture volume. This ratio is defined as the "dilution rate (D)." At steady state the specific growth rate is equal to the dilution rate and the following equation is valid.

$$D = \frac{F}{V} = \mu \qquad (19)$$

The reciprocal of the dilution rate is known as the retention time.

$$\Theta = \frac{1}{D} \qquad (20)$$

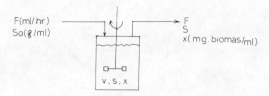

Figure 6. Continuous culture in single-stage chemostat.

Problem 5. Using the material balances on the substrate and microbial biomass which can be written for the single stage fermentor shown below, derive the following equations:

$$\frac{S_o - S}{\Theta} \cdot \frac{1}{x} = \frac{q_m S}{K_S + S} \tag{21}$$

or

$$\frac{x}{S_o - S} \Theta = \frac{K_S}{q_m} \cdot \frac{1}{S} + \frac{1}{q_m} \tag{22}$$

Considering the maintenance factor

$$\frac{D(S_o - S)}{x} = \frac{1}{Y_G} D + m \tag{23}$$

or

$$\frac{D(\Delta S)}{x} = \frac{1}{Y_G} D + m \tag{24}$$

Problem 6. The following data were obtained under optimum conditions for substrate uptake of a mix-culture growing in a continuous waste treatment vessel. Using one of the equations obtained in Problem 5., calculate the average K_S and q_m values for this microbial population.

S_o (Substrate in the feed) (mg/l)	S (Substrate in the fermentor) (mg/l)	Θ (day)	x (mg/l)
300	7	3.2	128
300	13	2.0	125
300	18	1.6	133
300	30	1.1	129
300	41	1.1	121

Results: $q_m = 3.125$ day^{-1} ; $K_S = 24.0$ mg/l

WASTE FOR PRODUCTION OF MICROBIAL BIOMASS AND BIO-ENERGY

Problem 7. The following data were obtained for the growth of A. aerogenes on a glycerol limited growth medium.

D, hr^{-1} (Dilution rate)	$\Delta S = S_9 - S$ (mg/ml)	x(mg-dry biomass/ml)
0.05	9.988	3.2
0.10	9.972	3.7
0.20	9.950	4.0
0.30	9.925	4.2
0.40	9.900	4.4
0.50	9.876	4.6
0.60	9.850	4.75
0.70	9.824	4.9
0.80	9.200	4.5

Estimate the values of K_s, Y_G, μ_m and m for this system.

Bauchop and Elsden (1960) have found that growth of some anaerobic microorganisms growing in complex media was directly proportional to the moles of (ATP) produced by their catabolism. Thus, Y_{ATP} can be defined by:

$$Y_{ATP} = \frac{\Delta x}{\Delta(ATP)} \tag{25}$$

or

$$Y_{ATP} = \frac{\Delta x/-\Delta S}{\Delta(ATP)/-\Delta S} = \frac{Y_{x/s}}{Y_{A/S}} \tag{26}$$

Where Y_{ATP} = growth yield based on ATP generation (g/mole). $Y_{A/S}$ = ATP yield from catabolized energy sources in media.

The average values of Y_{ATP} for some microorganisms are given in Table 6.

For the assessment of a maintenance coefficient, mass balance on ATP utilized (= formed), in a small interval of time can be written as:

$$\Delta(ATP)_F = \Delta(ATP)_m = \Delta(ATP)_G \tag{27}$$

Where F, m and G stand for formed, maintenance and growth respectively.

We can assume that the amount of ATP utilized for the maintenance metabolism is proportional to the cell concentration at an arbitrarly given time. Therefore, the following relation is valid:

$$\Delta(ATP)_m = M_A X \Delta t \tag{28}$$

Where M_A = maintenance coefficient for ATP (mole.g.$^{-1}$h^{-1}) Δt = time interval and X = cell concentration.

The amount of ATP utilized for growing microorganisms is given below:

$$\Delta(ATP)_G = \frac{\Delta x}{Y^{max}_{ATP}} \tag{29}$$

Where Y^{max}_{ATP} is maximum growth yield for ATP (g.mole^{-1}). Substituting Eqns. 28 and 29 into Eqn. 27 we obtain

$$\Delta(ATP)_F = M_A X \Delta t \quad \frac{\Delta x}{Y^{max}_{ATP}} \tag{30}$$

or, for a small interval of time

$$q_{(ATP)} = M_A \quad \frac{1}{Y^{max}_{ATP}} \quad \frac{1}{x} \quad \frac{dx}{dt} \tag{31}$$

Since, $\frac{1}{x}\frac{dx}{dt}$ we can write Eqn. 31 as follows:

$$q_{(ATP)} = M_A \quad \frac{1}{Y^{max}_{ATP}} \mu \tag{32}$$

Table 6. Growth yields based on ATP generation of microorganisms in complex media under anaerobic conditions. (Some data were taken by permission from S. Nagai 1979).

Microorganism	Substrate	Y_{ATP}, g.mole^{-1}
Lactobacillus plantarum	Glucose	9.4
Aerobacter aerogenes	Glucose	10.3
" "	Fructose	10.7
" "	Mannitol	10.0
" "	Gluconic acid	11.0
Saccharomyces cerevisiae	Glucose	10.5
Escherichia coli	Glucose	9.4
Zymomonas mobilis	Glucose	8.3
Streptococcus faecalis	Glucose	11.5
" "	Ribose	12.6
" "	Arginine	10.0

Where, q_{ATP} is specific rate of ATP formation (mole.g.$^{-1}$h^{-1}) and is expressed as:

$$q(ATP) = \frac{1}{x} \frac{d(ATP)}{dt} \qquad (33)$$

The growth yield based on total energy available from the medium can be written as:

$$Y_{Kcal} = \frac{\Delta_x}{\Delta H_a \Delta x + \Delta H_c} \qquad (34)$$

Where Y_{KCal} = growth yield based on total energy available (g.KCal^{-1}); ΔH_a = heat by combustion of dry cell (KCal.g.$^{-1}$) ΔH_c = heat generation by catabolism (KCal.l.$^{-1}$).

ENZYMES AND ENZYME KINETICS

Enzymes are proteins possessing catalytic functions. Therefore, similar to microorganisms they are termed as biocatalysts. Their molecular weights range from about 15,000 to over a million. Enzyme extracts, like microorganisms, have been used since before recorded history, but great developments have come during the last 45 years.

The first clear recognition of an enzyme was made by Payen and Persoz in 1833. They found that an alcohol precipitate of malt extract contained a thermolabile substance which converted starch into sugar and named this substance as "diastase (now called amylase)." Later, many scientists have observed a parallel behaviour between the action of enzymes and that of yeast during fermentation. Therefore, the name "ferment" was consequently used for enzymes.

In 1878 Kühne proposed the word "enzyme (Yeast in Greek)" instead of ferment. Several other enzymes were discovered during the later half of the 19th century.

Emil Fischer (the great German chemist) developed the concept of enzyme specifity and of the close steric relationship between an enzyme and its substrate. Today we know that most enzymes act on a limited group of compounds, but some of them have an almost absolute specifity for one substrate.

As is shown in Table 7 enzymes are classified according to the reaction which they catalyze.

In 1926, urease was crystallized by Sumner and shown to be a protein. Today more than 2000 enzymes are known and about 120 of them are used in industries and sold on commercial scale. In 1969 the U.S. fermentation industry produced products valued at $2 billion, of which $64 million was accounted for by enzymes.

Table 7. Classification of enzymes.

Major class number	Class	Reaction catalyzed
1	Oxidoreductases	Oxidation-reduction
2	Transferases	Group transfer
3	Hydrolases	Hydrolysis
4	Lyases	Removal of group leaving double bond. Addition of group to double bond.
5	Isomerases	Isomerization
6	Ligasses (Synthetases)	Joining of two molecules coupled with clevage of a pryophosphate bond of ATP or similar triphosphate.

There are three sources of enzymes: These are 1. Plant, 2. Animal and 3. Microbial sources. Since microorganisms have a shorter doubling time compared with plants and animals, microbial enzymes are most important and they can be easily scaled-up.

Enzymes are also classified as 1. Extracellular and 2. Intracellular according to the localization in living organisms. The extracellular enzymes excreted by the cell. Therefore, the isolation of them is easier than intracellular enzymes.

Recently, Pollock (1962) laid down a classification of exoenzymes dividing them 1. Cell-bound enzymes which could be a. Intracellular or b. Surface-bound, and 2. Cell-free or truly extracellular enzymes.

Truly extracellular enzymes are excreted during normal active growth. Microorganisms can also release intracellular enzymes as a result of autolysis or cell damage. As we mentioned before microbial growth on an insoluble substrate is known as "solid state fermentation." Insoluble substrates (both solid and liquid) or high molecular-weight soluble substances of polymeric structure can not penetrate into cells. Much of the waste materials found in nature are polymeric in structure such as polysaccharides, proteins, fats, lignins, nucleic acids, etc. Microorganisms (excluding protozoa) release into the environment either true-extracellular enzymes or cell (surface)-bounded enzymes to degrade polymeric or insoluble materials. Cell-free enzymes may strongly adsorb on these materials and thus be undetectable in the culture filtrate. For a cell-bound enzyme to act on the insoluble substrate the cell must be close to the substrate.

There are some growth conditions under which the amount of enzymes in microorganisms can be increased. For example, growing bacteria in a culture rich in protein or in carbohydrates may induce the production of some proteolytic enzymes and carbohydrates, respectively; or similarly, growth on maltose as the sole carbon source may induce high production of maltase in some growing bacteria. Also by using some genetic engineering methods, mutants of microorganisms which have high concentrations of some specific enzymes can be obtained. The level of enzyme is not the same throughout the life cycle of a microorganism. The enzyme level may be absent or very low at an early stage of growth, later rise to a maximum and decrease again.

Two types of culture processes may be used to produce enzymes: 1. Solid culture, or Koji and 2. Submerged-liquid culture methods.

In Figure 7. the Koji Culture Technique is shown schematically. As shown in this flow diagram, Koji process microorganisms are cultivated in trays containing the solid substrate mixed with water and other ingredients such as buffers and supplementary nutrients. In the case of cultivation of Trichoderma viride, wheat bran is used as the solid substrate. The surface culture is washed by water or buffer and crude extracts of extracellular-cellulase are obtained.

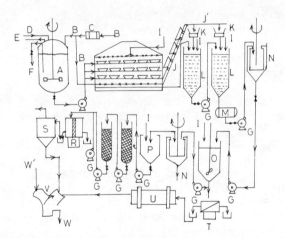

Figure 7. Cellulase production by T. viride, using the Koji technique
A- submerged seed culture of Trichoderma viride; B- oil-free compressed air; C- air filter; D- inoculum; E-exhaust air; F- sample collection; G- centrifugal pump; H-automatic wheat bran culture of T. viride; I- water spray; I-ammonium sulphate, acetone, alcohol and water: J- belt conveyor; J- screw conveyor; K- hopper; L- extraction column; R- membrane concentrator; S- spray dryer; T-filter press: U- rotary dryer; V- mixer; W-cellulase preparation: W- salt stabiliser. (Reprinted by permission from T.K. Ghose and A.N. Pathak, 1973).

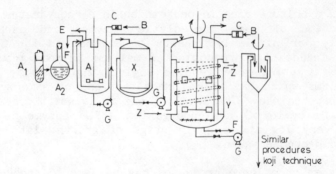

Figure 8. Submerged-culture process for cellulase production from T. viride: A_1 - culture of T. viride; A_2 - spore suspension of T. viride; A- seed; B- oil-free compressed air; C- air filter: E- exhaust air; F- sample collection: G- centrifugal pump; X- substrate storage tank; Y- main fermenter; Z- water line; N- centrifuge. Further processes are similar to Figure 7. (Reprinted by permission from T.K. Ghose and A.N. Pathak 1973).

The crude extract is purified further. In Figure 8, the submerged-liquid culture technique is shown. This a modern technique and is used to produce both extracellular and intracellular enzymes. In this method, the isolation and purification procedures are similar to the Koji Technique.

The kinetic mechanism for the conversion of a single substrate (S) to a product (P) by an enzyme, is represented as follows:

$$S + E \underset{k_{-1}}{\overset{k_1}{\rightleftarrows}} ES \xrightarrow{k_2} E + P$$

Where k_1, k_{-1} and k_2 are rate constants and ES is intermediate enzyme-substrate complex.

The rate of this reaction is given and calculated by the Michaelis-Menten equation.

$$V = -\frac{ds}{dt} = \frac{k_2 E_o S}{K_m + S} \qquad (25)$$

Where E_o is the total enzyme concentration and K_m is the Michaelis-Menten constant and they are represented as follows:

$$E_o = [A] + [ES] \qquad (26)$$

$$K_m = \frac{k_{-1} + k_2}{k_1} \qquad (27)$$

When, $S \gg K_m$, the rate of reaction approaches the limiting maximum value (V_m) which is given by

$$V_m = k_2 E_o \qquad (28)$$

Similar to the Monod equation, Eqn.25 may be re-arranged as follows

$$\frac{1}{V} = \frac{1}{V_m} + \frac{K_m}{V_m} \frac{1}{S} \qquad (29)$$

A plot of $1/V$ against $1/S$ gives a straight line. This is usually referred to as Lineweaver-Burk diagram. From this plot, K_m and V_m values can be calculated.

Enzymes are immobilized on water-insoluble materials. Similar to immobilized enzymes, the application of immobilized microorganisms represents a new and very useful technique in modern biotechnology. When compared with free enzymes or free cells, immobilized biocatalysts have many advantages. For example, free cells or enzymes are very difficult to re-use or to use continuously, they are too small to filter, and recovery by centrifugation is too expensive. Immobilized systems are easily added to or removed from the reaction mixture, and re-used, their reactions are easily controlled by automatic procedure. They can be used for continuous processing and product inhibition can be minimized, etc.

Different immobilization methods of enzymes (or microorganisms) are shown schematically in Figure 9.

In immobilized systems, external mass transfer limitation effects are observed. Therefore, the rate of mass transfer from the bulk solution to the surface can be expressed as

$$V' = k_L (S_b - S_s) \qquad (30)$$

Where,

S_s = Surface concentration; S_b = Bulk phase substrate concentration; a = Surface area per unit volume and k_L = Mass transfer coefficient (dimensional units of length/time).

At steady state, the mass transfer rate must be equal to the reaction rate at the surface or to the apparent reaction rate within the solid. In this case the internal diffusion limitations are not significant and simple Michaelis-Menten kinetics are applied; therefore the following steady-state expression is valid:

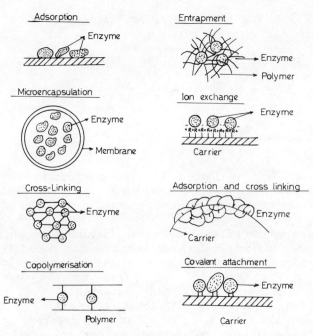

Figure 9. A schematic representation of immobilization techniques of enzymes (or microorganisms). Reprinted by permission from H.H. Weetall 1975.

$$V = V' = k_L a(S_b - S_s) = \frac{V_m S}{k_m + S} \quad (31)$$

Solving Eqn.30 for S_s, we obtain

$$S_s = S_b - \frac{V'}{k_L a} \quad (32)$$

When this expression is substituted into Eqn.31 and rearranged, the following equation is obtained

$$\frac{1}{V} = \frac{1}{V_m} + \left(\frac{K_m}{V_m}\right)\left(\frac{1}{S_b}\right)\frac{1}{\left(1 - \frac{V}{S_b k_L a}\right)} \quad (33)$$

If $S_b \longrightarrow \infty$, then $V \longrightarrow V_m$ and Eqn.33 approaches

$$\frac{1}{V} = \frac{1}{V_m} + \frac{K_m}{V_m}\frac{1}{S_b} \quad (34)$$

As $S_b \longrightarrow 0$, Eqn.33 approaches

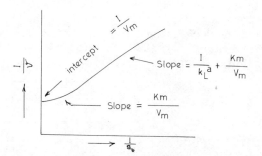

Figure 10. Effect of external mass transfer limitation for immobilized enzyme systems. (Reprinted by permission W.H. Pitcher, Jr. 1975).

$$\frac{1}{V} = (\frac{1}{k_L a} + \frac{K_m}{V_m}) \frac{1}{S_b} \tag{35}$$

Figure 10 shows a Lineweaver-Burk plot for a reaction with Michaelis-Menten kinetics limited by external mass transfer. At high values of S_b, the slope approaches k_m, while at low values of S_b, it approaches $(1/k_L a) + (K_m/V_m)$. From these two slopes and the intercept the value of $K_L a$ and k_m can be determined. In Figure 10 the effect of external mass transfer is represented.

Immobilized enzyme reactors may be classified as 1. Batch reactor. 2. Continuous stirred tank reactor (CSTR), 3. Fixed-bed and 4. Fluidized-bed reactors. These are shown in Figure 11.

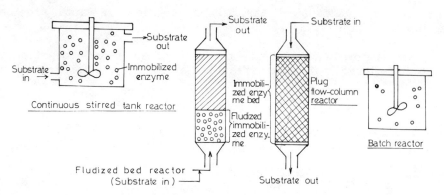

Figure 11. Reactor types. (Reprinted by permission from W.H. Pitcher, Jr. (1975).

Problem 8. Thornton et al. (1975) studied the hydrolysis of sucrose at 25°C and pH=4.5 using free and immobilized crude invertase obtained from bakers' yeast. The following data were obtained with E_0 = 408 units* of free crude enzyme and immobilized form.

Initial Sucrose Concentration (S)	$v_0 = -\dfrac{dS}{dt}$ (m mole/liter-min)	
moles/liter	Free Enzymes	Immobilized Enzymes
0.01	0.083	0.055
0.02	0.1429	0.0962
0.03	0.1876	0.1266
0.04	0.2222	0.1493
0.05	0.2500	0.1680
0.10	0.3333	0.2273
0.29	0.4084	0.2899

Using the above data,**

a) Determine the K_m and V_m for this reaction for both free and immobilized enzymes.

b) Plotting a graph of V_0 against S and $1/V_0$ against $1/S$ explain the immobilization effect of the substrate diffusion rate to the enzyme reaction rate.

* 1 unit (Enzyme unit) = Quantity of enzyme hydrolyzing 1 μ mol of sucrose/min when incubated with 0.29 M. sucrose in buffer at pH = 4.5 and 25°C.

**From Biochemical Engineering Lectures of Dr. A.E. Humphrey at The University of Pennsylvania during 1975-76 term.

Some applications of immobilized enzymes in waste treatment are shown in Table 8.

Some compounds that reduce the rate of an enzymic reaction are called inhibitors. The classification of inhibitors is shown below:

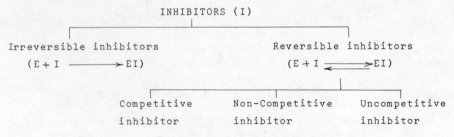

Scheme of inhibitors.

Problem 9. Weetall and Havewala (1972) report the following data for the production of dextrose from corn starch using both soluble and immobilized glucoamylase in a fully agitated CSTR system.*
From this data determine the initial rate of the reaction.

	Conversion of starch to dextrose (mg/ml)	
Time (Minute)	Free enzyme	Immobilized enzyme
0	12.0	18.4
15	40.0	135
30	76.5	200
45	94.3	236
60	120.0	260
75	135.5	258
90	151.2	262
120	155.7	278
150	164.9	310
165	170.0	306

*Data obtained at 60 °C for S_o = 336 mg starch/ml and E = 46.400 units in a reactor with an operating volume of 1800 ml.

Table 8. Some applications of immobilized enzymes in waste treatment (Adapted from H.H. Weetall 1975).

Application	Specific enzymes
Hydrolysis of cheese whey	Lactase
Hydrolysis of starches	Amylases (alpha and beta)
Dextrose production (from starches)	Amyloglucosidase
Hydrolysis of cellulose	Cellulase
Convert dextrose (glucose) to fructose	Glucose isomerase
Hydrolysis of raffinose in beet molasses	α-galactosidase
Hydrolysis of lipids	Lipases
Hydrolysis of pectins	Pectinases

Accepting steady state assumption and using simple Michaelis-Menten kinetics, the following rate equations can be derived for each class of inhibitors.

$$\frac{1}{V} = \frac{K_m(1 + K_i I)}{V_m} \frac{1}{S} + \frac{1}{V_m} \text{(For Competitive Inhibitors)} \quad (36)$$

$$\frac{1}{V} = \frac{1 + K_i I}{K V_m} \frac{1}{S} + \frac{1 + K_i I}{V_m} \text{(For Non-Competitive Inhibitors)} \quad (37)$$

$$\frac{1}{V} = \frac{1}{KV_m}\frac{1}{S} + \frac{1 + K_i I}{V_m} \quad \text{(For Uncompetitive Inhibitors)} \tag{38}$$

Where, K_i is equilibrium constant between E, I, and EI; K is equilibrium constant between E, S, and ES.

Problem 10. Plot a Lineweaver-Burk diagram between $1/V$ and $1/S$ for Eqns. 36, 37 and 38. Explain the graphs for three kinds of inhibitors.

Problem 11. The experimental data on the hydrolysis of starch to maltose by α-amylase are given below. (Adapted from S. Aiba, A.E. Humphrey, N.F. Millis 1973 and B. Pekin 1979).

Starch Concentration S (mg/ml)	12.56	11.24	9.00	8.12	6.33	5.61	4.28	3.56	2.34	1.00
Relative rate of Hydrolysis (V)	101	98.2	92.4	90.0	82.7	79.1	70.9	65.0	51.7	28.8

When 3.34 mg/ml α-dextrin is added into the reaction mixture, the following data are obtained for the same reaction.

S (mg/ml)	33.0	20.0	12.9	10.0	6.06	3.64	2.82	1.84	1.60	1.43
V	116	109	102	85.5	71.5	55.6	47.6	35.6	32.2	28.5

From these data explain the effect of α-dextrin on α-amylase and calculate K_i value for α-dextrin.

Some compounds called cofactors may be required for the activity of a particular enzyme. The classification of cofactors is shown below.

COFACTORS

Enzyme bound
(Prosthetic groups)
i- Organic
ii- Inorganic (Metalloenzymes)

Generally free in Solution
i- Organic (Coenzymes)
ii- Metal ions (Activators)

Scheme of cofactors.

In the presence of activators, the enzymic reaction rate can be expressed as:

$$\frac{1}{V} = \left(\frac{\overline{K}_1 \overline{K}_A}{V_m A} + \frac{\overline{K}_1}{V_m} \right) \frac{1}{S} + \frac{1}{V_m} \tag{39}$$

Here, A is the concentration of activator, K_1 and K_A are equilibrium constants depending on the equilibrium between substrate, activator and enzyme.

PART II: GENERAL CONCEPTS OF WASTE UTILIZATION

INTRODUCTION

As was mentioned in Part I, treatment and utilization of wastes are very important from the point of economics as well as environmental pollution problems.

The agricultural and forestry wastes, or raw materials such as farm crops, bagasse, woods, etc., and the wastes or by-products from industries such as cane and beet molasses, whey, sulphite waste liquor, vinasse, some starchy materials and so on or, urban and domestic wastes are extremely important sources for the production of either microbial biomass (mainly bakers' yeast and single cell protein) or bio-energy as well as for the production of some fine chemicals and pharmaceuticals.

There are more than 2 billion tons of solid waste per year in the world, half of them being cellulosic materials. Agricultural and food wastes, urban refuse and manure are important sources of cellulosic wastes. Solid wastes in the U.S. are shown in Table 9 below.

It is estimated that 2 million tonnes of single-cell protein, mostly yeast, are produced annually in the world (Rose 1979).

The ecological carbon balance is shown in Figure 12.

Table 9. U.S. solid wastes (From A.E. Humphrey 1975).

Waste type	Ton/year x 10^6
Agricultural and food wastes	400
Manure	200
Urban refuse	150
Logging and other wood wastes	60
Industrial wastes	45
Municipal sewage solids	15
Miscellaneous organic wastes	70
Total :	940

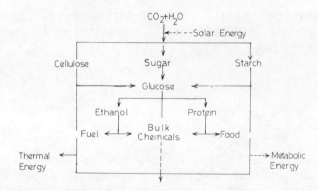

Figure 12. The ecological carbon balance (From E.L. Gaden, Jr. 1975).

As was mentioned in Part I, bacteria, yeasts, molds and algae used as a source of protein is called single-cell protein (SCP) and can be utilized as human or animal food. A general formula for the production of SCP from wastes is described below:

$$(CH_2O)_x + O_2 + NH_3 \longrightarrow C_5H_7NO_2 + CO_2 + H_2O$$

Wastes: Bacteria
Cellulose   Yeast
Starch   Mold
Sugar   Algae

High growth rates of microorganisms make their utilization as SCP economically feasible.

Comparison of the doubling times of some organisms is given below (Table 10).

Besides yeasts and bacteria, more recently, algae are also becoming a very important source for SCP production called algae-SCP.

Table 10. Doubling time of some organisms (From A.E. Humphrey 1975).

Organisms	Time for one mass doubling
Bacteria and yeast	10-120 minutes
Mold and algae	2-6 hours
Grass and some plants	1-2 weeks
Chickens	2-4 weeks
Pigs	4-6 weeks
Cattle	1-2 months
People	0.2-0.5 years

WASTE FOR PRODUCTION OF MICROBIAL BIOMASS AND BIO-ENERGY

Their desirable properties are: 1. High growth rate; 2. High nutritive value; 3. High protein content; and 4. Easy to harvest and process; 5. Ability to grow on sewage (in the case of high-rate protein ponds). The following algae have been tested for mass cultivation: <u>Chlorella</u>, <u>Scenedesmus</u>, <u>Spirulina</u>, <u>Coelastrum</u>, <u>Uronema</u>, <u>Dunaliella</u>; in sewage and waste-water cultures: <u>Micractinium</u>, <u>Docystis</u>, <u>Oscillatoria</u>, <u>Chlamydomonas</u>, <u>Euglena</u>, <u>Ankistrodesmus</u>.

Table 11 shows the chemical composition of different algae compared with soya.

As an example, a flow chart of the cultivation and processing of microalgae is shown in Figure 13.

Agricultural, industrial and municipal wastes are also utilized in the production of bioenergy such as ethanol, biogas and hydrogen.

In the following sections utilization of different types of wastes, microbial biomass (SCP) and bio-energy production will be described briefly.

PRODUCTION OF MICROBIAL BIOMASS FROM CELLULOSIC MATERIALS

Cellulose, in waste or raw materials, is usually associated and complexed with other polymers, such as hemicelluloses and lignins. Hemicelluloses are polysaccharides occurring in the plant cell walls associated with cellulose in lignified tissues.

They are composed of various hexoses, pentoses, uronic acids and other minor sugars.

Hydrolysis of cellulose can be effectively carried out by both acid or cellulase enzyme.

The presence of lignin in cellulosic materials decreases the susceptibility of cellulose to enzymic attack. Many different methods have been improved to remove lignin and change the physico-chemical properties of cellulose such as pore structure and crystallin-

Table 11. Chemical composition of different algae compared with soya (% dry matter) (From E.W. Becker 1981).

Component	Scenedesmus	Spirulina	Chlorella	Soya
Crude protein	50-55	55-65	40-45	35-40
Lipids	8-12	2- 6	10-15	15-20
Carbohydrates	10-15	10-15	10-15	20-35
Crude fibre	5-12	1- 4	5-10	3- 5
Ash	8-12	5-12	5-10	4- 5
Moisture	5-10	5-10	5-10	7-10

As an example a flow chart of the cultivation and processing of microalgae is shown in Fig.13.

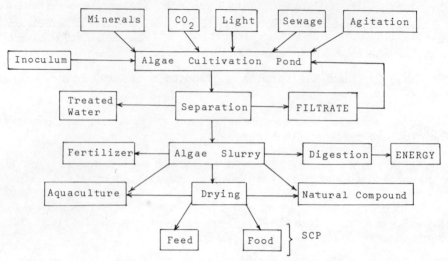

Figure 13. Flow diagram of the cultivation and processing of microalgae (Reprinted by permission from E.W. Becker 1981).

ity. This kind of modification of cellulose structure improves its swelling ability and increases the hydrolysis rate.

Amorphous cellulose is relatively easy to hydrolyze by enzymes or acid while crystalline cellulose has more resistance to hydrolysis. The ratio of crystalline to amorphous cellulose is variable in nature, the average being 85 percent crystalline.

The method of pretreatment includes the following processes: sodium hydroxide, ammonia, acids, kraft, bisulphite, sulphur dioxide, peracetic acid, dry heat steaming (under pressure) and irradiation. These processes have been used singly or in combination with each other, or with milling.

Intense research work has been devoted to cellulose hydrolysis aiming at the utilization of this renewable source of raw and waste material for SCP, ethanol and other chemicals.

Products from cellulosic waste degradation are shown in the following diagram (Figure 14).

The enzymic action leading to an effective breakdown of the native cellulose to glucose requires several basic cellulase (enzyme) components. These are the endoglucanase (C_x), the exoglucanase (C_1) and cellobiase (C_b). C_x is an endo and random-cutting enzyme and it is characterized by the release of free fiber from filter paper or production of glucose from the carboxymethyl cellulose (CMC).

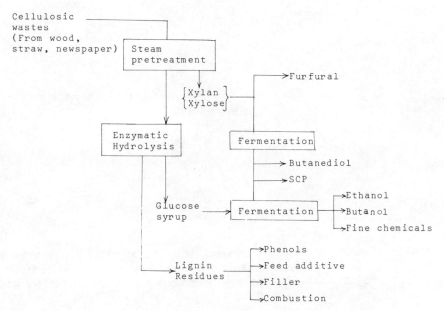

Figure 14. Products from cellulosic waste degradation (Reprinted by permission from K. Buchholz et al. 1980/81).

The endogluconase (C_x) has been assumed to be capable of cutting the cellulose molecule in the middle of a chain to create reactive ends for the subsequent action of exogluconase. Exogluconase (C_1) is an end-cutting enzyme with a nickname "Avicellase". This enzyme is mainly responsible for the production of the cellobiose from the crystalline cellulose. C_b is an enzyme specific for cellobiose and hydrolyses it to glucose. Acting together C_1 and C_x can solubilize native cellulose to cellobiose but neither C_1, nor C_x can do this job alone. This is the so-called synergistic effect. A schematic diagram showing action of the three-enzyme cellulase complex is shown in Figure 15.

The time course of sugar production can be derived from Michaelis-Menten kinetics and be expressed by the following formula:

$$t = \frac{K_m}{V_m} \ln \frac{S_o}{S_o - P} + \frac{P}{V_m} \qquad (40)$$

Where S_o is the initial substrate concentration, P is the soluble product concentration, $S = S_o - P$ and

$$V_m = k_2 E_o \qquad (41)$$

Various types of product inhibition have been included by many workers to modify the basic equation 40. For instance, Howell and

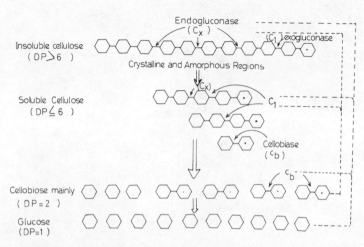

Figure 15. Schematic diagram showing the action of three-enzyme cellulase complex. (Dashed lines show possible feedback inhibition. DP = degree of polymerization) (Adapted with some alteration from J.E. Bailey 1981).

Stuck (1975) assumed a non-competitive inhibition and derived the time-course equation in the following form:

$$t = \frac{K_s}{V_m}(1 + \frac{S_o}{K_i})\ln\frac{S_o}{S_o - P} + (1 - \frac{K_s}{K_i})\frac{P}{V_m} + \frac{P^2}{2K_i V_m} \qquad (42)$$

where K_s, K_i are dissociation constants for the ES and EP complexes Huang (1975) considered a fast adsorption of the enzyme followed by a slow hydrolysis and subsequent product inhibition. The time-course was obtained in the following form:

$$t = \frac{1 + KE_o + K'S_o}{K_2 X_m KE_o} \ln\frac{S_o}{S_o - P} + \frac{X_m K - K'}{K_2 X_m KE_o} P \qquad (43)$$

where K, K', K_2 are constant and X_m is the adsorption parameter.

Some information on growth of microbes on cellulosic materials and production SCP are given in Table 12.

MICROBIAL BIOMASS FROM SULPHITE-WASTE LIQUOR

Sulphite-waste liquor is waste from wood pulp mills. A strong sulphite solution is used to dissolve lignin and other components from wood to release the cellulose fibres. It has a very high BOD ranging from 25,000 to 50,000 mg.l^{-1} and therefore presents a serious environmental pollution problem. The typical composition of sulphite-waste liquor, dry basis, is lignin sulphonate 50%, mixed sugars 18%, sugar sulphonic acids 8%, calcium salts 8%, total SO_2 10% and total solids 8-14% (Webb 1964).

A large proportion of mixed sugars of sulphite-waste liquor are pentoses. They can not be directly assimilated by human beings nor by most species of microorganisms. But the Candida group of yeasts can be adapted to make efficient use of pentoses; therefore, they can be used to produce SCP from sulphite-waste liquor.

MICROBIAL BIOMASS FROM STARCHES AND STARCHY MATERIALS

The cost of food-grade starches make them economically unrealistic to use for SCP, but plants processing potatoes, corn and other related food factories have some quantities of starch in their effluents. Some microorganisms such as Aspergillus oryzae (fungus) or another thermophilic strain. of A. fumigatus and a symbiotic culture of the yeast Endomycopsis fibulizer and Candida utilis or some others are used to produce SCP from starch-containing effluents.

Enzymic processes are used to modify starchy foods. The mechanism of action of various types of amylases on starch is shown in Figure 16.

As is shown in Figure 16, α-amylase which is an endo-enzyme, cleaves the starch within the molecule to form dextrins. β-amylase, an exo-enzyme, acts only on the end molecule chain and cleaves off sets of two glucose molecules (maltose). Amyloglucosidases (or gluco-amylase) are capable of splitting off glucose molecules from the end of a chain. α - 1,6 - glucosidase (or R-enzyme) can only break down the branch points with α - 1,6 - linkage.

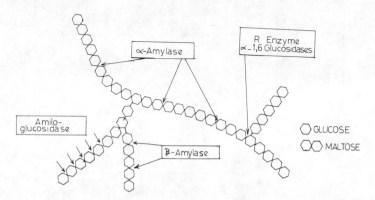

Figure 16. Action of various types of amylases (Reprinted by permission from E. Ter Haseborg 1981).

Table 12. Some information of growth of microbes on cellulosic materials and production SCP. (Some parts reprinted by permission from J.S. Knapp and J.A. Howell 1980).

Microorganisms	Cellulosic Materials	Pretreatment	Culture Time (hr)	Rate of Protein Production (g.protein/l/h)	Yield $Y_{Protein/Substrate}$
Trichoderma viride	Barley straw	5.7 % NaOH	48	0.0383	0.184
T. viride (QM 9414)	Cellulose	—	36	0.0306	0.11
Cellulomonas SP.	Cellulose powder	NaOH	96	0.024	0.115
Myrothecium verrucaria	Newspaper	Ballmill	144	0.023	0.0825
M. verrucaria	Wood pulp	Ground to 60	120	0.0124	0.1
Thermoactinomyces SP.	Micro crystalline cellulose (Avicel)	—	16	0.0343	0.11
Thermoactinomyces SP.	Cellulose	—	13	0.135	0.175
Thermonospora fuseu	Wood fibers (Pulping fines)	—	96	0.004	0.08
Sporotrichum pulverulentum	Powdered cellulose	NH_4 (as-N-Source)	144	0.0035	0.05
Chaetomium cellulolyticum	Solkafloc cellulose	—	36	0.0306	0.11
Chaetomium cellulolyticum	Sawdust	Ground to 40 mesh 1 % NaOH	36	0.05	0.18

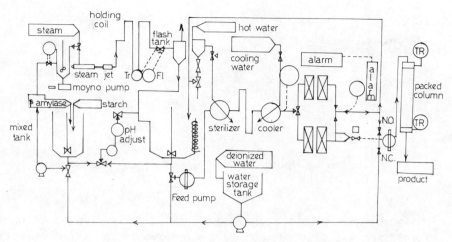

Figure 17. Process flow diagram of immobilized glucoamylase pilot plant (Reprinted by permission from B. Solomon 1978).

The hydrolysis of starch to glucose makes starch and starch wastes a better substrate for growing microbial biomass (for SCP) or to produce alcohol which will be explained later. Glucose from starchy materials is also used for production of fructose.

Recently, immobilized glucoamylase (enzyme attached to alkylamine porous silica glass using glutaraldehyde) was employed in a pilot plant for the production of glucose. The flow diagram of the process is shown above (Figure 17).

Glucose isomerase (bacterial or fungal origin mainly of <u>Streptomyces</u> species) converts glucose to fructose with about a 50 percent yield, thus producing invert sugar from glucose. Fructose is about 2.8 times sweeter than glucose and may be suitable as a substitute for sucrose (which is more expensive).

PRODUCTION OF MICROBIAL BIOMASS FROM MOLASSES

Molasses is a by-product of beet and cane sugar refining. Both sugar beet and sugar cane molasses are used for the production of the yeasts such as <u>Saccharomyces cerevisiae</u>, <u>Candida utilis</u>, and bakers' yeast.

Molasses is the source of carbohydrates as well as other organic and inorganic nutrients and many vitamins. The composition of molasses changes according to country of origin, season, factory, etc. The gross composition of both beet and cane molasses is given in Table 13.

Composition	Beet Molasses	Cane Molasses
Sucrose	48.5	33.4
Raffinose	1.0	-
Invert	1.0	21.2
Ash	10.8	9.8
Organic (non-sugar)	20.7	19.6
Nitrogen	1.5-2.0	16.0
Water	18.0	

Table 13. The gross compositions of beet and cane molasses. (From A.H. Roseled 1979.)

Molasses requires little or sometimes, no pretreatment before it is used for SCP production. In some processes the mineral content of molasses is lowered by acid clarification; it is heated and aerated to remove sulphur dioxide which can inhibit yeast growth. Generally, molasses liquor is diluted to 4-6 percent sugar concentration and sufficient nitrogenous and phosphate nutrients are added. Besides SCP production, molasses can also be used for ethanol production, extensively.

PRODUCTION OF MICROBIAL BIOMASS FROM WHEY

Whey is a by-product of cheese manufacture containing mostly lactose plus small amounts of non-coagulatable protein, citric and lactic acids, minerals, vitamins and trace amounts of fat. According to the types of cheese manufacture, low acidic (or sweet) whey and high acidic (or sour) whey can be obtained. In Table 14, the gross compositions (sweet and sour whey) are shown.

Different types of microorganisms have been used to produce SCP from whey. Some of them are: Saccharomyces fragilis (now Kluyveromyces fragilis), K. lactis, Candida krusei, C. pseudotropicalis, C. utilis, C. intermedia, Torulopsis cutaneum, Oidium lactis, etc.

Lactase (β-galactosidase) hydrolyses the lactose into glucose and galactose. Therefore, this enzyme makes waste whey a better substrate for growing microbial biomass for SCP.

Lactase is produced from yeast such as S. fragilis, Z. lactis and C. pseudotro picalis grown on lactose.

Immobilized lactase has found a very important application in the treatment of waste whey. A flow sheet for lactose hydrolysis by immobilized lactase is shown in Figure 18.

Table 14. The gross compositions of sweet and sour whey (From A.H. Rose 1979).

Components	Sweet Whey (%)	Sour Whey (%)
Dry matter	6-7	5-6
Ash	0.5-0.7	0.7-0.8
Protein	0.8-1.0	0.8-1.0
Lactose	4.5-5.0	3.8-4.2
Lactic acid	Traces	up to 0.8
Citric acid	0.1	0.1

The fermentation of whey lactose to lactic acid by Lactobacillus bulgaricus is also used for the production of lactic acid.

Most of the fruit and vegetable processing factories have also large quantities of trimmings which often present considerable disposal problems. Much of this material has a large sugar content and can be converted quite readily to fodder yeast and SCP or ethanol.

PRODUCTION OF ALCOHOL FROM WASTES

Increase in the cost of oil since 1972 added special significance to the production of ethanol from sugars, starchy or cellulosic materials and wastes. The total industrial ethanol production in the world is approximately 3.2 billion liters (800 million gallons) per year (Maiorella et al. 1981). Three-quarters of this production uses the same techniques developed prior to the 1940's; a slow batch process, followed by an efficient multistep distillation. Recently, some alternative processes such as CSTR, or series of CSTR's, are under development. These are termed high rate processes. In these processes some modified fermenters such as tower, dialysis, plug, flow, hollow fiber and pressure rotorfermenter are being considered and used.

The microorganisms most suitable for the production of ethanol from sugars are yeasts of the genera Saccharomyces and Kluyveromyces. Clostridium thermosaccharolyticum and other thermophilic bacteria, and Zymomonas mobilis (bacterium) are also under development to use in ethanol production. A genetic modification of the bacterium Z. mobilis to give a high product concentration without inhibition is currently being investigated.

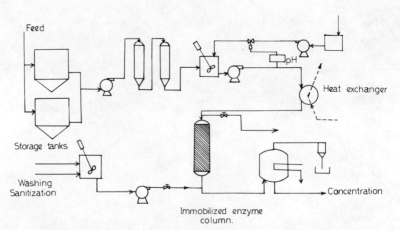

Figure 18. A flow sheet of lactose hydrolysis by immobilized lactase (Reprinted by permission from W.H. Pitcher, Jr. 1975).

A schematic representation of the metabolic pathways for anaerobic and aerobic yeasts are shown in Figure 19.

As is shown in Figure 19, under anaerobic conditions, glucose is converted to ethanol and CO_2 by glycolysis. The overall reaction liberates energy for biosynthesis by producing 2 moles of ethanol and CO_2 for every mole of glucose consumed. Thus, the following equation can be written, theoretically.

$$C_6H_{12}O_6 \xrightarrow{\text{Yeast}} 2C_2H_5OH + 2CO_2 + \text{Energy (Stored as ATP)}$$

180 g.   92 g.   88 g.

In practice actual ethanol yield is about 90 percent of the theoretical production because some portion of glucose is being used for biosynthesis of new cellular material.

Under aerobic conditions, glucose is converted completely to CO_2 and water and no ethanol is formed.

Sugars, starches, molasses, cellulosic materials, agricultural and forestry, urban and some industrial wastes can be used for ethanol production.

The ability of microorganisms to assimilate carbohydrates depends upon the substrate used as well as the strains of yeast and fermentation conditions. The abilities of various yeast species to ferment sugars are given in Table. 15.

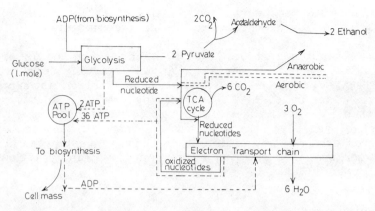

Figure 19. Simplified chart of anaerobic and aerobic catabolism by S. cerevisiae (Reprinted by permission from B. Maiorella et al. 1981).

Table 15. The ability of Saccharomyces and Kluyveromyces species to ferment some sugars. (Reprinted by permission from R. P. Jones et al. 1981.)

Sugar	Basic unit	YEASTS S.Cerevisiae	S.Uvarium	K.Fragilis
Glucose	Glucose	+	+	+
Maltose	Glucose	+	+	-
Maltotriose	Glucose	+	+	-
Cellobiose	Glucose	-	-	-
Trehalose	Glucose	+/-	+/-	-
Galactose	Galactose	+	+	+
Mannose	Mannose	+	+	+
Lactose	Glucose, Galactose	-	-	+
Mellobiose	Glucose, Galactose	-	+	
Fructose	Fructose	+	+	+
Sorbose	Sorbose	-	-	-
Sucrose	Glucose, Fructose	+/-	+	+/-
Rhamnose	6-Deoxymannose	-	-	-
Deoxyribose	2-Deoxyribose	+/-	+/-	+/-
Aldose	Arabinose	-	-	-
Xylose	Xylose	-	-	-

In Table 16 some waste and raw materials containing carbohydrates suitable for alcohol fermentation are summaried.

As was mentioned above, either batch or continuous processes are used for ethanol fermentation. The starting temperature is mostly 21-27° C. Since the fermentation reaction is exothermic, temperature rises up to 32-40°C during the process. Initial pH must be less than 5. The main limiting factor for alcohol yield is the inhibition of yeast growth by ethanol. Ethanol inhibition of yeast growth and ethanol production have been investigated by some research workers through some mathematical models. A summary of these is given in Table 17.

PRODUCTION OF METHANE FROM WASTES

Methane gas biologically produced from waste is becoming very important among the new alternative sources of energy. Traditionally, anaerobic digestion has been used as a method of treating municipal sewage, but more recently it is being used to produce methane gas, so-called "biogas," from garbage, food and industrial wastes, manures and crop-residues (Clausen et al. 1981).

Table 16. Carbohydrate raw materials suitable for fermentation to ethanol. (Reprinted by permission from R.P. Jones 1981).

Waste raw materials	Hydrolysis products
Sucrose: Cane Sugar, Beet Sugar, Molasses, Cannery Wastes	Glucose + Fructose
Starch: Corn and other Grains, Cassava, Potatoes, Jerusalem Artichokes, Sago, Toro etc.	Glucose + Mannose + Maltotriose + Highes M.W. Dextrins
Cellulose: Wood and Wood Wastes, Agricultural Residues (Bagasse, straw etc.), Municipal Wastes	Glucose + Mannose + Galactose + Arabinose + Xylose
Lactose: Whey	Glucose + Galactose

Table 17. Models for ethanol inhibition of yeast growth and ethanol production. (Reprinted by permission from R.P. Jones 1981).

Investigators	Holzberg et al	Aiba et al	Bazua and Wilkie
Kinetic equation (Growth)	$\mu = \mu_o - k_1(P - k_2)$	$\mu = \mu_o e^{-k_1 P} \dfrac{S}{K_S + S}$ or $\mu = \dfrac{\mu_o}{1 + P/K_P} \dfrac{S}{K_S + S}$	$\mu = \mu_o \dfrac{k_1 P}{k_2 - P} \dfrac{S}{K_S + S}$
Kinetic equation (Ethanol production)	$\dfrac{dP}{dt} = k_3\left(\dfrac{\ln X}{\mu}\right) - k_4 P - k_5$	$\nu = \nu_o e^{-k_2 P} \dfrac{S}{K'_S + S}$ or $\nu = \dfrac{\nu_o}{1 + P/K_P} \dfrac{S}{K'_S + S}$	$\nu = \nu_o \dfrac{k_3 P}{k_4 - P} \dfrac{S}{K'_S + S}$
Maximum ethanol production	68.5 g/l	76.4 g/l	93 g/l
Strain	S. Cerevisiae ellipsodeus (New York Agric. Expt. Station No. 223)	S. Cerevisiae Japan Sugar Refinery Ltd. H-1	S. Cerevisiae (ATCC No. 4126)
Conditions	21 °C; pH = 3.6 Enriched grape juice	30 °C; pH = 4.0 Glucose	30 °C; pH = 4.0 Glucose
Comments	Equation only apply where fermentation is not nutrient or substrate limited	Equation predict continuous cell growth and alcohol productivity	Productivity is affected by alcohol even at low concentrations growth and ethanol production cease at finite ethanol concentration

S = Substrate concentration, g/l
P = Product concentration, g/l
X = Biomass concentration, g/l
t = Time
μ = Specific growth rate, 1/hr
ν = Specific rate of product formation, g/g-hr
All other symbols represent parameters.

According to the oxygen demand, microorganisms are classified into three groups. (Cowley and Wase 1981): 1. Aerobes: These microorganisms use molecular oxygen as the ultimate electron acceptor in their respiratory chain for obtaining utilizable chemical energy. Therefore, they require the presence of oxygen for growth. 2. Anaerobes: These microorganisms are unable to grow in the presence of oxygen and are sometimes termed "obligate anaerobes." They obtain their energy from the partial oxidation of complex organic matter and utilize compounds other than dissolved oxygen. The process by which insoluble organic material is degraded in the absence of oxygen is often called "anaerobic fermentation" or "digestion." 3. Facultative anaerobes: These are able to adjust themselves rapidly to live and grow in an oxygen-rich or oxygen-deficient environment. Relatively, few microorganisms are obligate anaerobes. A large number, however, are facultative anaerobes.

The anaerobic digestion process is very complex and can be divided into two stages. The first stage is subdivided into two parts (Eckenfelder 1976; Dowing and Kell 1976).

1. Digestion by acid-forming bacteria: This stage is carried out by acid-forming bacteria, such as <u>Clostridium</u> spp., <u>Peptococcus anaerobus</u>, <u>Actynomyces</u>, <u>Staphylococcus</u> and <u>Escherichia coli</u>, etc. It occurs in two parts:

a. Liquefaction

This process is carried out by exo-enzyme of the bacteria. During this phase, proteins are broken down into amino acids. Carbohydrates are first hydrolysed into smaller soluble compounds and then broken down further to produce mostly short chain fatty acids. Fats are hydrolysed to a mixture of long chain fatty acids and glycerol. This is a rapid process and occurs in less than five days.

b. Acidification or acid-fermentation

The solubilized organic matter is rapidly converted to organic acids by both anaerobic or facultative acid-forming bacteria, under anaerobic conditions. In this phase, the primary acids produced are acetic, propionic, and butyric with trace amounts of formic, valeric, isovaleric and caproic. During this process both CO_2 and H_2 gases are formed and sufficient energy is released for bacterial growth. This phase is characterized by a drop in pH from near neutral to about pH 5.0: After this process the conversion of acids into methane and carbon dioxide occur and the pH rises to 6.8-7.4.

2. Gasification or methane fermentation: In this stage of the process, methane-forming bacteria such as Methanobacterium, Methanobacillus, Methanococcus and Methanosarcina break down the products (which occurred in the acidifaction phase) to CH_4 and CO_2.

In Figure 20 anaerobic digestion process is shown, schematically.

Although, very small amounts of hydrogen and H_2S are produced depending on the composition of waste materials, the overall conversion of organic matter into methane and carbon dioxide can be repsented stoichiometrically as

$$C_nH_aO_b + (n - \frac{a}{4} - \frac{b}{2})H_2O \longrightarrow (\frac{n}{2} - \frac{a}{8} + \frac{b}{4})CO_2$$
$$+ (\frac{n}{2} + \frac{a}{8} - \frac{b}{4}) CH_4$$

Anaerobic digestion occurs at a temperature between 5-55°C (41-131°F). Two sets of bacteria, the mesophiles 5-40°C (41-104°F), and the thermophiles 40-55°C (104-131°F) are effective within this range.

Optimum gas production occurs at 35°C (mesophilic digestion) or 55°C (thermophilic digestion).

The effect of temperature on methane production is shown in Figure 21.

The best example for anaerobic degradation, in living systems, is rumen microflora. In this "living fermenter" a multispecies microflora digest insoluble substrates like cellulose to produce carbon compounds (mainly volatile fatty acids such as acetic and propionic) which can be assimilated by the animal. During the process, methanogenic bacteria act as hydrogen utilizers. Figure 22 illustrates these processes.

MICROBIAL AND ENZYMATIC PRODUCTION OF HYDROGEN

Hydrogen gas is formed via different metabolic pathways in several photosynthetic and non-photosynthetic bacteria and in algae cultures. Hydrogen is a real clean and ideal energy source. The mechanism of hydrogen production varies according to types of microorganisms present in the process. For example, Clostridium and E. coli help to increase the rate of the following reversible reaction to form hydrogen, in different pathways:

$$2H^+ + e^- \rightleftharpoons H_2$$

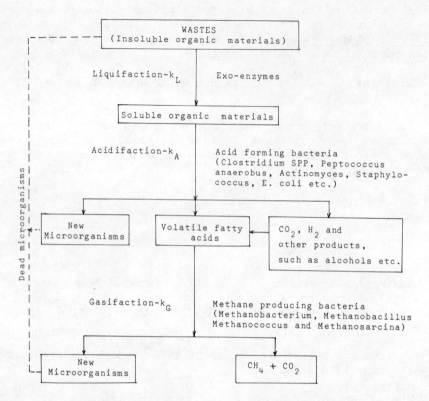

Figure 20. Sequence of anaerobic digestion.

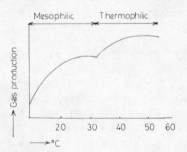

Figure 21. Effect of temperature on gas production in anaerobic digestion (From T.H.Y. Tebbutt, 1977.)

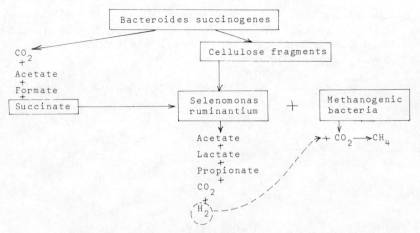

Figure 22. Degradation of biopolymers (cellulose) by typical rumen microorganisms and illustration of hydrogen transfer. (By permission from C. Rolz and A. Humphrey 1982.)

As it is known, pyruvate is a major intermediate in the carbohydrate metabolism. The <u>Clostridia</u> and <u>Coliaerogenes</u> bacteria convert pyruvate into acetic acid and carbon dioxide with the formation of hydrogen gas. This process occurs in the presence of an hydrogen acceptor under anaerobic conditions. In this reaction several enzymes and electron-carrying elements such as ferredoxin and cytochrome-c play an important role. Biochemistry of the production of hydrogen formation in the <u>E. coli</u> and clostridial system are represented (Zajic et al. 1978).

<u>Escherichia coli system:</u>

1- Pyruvate + E $\rightleftharpoons$ Acetyl-E + Formate (HCOOH)

    E = Formate lyase

2- Acetyl-E + CoA $\rightleftharpoons$ E + Acetyl-CoA

3- $\underbrace{CH_3CO\text{-}S\text{-}CoA}_{\text{Acetyl-CoA}} + HPO_4^2 \rightleftharpoons \underbrace{CH_3COPO_4}_{\substack{\text{Phospho-}\\\text{transace-}\\\text{tylase}}} + \underbrace{HS\text{-}CoA}_{\substack{\text{Acetyl-}\quad\text{CoA}\\\text{phosphate}}}$

4- HCOOH + Fd_o $\rightleftharpoons$ CO_2 + Fd_r

             Formate
             dehydrogenase

    Fd_o = Oxidized form of ferredoxin

    Fd_r = Reduced form of ferredoxin

5- Fd_r + 2H $\rightleftharpoons$ Fd_o H_2

             Hydrogenase

In the mechanisms 4-5, the enzyme systems "formate dehydrogenase" and hydrogenase" are expressed as "formic hydrogenlyase" and during the decomposition of formic acid, electrons are transferred via cytochrome C_{552}. This can be shown as:

$$\text{Chtochrome } C_{552} \longrightarrow \text{Flavodoxin} \longrightarrow \text{Ferredoxin } (Fd_r)$$

Clostridial system:

1- Pyruvate + TPP - E_o ⇌ Hydroxyethyl - TPP - E_o + CO_2

   E = Ferredoxin oxidoreductase (O = Oxidized form; r = reduced form)

   TPP = Thiamine pyrophsphate

2- Hydroxyethyl - TPP - E_o + CoASH ⇌ Acetyl - S - CoA + TPP - E_r

3- TPP - E_r + Fd_o ⇌ TPP - E_o + Fd_r

4- $\underbrace{CH_3CO-S-CoA}_{\text{Acetyl-CoA}}$ + HPO_4^{2-} ⇌ $\underbrace{CH_3COPO_4}_{\text{Acetyl-phosphate}}$ + $\underbrace{HS-CoA}_{\text{CoA}}$

   Phospho-transacetyase

The mechanisms of hydrogen gas production from <u>E. coli</u> and <u>Clostridium</u> are shown comparatively in the following diagram (Figure 23).

The hydrogenase enzyme is present in many microorganisms and it catalyzes the following reaction, reversibly.

$$2H^+ + 2e^- \xrightleftharpoons{\text{Hydrogenase}} H_2$$

The mechanism of this reaction is shown in Figure 24.

In this mechanism some reductants such as ferredoxin, cytochrome C_3 and reduced methyl viologen play an important role as immediate electron donors for hydrogenase. Properties of some bacterial hydrogenases are shown in Table 18.

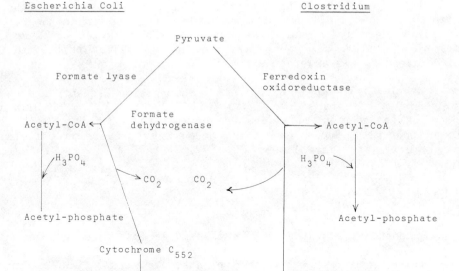

Figure 23. Comparative diagram showing the production of hydrogen gas from the anaerobic breakdown of pyruvate by saccharolytic clostridia and E. coli. (Reprinted by permission from J.E. Zajic et al. 1978.)

In the green plants and alage (such as **Chlorella**, **Scenedesmus**, **Chlamydomonas**, etc.) under natural circumstances, solar energy of radiation causes the oxidation of water and an oxygen molecule is released during this process. The released electrons are utilized in the conversion of carbon dioxide and production of hydrogen molecules via several carriers. The resultant equation describing this phenomenon taking place in plants and algae are as follows:

$$2H_2O \longrightarrow 4H^+ + 4e^- + O_2$$

$$4H^+ + 4e^- \longrightarrow 2H_2$$

This process is known as water "biophotolysis." The energy conversion efficiency (η) of this process is:

$$\eta = \eta' \left[\frac{\Delta G}{nE_o} \right] \tag{44}$$

Where is the solar light-fraction absorbed by photosynthetic pigments, G= the free energy of hydrogen oxidation, n=the number of quanta needed for the formation of 1 mol of oxygen (or 2 mol of hydrogen), and E_o=the energy per quantum at maximum photosynthetic effectiveness. As an example for photosynthesis:

G = 113.4 KCal/mol-O_2; n = 8 (suspposing 2 quanta are needed per electron transfer); E_o=40 KCal/mol (at λ_o = 700 nm) and η' = 0.4 (because green plants are capable of absorbing 40-50 % of solar energy). From these data, the energy conversion efficiency can be calculated as follows:

$$\eta = 0.4 \; \frac{113.4}{8 \times 40} = 14\% \tag{45}$$

Biophotolytic systems may be classified into three groups:

1. Chloroplasts of higher plants, ferredoxin, and bacterial hydrogenase. The mechanism of this system is as follows (Beremann et al. 1973; Rao et al. 1976).

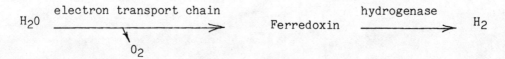

2. Chloroplasts, a low molecular-weight electron carrier (mediator) and bacterial hydrogenase (Brezin and Varfolomeev 1979).

$H_2O \xrightarrow[O_2]{\text{electron transport chain}} M_r \xrightarrow{\text{hydrogenase}} H_2$

Where M_r is the reduced form of the low-molecular weight mediator.

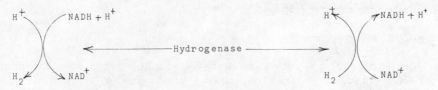

Figure 24. The mediation of the reversible reaction between hydrogen and NAD+ by the enzyme hydrogenase. (By Permission from J.E. Zajic et al. 1978.)

3. Systems built around hydrogen-producing microbial cells with a process sequence of

$$H_2O \xrightarrow[O_2]{\text{algae in light}} \text{Energy-Consuming metabolites (in light)}$$

$$\text{Metabolites} \longrightarrow H_2 \text{ (in dark)}$$

The electron transport mechanism in plants during photosynthesis is a rather complex phenomenon. A tentative diagram for electron transport in plant photosynthesis is shown in Figure 25.

Purple bacterium is a poor system to use in bio-solar energy conversion processes. Hydrogen photoproduction by green or blue-green algae is apparently due to an electron transport driven by direct light. However, the situation in vivo is very complicated. Some light and dark reactions in these systems are shown below (NSF 1974):

Light-Dependent Reactions:

1- Photosynthesis: $CO_2 + 2H_2O + \text{Light} \longrightarrow (CH_2O) + O_2 + H_2O$

2- Photo-reduction: $CO_2 + 2H_2 + \text{Light} \xrightarrow{\text{Hydrogenase}} CH_2O + H_2O$

3- H_2-Photoproduction: $RH_2 + \text{Light} \xrightarrow{\text{Hydrogenase}} R + H_2$

Dark Reactions:

Dark Reactions:

4- Respiration $1/2 O_2 + RH_2 \longrightarrow H_2O + R + \text{Energy}$

5- Dark H_2-Production: $RH_2 \xrightarrow{\text{Hydrogenase}} R + H_2$

6- H_2-Absorption: $H_2 + R \xrightarrow{\text{Hydrogenase}} RH_2$

7- Oxy-Hydrogen reaction: $O_2 + 2H_2 \xrightarrow{\text{Hydrogenase}} 2H_2O + \text{Energy}$

Overall Reaction: $CO_2 + 2H_2 + \text{Energy} \xrightarrow{\text{Hydrogenase}} (CH_2O) + H_2O$

Table 18. Properties of some pure bacterial hydrogenases. (Reprinted by permission from Ilia V. Berezin and S.D. Varfolomeev 1979.)

Microbial source	Molecular weight	Groups in active center	Specific activity ($molH_2$/min/mg. protein)	Catalytic rate constant[x] (sec^{-1})
Clostridium pasteurianum	60,000	$4Fe^{2+}$, $4S^{2-}$ $12Fe^{2+}$, $12S^{2-}$	320	310
Chromatium	98,000	$4Fe^{2+}$, $4S^{2-}$	82,1	130
Alcaligenes entrophus (H 16)	205,000	___	48	160
(Z1)		___	54	180
Desulfovibrio	89,000	$7-9Fe^{2+}$, $7-9S^{2-}$		
Thiocapsa roseopersicina	68,000	$3.9 \pm 0.2Fe^{2+}$ $3.9 \pm 0.2S^{2-}$	3.4	3.8

(x) Computed from molecular weight and specific activity.

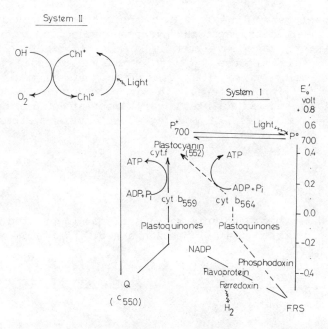

Figure 25. Tentative diagram for electron transport in plant photosynthesis (Reprinted by permission from J.E. Zajic et al. 1978). FRS = Ferredoxin reducing substance; Cytochrome f = C-type cytochrome (C_{552}) in contact with a copper protein plastocyanin, P_{700} = Pigment established as the photoreaction centre of system-1 and contains chlorophyll, phosphodoxin substance of unknown nature, Q = Substance of unknown nature although not a cytochrome, E_o = Oxidation reduction scale.

Heterocystous, filamentous blue-green algae are probably the most promising organisms for biophotolysis because of the separation of O_2 and H_2 production sites into different cells. The filament of blue-green algae contains two types of cells, so-called heterocysts and vegetative cells. The vegetative cells use light as an energy source and water as an electron donor to reduce CO_2 through the photosynthesis. In the process, oxygen is evolved and CO_2 is reduced to carbohydrates. Carbohydrate is transferred to the heterocyst where it serves as an electron donor to reduce either N_2 or H^+ to NH_3 or H_2 through the action of nitrogenase. This system is shown in Figure 26, schematically.

A three stage pilot plant system has been developed and constructed for cultivation of algae in ponds during waste water treatment for hydrogen and methane production. The diagram of this process is shown in Figure 27.

In recent years, besides the production of energy from the above mentioned living systems, hydrogen production by biophotocatlytic systems *in* vitro have been improved. In order to produce the hydrogen by this process, the chloroplasts of some plants like spinach are being mixed with ferredoxin and enzymes of acterial hydrogenase *in* vitro systems. These processes are also receiving wide attention and constitute a very important field of research. In Table 19 some models of complex biocatalysis systems for water biophotolysis are shown.

Table 19. Water biophotolysis model systems. (Reprinted by permission from Ilia. U. Berezin and S.D. Varfolomeev 1979).

System	Hydrogen photoproduction rate (mol/sec/mg) (Chlorophil II)
1. a- Spinach chloroplasts, ferredoxin, and hydrogenase	$5.8 \cdot 10^{-10}$
b- In the presence of glucose and glucoseoxidase	$33.4 \cdot 10^{-10}$
2. a- Spinach chloroplasts, spinach ferredoxin, and hydrogenase	$2.6 \cdot 10^{-10}$
b- In the presence of glucose and glucoseoxidase	$14.3 \cdot 10^{-10}$
3. a- Spinach chloroplasts, spinach ferredoxin, and hydrogenase in the presence of glucose and glucoseoxidase	$7.8 \cdot 10^{-10}$
b- In the presence of glucose, glucoseoxidase, catalase, and ethanol	$22.6 \cdot 10^{-10}$
4. Pea chloroplasts, glucose, glucoseoxidase, catalase, plastocyanine, ferredoxin, and hydrogenase	$5.0 \cdot 10^{-11}$

WASTE FOR PRODUCTION OF MICROBIAL BIOMASS AND BIO-ENERGY   361

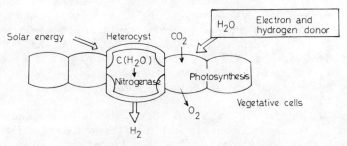

Figure 26. Production of O_2 and H_2 from filamentous blue-green algae (From J.W. Jeffries et al. 1976.)

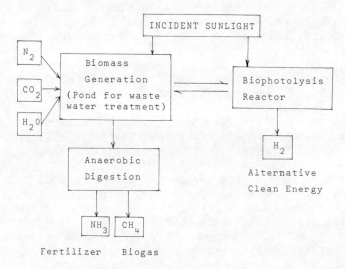

Figure 27. Diagrammatic representation of system inputs (N_2, CO_2, and H_2O) and outputs H_2, CH_4 and NH_3.

REFERENCES

Aiba, S., A.E. Humphrey and N.F. Millis. 1973. Biochemical Engineering, 2nd Ed. University of Tokyo Press.
Bailey, J.E. (personal communication, 1981). Biochemical Reaction. Engineering and Biochemical Reactors, Dept. of Chemical Engineering, University of Houston, Texas 77004, U.S.A.
Bailey, J.E. and D.F. Ollis. 1977. Biochemical Engineering Funamentals. McGraw-Hill Book Co.
Bauchop, T. and S.R. Elsden. 1960. J. Gen. Microbiol. 23:457.
Becker, E.W. 1981. Algae mass cultivation -- Production and Utilization. Process Biochemistry, 16:(5). Aug./Sept.
Beremann, J.R., Berenson, J.A., Kaplan, N.O. and M.D. Kamen. 1973. Proc. National Acad. Sci., U.S.A. 70:2317.

Berezin, I.V. and S.D. Varfolomeev. 1979. Energy related application of immobilized enzymes. In: Applied Biochemistry and Bioengineering, Wingard, L.B., Jr., Katchalski-Katzir, E and L. Goldstein, eds. Vol. 2, Academic Press.

Buchholz, K., Puls, B., Godelmann, B. and H.H. Dietrichs. 1980/1981. Hydrolysis of cellulosic wastes. Process Biochemistry, 16(1), Dec./Jan.

Clausen, E.C., Ford, J.R. and A.H. Shah. 1981. Importance of start-up in the anaerobic digestion of crop materials to methane. Process Biochemistry 16(2), Feb./March.

Cowley, I.D. and D.A.J. Wase. 1981. Anaerobic digestion of farm wastes, a review. Part 1. Process Biochemistry, 16(5), Aug./Sept.

Dowing, A.L. and A.D.K. Kell. 1976. In: Theory and Practice of Biological Wastewater Treatment, NATO ASI directed by K. Curi at Bogazici University School of Engineering, Istanbul, Turkey, July 11-23.

Eckenfelder, W.W., Jr. 1976. In: Theory and Practice of Biological Wastewater Treatment, NATO ASI directed by K. Curi at Bogazici University School of Engineering, Istanbul, Turkey, July 11-23.

Gaden, E.L., Jr. 1975. Biotechnology as old solution to new problems. Chemical Engineering Education, p.40-47, Winter, 1975.

Ghose, T.K. and A.N. Pathak. 1973. Cellulase I: Sources, Technology. Process Biochem., 35, April, 1973.

Harima, T. 1978. Energy from the sun -- Bioconversion. 25th Anniversary of Biochemical Engineering Education. The College of Engineering and Applied Science, Univ. of Pennsylvania, Philadelphia, PA. 19174. U.S.A.

Haseborg, E. ter. 1981. Enzymes in flour and baking applications, especially waffle batters. Process Biochemistry, (16)5, Aug. Sept.

Howell, J.A. and J.E. Stuck. 1975. Kinetics of Solka Floc cellulose hydrolysis by _Trichoderma viride_ cellulase. Biotech. Bioeng. 17:873.

Huang, A.A. 1975. Kinetic studies on insoluble cellulose-cellulase system. Biotech. and Bioeng. 17:1421-1433.

Humphrey, A.E. 1975. Economics and utilization of enzymatic hydrolized cellulose. Symposium on Enzymatic Hydrolysis of Cellulose, Aulanko, Finland, March 12-14, 1975.

Jeffries, J.W. et al. 1976. Biosolar production of fuels from algae. Lawrence Livermore Lab., U.S. Dept. of Commerce National Technical Information Service UCRL-52177, Nov.

Jones, R.P., Pamment, N. and P.F. Greenfield. 1981. Alcohol fermentation by yeasts -- The effect of environmental and other variables. Process Biochemistry (16)3, April/May.

Knapp, J.S. and J.A. Howell. 1980. Solid substrate fermentation. In: Topics in Enzyme and Fermentation Biotechnology, Wiseman, A., Ed., Vol. 4, Chapt. 4, pp. 85-143. Halsted Press: a division of John Wiley and Sons.

Mairolla, B., Wilke, Ch. R. and H.W. Blanch. 1981. Alcohol production and recovery. In: Advances in Biochemical Engineering, Fiechter, A., Ed. Vol. 20. Springer-Verlag.

Mateles, R.I. and S.R. Tanenbaum, Eds. 1968. Single Cell Protein, Vol. 1. MIT Press, Cambridge, Mass.

Moreira, A.R. 1978. Biological Conversion of Waste Materials into Useful Products. 25th Anniversary of Biochemical Engineering and Applied Science, Univ. of Pennsylvania, Philadelphia PA., 19174, U.S.A.

Nagai, S. 1979. Mass and energy balances for microbial growth kinetics. In: Advances in Biochemical Engineering, Ghose, T. K., Fiechter, A. and N. Blakebrought, Eds. Vol. II, Springer-Verlag.

National Science Foundation (NSF). 1974. Proceedings of the Workshop on Bio-Solar Conversion. U.S. Depart. of Commerce National Technical Information Service. (Prepared by Indiana Univ.), PB-236 142.

Pekin, B. 1979/80. Biyokimya Muhendisligi (Temel Ilkeler) Birinci Kitap, Kisim I-II (in Turkish), Ege University Kimya Fakultesi Yayini No. 3, Ege University Press.

Pekin, B. 1982. Biyoteknoloji 1. inci Baski (in Turkish), Ege Universitesi Kimya Fakultesi Ders Notlari Yayin Cogaltma No. 29.

Pekin, B. 1983. Biyokimya Muhendisligi (Biyoteknoloji) Ikinci Kitap (in Turkish), Ege Universitesi Kimya Fakultesi Yayini No. 3, Mas Ambalaj Press.

Pirt, S.J. 1975. Principles of Microbe and Cell Cultivation. A Halsted Press Book, John Wiley and Sons.

Pitcher, W.H., Jr. 1975. Design and operation of immobilized enzyme reactions. In: Immobilized Enzymes for Industrial Reactors. Messing, Ralph, Ed. Academic Press.

Pollock, M.R. 1962. Chapter IV. The Physiology of Growth, pp. 121-178. In: The Bacteria, Vol. IV. Gunwalus, I.C. and R.Y. Stanier, Eds. Academic Press.

Rao, K.K., Rosa, L. and D.O. Hall. 1976. Subject? Biochem. Biophys. Res. Commun. 68:21.

Rolz, C. and A. E. Humphrey. 1982. Microbial Biomass from Renewables: Review of Alternatives, pp. 1-53. In: Advances in Biochemical Engineering. A. Fiechter, Ed. Vol. 21. Springer-Verlag.

Rose, A.H. 1979. Economic Microbiology -- Microbial Biomass, Vol. 4 Academic Press.

Solomon, B 1978. Starch hydrolysis by immobilized enzyme/industrial applications. In: Advances in Biochemical Engineering. Ghose, T.K., Fiechter, A. and N. Blakebrough, Eds. Vol 10. Springer-Verlag.

Tebbutt, T.H.Y. 1977. Principles of Water Quality Control. Second Edition. Pergamon Press.

Thornton, D., Flynn, A., Johnson, D.B. and P.D. Ryan. 1975. The preparation and properties of hornblende as a support for immobilized invertase. Biotech. and Bioeng. 17:1679-1693.

Wang, D.I.C., Cooney, C.L., Demain, A.L., Dunnill, P., Humphrey, E.A. and M.D. Lilly. 1979. Fermentation and Enzyme Technology. John Wiley and Sons.
Webb, F.C. 1964. Biochemical Engineering. D. Van Nostrand Co., Ltd.
Weetall, H.H. 1975. Application of immobilized enzymes. In: Immobilized Enzymes for Industrial Reactors. Messing. R.A., Ed. Academic Press.
Weetall, H.H. 1975. Immobilized enzymes and their application in food and beverage industry. Process Biochem. p.3. July/Aug.
Weetall, H.H., and N.B. Havewala. 1972. Continuous production of dextrose from cornstarch. In: Enzyme Engineering, Wingard, L.B., Jr. Ed. Interscience Publishers.
Zajic, J.E., N. Kosaric and J.D. Brosseau. 1978. Microbial production of hydrogen. In: Advances in Biochemical Engineering. Ghose, T.K., Fiechter, A. and N. Blakebrough, Eds. Vol. 9. Springer-Verlag.

THE FERMENTATION OF BIOMASS - CURRENT ASPECTS

Douglas E. Eveleigh

Department of Biochemistry and Microbiology
Cook College
New Jersey Agricultural Experiment Station
Rutgers University
New Brunswick, New Jersey 08903
U.S.A.

INTRODUCTION

Biomass is a major resource from which man can harvest solar energy and offers a renewable solution to the world's deficiencies of energy, food and chemicals. The availability and composition of biomass has been detailed at this Study Institute. Its uses are diverse. This presentation addresses the application of microbes for the production of chemicals. Microorganisms are well known for their ability to grow rapidly on a wide range of biomass substrates, to effect a broad spectrum of specific chemical transformations, for their ease of culture and their reasonably high rates of synthesis of organic materials. For these reasons, bacteria and fungi have been used for the past century for the large scale production of a range of chemicals: citric acid, ethyl alcohol, enzymes and polysaccharides (Laskin et al. 1980; Office of Technology Assessment 1981; Peppler and Perlman 1979; Perlman and Tsao 1977-1982; Rose 1977).

Although microbes produce such chemicals efficiently, the fermentation industry has had a chequered history, as the success of a fermentation is dictated by numerous extrinsic economic factors. A major one is the cost of the substrate. The demise of the ethanol and butanol fermentation industries in the 1940's was due to competition from the petro-chemical industry, as the latter had an inexpensive and abundant substrate - oil. The fermentation industries have therefore focused on chemicals which were amendable to production mediated by the highly stereospecific action of microbial enzymes, e.g., L-amino acids. In other instances, where complex organic materials such as enzymes are the product of interest, biological production has been the only practical means of synthesis. Today the microbial production of chemicals is a highly viable industry, and the forecast for the fermentation industry is exceptionally bright. This is perhaps enigmatic for although microbes produce perhaps two hundred useful primary products, only six or seven of them are currently made by industrial fermentation, e.g., ethanol, isopropanol, n-butanol, acetone, glycerol and acetic, citric and lactic acids. However, the recent rapidly and disproportinately escalating price of crude oil now means that the petrochemical industry no longer has the advantage of inexpensive feedstocks for production of chemicals. The need for alternate renewable energy resources can be fulfilled in part through use of biomass. This paper addresses the microbiological conversion of biomass. The different types and components of biomass and the variety of microorganisms suggest several approaches to biomass conversion (Figure 1). Rather than attempt a complete review, four important topic areas are discussed: cel-lulase, pentose sugar utilization, the application of thermophilic microorganisms and a discussion of speciality chemicals.

CELLULOSIC BIOMASS AND CELLULASES

Biomass is primarily thought of as plant matter (wood, straw, grain*) and includes agricultural, forestry and municipal wastes. It is ubiquitously available and relatively inexpensive. In the context of an energy feedstock material, biomass is composed of cellulose, hemicellulose and lignin. The cellulose, hemicellulose and lignin components of biomass occur in roughly equivalent amounts, dependent to a degree on the plant species. Cellulose is more often the major component (Brown 1982; Côté 1983).

Native cellulose consists of linear chains of B-1,4-linked glucose units strongly hydrogen bonded to form a crystalline lattice (Figure 2; see Brown 1982 and Cote 1983 for review). It is thus resistant to hydrolysis either by acid or enzymes. Furthermore, with most biomass, the crystalline cellulose is cemented by lignin and this results in a highly stable and insoluble lignocellulose complex. Recently, a number of chemical and physical pretreatments

---

* Grains are not focused on in this discussion of production of chemical feedstocks, as they are a major food resource. However, substrates such as corn should not be absolutely excluded. Ninety-five percent of the U.S. corn crop is used as animal feed (Venkatasubramanian and Keim, 1980). In time of energy crisis, a fraction of this substrate could well serve as a further energy resource. Spoiled crops are also a potential resource (Nofsinger and Bothast 1981).

FERMENTATION OF BIOMASS-CURRENT ASPECTS

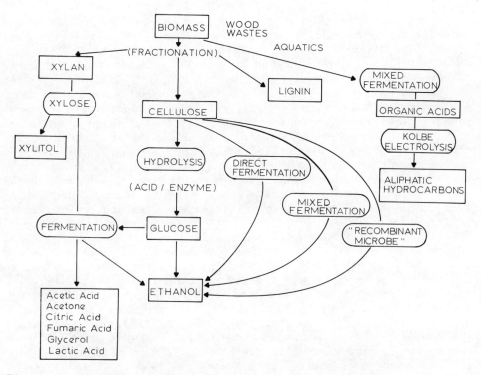

Figure 1. Primary routes for the microbial fermentation of biomass.

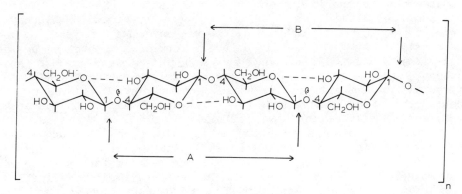

Figure 2. Cellulose - primary structure consisting of β-(1 $\rightarrow$ 4)-D-glucose. The alternate sugar residues are rotated 180° and hence yield two cellobiose conformations for enzymatic attack (A & B), in unidirectional attack of the insoluble substrate.

which disassociate this complex and make the cellulose more readily amenable to hydrolysis by either enzymes or acid, have been evaluated, e.g., steam explosion, solvent extraction (Bungay 1981; Chang et al. 1982). In this regard it should be noted that the direct conversion of cellulose and biomass via acid hydrolysis to a variety of feedstocks and commodities is one major potential approach (Goldstein 1983). However, acid hydrolysis of cellulose has a number of disadvantages which include: high capital costs, lower overall conversion efficiency 65-72 percent with dilute acid; 90 percent with concentrated acid), degradation of the products (glucose) to yield toxic impurities and equipment corrosion (Table 1). On the other other hand, enzymatic hydrolysis of cellulose yields pure syrups devoid of degradation products, requires low energy input as enzymatic reactions occur at moderate temperatures and pressures, and is far more efficient in terms of percent conversion (95 percent). Unfortunately, the enzymatic process is slow and the price of cellulase is high. Research has focused on means of reducing the cost of enzyme production through selection of hypercellulolytic mutants and their use in optimized fermentor mode (Cuskey et al. 1982). In this approach, it is envisioned that cellulase technology will become as refined as that of enzymatic conversion of starch, an extremely rapidly growing industry (Aunstrup 1978). The potential of the combined use of acid and enzyme hydrolysis to gain efficient conversion to give glucose in high yields is also a further research avenue (Knappert et al. 1980; Tanaka et al. 1980). These approaches using a discrete enzymatic stage for the conversion of cellulose to glucose, permit subsequent fermentative conversion of the sugar to a variety of solvents and feedstocks (Figure 1).

One system for the conversion of cellulose to ethanol that is at a relatively advanced development stage, utilizes the concept of simultaneous saccharification and fermentation (SSF). In this approach, cellulase from _Trichoderma reesei_ is used to hydrolyse municipal solid waste cellulose, in the presence of yeast. The hydrolysis product glucose, is directly fermented to ethanol by the yeast (Emert et al. 1982; Takagi et al. 1978). No end-product inhibition of the cellulase complex occurs, as the products are maintained at low concentration through fermentation of glucose by the yeast. This fermentative mode of operation in which both _T. reesei_ mycelium and culture both are utilized, circumvents the problem of the relatively low amounts of secreted _Trichoderma_ β-glucosidase, as the mycelium has most adequate cell-bound levels of this enzyme. Inherent problems in this process are: requirements for two independent stages [(i) enzyme production, and (ii) saccharification/fermentation], recycle and stability of the cellulases, and the ethanol tolerance of the yeast.

An alternate approach for ethanol production is to utilize a cellulolytic ethanol producing microbe. Unfortunately such microorganisms are relatively rare and suffer major restrictions with

Table 1. Hydrolysis of cellulose to glucose: Relative advantages of the use of enzyme or acid.

	Hydrolysis by	
	Cellulase	Acid
Efficiency of Conversion	++++	+++
Purity of Product	++++	++
Mild Conditions	++++	++
Non-Polluting	+++	-
Reusable	++	-
Conversion Rate	+	++++
Low Scale Technology	+	+++
Low cost of capital equipment	+++	+
Low cost of operation	++	+++

regard to industrial application. Thus certain Fusaria and Mucorales are efficient ethanol producers but grow relatively slowly (Gong et al. 1981; Rosenberg 1980; Ueng and Gong 1982). Clostridium thermocellus in contrast is a cellulolytic bacterium, that grows rapidly at 55° (Avgerinos and Wang 1980; Zeikus 1980; Zertuche and Zall 1982). Unfortunately, it is tolerant of relatively low concentrations of ethanol and also produces a co-product, acetic acid, which necessitates an extra separation step. Despite these deficiencies, alcohol production from cellulose is rapidly approaching commercial feasibility, with theoretical costs for ethanol already being close to gasoline prices when appropriate by-product credits are taken into account. However, the rate of hydrolysis of cellulose is a rate limiting step and furthermore it has been estimated that 43.4 percent of the total cost of ethanol from cellulose resides in enzyme production (Ryu and Mandels 1980). Thus there is a need for more effective cellulases and for their production in greater yields. The cellulase system is therefore discussed in greater detail.

Cellulases

A diverse range of microbes can degrade crystalline cellulose; fungi, bacteria and actinomycetes. Fungi give exceptionally high yields of cellulase. However, though many microbes are efficient cellulose degraders in vivo, such as ruminant bacteria, they fail to yield active cell-free cellulase preparations. Further examples are given in Table 2. The cellulase enzymes comprise a complex of endo- and exo-splitting hydrolases, in association with a variety of enzymes that transform the intermediary product cellobiose to sugar in a utilizable form (Figure 3).

A broad overview of microbial cellulases is presented, as they have been the subject of several recent reviews (Bailey et al. 1975;

Table 2. Cellulolytic microbes (in vivo).

GROUP I

ENZYME ACTIVE IN VITRO TOWARDS COTTON

Fusarium solani
Penicillium funiculosum
Phanerochaete chrysosporium
Talaromyces emersonii
Thielavia terrestris
Trichoderma reesei

Acetivibrio cellulolyticus
Cellulomonas spp.
Cellvibrio spp.
Clostridium thermocellum
Streptomyces spp.
Thermomonospora sp.

GROUP II

ENZYME INACTIVE IN VITRO TOWARDS COTTON

Gliocladium roseum
Memnoniella echinata
Myrothecium verrucaria
Stachybotrys atra

Bacteroides succinogenes
Butyrivibrio fibrisolvens
Clostridium lochheadii
Ruminococcus albus
Ruminococcus flavifaciens

Bisaria and Ghose 1981; Brown and Jurasek 1979; Ekman Days 1981; Enari and Markkanen 1977; Eriksson 1978, 1981; Gaden, et al. 1976; Ghose 1977, 1981; Gong and Tsao 1979; Goksøyr and Eriksen 1980; Halliwell 1979; Lee and Fan 1980; Linko 1977; Mandels 1982; Ryu and Mandels 1980; Wood 1975, 1981.

Fungal Cellulases

The cellulase complex of Trichoderma has been most extensively studied as it has been the best available source of an extracellular enzyme complex that is capable of attacking highly ordered crystalline cellulose (Gaden et al. 1976; Halliwell 1979; Ryu and Mandels 1980). The complex is representative of fungal cellulases and contains three major components:

    (i)   endoglucanase (1,4- β-D-glucano-hydrolase)
    (ii)   cellobiohydrolase (1,4- β-D-glucan cellobiohydrolase)
    (iii) β-glucosidase (cellobiase)

(i) Endo-glucanases. These enzymes randomly hydrolyse cellulose glucan chains to yield oligosaccharides. Several endoglucanases exist. Some act preferentially towards the larger glucan chains while others act on shorter chains. (Gritzali and Brown 1979; Shoemaker et al. 1981; and see the reviews cited above).

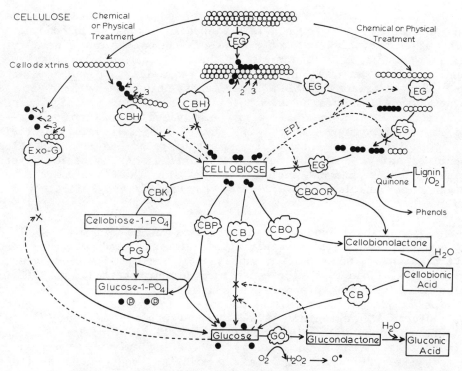

Figure 3. Generalized scheme of enzymes involved in the degradation of cellulose. EG - endoglucanase; CBH - cellobiohydrolase; EXO-G- exoglucanase; CB - cellobiase; CBO - cellobiose oxidase; CBQR - cellobiose quinone oxido-reductase; CBP - cellobiose phosphorylase; CBK - cellobiose kinase; PG - phosphoglucosidase; GO - glucose oxidase; end-product inhibition.

(ii) <u>Cellobiohydrolase</u>. Cellobiohydrolase hydrolyzes the glucan chain by sequentially removing cellobiose units from the non-reducing end.

(iii) <u>β-Glucosidase</u>. β-Glucosidase hydrolyses cellobiose and small oligomers to glucose (Woodward and Wise 1982).

These three enzymes act synergistically (Wood 1981). Endoglucanase initiates the attack of crystalline cellulose at internal sites, and this is followed by a co-ordinate action of cellobiohydrolase (Eriksson 1981; Wood 1981). The -glucosidase cleaves the major product cellobiose to glucose, thus circumventing end-product inhibition of the endo-glucanase and cellobiohydrolase by cellobiose. This synergistic and sequential action of three hydrolases, is the current, more generally accepted model for the fungal cellulase complex. The sequence is reviewed in Figure 3, in conjunction with other microbial enzymes of the cellulase complex. These three major types of enzymes comprising the cellulase complex, have been purified from several fungi (Fusaria, Penicillia, Trichoderma spp.) and through reconstitution experiments, have been shown to act synergistically, even between species. The action of endoglucanase and cellobiohydrolase has also been visualized through use of electron microscopy (White and Brown 1981). Using bacterial (Acetobacter xylinum) cellulose as a substrate, endoglucanase alone promoted splaying of the substrate fibrils, whereas cellobiohydrolase showed no apparent action. However when these two enzymes were used in combination, the cellulose was rapidly and completely dissolved.

In an initial hypothesis of cellulase action (Reese et al. 1950) it was suggested that enzymatic attack was initiated by a non-hydrolytic fator C_1, whose action subsequently permitted attack by endoglucanases. This C_1 factor has yet to be demonstrated. However, in view of the current proposed mechanism for cellulase (Figure 3) it has been inferred that cellobiohydrolase is infact the hypothetical C_1 factor. As Reese has rejected this suggestion, the term C_1 should not be confused with the concept of a cellobiohydrolase. He considers that there is possible a "disordering enzyme" initiates the reaction sequence and could be a specialized endo-glucanase (Reese 1977). Thus he postulates that the action of C_1 "a covalent linkage is split and that this act is accompanied by splitting of hydrogen bonds. Thus, C_1 becomes a member of the endo-β-glucanases (Cx). But it is a very special member having properties not possessed by most endo-glucanses, e.g. activity on crystalline cellulose; disruption of H-bonds; lack of action CMC, and inability to act on products of it own action."

It has been emphasized that the repeating unit of the $\beta-(1,4)$ glucan is cellobiose (Wood 1981). Hence there are two conformations for cellobiose within cellulose (Figure 2, A and B) and one can predict the existence of two forms of cellobiohydrolase. Two distinct cellobiohydrolases have been characterized, although it is currently unknown if their differences include individual attack of these two discrete cellobiose conformations.

Fungal β-glucosidases differ with regard to their yield, location (cellbound or extracellular), substrate specifity and reaction kinetics (Woodward and Wiseman 1982). If high concentrations of glucose syrups are integral to biomass transformation schemes, their susceptibility to product inhibition is a major limitation.

Further enzymes are now realized to comprise certain fungal cellulase complexes. Cellobiose quinone oxido-reductase occurs in some Basidiomycetes (Eriksson 1981) and Monilia sp. (Dekker 1981).

It has been proposed that this enzyme can facilitate lignin degradation by prevention of polymerization of phenolic residues due to direct reduction of the phenolic quinoidal moieties (Eriksson 1981). Cellobiose oxidase and glucose oxidase have also been recorded in Basidomycetes (Eriksson 1981). Traditionally, fungal cellulases have been considered to be extracellular proteins. However, it is becoming increasingly clear that some, if not all, of the enzymes in the complex are either periplasmic or wall bound during the early stages of development (Inglin et al. 1980; Kubicek 1981; Montenecourt et al. 1981; Vaheri et al. 1979).

Through a consideration of the specifity of the individual cellulase components and of the regulation of their synthesis, significant increases in productivity have been achieved by the isolation of mutant strains in several laboratories (Figure 4) (Bailey and Nevalainen 1981; Cuskey, et al. 1982; Durand and Tiraby 1980; Gallo 1982; Gallo et al. 1979; Shoemaker et al. 1981). Optimization of fermentor conditions has resulted in further increased in productivity (Allen and Andreotti 1982; Gottvaldova et al. 1982; Tangnu et al. 1981). Yields of one hundred filter paper units (fpu)/l/hr and 2 percent extracellular protein have been reported in comparison to the wild type strain QM6a of 15 fpu/l/hr and 0.1 percent extracellular protein in flask culture. Immobilization of the hypercellulolytic mutant T. reesei RUT-C30 on Celite, (diatomaceous earth) has resulted in productivities of 200 fpu/l/hr. (M. Frein, unpublished results). The high cellulase productivity of this mutant is related to enhanced synthesis of endoplasmic reticulum (Ghosh et al. 1982a).

From this brief review, it is seen that in spite of the complexity of the fungal cellulase system, current technologies have resulted in considerably increased productivities and when coupled with greater efficiency of conversion, will allow enzymatic hydrolysis to favorably compete with alternative chemical technologies.

Bacterial Cellulases

The bacterial cellulase complex differs from the fungal system in that only two principle components, endo-glucanase and β-glucosidase have to date, been adequately described. A few examples are cited in Table 3.

Ramasamay and Verachtert, 1980 have proposed that in their Pseudomonad system, the cellulase complex contains only endoglucanases. Certain endoglucanases act optimally against long chain cellulose chains and the subsequent further hydrolysis of the smaller oligomers occurs as a result of endoglucanases with greater specifity towards the shorter substrate molecules. However, there is also mounting evidence for the occurrence of bacterial cellobiohydrolases e.g. in Acetivibrio cellulolyticus (Saddler and Khan 1981).

Figure 4. Genealogical chart of hypercellulolytic mutants derived from Trichoderma reesei QM6a.

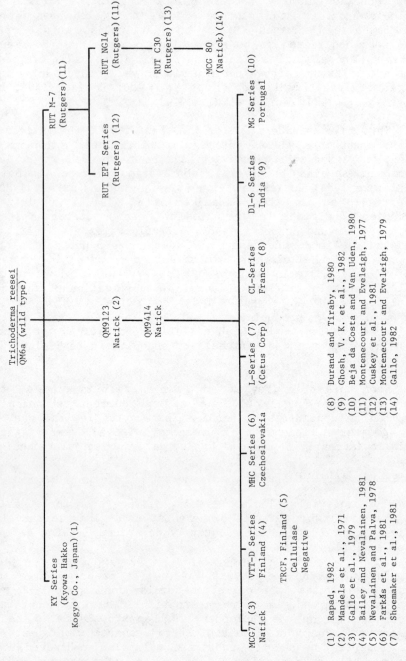

(1) Rapad, 1982
(2) Mandels et al., 1971
(3) Gallo et al., 1979
(4) Bailey and Nevalainen, 1981
(5) Nevalainen and Palva, 1978
(6) Farkas et al., 1981
(7) Shoemaker et al., 1981
(8) Durand and Tiraby, 1980
(9) Ghosh, V. K. et al., 1982
(10) Beja da Costa and Van Uden, 1980
(11) Montenecourt and Eveleigh, 1977
(12) Cuskey et al., 1981
(13) Montenecourt and Eveleigh, 1979
(14) Gallo, 1982

Table 3. Cellulolytic bacteria.

Acetivibrio cellulyticus	Saddler and Kahn 1981.
Bacteroides succinogenes	Groleau and Forsberg 1981.
Cellulomonas spp.	Beguin and Eisen 1978; Choudhury et al. 1980; Hitchner and Leatherwood 1980; Rickard and Peiris 1981.
Cellvibrio spp.	Berg 1975.
Clostridia	Ait et al. 1979; Johnson et al. 1982; Ng et al. 1977; Zertuche and Zall 1982.
Pseudomonads	Ramasamy and Verachtert 1980; Thayer 1978; Yamane et al. 1970; Yoshikawa et al. 1974.
Thermomonospora	Hagerdal et al. 1979.

Bacteria possess a variety of mechanisms for transforming cellobiose. These include β-glucosidases, cellobiose phosphorylase (active also against cellodextrins), and even β-glucosidase kinase plus cellobiose phospho- β-glucosidase in the non-cellulolytic Enterobacter aerogenes (Palmer and Anderson 1972a,b) (see Figure 3). The potential of all of these enzymes in biomass conversion schemes remains to be clarified.

HEMICELLULOSE AND PENTOSE UTILIZATION

Hemicelluloses comprise amorphous polysaccharides associated with the plant cell-wall. These pentosans or hexosans are classified according to the major component of the polysaccharide chain; arabans, galactans, mannans, xylans. The most common pentosan is xylan. It can occur as a homopolymer or with side-chain substituents (Figure 5). Grasses and cereals contain a major xylan component typically with single L-arabinose side chains. Wood xylans also contain 4-O-methyl-D-glucuronic acid substituents, with greater amounts of xylan occuring in hard woods. Hexosan hemicelluloses are less common; for instance glucomannans of gymnosperms (up to 11 percent) of the wood).

Hemicelluloses have limited uses. Currently, xylans are used for the preparation of furfurals. Sugar cane bagasse which contains a high proportion of xylan (30-40 percent), is used for energy generation through direct combustion. Bagasse per se also has limited use in newsprint production in Argentina, India, Indonesia, Peru, Mexico and the United States. To what further uses can pentoses be put?

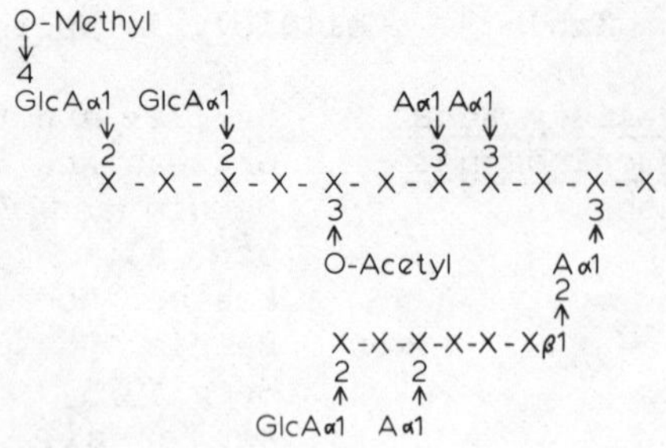

```
                    O-Methyl
                       ↓
                       4
              GlcAα1  GlcAα1            Aα1 Aα1
                ↓       ↓                ↓   ↓
                2       2                3   3
              X - X - X - X - X - X - X - X - X - X - X
                              3                   3
                              ↑                   ↑
                           O-Acetyl             Aα1
                                                 2
                                                 ↑
                              X-X-X-X-X-Xβ1
                              2       2
                              ↑       ↑
                           GlcAα1    Aα1
```

Figure 5. Generalized xylan structure to illustrate the main chain and attached substituent sugars.

It is noteworthy, that in the recent U. S. "Fuels-from-Biomass" programs the initial research focus was singularly directed towards the cellulose component of biomass (Bungay 1981; Douglas 1982). However, the cost of producing feedstocks such as ethanol is particularly sensitive to the price of biomass (Righelato 1980). As agricultural residues and wood can contain up to 20-40 percent xylan (Flickinger 1980; Rydholm 1956; Timell 1964) it is essential to find methods for the efficient utilization of xylan. The potential fermentation of xylan is attractive as it is readily extracted from biomass and simultaneously hydrolysed to xylose via dilute acid treatment. Bothast (1981) has estimated that 4-5 billion gallons of ethanol potentially can be produced in the U.S.A., from the fermentation of the $50-125 \times 10^6$ tons of xylose annually available from crop residues. Acid extraction and hydrolysis of xylan results in the availability of a single component feedstock, in contrast to the mixed heteropolymeric and recalcitrant lignocellulosics in the residual biomass. This simple extraction is especially noteworthy in that sugar cane factories routinely burn bagasse as a source of energy. Even so, there is an excess of bagasse (30 percent) that still has to be disposed. The concept from the fermentation viewpoint, is therefore to acid extract and ferment the xylan component (one third of the bagasse), burn the residual lignocellulose for the production of energy and thus have no excess biomass remaining. The ready availability of xylose from bagasse and straw, and the massive amounts from wood in sulfite waste liquor, have prompted considerable research in the fermentation of xylose to ethanol. However, until recently although yeasts were known that could oxidatively utilize pentoses, none were known to ferment xylose. Three major research approaches are under active investigation:

(a) indirect fermentation of xylose, via prior conversion to xylulose.
(b) screening for yeasts that ferment pentoses.
(c) construction of yeasts through recombinant DNA approaches that can ferment xylose.

(a) <u>Indirect fermentation of xylose</u>. Yeasts were focussed on as they are highly tolerant of their product ethanol (10 percent), in contrast to many bacteria that are sensitive to 1-2 percent. It was realized that oxidative utilization of pentoses in yeast proceeded via xylitol and thence to xylulose (Figure 6). Furthermore, as <u>Saccharomyces cerevisiae</u> could ferment xylulose directly, it was proposed to initially convert xylose to xylulose via the use of an immobilized bacterial enzyme, xylose isomerase, and subsequently ferment the xylulose product to ethanol using <u>Saccharomyces cerevisiae</u> (Wang et al. 1980) or <u>Schizosaccharomyces pombe</u> (Ueng et al. 1981). The results were promising with gradual increases in efficiency occurring through enhancement of the initial isomerization in favor of xylulose through use of borate (Hsiao et al. 1982). Ethanol yields of 3.2 percent from 10 percent xylose were achieved (64 percent efficiency).

(b) <u>Direct fermentation of xylose by yeast</u>. Although <u>Saccharomyces cerevisiae</u> and four hundred other yeasts did not ferment pentoses, the potential usefulness of a fermenting species prompted considerable screening for such a fermentative yeast. Three species have been found to date, within a relatively short period of time: <u>Pachysolen tannophilus</u> was initially discovered at the U.S.D.A. Laboratory at Peoria, IL (Slininger et al. 1982) and also at the National Research Council, Ottawa, Canada (Schneider et al. 1981). Jeffries noted ethanol production by <u>Candida tropicalis</u> (Jeffries 1981), while at Purdue University a mutant strain of a <u>Candida</u> species XF217 has been described (Gong et al. 1981). All are potentially useful organisms. <u>C. tropicalis</u> requires low levels of oxygen and has not given high ethanol yields. <u>P. tannophilus</u> also requires a supply of oxygen but given better ethanol yields (Slininger et al. 1982). <u>Candida</u> sp. XF217 is capable of completely anaerobic fermentation, and xylose (12 percent) has been fermented to yield 4.8 percent ethanol.

(c) <u>Construction of a fermentative yeast through a recombinant DNA approach</u>. Utilization of xylose by <u>S. cerevisiae</u> is via xylitol, and this product may accumulate (Figure 6). Xylulose can be directly fermented to ethanol. It should therefore be possible to block xylitol formation, and concomitantly insert a bacterial xylose isomerasegene to allow conversion of xylose to xylulose and hence to ethanol in the yeast (Figure 6). Several groups in North America are attempting this aproach. To date, the cloning of the bacterial glucose isomerase in <u>E. coli</u> has been reported and is an initial step towards this goal (Maleszka et al. 1982; Polaina et al. 1982).

THERMOPHILIC MICRO-ORGANISMS

Thermophilic microorganisms have recently attracted considerable attention for their potential industrial application. The

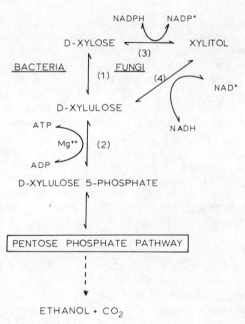

Figure 6. Major pathways of bacterial and of fungal catabolism of xylose. (1) D-xylose isomerase; (2) D-xylulokinase; (3) aldoreductase; (4) xylulose reductase.

basis for this is the rapid growth of these microbes, the relatively high stability of their enzymes, their lack of pathogenicity, and in fermentations for solvent production the facilitation of distillation for product recovery at the higher temperature. Thermophilic microorganisms can be categorized in relation to their optimal growth temperature and range. Fungal thermophiles have growth optima between 45-55°C. Bacterial thermophiles are more generally referred to as growing well from 50-65°C, though caldoactive strains occur which can grow at temperatures as high as 97°C. However, at the latter extreme temperatures, the microbes are under severe stress and growth rates are relatively slow. Hence for fermentative application, bacteria with growth optima in the 55-65°C range have been studied to a greater degree (Ljundahl et al. 1981; Zeikus 1979; Zeikus et al. 1981; Zeikus and Ng 1982).

Ethanol has been focussed on as a product of direct interest. Wiegel (1980) points out that of the 133 ethanol producing bacteria, only twelve are major producers. Six thermophilic species of current interest are ethanol fermentors (Table 4). All six strains ferment a variety of sugars to essentially the same products, but in vastly different proportions.

With regard to biomass utilization, the cellulolytic Clostridium thermocellum is of prime importance though it produces acetate

Table 4. Saccharolytic thermophilic bacteria[a].

	Clostridium thermocellum LQRI	Clostridium Thermohydro-sulfuricum 39E	Clostridium thermo-saccharolyticum HG-3	Thermo-anaerobacter ethanolicus[b] JW200	Thermo-anaerobium brockii HTD4	Thermo-bacteroides acetoethylicus HTB2
Temp. Range (°C)	40-70	38-78	55-64	37-78	40-80	40-80
Temp. Optimum	62	65-70		69	70	65
Growth substrates						
Glucose	+	+	+	+	+	+
Cellobiose	+	+	+	+	+	+
Cellulose	+	−	−	−	−	−
Starch	−	+	+	+	+	N.T.
Xylose	−	+	+	+	−	+
Xylan	−[c]	−	+	−	−	N.T.
Sucrose	−	+	+	+	+	+
Products (from cellobiose at 60-65°C)			Relative moles formed			
Ethanol	157	543	4.2d	78.4	224	139
Acetate	165	31	2.28	4.5	48	134
L-Lactate	24	50	0.6	4.0	352	0
CO_2	346	580	N.T.	89.2	230	190
H_2	286	31	N.T.	4.3	20	29
EtOH yield (mol/mol sugar)		1.9		1.8		

[a] Avgerinos and Wang 1981; Ljundahl et al. 1981; Wiegel and Ljungdahl 1981; Zeikus and Ng 1982.
[b] Culture grown on glucose (44 moles).
[c] Cultures grown on cellulose, produce a cellulase that degrades xylan. Note + = metabolic utilization of the substrate.

and ethanol in roughly equal proportions. By selective isolation, strains have been selected that now produce ethanol predominantly and that are also more tolerant of it, e.g., strain HG-4 (Avgerinos and Wang 1981). Although the cellulase of this bacterium degrades xylan to xylose, it is unable to utilize this pentose. Hence, C. thermohydrosulfuricum, C. thermosaccharolyticum and Thermoanaerobacter ethanolicus assume importance in biomass fermentations as they can ferment this pentose. The efficiency of ethanol fermentation is also critical and here again C. thermohydrosulfuricum and T. ethanolicus are seen to be effective for biomass transformations. Efficiency of transformation of biomass to ethanol of C. thermocellum has been increased through coculture with these efficient ethanol producing strains.

Obviously these rapidly growing and effective thermophilic fermentors are of great future interest. They have only recently received intensive study, and already novel characteristics are being found, e.g., the rare optically active synthesis of 2-pentanol via T. brockii alcohol dehydrogenase (Zeikus and Ng 1982). Their importance will increase rapidly.

HIGH VALUE SPECIALITY CHEMICALS

The recent research directions on fermentation of biomass have emphasized "fuels-from-biomass". The economics of these proposed processes have been under strong criticism, although the fermentative production of ethanol from starch in the U. S. A. is currently less expensive than alcohol production via petroleum derived ethylene. The conversion of biomass into high value speciality chemicals is now also receiving focus. Two examples in which biomass may also be used as a substrate are in enzyme and amino-acid production.

Enzymes have been utilized by man for centuries; well known applications include the malting of beer and the use of rennin in the preparation of cheese. Although the attributes of enzymes for effecting chemical transformations are well known, namely their high specific action and stereospecificity, the industrial application of microbial enzymes has been slow. The first commercial microbial enzyme production in the U.S.A. was by Jokichi Takamine who introduced solid substrate fermentation in 1894 for the preparation of fungal amylases. This process was based on the traditional koji fermentation used for saké production in Japan. Initially, these enzymes were expensive to produce and were relatively unstable, and with these restraints there has been a gradual but steady development of the use of microbial enzymes. Production costs have now plummeted through the use of large scale submerged fermentations (50,000 gallons) and the operational stability of enzymes has been vastly improved by immobilization techniques. The result has been a recent dramatic growth of the enzyme industry, though in limited spheres (Table 5). The four major enzymes (Table 5) represent a current world market of around $300 million. Forecasts for the enzyme industry are for continued growth through the next decade including both greater sales of bulk enzymes as well as stronger development of diagnostic and analytical enzymes. Each area of development will be nurtured by recombinant DNA techniques.

Amino acids: Proteins, one of the essential components of living matter, are composed of about twenty amino acids. Animals, including man, require balanced amounts of these amino acids in their diet. As certain major foodstuffs (cereals) contain low amounts of two of them, L-methionine and L-lysine, the latter are added to animal feed as supplements. L-Methionine is produced commercially through chemical synthesis, while L-lysine is mainly produced (80 percent) via fermentation. Another amino acid, monosodium glutamate (MSG), a widely used flavor enhancer, is also obtained through microbial fermentation. In vitro enzymatic conversions can also be used advantageously for synthesis for a few amino acids. The fermentative production of L-lysine and MSG are success stories of industrial microbiology (Table 6), and it is projected that other amino acids will also soon be produced in similar manner.

Epilogue

It is obvious that the technology exists to permit biomass to contribute significantly as a raw material for the fermentation production of chemical feedstocks. However with the recent introduction of recombinant (R)-DNA technology, this is "just the tip of the iceberg" (Office of Technology Assessment 1981).

Classical microbial genetics has allowed the selection of more rapidly growing mutants and through excision or modification of regulatory controls, has yielded strains with significantly greater productivity. The major limitation in this approach was that the microbial geneticist had a finite genome to manipulate. The R-DNA technology has lifted this restriction and permits design and construction of chimeric organisms with completely new characteristics and capabilities. For instance, the construction of bacteria that can utilize a broad range of "waste materials" is envisaged. A. M. Chakrabarty, while at the General Electric Co., Schenectady, N.Y. conceived the notion of assembling "degradative-genes" into a so-called "super-bug" through genetic means. He was able to construct a "novel" bacterium that could utilize naphthalene, xylene, alkanes and camphor. R-DNA technology will also aid in the development of new strains capable of using inexpensive fermentation substrates, e.g., construction of cellulose utilizing, ethanol producing strains through recombinant procedures has been proposed (see Figure 1). In B.C. (before cloning) genetics, a regulatory pathway could be modified in order to gain greater metabolite yields. In contrast, the R-DNA approach can include replacing the

Table 5. Examples of world production of enzymes[1].

Enzymes	Current Production[2] (tonnes/year)	Cost (bulk) ($/lb)	Sales 1980 ($ x 10^6)	Application
Bacterial protease	530	62	66	Cleaning aid in detergents
Amylase[3]	320	19	12	Production of starch dextrins in the preparation of high fructose corn syrups.
Amylase				
Fungal amyloglucoside[3]	350	51	36	Conversion of starch dextrins to glucose.
Glucose Isomerase	70	400	56	Conversion of glucose to fructose - a sweetner.
Rennet (calf)[4]	10		42	Cheese manufacture.
Rennet (cow)[4]	8		10	
Rennet (microbial)[4]	8		12	
Glucose Oxidase	2.5	160	0.8	Blood glucose determinations.
Urokinase[5]			50	Therapeutic for blood clot removal.

[1]Data based on the office of Technology Assessment's "Impacts of Applied Genetics" and personal communication with L. Glick, Genex Corporation.

[2]The industry is projected to grow at 8% annually through 2000 A.D.

[3]In-house use of plant malt amylase for beer and liquor production, and of fungal koji amylases for sake preparation, may exceed 15,000 tons/annum.

[4]In the U. S. A., microbial production of rennet for cheese manufacture accounts for 50 percent of rennet use. In Europe, only 10-20 percent microbial rennet is used in cheese manufacture. Calf and cow rennet are obtained as water extracts from the abomasum stomach. As protein content will vary, cost/lb is not cited. The genes for bovine rennet production have recently been cloned in E. coli.

[5]Urokinase is used for removal of blood clots and in Japan in conjunction with anti-cancer agents. Only a few pounds are prepared each year from either human urine or from tissue culture. The market price reflects the cost of an experimental therapeutic agent.

Table 6. Aminoacids - world production 1981.[1]

Aminoacid	Current Production (tonnes/year)	Synthesis (Chemical, C, or fermentation, F)	Cost ($/lb)	Sales ($x10⁶)	Role of Recombinant-DNA Techniques
Glutamic acid (A flavor enhancer as monosodium glutamate)	300,000	F	1.80	1,080	Current production is completely through microbial fermentation, R-DNA techniques will be used to gain more efficient operation.
Lysine (An essential aminoacid)	50,000	F	2.10	258	Five year prediction for complete production by fermentation, with the advantages of R.DNA techniques being exploited in analogous manner to those for glutamic acid production.
Methionine (An essential amino aminoacid)	105,000	C	1.20	246	Currently synthesized from acrolein as a racemic D,L mixture. Now that the regulatory controls of the biosynthetic pathways are known. R.DNA technology can yield a microbial strain that will produce only the L-form and in economically attractive yield. 10 year prediction.

[1] Based on the Office of Technology Assessment's "Impacts of Applied Genetics" and personal communication with L. Glick, Genex Corporation.

regulatory pathway with a new one that is more conducive to gaining higher yields. Faster biosynthetic rates are possible to achieve through the introduction of components which relax the control of reaction rates. For instance, promoter sites on DNA regulate the rate of attachment of RNA polymerase. As certain "up-promotors" have especially high affinity for RNA polymerase, their presence results in more effective transcription of mRNA, and results in more rapidly gaining greater yields of a gene product. Greater yields of enzymes can also be obtained by increasing the copy number of the gene of interest through R-DNA techniques.

In summary, by focusing on the major biomass substrates (cellulose and xylose), through the use of rapidly growing thermophiles and by applicaton of recombinant DNA techniques, the fermentation industries can look forward to a progressive new era.

ACKNOWLEDGEMENTS

This work is a New Jersey Agricultural Experiment Station publication No. D-01111-2-82 supported by state funds, and the U.S. Department of Energy Contract No. ET-78-S-02-4591-A000.

REFERENCES

Ait, N., N. Creuzet and P. Forget. 1979. Partial purification of cellulase of Clostridium thermocellum. J. Gen. Microbiol. 113:399-402.
Allen, A.L. and R.E. Andreotti. 1982. Cellulase production in continuous and feed batch culture, e.g., Trichoderma reesei MCG80 (preprint Gatlinburg Energy Conference, 1982).
Aunstrup, K. 1978. Enzymes of industrial interest; traditional aspects. Ann. Rep. Fermentation Processes. 2:125-154.
Avgerinos, G.C. and D.I.C. Wang. 1980. Direct microbiological conversion of cellulose to ethanol. Ann. Reports Ferm. Processes. 4:165-191. (G. Tsao, Ed.). Academic Press, New York N.Y.
Bailey, M., T.-M. Enari and M. Linko (Eds.). 1975. Symposium on Enzymatic Hydrolysis of Cellulose. Finnish National Fund for Research and Development, Helsinki.
Bailey, M.J. and K.M.H. Nevalainen. 1981. Induction, isolation, and testing of stable Trichoderma reesei mutants with improved production of solubilizing cellulase. Enzyme Microbial Tech. 3:153-157.
Beguin, P. and H. Eisen. 1978. Purification and partial characteristics of three extracellular cellulases from Cellulomonas sp. Europ. J. Biochem. 87:525-531.
Beja da Costa, M. and N. Van Uden. 1980. Use of 2-deoxglucose in the selective isolation of mutants of Trichoderma reesei with

enhanced β-glucosidase production. Biotechnol. Bioeng. 22: 2429-2432.
Berg, B. 1975. Cellulase location in Cellvibrio fulvus. Can. J. Microbiol. 21, 51-57.
Bisaria, V.S. and T.K. Ghose. 1981. Biodegradation of cellulosic materials: substrates, microorganisms, enzymes and products. Enzyme Microbial Technol. 3:90-104.
Bothast, R. 1981. Cited in Biotechnology News. 1:No. 12. June 15.
Brown, Jr., R.D. and L. Jurasek. 1979. Hydrolysis of Cellulose: Mechanisms of Enzymatic and Acid Catalysis. Advances in Chemistry Series 181. Pub. Amer. Chem. Soc., Washington, D.C.
Brown, Jr., R.M. (Ed.) 1982. Cellulose and other natural polymer systems: Biogenesis, structure and degradation. pp. 519. Plenum Press, New York, N.Y.
Bungay, H.R. 1981. Energy, the Biomass Options. John Wiley and Sons, New York. 347 pp.
Chang, M.M., T.Y.C. Chou and G.T. Tsao. 1982. Structure, pretreatment and hydrolysis of cellulose. Adv. Biochem. Eng. 20:16-42.
Choi, W.Y., K.D. Haggett and N.W. Dunn. 1978. Isolation of a cotton wool degrading strain of Cellulomonas: mutants with altered ability to degrade cotton wool. Aust. J. Biol. Sci. 31:553-564.
Choudhury, N., P.P. Gray and N.W. Dunn. 1980. Saccharification of sugar cane bagasses by an enzyme preparation from Cellulomonas: Resistance to product inhibition. Biotech. Lett. 2:427-428.
Côté, W. 1983. this symposium.
Cuskey, S.M., E.M. Frein, B.S. Montenecourt and D.E. Eveleigh. 1982. Overproduction of Cellulase: Screening and Selection. Proceedings from FEMS Symposium on the Overproduction of Microbial Enzymes. Prague, Czechoslovakia, Chapt. 32. pp. 405-416.
Cuskey, S.M., B.S. Montenecourt and D.E. Eveleigh. 1981. Approaches for the isolation of β-glucosidase end-product inhibition resistant mutants. Symposium Second International Course cum Symposium on Bioconversion and Biochemical Engineering. IIT Delhi, March, 1980, pp. 63-73.
Dekker, R.F.H. 1980. Induction and characterization of a cellobiose dehydrogenase produced by a species of Monilia. J. Gen. Microbiol. 120:309-316.
Douglas, L. 1982. The chemistry and energetics of biomass conversion: An overview. In: Chemistry in Energy Production. (Eds.), R.G. Wymer and O.L. Keller. Southeast-Southwest Regional Amer. Chem. Soc. Meeting. Pub. Oak Ridge Natl. Lab., Oak Ridge, Tenn.
Durand, H. and G. Tiraby. 1980. Abstract (poster session) 2nd Amer. Soc. Microbiol. Conf. on Genetics and Molecular Biology of Industrial Microorganisms. Indiana University, Bloomington, IN.

Ekman Days. 1981. Biosynthesis and Biodegradation of Wood Components. Int. Symp. Wood and Pulping Chemistry. Stockholm, Sweden.
Emert, G.H., R. Katzen, R.E. Fredrickson, K.F. Kaupisch and C.E. Yeats. 1982. Design of the 45 MT/D cellulose to ethanol plant. In: First Pan-Pacific Synfuels Conference, pp. 447-454. The Japan Petroleum Institute, Tokyo, Japan.
Enari, T.M. and P. Markkanen. 1977. Production of cellulolytic enzymes by fungi. Adv. Biochem. Eng. 5:1-24.
Eriksson, K.-E. 1978. Enzyme mechanism involved in cellulose·hydrolysis by the rot fungus Sporotrichum pulverulentum. Biotech. Bioen. 20:317-332.
Eriksson, K.E. 1981. Cellulases of fungi In: Trends in the Biology of Fermentations (Ed. A. Hollaender). pp.19-32. Plenum Press, New York, N.Y.
Farkas, V., I. Labudova, S. Bauer and L. Ferenczy. 1981. Preparation of mutants of Trichoderma viride with increased production cellulase. Folia Microbiol. 26:129-132.
Flickinger, M.C. 1980. Current biological research in conversion of cellulosic carbohydrates into liquid fuels: how far have we come? Biotechnol. Bioeng. 22: Suppl. 1, 27-48.
Flickinger, M.C. and G.T. Tsao. 1978. Fermentation substrates from cellulosic materials. In: Annual Reports on Fermentation Processes. 2:23-42. (D. Perlman, Ed.) Academic Press, New York, N.Y.
Gaden, E.L., Jr., M.H. Mandel, E.T. Reese and L.A. Spano, Eds. 1976. Enzymatic conversion of cellulosic materials: technology and applications. Biotechnol. Bioeng. Symp. No. 6. John Wiley & Sons, New York, N.Y.
Gallo, B.J. 1982. Cellulase production by the new hyperproducing strain of Trichoderma reesei MCG 80 presented at American Institute of Chem. Engineers, Orlando, Fla. (February, 1982).
Gallo, B.J., R. Andreotti, C. Roche, D. Ryu and M. Mandels. 1979. Cellulase production by a new mutant strain of Trichoderma reesei MCG-77. Biotech. Bioeng. Symp. 8. Biotechnology in Energy Production and Conversion. pp. 89-101. (C.D. Scott, Ed.).
Ghose, T.K. 1977. Cellulase biosynthesis and hydrolysis of cellulosic substances. Adv. Biochem. Eng. 6:39-76.
Ghose, T.K., (Ed.). 1981. Bioconversion and Biochemical Engineering Symposium 2, Indian Institute of Technology New Delhi.
Ghosh, A., S. Al-Rabiai, B.K. Ghosh, H. Trimino-Vazquez, D.E. Eveleigh, and B.S. Montenecourt. 1982a. Increased endplasmic reticulum content of a mutant of Trichoderma reesei RUT-C30 in relation to cellulase synthesis. Enzyme Microbial Tech. 4:110-114.
Ghosh, V.K., R.K. Ghose and K.S. Gopalkrishnan. 1982b. Improvement in T. reesei strain through mutation and selective screening techniques. Biotech. Bioeng. 24:241-243.

Goldstein, I.S. 1983. This symposium.
Gong, C.-S. L.F. Chen, M.C. Flickinger and G.T. Tsao. 1981. Conversion of hemicellulose carbohydrates. Adv. Biochem. Eng. 20: 93-118.
Gong, C.-S., L.D. McCracken and G.T. Tsao. 1981. Direct fermentation of D-xylose to ethanol by a xylose fermenting-yeast mutant, Candida sp. XF 217. Biotechnol. Letters. 3:245-250.
Gong, C.-S. and G.T. Tsao. 1979. Cellulase and biosynthesis Regulation. Ann. Reports Fermentn. Process 3:111-140.
Goksøyr, J. and J. Eriksen. 1980. Cellulases. In: Economic Microbiology. 5:283-330. (Ed. A.H. Rose). Academic Press, New York, N.Y.
Gottvaldova, M., J. Kucera and V. Podrazky. 1982. Enhancement of cellulase production by Trichoderma viride using carbon/nitrogen double-fed-batch. Biotech. Lett. 4:229-232.
Gritzali, M. and R.D. Brown, Jr. 1979. The cellulase system of Trichoderma. Relationships between purified extracellular enzymes from induced or cellulose-grown cells. In: Hydroloysis of Cellulose: Mechanisms of Enzymatic and Acid Catalysis (Eds. R.D. Brown, Jr. and L. Jurasek.) Amer. Chem. Soc. Ser. 181, Washington, D.C.
Groleau, D. and C.W. Forsberg. 1981. Celluloytic activity of the rumen bacterium Bacteroides succinogenes. Can. J. Microbiol. 27:517-535.
Hägerdal, B., J. Ferchak, E.K. Pye and J.R. Forro. 1979. The cellulolytic enzyme system of Thermoactinomyces. In: Hydrolysis of Cellulose: Mechanisms of Enzymatic and Acid Catalysis. (Eds. R.D. Brown, Jr. and L. Jurasek.) Pub. Amer. Chem. Soc. pp. 331-345.
Halliwell, G. 1978. Microbial β-glucanases. Prog. Indust. Microbiol. 15:1-58.
Hitchner, E.V. and J.M. Leatherwood. 1980. Use of a cellulase-derepressed mutant of Cellulomonas in the production of a single-cell protein product from cellulose. Appl. Environment. Microbiol. 39:382-386.
Hsiao, H.-Y., L.-C. Chiang, L.F. Chen and G.T. Tsao. 1982. Effects of borate on isomerization and yeast fermentations of high xylulose solution and acid hydrolysate of hemicellulose. Enzyme Microbial Technol. 4:25-31.
Inglin, M., B.A. Feinberg and J.R. Loewenberg. 1980. Partial purification and characterization of a new intracellular -glucosidase of Trichoderma reesei. Biochem. J. 185:515-519.
Jeffries, T.W. 1981. Conversion of xylose to ethanol under aerobic conditions by Candida tropicalis. Biotech. Lett. 3:213-218.
Johnson, E.A., M. Sakajoh, G. Halliwell, A. Madia, and A.L. Demain. 1982. Saccharification of complex cellulosic substrates by the cellulase system from Clostridium thermocellum. Appl. Environmen. Microbiol. 43:1125-1132.
Knappert, F., H. Grethlein and A. Converse. 1980. Partial acid hydrolysis of cellulosic materials as a pretreatment for enzymatic hydrolysis. Biotech. Bioeng. 22:1149-1463.

Kubicek, C.P. 1981. Release of carboxymethyl-cellulase and β-glucosidase from cell walls of Trichoderma reesei. Europ. J. Appl. Microbiol. Biotech. 13:226-231.

Laskin, A.I., M.C. Flickinger and E.L. Gaden, Jr. (Eds.) 1980. Fermentation: Science and Technology with a Future. Biotechnology and Bioengineering. Vol. 22. Supplement 1980.

Lee, Y.-H. and L.T. Fan. 1980. Properties and mode of action of cellulase. Adv. Biochem. Eng. 17:101-129.

Linko, M. 1977. An evaluation of enzymatic hydrolysis of cellulosic materials. Adv. Biochem. Eng. 5:25-48.

Ljungdahl, L.G., F. Bryant, L. Carreira, T. Saiki and J. Wiegel. 1981. Some aspects of thermophilic and extreme thermophilic anaerobic microorganisms. In: Trends in the Biology of Fermentation for Fuels and Chemicals (A. Hollaender, Ed.) pp. 397-419. Plenum Press, New York, N.Y.

Maleszka, R., P.Y. Wang and H. Schneider. 1982. A Col El hybrid plasmid containing Escherichia coli genes complementing D-xylose negative mutants of Escherichia coli and Salmonella typhimurium. Can. J. Biochem. 60:144-151.

Mandels, M. 1982. Cellulases. Ann. Rept. Ferm. Proc. 5:35-78.

Mandels, M., J. Weber and R. Parizek. 1971. Enhanced cellulase production by a mutant of Trichoderma viride. Appl. Microbiol. 21:152-154.

Montenecourt, B.S., and D.E. Eveleigh. 1977. Preparation of mutants of Trichoderma reesei with enhanced cellulase production. Appl. Environ. Microbiol. 34:777-782.

Montenecourt, B.S. and D.E. Eveleigh. 1979. Production and characterization of high-yielding mutants of Trichoderma reesei. TAPPI 28:101-108.

Montenecourt, B.S., S.D. Nhlapo, H. Trimiño-Vazquez, S. Cuskey, D.H.J. Schamhart, and D.E. Eveleigh. 1981. Regulatory controls in relation of over-production of fungal cellulases. In: Trends in the Biology of Fermentation of Fuels and Chemicals. 18:33-53, Ed. A. Hollander, Plenum Press, New York.

Nevalainen, K.M.H. and E.T. Palva. 1978. Production of extracellular enzymes in mutants isolated from Trichoderma viride unable hydrolyze cellulose. Appl. Environ. Microbiol. 35:11-16.

Ng, T.K., P.J. Wiemer and J.G. Zeikus. 1977. Cellulolytic and Physiological properties of Clostridium thermocellum. Arch. Microbiol. 114:1-7.

Nofsinger, G.W. and R.J. Bothast. 1981. Ethanol production by Zymomanas mobilis and Saccharomyces uvarum on aflatoxin-contaminated and ammonia-detoxified corn. Can. J. Microbiol. 27:162-167.

Office of Technology Assessment. 1980. Energy from biological processes. (T.E. Bull, Project Director). Library of Congress, Catalog Card N. 80-600118.

Office of Technology Assessment. 1981. Impacts of applied genetics and microorganisms, plants and animals. (J.N. Gibbons,

Director). pp. 331. Library of Congress, Catalog Card No. 81-600046.
Palmer, R.E. and R.L. Anderson. 1972a. Cellobiose metabolism in Aerobacter aerogenes II. Phosphorylation of cellobiose with a cellobiose kinase. J. Biol. Chem. 247:3415-3419.
Palmer, R.E. and R.L. Anderson. 1972b. Cellobiose metabolism in Aerobacter aerogenes III. Clevage of cellobiose monophosphate by a phospho-β-glucosidase. J. Biol. Chem. 247:3420-3423.
Peppler, H.J. and D. Perlman. 1979. Microbial Technology. Volumes 1 and 2, 2nd Edition. Academic Press, New York, N.Y.
Perlman, D. and G.T. Tsao (Eds.). 1977-82. Annual Reports of Fermentation Processes. Vols. 1-5. Academic Press, New York, N.Y.
Polaina, J., J. Wiggs, R.H. Villet and K. Grohmann. 1982. Recombinant technology for ethanolic xylose fermentations. Amer. Chem. Soc. 183rd Nat. Meeting, Las Vegas (Abstract).
Ramasamy, K. and H. Verachtert. 1980. Localization of cellulase components in Pseudomonas sp. isolated in activated sludge. J. Gen. Microbiol. 117:181-191.
Rapad (Research Association for Petroleum Alternatives Development). 1982. Research and Development on Synfuels Annual Report, Tokyo, Japan.
Reese, E.T., R.G.H. Siu and H.S. Levinson. 1950. The biological degradation of soluble cellulose derivatives and its relationship to the mechanism of cellulose hydrolysis. J. Bacteriol. 59:485-497.
Reese, E.T. 1977. Degradation of polymeric carbohydrates by microbial enzymes. In: Recent Adv. Phytochemistry Vol. II. Edit. F. Loewus and V.C. Runneckles.
Rickard, P.A.D. and S.P. Peiris. 1981. The hydrolysis of bagasse hemicelluloses by selected strains of Cellulomonas. Biotech. Lett. 3:39-44.
Righelato, R.C. 1980. Anaerobic fermentation: Alcohol production. Phil. Trans. R. Soc. London. B290:303-312.
Rose, A.H. 1977. Economic Microbiology. Vols. 1-5. Academic Press, New York, N.Y.
Rosenberg, S.L. 1980. Fermentation of pentose sugars to ethanol and neutral products by microorganisms. Enzyme Microbial. Tech. 2:185-193.
Rydholm, S.A. 1956. Pulping processes. Interscience Publishers, New York, N.Y.
Ryu, D. and M. Mandels. 1980. Cellulases: Biosynthesis and applications. Enzyme Microbial Technol. 2:91-102.
Saddler, J.N. and A.W. Khan. 1981. Cellulolytic enzyme system of Acetivibrio cellulolyticus. Can. J. Microbiol. 27:288-294.
Schneider, H., P.Y. Wang, Y.K. Chan and R. Maleszka. 1981. Conversion of D-xylose into ethanol by the yeast Pachysolen tannophilus. Biotechnol. Lett. 3:89-92.

Shoemaker, S.P., J.C. Raymond and R. Bruner. 1981. Cellulases: diversity amongst improved Trichoderma strains. In: Trends in the Biology of Fermentations for Fuels and Chemicals. (A. Hollaender, Ed.). pp.89-109, Plenum Press, New York, N.Y.

Slininger, P.J., R.J. Bothast, J.E. van Cauwenberge and C.P. Kurtzman. 1982. Conversion of D-xylose to ethanol by the yeast Pachysolen tannophilus. Biotech. Bioeng. 24:371-384.

Takagi, M., S. Abe, S. Susuke, G.H. Emert and Y. Nata. 1978. Proc. Bioconversion. Symp. IIT (Ed. T.K. Ghose). pp. 551-571.

Tanaka, M., T. Morita, M. Taniguchi, R. Matsuno and T. Kamikubo. 1980. Saccharification of cellulose by combined hydrolysis with acid and enzyme. J. Ferm. Tech. 58:517-524.

Tangnu, S.K., H.W. Blanch and C.R. Wilke. 1981. Enhanced production of cellulase, hemicellulase and β-glucosidase by Trichoderma reesei (RUT-C30). Biotech. Bioeng. 23:1837-1849.

Thayer, D.W. 1978. Cellulolytic and physiological activities of bacteria during production of single cell protein from wood. AICHE Symp. Ser. 74:126-135.

Timell, T.E. 1964. Hemicelluloses. Adv. Carbohydrate Chem. 19:247-302.

Tsao, G.T. 1978. Cellulosic material as a renewable resource. Process Biochem. 13(10):12-14.

Ueng, P.P. and C.-S. Gong. 1982. Ethanol production from pentoses and sugar-cane bagasse hemicellulose hydrolysate by Mucor and Fusarium species. Enzyme Microbial Technol. 4:169-191.

Ueng, P.P., C.A. Hunter, C.-S. Gong and G.T. Tsao. 1981. D-Xylulose fermentation in yeast. Biotechnol. Lett. 3:315-320.

U.S. National Alcohol Fuels Commission. 1981. Fuel Alcohol. An energy alternative for the 1980's. Final Report. pp.146. ISBN 0-9605762-0-7.

Vaheri, M.P., M.E.O. Vaheri and V.S. Kauppinen. 1979. Formation and release of cellulolytic enzymes during growth of Trichoderma reesei on cellobiose and glycerol. Eur. J. Appl. Microbiol. Biotechnol. 8:73-80.

Venkatasubramanian, K. and C.R. Keim. 1981. Gasohol: A commercial perspective. New York Acad. Sciences. 369:187-204.

Wang, P.Y., B.F. Johnson and H. Schneider. 1980. Fermentation of D-xylose by yeasts using glucose isomerase in the medium to convert D-xylose to D-xylulose. Biotechnol. Lett. 2:273-278.

White, A.R. and R.M. Brown, Jr. 1981. Enzymatic hydrolysis of cellulose: Visual characterization of the process. Proc. Natl. Acad. Sci. 78:1047-1051.

Wiegel, J. 1980. Formation of ethanol by bacteria. A pledge for the use of extreme thermophilic anaerobic bacteria in industrial ethanol fermentation processes. Experientia. 36:1434-1446.

Wiegel, J. and L.G. Ljungdahl. 1981. Thermoanaerobacter ethanolicus gen. nov., spec. nov. A new extreme thermophilic anaerobic bacterium. Arch. Microbiol. 128:343-348.

Wood, T.M. 1981. Enzyme interactions involved in fungal degradation of cellulosic materials. Proceedings of the International

Symposium on Wood and Pulping Chemistry. The Ekman Days. Stockholm, Sweden. 3:31-38.
Wood, T.M. and S.I. McCrae. 1975. The cellulase complex of Trichoderma koningii. In: Symposium on Enzymatic Hydrolysis of Cellulose. (M. Bailey, T.-M. Enari and M. Linko, Eds.). pp. 231-254. SITRA, Aulanko, Finland.
Woodward, J. and A. Wiseman. 1982. Fungal and other β-D-glucosidases - their properties and applications. Enzyme Microbial Technol. 4:73-79.
Yamane, K., H. Suzuki and K. Nisizawa. 1970. Purification and propties of extracellular and cell-bound cellualse components of Pseudomonas fluorescens var. cellulosa. J. Biochem. (Tokyo) 67:19-35.
Yoshikawa, T., H. Suzuki and H. Nisizawa. 1974. Biogenesis of multiple cellulase components of Pseudomonas fluorescens var. cellulosa. J. Biochem. 75:531-540.
Zeikus, J.G. 1979. Thermophilic Bacteria: Ecology, Physiology and Technology. Enzyme Microbial Technol. 1:243-252.
Zeikus, J.G. 1980. Chemical and fuel production by anaerobic bacteria. Ann. Rev. Microbiol. 34:423-464.
Zeikus, J.G., A. Ben-Bassat, T.K. Ng and R.J. Lamed. 1981. Thermophilic Ethanol Fermentations. In: Trends in the Biology of Fermentations, (Ed. A. Hollaender.) pp. 441-461. Plenum Press, New York, N.Y.
Zeikus, J.G. and T. Ng. 1982. Thermophilic saccharide fermentations. Ann. Report Ferm. Proc. 5:263-289.
Zertuche, L. and R.R. Zall. 1982. A study of producing ethanol from cellulose using Clostridium thermocellum. Biotech. Bioeng. 24:57-68.

BIOTECHNOLOGICAL UPGRADE OF RICE HUSK:

I. THE INFLUENCE OF SUBSTRATE PRE-TREATMENTS ON

THE GROWTH OF Sporotrichum pulverulentum

José M. C. Duarte, A. Clemente, A. T. Dias, M. E. Andrade

Biology/Institute of Energy
P. O. Box 21
2686 Sacavem Codex
Portugal

INTRODUCTION

Everybody knows that rice is one of the most widespread and valuable sources of food for mankind. Annual world paddy rice production by 1978 was calculated at 300 million tons. As shown on the simplified diagram below, after harvesting, rice is treated in industrial mills giving rise to two main by-products: one designated "huller bran" is valuable for direct incorporation in animal feeding due to its relatively high content of fat and protein (ca. 12.5%); the other is the husk and represents one-fifth of the paddy by weight.

The morphological development of the rice plant has resulted in the evolution of a protective envelope, the husk, which must be removed before the grain can be used for food. These husks represent therefore a huge waste of the rice-milling industries (ca. 60 million tons worldwide) and its utilization would have a major impact on the economies of both the rice milling industry and the countries concerned.

In Portugal, rice is still today an important agricultural and industrial activity; it was first introduced there probably by the Arabs (Vianna e Silva 1969) who called it al roz (arroz, in Portuguese), in the eighth century. Despite the fact that only one-third of the rice consumed is of domestic origin (imported rice is husk-free) there are available some thirty thousand tons of husk which in most of the places constitute a serious disposable problem.

Rice Chain Diagram

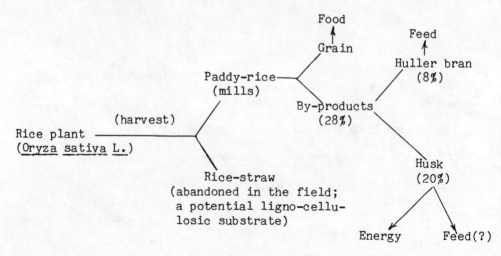

The husks are mainly of ligno-cellulosic nature. Therefore their utilization (even because there are, nationally, large amounts of other wastes of the same nature) could free more precious resources for applications where they can be most efficiently utilized domestically or exported in order to generate foreign exchange. A reappraisal of the most beneficial uses of all energy and resources is therefore necessary.

RICE-HUSK UTILIZATION

Only a small percentage of the available husks are being utilized in any manner. Their traditional use has been as a fuel to provide energy for the mill operation and this goes on being the main application. The conversion of rice husks into energy was the subject of a recent FAO report (Beagle 1978). Among the factors which may influence the use of rice husk as a raw-material are: available quantity, available area, seasonal considerations, transportation costs, and the availability of competitive fuels or power sources. Burning a ton of husk liberates 3.4×10^6 kcal with an equivalent heating oil value of ca. U.S. $19.00 (at 1978 prices); the residue left may also be of some value mainly due to its silica content. However, burning methods need research into producing higher grade and more uniform residues (Beagle 1978).

The opportunities for utilization of the husks are influenced by their physical and chemical properties which affect handling, storage and application. The chemical composition of the husk is shown in Table I. However, this composition may vary or with the place and the year or with the variety of the plant.

Table I. Chemical composition of rice husks.

Water	10%
Ash	19%
Protein	2%
Fat	1%
Cellulose	41%
Hemicellulose	17%
Nitrogen Free Extract	10%

Note the high ash content of the husk; this is normally more than 95% silica. The silica-cellulose structural arrangement makes it difficult to handle, and it is also responsible for its quite low apparent density (about 0.01 g/cm3).

As an alternative to combustion, pyrolysis can be chosen for production of char and oil; by gasification you can obtain a type of "producer gas" (Beagle 1978) consisting principally of flammable carbon monoxide and hydrogen and the non-flammable gases, carbon dioxide and nitrogen.

Acetic acid, methanol and tar may be obtained by distillation. After some chemical treatment rice husk derivatives may have chemical uses or be used for rubber reinforcement and ceramic additive. It has also been used in furfural production (Vianna e Silva 1969) after acid hydrolysis.

Biotechnological utilization of rice husk has been practically non-existent. However, it can be used as a compost or mulching agent (Beagle 1978) due to its moisture retention and soil sterilizing effects. Fermentation of the pentose sugars obtained from agricultural residues, including rice husk, by Clostridium acetobutylicum, is referred to in the USDA-NRRL process (Tsuchiya 1949), with production of liquid fuels or solvents. Acid and enzymatic hydrolysis to produce fermentable sugars is also a possibility. These sugars can be fermented to produce ethyl alcohol, citric acid, animal fodder or other useful substances. Two processes to produce SCP directly by fermentation are referred in the FAO report (Beagle 1978). Anaerobic digestion (with biogas production) could also be a possibility, mainly if composted with other wastes with a low C/N ratio and high water content; however, rice husk may not be sufficiently biodegradable to be of significance for such production.

In conclusion, it looks as if rice husks have not attracted much attention for biotechnological studies; however, their high carbohydrate content makes them an attractive substrate for such studies and successful bio-utilization could be of significance mainly in those places where they go on creating a disposal problem.

LIGNOCELLULOSICS AS FERMENTATION SUBSTRATES

From Table I it can be appreciated that about 70% of the husk is lignocellulosic material.

Lignocellulosic substrates have been receiving much attention in the recent years as a substrate for bacteria and fungal fermentations (Han 1975; Eriksson and Larsson 1975; Peitersen 1975; Tsao 1978; Moo-Young et al. 1977; Ladrazil 1980; Kirk et al. 1978; Peck and Odon 1981). The ability of microorganisms to attack these substrates differs widely in relation to time, optimal temperature and activity. Bacteria are not much utilized for aerobial degradation of cellulose; on the contrary, anaerobic degradation of cellulose (Peck and Odon 1981) is almost exclusively effected by bacteria, especially Clostridia sp. Lignin is decomposed by fungi and bacteria (Crawford et al. 1973) but whiterot fungi are by far the most active lignolytic organisms isolated to date. Because the rate of lignin degradation is known to be highly dependent on the oxygen concentration, (Kirk et al. 1978) we have looked for an organism that could degrade both cellulose and lignin aerobially. In Table II are shown some of the process characteristics of microorganisms which utilize cellulose or both cellulose and lignin. When microorganisms utilize only cellulose, lignocellulosic substrates must be pretreated to make cellulose accessible to the action of the hydrolytic enzymes. However, these pre-treatments reflect heavily on the final economics of the process (Ladisch 1979).

Among the microorganisms of Table II, Phanerochaete chrysosporium was chosen by us to study the effect of pre-treatments on the degradability of rice husk.

Table II. Microorganisms utilizing lignocellulosic substrates.

Microorganism	Substrate	Temperature	Time Required
-Pleurotus sp.	Lignocellulose	Ambient	weeks
-Phanerochaete chrysosporium	Lignocellulose	Ambient to 40°C	weeks
-Trichoderma reesei	Cellulose	Ambient	days
-Thermoactinomycetes sp.	Cellulose	55 - 65°C	days
-Clostridium thermocellum	Cellulose	55 - 65°C	days
-Pseudomonas fluorescens var.cellulosae	Cellulose; lignocellulose	Ambient	days

There were two main advantages of using this Basidiomycete: it has a strong lignolytic action and grows over a wide range of temperature from ambient to more than 40°C. It is however slower than other typical cellulolytic microorganisms. P. chrysosporium had earlier been classified among the Fungi Imperfecti and called either Chrysosporium lignorum or Sporotrichum pulverulentum. In this work we will refer to it as Sporotrichum pulverulentum, the name of its imperfect stage, as did Westermark and Eriksson (1974).

Another reason to choose this organism is that its cellulolytic complex has been exhaustively characterized by Eriksson (1978). The fungus utilizes three different types of hydrolytic enzymes for the degradation of cellulose: five endo-1, 4-β-glucanases, one exo-1, 4-β-glucanase and two 1, 4-β-glucosidases. An oxidative enzyme, cellobiose oxidase was isolated and shown to be of importance in in vitro cellulose . . . degradation (Ayers et al. 1978). Another enzyme, cellobiose quinone oxidoreductase, is of importance both in cellulose and in lignin degradation. The regulatory mechanisms involved in cellulose degradation such as catabolite repression of the endo 1 , 4-β-glucanases by monosugars was also studied (Eriksson 1978).

Even with this lignin-degrading organism the fermentation process will be very slow when the lignin content is high. Lignin metabolism seems to require large amounts of energy which necessitates that simultaneously an easier co-substrate is being utilized (Kirk et al. 1978) therefore reducing the efficiency of the wood cellulose degradation. This attack of the lignin by the fungus is in itself a pre-treatment and as such it was shown to cause a decrease in the energy demand for production of thermomechanical pulp from wood chips (Eriksson et al. 1980).

Pre-treatments of most of these lignocellulosic materials are needed to obtain acceptable enzymatic activities; the effect of pre-treatments is basically to disrupt or destroy the lignin barrier which also permits that a broader range of microorganisms is used. In this work we attempted the use of physical, mechanical and chemical pre-treatments on rice husk as a potential substrate for S. pulverulentum growth.

GROWTH OF S. PULVERULENTUM

Materials and Methods

Sporotrichum pulverulentum, Novobranova obtained from K. E. Eriksson is maintained in Petri plates and tubes on both malt and potato dextrose agar; new cultures are made every two months. Fresh cultures are obtained by replicating from these to malt-agar plates and left to grow for seven days. These are used to inoculate liquid medium with a modified Norkran's (Eriksson and Larsson 1975) composition; this inoculum is grown with 0.5% glucose as the carbon source during two days.

For the experiments with the pre-treated substrates the inoculum was grown with 2.5% cellulose during five days and 1 ml of the final suspension was used.

The experiments were in 250 ml shake flasks with 50 ml of medium; final pH was 5.4. The medium together with the substrate (treated or not) was previously sterilized at 115°C for 15 minutes. These flasks were agitated at 150 rpm in an orbital shaker (New Brunswick) with the temperature controlled at 30°C. Experiments were done with duplicates.

Glucose was analyzed by the Dubrowsky method; reducing sugars by the method of Somogyi-Nelson. Dry weight was determined at 115°C for 6 hours and used as a measure of the biodegradability of the rice husks (Han 1975; Eriksson et al. 1980; Zadrazil and Brunnert 1981). Carboxy methyl cellulase activity was measured by the production of reducing sugars.

Growth in Glucose and Cellulose

When growing S. pulverulentum in lignocellulosic substrates, cellulose is usually the main source of energy. Cellulose is hydrolyzed to glucose and cellobiose, the latter being mainly hydrolyzed to glucose by the action of β-glucosidases (Eriksson 1978).

Therefore, it was important to know the yield obtained in our experimental conditions when growing S. pulverulentum in powdered cellulose (for chromatography). The results are shown in Figure 1; in Figure 2 are calculated the yields when glucose was used instead of cellulose. The biomass yields from cellulose and glucose do not seem to be much different; at the very low concentrations (less than 0.5%) yields seem to be larger but this could be due to the influence of inoculum dry weight and composition. In both cases the yield quickly goes down until it reaches 0.2. Dry weights are practically proportional to the substrate consumed for a wide range of concentrations; however, the increase in dry weight slows down at lower concentrations for cellulose than for glucose. This may be due to the production of large amounts of extracellular enzymes when cellulose concentration increases. This is partially confirmed in Figure 3 where the production of endo-glucanases is shown for cultures with different concentrations of cellulose at days seven and eighteen of the cultures. While activity is already disappearing at the low cellulose concentrations it is still increasing for the larger concentrations as observed from Figure 4. The endo-glucanase activity seems to increase when the cellulose concentration is larger than 0.1% although this is only visible after 18 days of growth.

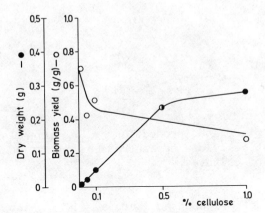

Figure 1. Growth of <u>Sporotrichum pulverulentum</u> on cellulose: dry weight and fungal yield as function of initial cellulose percentage in the medium. Agitation: orbital shaker at 150 rpm; T = 30° C.

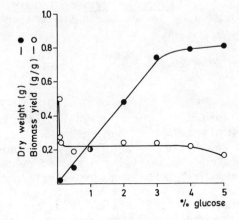

Figure 2. Growth of <u>S. pulverulentum</u> on glucose: dry weight and fungal yield as function of inital glucose concentration in the medium. Agitation: orbital shaker at 150 rpm; T = 30° C.

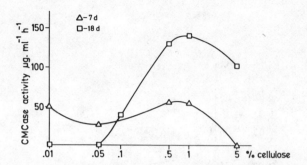

Figure 3. Endo-glucanases production by S. pulverulentum with different cellulose concentrations and at seven and eighteen days of culture.

In the work of Eriksson and Larsson (1975), activities also increase with the (powdered) cellulose concentration (from 4 to 10.0 g/l); the extracellular protein concentration increased with the cellulose concentration making as much as 30 percent of the total protein. It was concluded that degradation of a more complex carbon source requires a higher amount of enzyme; the more crystalline the cellulose the more difficult is its degradation by the fungus. This may be the case with the cellulose we used. The low rate of degradation would be responsible for the continuous enzyme production at the higher cellulose concentrations even after 18 days of growth. A similar large increase of endo-glucanase activity with time (6 to 16 days) was observed with Sporotrichum pruinosum (most probably an identical organism) growing on laminarin (Merril and French 1966).

RICE HUSK PRE-TREATMENTS

Grinding

Previously (Clemente and Andrade 1981; Clemente et al. 1982) it was observed that the grinding of the husks was essential for a good involvement of the substrate by the fungus. Gracheck et al. (1981) found that unpretreated lignocellulosic wastes produced unacceptable enzyme activities and that mechanical pre-treatments enhanced enzyme production; the smaller the particle size, the better the substrate. However, Han and Callihan (1974) did not find any effect of particle size reduction on the growth of microorganisms until reaching a 60-mesh size for their substrates.

We used a Max Fritsch knife mill with a 1.5 mm screen to prepare the husk samples for growth of S. pulverulentum. The results are shown in Figure 5.

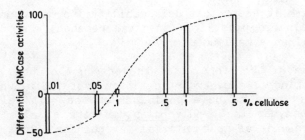

Figure 4. Pattern of change (between days 7 and 18 of culture) of endoglucanases production by S. pulverulentum as a function of initial cellulose in the medium. Negative values represent a decrease in the activity in solution.

As protein measurement in these systems is difficult (Eriksson and Larsson 1975) here we use the reduction in dry weight as a measure of the degradability (organic matter solubilization) of the substrate and indirect measure of fungal growth. We oberved that even with pure cellulose (between 1 and 10%) dry weight reduction was proportional to the cellulose concentration.

From Figure 5 it may be observed that ground husk degradability is only a few percent smaller (42% compared to 52%) than that of the cellulose powder at a concentration of 50 g/l. At 10 g/l almost no degradation occurred with rice husk and it was very small for cellulose. When cellulose (10%) was partially substituted for the same

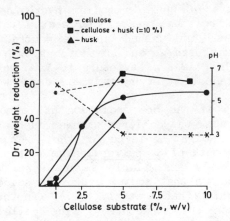

Figure 5. Growth of s. pulverulentum: the influence of both cellulose, husk and cellulose plus ground husk concentration in the dry weight reduction, after 10 days of culture. Dash line show the final pH of cultures: X - culture with cellulose; * cultures with rice husk.

amount of husk it seems that degradability increased until a mixture of 5% of cellulose plus 5% of husk was reached, when dry weight reduction attained a maximum of 66%.

It appears, therefore, that the presence of both cellulose and husk have a kind of a synergistic action over their own degradation. Cellulose is known to be a good inducer for both cellulolytic and lignolytic enzymes in S. pulverulentum (Kirk et al. 1978; Eriksson 1978) and presence of husk material may reinforce this action.

A remarkable difference between growth in cellulose and husk is observed in the final pH of the system (initial pH = 5.4). With cellulose the pH decreases to 3.0 corresponding to a strong mycelium development, while with the husk pH has a tendency to rise (pH 6.0 with 5% of husk) with low mycelium development, and substrate involvement.

Based on these results, a 5% concentration was chosen for use in the other experiments.

Sodium Hydroxide Treatment

Sodium hydroxide has been widely used for chemical pre-treatment of lignocellulosics (Han 1975; Han and Callihan 1974; Chahal et al. 1981; Miron and Ben-Ghedalia 1981; Cabello et al. 1981; Vandecasteel and Pourquie 1981). By using sodium hydroxide Chahal et al. (1981) increased the protein content by 17.3% with a cellulose utilization of 78.5%. However, they used a high temperature treatment for 30 minutes. We decided to favour first a treatment at room temperature using a high sodium hydroxide concentration. A 12.5% (w/v) mixture of husk in sodium hydroxide (10N) was used. Several times of treatment were experimented with; afterward the mixture was filtered and washed until a neutral pH was obtained. The treated husk was used as substrate for growth of S. pulverulentum (Figure 6). The times of exposure to NaOH were 6 and 24 hours, when significant reduction in dry weight and ash were obtained.

From Figure 6 it can be concluded that almost complete solubilization of the husk was achieved with as much as 90% of reduction of ground husk dry weight achieved for the 24-hour-treated samples. The visible weak growth of S. pulverulentum is probably related with the fact that readily assimilable sugars were solubilized by the treatment; it is known that sodium hydroxide solubilized hemicelluloses and some lignin (Chahal et al. 1981). Another effect of sodium hydroxide is on the structure of the cellulose fiber, which caused a growth increase of a Cellulomonas Sp. in alkali-treated sugar can bagasse (Han and Callihan 1974). It is possible that better Sporotrichum pulverulentum yields are obtained if the solubilized carbohydrates are also used in the growth medium after pH neutralization. This is confirmed when cellulose was used instead

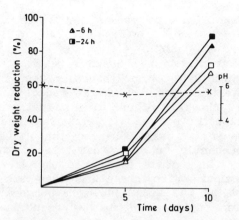

Figure 6. Dry weight reduction of liquid cultures of S. pulverulentum on rice husk pre-treated with sodium hydroxide, (10 N). Full symbols: ground husk; Empty symbols: whole husk.

of the husk since the treated cellulose was neutralized in situ. Better growth was obtained which is probably responsible for the smaller dry weight reductions obtained (Figure 7).

Steam Treatment

Steam is another common pretreatment (Han and Callihan 1974; Chahal et al. 1981; Vandecasteel and Pourquie 1981) used for increasing efficiency of lignocellulosic substrates utilization by microorganisms or ruminants, with the advantage that the economics

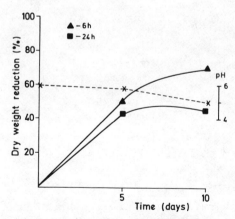

Figure 7. Dry weight reduction of S. pulverulentum cultures on cellulose pre-treated with sodium hydroxide.

of steam pretreatment seem to be on the favourable side (Ladisch 1978).

To test this treatment the husks were pressure cooked in an autoclave at 124°C for 0.5, 1 and 3 hours, respectively. The effect of this pre-treatment on the dry weight reduction when grown with S. pulverulentum is shown in Figure 8. The pattern of variation of dry weight reduction is not easily explained due to the lack of other data. However, it was observed that mycelial growth was occurring more abundantly on the steam-treated than on the other pre-treated husks. This could be due to the way steaming acts on the husk, by disrupting its structure (Han and Callihan 1974), while leaving most of the assimilable carbohydrates intact. The effect was much more noticeable on the ground husk than on whole husk.

Therefore, the slow increase in dry weight reduction and even its decrease observed mainly between the 5 and 10 days of growth could be due to intense growth of the fungus. The strong involvement of the husk also made it difficult to sample for dry weight determinations and could be the reasons why practically zero dry weight reductions were observed in a number of cases.

In the work of Han and Callihan (1974) pressure cooking of rice straw and sugar cane bagasse did not increase the digestibility of these ligno-cellulosic substrates when a pressure of about 28 bar (during 90 s) was used. When ground aspen wood (1 mm) was used as a substrate for C. cellulolyticum and pre-treated with steam at atmospheric pressure for 1 hour (Chahal et al. 1981), a product with

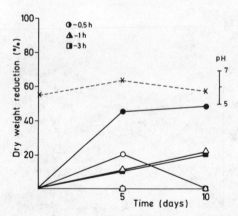

Figure 8. Dry weight reduction of liquid cultures of S. pulverulentum on rice husk pre-treated by steaming on an autoclave at 124° C. Full symbols: ground husk; Empty symbols: whole husk.

3.6% protein was obtained (dry weight) with a cellulose utilization of 12.4; but when steam at higher pressure (20 bar for 4 min) was used the percentage in the final product increased to 21.4% with 83.3% of cellulose utilization. It may then be concluded that the effect of steam depends on the particular substrate under use.

Use of γ-radiation

γ-radiation has been used in our Laboratory for sterilization of a variety of materials. It was also known (and observed) that high dosages of radiation will alter and destroy the degree of polymerization and crystallinity of these materials (the plastic support has to be frequently substituted). Therefore, we decided to irradiate the rice husks, before using it as substrate, with γ-radiation from a ^{60}Co source of 10 KCi*. Previously obtained results (Clemente and Andrade 1981) did not show a significant increase of microbial attack and therefore we increased the dose utilized from 0.05 MGy to 1 MGy. Four doses were applied: 0.1; 0.5; 0.75 and 1.0 MGy. The results obtained when growing S. pulverulentum on husk treated this way are shown in Figure 9. Only whole husk was used (results for ground husk, are not yet available) without any preparation. It was evident that the higher the dosage the better the involvement of the husk by the fungus; it is also evident that at the higher doses the reduction in dry weight was quicker although there is not much difference in the dry weight reduction obtained after 10 days. Han et al. (1981) also used γ-radiation to pretreat lignocellulosic materials; they used pre-swollen substrates and when

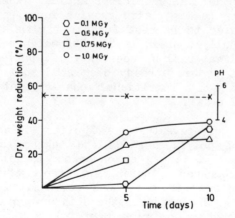

Figure 9. Dry weight reduction of liquid cultures of S. pulverulentum on whole rice husk irradiated in a γ ray source of ^{60}Co.

---

* 1 Ci (Curie) is equivalent to 3.7×10^{10} Bq (Becquerel).

NaOH (17%) was used sugar cane bagasse was completely solubilized after irradiation with 0.85 MGy from ^{137}Cs source; however, the degradation products contained little glucose due to combined effect of alkali and high dosage of γ-radiation.

The use of γ-radiation is probably not economical, but in combination with a chemical treatment it can reduce the dosage required by almost 90% (Han et al. 1981). This, together with the use of ^{137}Cs, a waste product of the nuclear industry, may make this process interesting for further studies.

PROSPECTS FOR FUTURE WORK

Results show that rice-husk may be used as a substrate for growth of S. pulverulentum; large dry weight reductions were observed in most of the cases. Among all treatments, sodium hydroxide was the one that caused most extensive solubilization of the husk, although its products were not yet well characterized. Steaming the husk at 124°C seemed to be beneficial for the attack of the husk by the fungus which involved the substrate almost completely. These treatments may be used preferentially whether the interest is mainly in producing enzymatic activities (for lignocellulosic degradation) or in producing fungal biomass. These studies are now in progress in our Laboratory to evaluate each of these possibilities.

The use of Sporotrichum pulverulentum may have some advantages over some of the other faster microorganisms. When the final product is destined for feeding, slow growing fungi have some advantages. They are easier to harvest (compared with unicellular organisms) and they usually contain little RNA which makes special processing of the cell material unnecessary. S. pulverulentum has also the advantage that it has already been subjected to experiments to evaluate its toxic and nutritional characteristics (Hofsten and Ryden 1975; Thomke et al. 1980).

Sporotrichum pulverulentum has a pleasant taste when grown on a wide variety of polysaccharides and sugars (Hofsten and Ryden 1975). From Table III it can be observed that its essential amino acid composition compares favourably with some of its usual competitors and is very near to the FAO recommendation. Of particular importance is the good content in the sulphur-containing amino acids (better than in soybean or rice protein) and tryptophan. Small scale experiments with the fungus (Hofsten and Ryden 1975) have not indicated that it is in any way toxic. Thomke et al. (1980) found that the amino acid digestibility of S. pulverulentum was inferior to values for soybean oil meal, and the more satisfactory results (82% digestibility of crude protein and a content of metabolizable energy of 70% of soybean oil meal) were obtained with sheep, indicating that it can be of potential interest for ruminant feeding.

On the other hand, Zadrazil (1980) found that solid state fermentation of rice husk by a number of other Basidiomycetes decreased the digestibility of rice husks. They attributed this effect to high incrustation of rice husks with silica. Treatments which may help in removing silica, such as the sodium hydroxide treatment, may however solve this problem.

Ek and Eriksson (1980) calculated that a process for conversion of lignocellulosic materials into biomass, using S. pulverulentum for protein feed, is only economic either if the material to be fermented has a certain negative value, either due to the combined effect of protein production and water purification (as it is the case in paper and fiber board mills).

Most of the uses today for rice husk are to expedite husk disposal which may not be the most beneficial; in our case (Portugal), it goes on being practically completely unused constituting a disposal problem and therefore a negative cost value can be attributed to it.

Table III. Essential amino acid composition of protein in different types of microorganisms and other conventional sources (in % of the sum of amino acids).

	S. pulverulentum		C. cellulolyticum	T. viride	Soybean	Rice	FAO Reference
	(1)	(2)	(3)	(4)			
Lys	6.6	6.2	6.8	4.4	6.5	3.7	4.2
Thr	5.7	4.9	6.1	4.9	4.0	4.1	2.8
Cys	2.6	2.5	0.31	1.4	1.4	1.1	2.0
Val	6.9	5.0	5.8	4.4	5.0	5.7	4.2
Met	2.7	1.9	2.3	1.3	1.4	2.4	2.2
Ile	5.2	3.9	4.7	3.5	5.4	3.9	4.2
Leu	9.3	6.9	7.5	5.8	7.7	8.0	4.8
Tyr	2.8	3.1	3.3	3.3	2.7	3.3	2.8
Phe	3.0	6.5	3.8	3.7	5.1	5.2	2.8
Trp	1.4	1.5	ND	ND	1.5	1.4	1.4

(1) Carbon source: spray-dried fiber board waste water (S. Thomke, et al. 1980).
(2) Carbon source: 2% barley flour (Hofsten and Ryden 1975).
(3) Carbon source: Solka-floc (Moo-Young et al. 1977).
(4) Carbon source: treated barley straw (N. Peitersen 1975).

The influence of cost factors must be reassessed and, it is not unrealistic to say, sometimes relegated to a less significant role, especially for developing countries.

With this work we hope to have shown that biotechnological solutions for utilization of rice husk are possible.

This work represents for our Laboratory the start-up of a more rational and application-oriented microbiology; the results presented show that one must go further. The importance of isolating fungal strains from the natural habitat of the plants was already recognized (Zomer et al. 1981). We have isolated a variety of fungi which utilizes lignocellulosic substrates and their potential must be assessed in a proper biotechnologically oriented way. From these few examples we must reiterate the statement of Eriksson (1978) "that there is a strong need for cooperation between biochemists, microbiologists, geneticists and engineers if the ultimate goal -- development of economically feasible large-scale bioconversion processes based on ligno-cellulosic waste material -- shall be a reality".

ACKNOWLEDGMENTS

We wish to thank our colleague, Mrs. Luisa Roda-Santos, for her help in mounting some of the analytical methods as well as in the prepartation of the final paper. We are also greatly indebted to Professor A. Xavier and his group at the New University of Lisbon for their help and encouragement throughout this work.

REFERENCES

Ayers, A.R., Ayers S.B., Eriksson K.E. 1978. Cellobiose oxidase, purification and partial characterization of a hemoprotein from S. pulverulentum. Eur. J. Biochem. 90: 171-181.
Beagle, E.C. 1978. Rice-husk conversion to energy. FAO Agricultural Services Bulletin no. 31, Rome.
Cabello, A., Conde J., Otero M. 1981. Prediction of the degradability of sugar cane cellulosic residues by indirect methods. Biotechnol. Bioeng. 23: 2737-3745.
Chahal, D.S., Moo-Young M., Vlach D. 1981. Effect of physical and physiocochemical pretreatments of wood for SCP production with Chaetomium cellulolyticum. Biotechnol. Bioeng. 23: 2417-2420.
Clemente, A. and Andrade, M. 1981. A preliminary study of the biodegradation of rice husk. VII National Congress of the Portuguese Society of Biochemistry, Povoa do Varzim.

Clemente, A., Duarte J., Roda-Santos, M. Andrade M. 1982. Microbial conversion of wastes for fuels and chemicals production. V National Meeting of the Portuguese Society of Chemistry, Porto.

Crawford, D.L. Kirk T.K., Harkin J.M., McCoy E. 1973. bacterial cleavage of an arylglycerol-β-aryl ether bond. Appl Microbio. 25: 322-324.

Ek, Mats and Eriksson, Karl-Erik. 1980. Utilization of the white-rot fungus Sporotrichum pulverulentum for water purification and protein production on mixed lignocellulosic waste-waters: Biotechnol. Bioeng. 22: 2273-2284.

Eriksson, K.E. 1978. Enzyme mechanisms involved in cellulose hydrolysis by the rot fungus S. pulverulentum. Biotechnol. Bioeng. 20: 317-332.

Eriksson, K.E. Grunewald A., Vallander L. 1980. Studies of growth conditions in wood for three white-rot fungi and their cellulaseless mutants. Biotechnol. Bioeng. 22: 363-376.

Eriksson, Karl-Erik and Larsson, Kjell. 1975. Fermentation of waste mechanical fibers from a newsprint mill by the rot fungus Sporotrichum pulverulentum. Biotechnol. Bioeng. 17: 327-348.

Gracheck, S.J., Rivers D.B., Woodford L.C., Giddings K.E., Emert G.H. 1981. Pretreatment of lignocellulosic to support cellulase production using Trichoderma reesei QM9414. Biotechnol. Bioeng. Symp. 11: 47-65.

Han Y.W. 1975. Microbial fermentation of rice straw: Applied Microbiology, 29: 510-514.

Han, Y.W., Callihan C.D. 1974. Cellulose fermentation: Effect of substrate pretreatment on microbial growth. Applied Microbiology, 27: 159-165.

Han, Y.W., Timpa J., Ciegler A., Courtney J., Curry W., Lambremont E. 1981. γ-Ray-induced degradation of lignocellulosic materials. Biotechnol. Bioeng. 23: 2525-2535.

Hofsten, B.V. and Ryden, A.L. 1975. Submerged cultivation of a thermotolerant basidiomycete on cereal flours and other substrates. Biotechnol. Bioeng. 17: 1183-1197.

Kirk, T.K., Schultz E., Connors W.J., Lorenz L.F., Zeikus J.G. 1978. Influence of culture parameters on lignin metabolism by Phanerochaete chrysosporium. Arch. Microbio. 117: 277-285.

Ladisch, M.R. 1979. Fermentable sugars from cellulosic residues. Process Biochemistry, 14. 1: 21-25, January.

Merril W. and French, D.W. 1966. Decay in wood and wood fiber products by Sporotrichum pruinosum: Mycologia 58: 592-596.

Miron, J. and Ben-Ghedalia, D. 1981. Effect of chemical treatments on the degradability of cotton straw by rumen microorganisms and by fungal cellulase. Biotechnol. Bioeng. 23: 2863-2873.

Moo-Young, M., Chahal D.S., Swan J.E., Robinson C.W. 1977. SCP production by Chaetomium cellulolyticum. Biotechnol. Bioeng. 19: 527-538.

Peck, H.D., Jr. and Odon, M. 1981. Anaerobic fermentations of cellulose to methane. In: Trends in the biology of fermentations for fuels and chemicals. Edited by Alexander Hollaender, Ed. 375-395. Plenum Publishing Corporation, N.Y.

Peitersen, N. 1975. Cellulase and protein production from mixed cultures of Trichoderma viride and a yeast. Biotechnol. Bioeng. 17: 1291-1299.

Thomke, S. Rundgren M., Eriksson S. 1980. Nutritional evaluation of the white-rot Fungus Sporotrichum pulverulentum as a feed-Stuff to rats, pigs and sheep. Biotechnol. Bioeng. 22: 2285-2303.

Tsao, G.T. 1978. Cellulosic material as a renewable resource. Process Biochemistry 13: 12-15, October.

Tsuchiya, H.M., Van Lanen J.M., Langlykke A.F. 1949. U.S. Patent 2,481,263, September 6.

Vandecasteel, J.P. and Pourquie J. 1981. Hydrolyse enzymatique de la paille de blé et des tiges de mais en vue de la production de carburants par fermentation. Colloque sur: "L'Hydrolyses des Produits Ligno-cellulosiques", Montreal, October.

Vianna e Silva, Manuel. 1969. Arroz. Fundacao calouste gulbenkian, Lisboa.

Westermark, Ulla and Eriksson, Karl-Erik. 1974. Cellobiose: Quinone oxidoreductase, a new wood-degrading enzyme from white-rot fungi. Acta. Chem. Scand. B28: 209-214.

Zadrazil, F. 1980. Conversion of different plant waste into feed by basidiomycetes. European J. Appl. Microbiol. Biotechnol. 9: 243-248.

Zadrazil, F. and Brunnert H. 1981. Investigation of physical parameters important for the solid state fermentation of straw by white-rot fungi. European J. Appl. Microbiol. Biotechnol. 11: 183-188.

Zomer, E., Klein D., Rozhanski M., Er-El Z. 1981. Winged bean haulm, a potential raw-material for single-cell protein production in the tropics. Biotechnology letters, 3: 513-518.

FOOD FROM BIOMASS

M. T. Amaral Collaço

Department of Food Research
LNETI
Rua Vale Formoso, 1
1900 Lisbon, Portugal

INTRODUCTION

The world is in a state of change. The following decades will be dominated by shortages in food and energy and also by environmental concerns. The so-called untapped or underemployed natural resources, forgotten until a short time ago, are now being listed by a large number of countries, giving rise to a certain competition as to their possible use as food or energy.

To alleviate food shortages (namely in proteins) production has been intensifying since the 1950's, but greater effort in this direction will be needed to feed the six billion world inhabitants in the year 2000. The world population increases about seventy million a year which means that the world area under cultivation is becoming insufficient for the feeding of humans and animals (Table 1).

In Portugal, only 53 percent of the total area of the country is used for agriculture with only 11 percent under irrigation (Figure 1).

A management policy for water resources could stimulate more intensive farming which is crucial, as Portugal is highly dependent on the foreign market to the order of 55 percent for human foodstuffs and up to 85 percent with imports of raw materials for animal feeding and 90 percent in energy costs.

The necessary intensive agriculture, or the "Green Revolution," which has been developing during the last few decades in the developed countries is, however, threatened by the "oil crisis" since the

Table 1. Farmland per person, 1970 and 1990. (From Allaby 1977.)

Continent	Area	Potentially arable (million hectares)	Cultivated 1965	Population 1970 (m)	Cult. ha/caput 1970	Population 1990 (m)	Cult. ha/caput 1990
AFRICA	3019	732,5	157,8	279	0,57	498	0,32
ASIA	2736	627,3	518,0	1986	0,26	3082	0,17
OCEANIA	822	153,8	16,2	15	1,08	21	0,77
EUROPE AND USSR	2711	530,2	380,4	702	0,54	812	0,47
N. AMERICA	2108	465,4	238,8	226	1,06	271	0,88
S. AMERICA	1752	679,9	76,9	284	0,27	489	0,16
TOTAL	13148	3189,1	1388,1	3621	0,38	5346	0,26

main support of the system lies in high energy input in the form of fertilizers, pesticides and mechanization.

An understanding of microbiology and the development of molecular, biological, genetic and enzymatic techniques led to hopes in

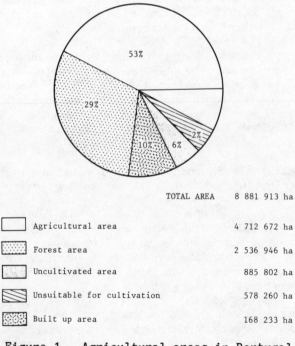

TOTAL AREA	8 881 913 ha
Agricultural area	4 712 672 ha
Forest area	2 536 946 ha
Uncultivated area	885 802 ha
Unsuitable for cultivation	578 260 ha
Built up area	168 233 ha

Figure 1. Agricultural areas in Portugal.

FOOD FROM BIOMASS

the biological revolution in which, through biotechnology, solutions will be found for a variety of areas in the production of new foods.

In the field of agriculture and the food industry the use of biological fertilizers and pesticides, genetic treatment of animal and vegetable species and the development of new foods and food supplements should be the solution to avoiding the catastrophe predicted by Malthus in 1798.

Non-conventional food will, of necessity, become the food of the future. Man is conservative by nature; adventure terrifies and attracts him at the same time. New types of food make him suspicious and only need will make him use them freely. That was the case with the yeast "Candida utilis" produced in sulphitic liquors and used for human nutrition by the Germans during the Second World War.

However, for centuries now, micro-organisms have been used by industry in fermentation processes. Bread, wine, beer, vinegar, cheese and yoghurt are examples of conventional foods that undergo a series of enzymatic reactions through microbiological intervention in which the final product has better organoleptic qualities, is more digestible and is easier to keep.

The impossibility of conventional agriculture being able to supply the world with food, makes the full use of existing resources essential. The use of superfluous biomass urgently needs the research focus of multidepartmental teams in the search for new and supplementary foods as well as further study into existing foods.

Several sources and processes in the production of new foods can be considered currently (Figure 2):

Without microbiological intervention:

- Protein concentrates
- Meat analogues

With microbiological intervention:

- Food with the use of micro-organisms
- Micro-organisms as food
- Constituents of micro-organisms as supplements to food

CONCENTRATED PROTEINS - MEAT ANALOGUES

Protein derivatives of vegetable origin should be taken from low cost, or indeed negative cost, raw materials that have a relatively high protein content and a balanced amount of amino acids,

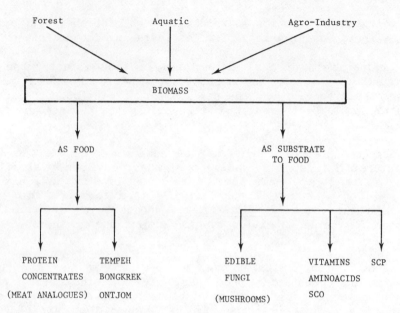

Figure 2. Sources and processes in the production of new foods.

atively high protein content and a balanced amount of amino acids, lack of anti-nutritional factors and allow for high extraction rates at a low cost. The proteins of soya beans, rapeseed, sunflower seed, peas, broad beans and cotton seed are already widely used in concentrates and isolated proteins. Recently considerable studies have focused on the leaves of green plants, a potential source that has yet to be explored. Green plants are made up of 1.5 - 3.5 percent proteins, 85 percent being found in the leaves.

Agricultural leaf by-products are extremely useful for feeding ruminants; however, large quantities are wasted, which results in an appreciable total annual quantity of protein waste.

Relatively similar extraction processes are used. In presses (Figure 3) the cell walls are broken, the juice is collected and the protein content then goes through various types of treatment - ultracentrifugation, preferential flocculation or chromatography. The proteins are then precipitated according to their isoelectric point.

According to Pirie, crude LPC contains 56-69 percent crude protein, 20-25 percent water, 5-10 percent starch, varying amounts of provitamin A ash and it is normally dark green in colour. The amino acid content of LPC is lower in nutritive value than meat or fish, higher than that in cereals with a slight deficit of total sulfur amino acids (Table 2).

Leaf protein concentrate (LPC) along with vitamins is used in India to supplement rice diets giving 2.5-5 percent extra protein.

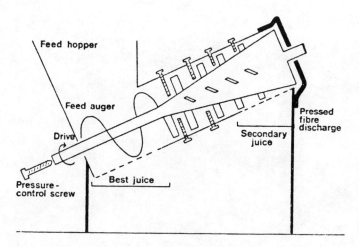

Figure 3. Extraction of LPC by press. (From Pirie 1976).

For the preparation of concentrates and isolated proteins, methods require making the protein soluble in a basic medium followed by precipitation in an acid medium. These concentrates usually appear in powder form, but to make them more appetizing an attempt has been made to make meat substitutes. In these the proteins are in fibre form along with lipids, traces of glucosides, mineral salts and flavouring, so that these substitutes resemble meat in flavour, texture and nutritive value.

FERMENTED FOODS

Asia is a vast storehouse of ancient culture, art and knowledge, including food fermentation technology. Oriental fermented foods are predominatly fermented soya products. The vast array of products can be appreciated by referring to Table 3.

Substrates also include copra, rice, maize and various agro-industrial residues such as peanut presscake and coconut presscake that are widely used in Indonesia, Japan, China and India. In certain cases yeasts and bacteria, and even mixed cultures are used besides moulds.

Bongkrek, Ontjom, Tempeh gembos, Oncom tahu and Tempeh mata kedele are fermented products in which Tempeh production technology is applied to different agro-industrial residues (Table 4).

Tempeh is a food produced by the solid substrate fermentation of whole soybeans by __Ryizopus oligosporus__. In Bongkrek the substrate is coconut presscake. In Tempeh gembos the substrate is

Table 2. Amino acids (essential) in leaf protein concentrate from different species. Samples, air-dried at 40°C-60°C, were hydrolysed and assayed microbiologically for essential amino-acids. (From PAG Nutrition Document R.10/Add.103 1966.)

Species	Protein, %	Lysine	Phenylalanine	Methionine	Mistidine	Leucine	Isoleucine	Valine	Tryptophane	Arginine	Threonine	Cystine
Lucerne	66	5.5	5.2	1.5	2.0*	4.0*	4.7*	6.0*	1.0	7.9	3.9	1.0
Banana	10.6	5.4	5.3	1.8	2.0	4.7	5.2	7.4	0.9	9.0	3.6	1.0
French-beans	44.5	4.4	6.3	2.1	2.0	5.9	6.9	8.1	1.0	16.0	4.8	0.7
Chicory	30	6.0	8.5	2.0	2.2	6.3	8.0	9.7	1.2	12.1	5.8	0.8
Water hyacinth	26.3	6.5	7.5	1.5	2.0	6.9	8.4	8.8	1.2	14.4	5.4	1.6
Dolichos lab-lab	51	4.9	6.3	1.3	1.7	5.6	7.0	8.2	1.5	10.0	4.9	0.4
Khol-khol	46	5.3	6.0	1.9	2.2	5.2	6.5	7.7	1.3	12.6	4.6	0.5
Maize	47	5.3	5.2	1.7	1.4	6.0	6.2	7.8	0.9	12.0	3.1	0.6
Carrot	33	4.5	5.2	1.4	1.7	5.3	5.6	8.7	1.2	12.6	4.5	0.5
Cauliflower	43	5.7	5.3	1.9	2.1	5.5	6.5	9.1	1.1	9.1	4.5	1.2

(* grms. amino acid per 100 g. protein).

Table 3. Fermented foods. (From Hesseltine and Wang 1980).

Name	Area or Country	Organisms Used	Substrate	Nature and Uses
Soy sauce (Chiang-Yu, Shoyu, Toyo, Kanjang, Kekap, See-Ieu)	The Orient	Aspergillus oryzae, Pediococcus halophilus, Lactobacillus deibrueckii, Torulopsis Versatilus, Saccharomyces rouxii.	Whole soybeans or defatted soy products, wheat	Dark reddish liquid salty taste resembling meat extract; a flavoring agent.
Miso (Chiang, Doenjang, Soybean paste, Tauco)	The Orient	Aspergillus oryzae, Saccharomyces rouxii, Torulopsis etchellsii, Pediococcus halophilus	Whole soybeans, rice or barley	Paste, smooth or chunky, light yellow to dark reddish brown, salty and strong flavored resembling soy sauce a flavoring agent.
Hamanatto (Toushih Tao-Si, Tao-Tjo)	The Orient	Aspergillus, Streptococcus, Pediococcus	Whole soybeans, wheat flour	Nearly black soft beans, salty flavor resembling soy sauce a condiment.
Sufu (Fu-Ru, Fu-Ju, Tou-Fu-Ju, Bean cake, Chinese cheese)	China	Actinomucor elegans, Mucor disperus	Soybean curd (Tofu)	Cream cheese-type-cubes, salty; a condiment, served with or without further cooking.
Tempeh (Tempe kedelee)	Indonesia and vicinity	Rhizopus oligosporus	Whole soybeans	Cooked soft beans bound together by mycelia as cake, a clean, fresh, yeasty odor; cooked and served as a meat substitute or snack.
Bongkrek	Indonesia	Rhizopus oligosporus	Coconut presscake	Similar to Tempeh
Ontjom (Oncom)	Indonesia	Neurospora intermedia	Peanut presscake	Similar to Tempeh

(Continued)

Table 3. Continued.

Bongkrek	Indonesia	_Rhizopus oligosporus_	Coconut presscake	Similar to Tempeh
Ontjom (Oncom)	Indonesia	_Neurospora intermedia_	Peanut presscake	Similar to Tempeh
Natto	Japan	_Bacillus natto_ (_B. subtilis_)	Whole Soybeans	Cooked beans covered with viscous, sticky polymers produced by the bacteria, ammonia odor, musty flavor; served with or without further cooking as a meat substitute.
Lao-Chao (Chiu-Niang, tape, ketan)	China, Indonesia	_Amylomyces rouxii_, _Rhizopus chinensis_ _Saccharomycopsis fibuleiger_, _S. malanga_	Glutinous rice (waxy variety of rice)	Soft, juicy, sweet and slightly alcoholic rice; served with or without cooking as a dessert.
Ang-Kak (Red rice, anka)	China, The Philippines	_Monascus purpureus_	Rice	Red color throughout the rice kernels; a coloring agent.
Idli	India (esp. South India)	Yeast, _Leuconostoc mesenteroides_	Rice, dehulled blackgram	Steamed cake
Doza (dosai)	India	Yeast, _Leuconostoc mesenteroides_	Rice, dehulled b.	Pancake

Table 4. Fermented products using Tempeh production technology.

Products	Substrate	Microorganisms
Tempeh	Whole soybeans	Rhizopus oligosporus
Bongkrek	Coconut presscake	Rhizopus oligosporus
Ontjom (Oncom)	Peanut Presscake	N. sitophila Neurospora intermedia
Tempeh gembos	Soybean curd	Rhizopus oligosporus
Oncom tahu	Soybean curd	Neurospora sitophila
Tempeh mata Kedele	Soybean Hypocotyledons	Rhizopus oligosporus

soybean curd. In Tempeh mata kedele the substrate is soybean hypocotyledons. Ontjom (Oncom) uses peanut presscake and the microorganism present is Neurospora sitophila or N. intermedia, which gives a red colour to the product. Oncom tahu also uses N. sitophila on soybean curd.

These fermented foods are of considerable social and economic importance. Solid substrate fermentation development is relatively advanced nowadays although controlling the conditions under which fermentation takes place in actually more difficult than controlling fermentation in liquid media. The advantages of solid substrate fermentation are the following:

1. Low energy consumption
2. Low growth rate ensuring minimum RNA content of microbes
3. Low water activity value, limiting growth of many contaminating agents
4. Facile processing before consumption

However, the basic requirements for these products, or for any product used as food, are: nutritional quality, palatability, digestibility and freedom from toxic compounds.

TEMPEH

In Indonesia, Tempeh is the main source of protein for the majority of the population. Annual production is about 75,600 t produced all over the country in small units. The biggest production plant produces 422 t/year.

Hesseltine, Sorenson, Steinkraus, Winarno, Shurtleff and Aoyagi are among the research workers who have done a great deal of research into this food product. Tempeh, the fermented soybean, after frying is a food with excellent organoleptic qualities and it is agreeable to Western tastes. Just as this technological process is applied to wastes such as peanut presscake to convert them into food in Indonesia, it would perhaps be worthwhile considering in the future the production of new products fermented by Mucorales using substrates of overproduced seeds or other raw materials.

At DTIA-LNETI, Tempeh has been made from chick-peas. Studies are still being conducted.

The Production of Tempeh

The micro-organisms used as a starter are stains of Rhizopus sp., mainly strains of Rhizopus oligosporus (Table 5). Sorenson and Hesseltine (1966) studied the metabolism of the strains used in these solid fermentations very thoroughly (Figure 5).

The rapid development of R. oligosporus followed by a drop in the pH means that probability of eventual bacterial contamination is lessened. Recent experiments have shown that Rhizopus sp. produce natural, heat-stable, antibacterial agents. However, mycotoxins, in particular aflatoxins, have never been reported as being produced by Mucorales.

Figure 4. Tempeh.

Two processing schemes for producing Tempeh are set out here:

Flowchart for Tempeh (1) Djien Process	Flowchart for Basic Indonesian Soy Tempeh Method (2)
Dry Soybeans ↓ Wash ↓ Soak overnight ↓ Seedcoats removed ↓ Boil (30 minutes) Drain ↓ Cool ↓ Inoculation with *Rhizopus* sp. spores ↓ Incubation 20-24h ↓ Temp. 31-32°C Finished Tempeh	Dry soybeans ↓ Wash ↓ First boil ↓ (1-30 minutes) Drain and dehull ↓ (Underfoot) Float off hulls ↓ Soak and prefermentation ↓ (24h at room temperature) Second boil ↓ (30-90 minutes in soak water) Drain ↓ Cool ↓ (To body or room temperature Inoculate ↓ (with starter grown on leaves) Put in Tempeh containers ↓ (Plastic bags, banana leaves or plastic-lined trays) Incubate ↓ (36-48 hours) Finished Tempeh

(1) From: Hesseltine 1965
(2) From: Shurtleff and Aoyagi 1979

Table 5. Summary of Rhizopus species which made acceptable Tempeh in pure culture. (From Hesseltine 1965).

Name	No. of Cultures
Rhizopus oligosporus Saito	25
R. stolonifer (Ehren) Vnill	4
R. arrhizus Fischer	3
R. oryzae Went & Geerligs	3
R. formosaensis Nakazawa	3
R. achlamydosporus Takeda	2
Total	40

Nutritional Characteristics of Tempeh

Although the fermentation time of soya by Rhizopus oligosporus is short, (one day) the changes in taste and smell, and the structural alterations coming from a change in the chemical composition, yield Tempeh with improved nutritional characteristics.

The protein value of a food depends to a great extent not only on the quantity of protein present, but mainly on the quality which is evaluated on the content of the essential aminoacids and by the digestibility of the protein present. To evaluate this quality we need five standards: PER (Protein Efficiency Ratio), BV (Biological Value), NPU (Net Protein Utilization), Chemical Score and Aminoacid Score.

During the fermentation of soya by R. oligosporus part of the protein is hydrolysed, freeing the amino acids. Although the total

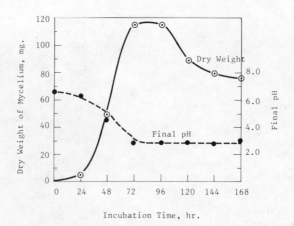

Figure 5. Typical growth curve of Rhizopus oligosporus in a glucose -- NH_4Cl medium. (From Sorenson and Hesseltine 1966.)

Table 6. Composition of nutrients in 100 grams of Tempeh (From Shurtleff and Aoyagi 1979).

Type of tempeh	Food Energy Cal.	Moisture %	Protein %	Fat %	Carbo-hydrates (incl.fiber) %	Fiber %	Ash %	Calcium Mg	Phos-phorus Mg	Iron Mg	Thiamine Vit.B$_1$ Mg	Riboflavin Vit.B$_2$ Mg	Niacin Nic.Acid Mg	Reference
Soy tempeh fresh	157	60,4	19,5	7,5	9,9	1,4	1,3	142	240	5,0	0,28	0,65	2,52	Dif.sources
Soy tempeh, dry basis	-	0	54,6	14,1	27,9	3,1	3,5	-	-	-	-	-	-	Lijas 1969
Okara tempeh, fresh	-	84,9	4,0	2,1	8,4	3,9	0,7	226	-	1,4	0,1	-	-	Gandjar 1977
Coconut-presscake tempeh, fresh	119	72,5	4,4	3,5	18,3	2,4	1,3	27	100	2,6	0,08	-	-	ITFC 1967
Peanut-presscake onchom	187	57,0	13,0	6,0	22,6	2,9	1,4	96	115	27,0	0,09	-	-	ITFC 1967

nitrogen remains constant, soluble nitrogen increases, consequently increasing the value of NPU. Tempeh shows a high value of lysine but is limiting with regard to methionine and cystine (Table 6, Table 7).

During this fermentation the reducing sugars are reduced giving Tempeh a carbohydrate content of about 9.9 percent starch and hemicellulose are absent or residual. Tempeh is an excellent diet food containing only 157 cal/100g of product. The saturated fatty acid content is low (14.5 percent in all) (Table 8).

The vitamin B_{12} content of Tempeh is about 1.5-6.3 g/100g of product, an amount that fulfills the U.S. recommended daily allowance (RDA) of 3 g/per adult per day (Table 9).

Toxins

One of the most important aspects of fermented foods is their freedom from toxins. Toxins produced by bacteria or molds can sometimes be very dangerous. In Tempeh bongkrek prepared from coconut presscake, Pseudmonas cocovenenans occasionally appears overgrowing the R. oligosporus mold and produces toxoflavin and bongkrek acid which are both poisonous and even responsible for many deaths. Micotoxins from fungi are also potentially dangerous as they are carcinogens. Aflatoxins from Aspergillus spp. have been most carefully studied. Their permitted level in food is 30 ppb.

Table 7. Comparative amino acid content of Tempeh. (From Shurtleff and Aoyagi 1979.)

Amino Acids	FAO/WHO Pattern	Tempeh (Soybean)	Tempeh as Percent of FAO/WHO Pattern	Soybeans	Eggs
Methionine-cystine	220	171	78	165	342
Threonine	250	267	107	247	302
Valine	310	349	113	291	.437
Lysine	340	404	119	391	417
Leucine	440	538	122	494	547
Phenylaninetyrosine	380	475	125	506	588
Isoleucine	250	340	136	290	378
Tryptophan	60	84	140	76	106
Total	2.250	2.627		2.460	3.117
Methionine		(71)		(84)	(192)
Cystine		(101)		(81)	(150)
Tyrosine		(170)		(165)	(240)
Phenylalanine		(305)		(341)	(348)

Table 8. Fatty acids in soy Tempeh. (From Shurtleff and Aoyagi 1979.

SATURATED TOTAL	0,595 grams (14,5% of total)
16,0(palmitic)	0,420 g
18,0(stearic)	0,175 g
MONOUNSATURATED TOTAL	0,713 grams (17,3% of total)
18,1 (oleic)	0,713 g
POLYNSATURATED, TOTAL	2,803 grams (62,2% of total)
18,2(linoleic)	2,510 g
18,3(linolenic)	0,293 g

Poorly stored peanut products have been contaminated, but R. oligosporus can degrade about 30-50 percent of the aflatoxin present in such Tempeh.

Table 9. Vitamins and minerals in soy Tempeh (From Shurtleff and Aoyagi 1979).

Nutrient	Amount Per 100 Grams Fresh Tempeh		U.S. Recommended Daily Allowance		100 Grams Tempeh As A Percent of RDA
Vitamin A	42	IU	5000	IU	1 %
Thiamine (B_1)	0.28	mg	1.5	mg	19 %
Riboflavin (B_6)	0.65	mg	1.7	mg	28 %
Niacin	2.52	mg	20		13 %
Pantothenic acid	0.52	mg	10.0	mg	7 %
Pyridoxine (B_6)	830	mcg	2000	mcg	42 %
Folacin (Folic acid)	100	mcg	400	mcg	25 %
Cyanocobalamin (B_{12})	3.9	mcg	3.0	mcg	130 %
Biotin	53	mcg	300	mcg	18 %
Calcium	142	mg	1000	mg	14 %
Phosphorus	240	mg	1000	mg	24 %
Iron	5	mg	18	mg	28 %

EDIBLE FUNGI

The edible fungi are Basidiomycetes of which the carpophore or fruiting body is edible. Known for their delicate flavour, throughout the centureis the mushrooms have been favoured by fairies and witches alike, some species being considered as having medicinal, pharmacological and occult properties.

At present, out of 45,000 species, 2,000 are edible, 25 are acceptable as food, but only 10 are commercially viable. The edible fungi are an example of an agreeable product derived from the use of biomass. Various substrates originating from the food industry, which are frequently a source of pollution, have been used experimentally to feed these saprophytic fungi and to convert such renewable resources into a quality food. At the same time the substrate is converted into fertilizer or can be converted into foodstuff through biotechnological techniques.

The economic importance of mushrooms is considerable and tends to increase each year. Production grows at an average of 10 percent a year. Delcaire (1978) estimated world production in 1975 at 670,000 t of <u>Agaricus bisporus</u>, 130,000 t of <u>Lentinusedodes</u> and 42,000 t of <u>Volvariella volvacea</u> (Figure 6).

The development of cultivation techniques has contributed to this increase in production and also to making the mushroom more popular and more generally available.

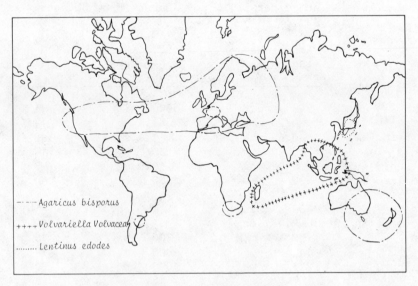

Figure 6. Mushroom culture in the world. (From Veder 1974.)

To grow mushrooms successfully various guidelines have to be taken into consideration depending on the species to be grown. Growing conditions to be considered are substrate, temperature, humidity and light which are fundamental factors for guaranteed production (Table 10).

Table 10. Mushroom cultivation conditions (From Kurtzman 1979).

Species	Temperature °C Spawn Running	Fruiting	Level of Environmental Control Required	Waste Substrate
Agaricus bisporus (Common mushroom)	20-27	10-20	+++	Composted Horse manure or rice straw
Agaricus bitorquis	25-30	20-25	+++	Composted Horse manure or rice straw
Auricularia spp. (Ear mushrooms)	25-35	20-30	+++	Sawdust-rice bran
Coprinus fimetarius	20-40	20-40	+	Straw
Flammulina velutipes (Winter mushroom)	18-25	3-8	+++	Sawdust-rice bran
Lentinus edodes (Shiitake mushroom)	20-30	12-20	++	Logs or sawdust-rice bran
Pholiota nameko (Nameko mushroom)	24-26	5-15	+++	Logs or sawdust-rice bran
Pleurotus ostreatus (Oyster mustroom)	20-27	10-20	+	Straw, paper, sawdust straw
Stropharia rugoso-annulata	25-28	10-20	++	Straw, paper, sawdust straw
Tremella fuciformis (White jelly mushroom)	20-25	20-27	+++	Logs or sawdust-rice bran
Volvariella volvacea (Straw mushroom)	35-40	30-35	+	Straw, Cotten wastes

Cultivation of Edible Mushrooms

Agaricus bisporus (Lange) Sing., (Figure 7) the common mushroom known the world over, is grown traditionally on a substrate of decomposed horse manure. Other substrates such as waste materials are used occasionally, as well as materials found locally, e.g., rice straw compost is used in Taiwan.

A. bitorquis, which is very similar to A. bisporus is an alternative species to the latter. This basidiomycete has a higher temperature for fructification (20-25°C), is virus dieback disease resistant and does not need as much attention as A. bisporus in cropping and storing. Although mushroom growing began in France in the seventeenth century, today production techniques are quite different. Much progress has been made in controlling the procedure so that nowadays the biochemical and technological mechanisms of the different phases are well understood. The following photographs (Figures 8-12) depict the cultivation procedure:

The preparation of compost involves the mixing of horse manure and straw which is followed by mixed fermentation dominated by thermophilic flora. The composting phase, illustrated in Figure 8, requires 15 days.

Figure 7. Agaricus bisporus (Lange) Sing.

Figure 8. Composting.

Pasteurization of the compost is accomplished by placing it on trays or shelves in a heated room at a decreasing temperature of 60°C, 55°C and finally 50°C where it is maintained until ammonia disappears under forced ventilation. This generally requires ± 72 hours. These final conditions call for a temperature of 26°C-28°C, with a humidity of 68 percent and a pH of 7.

The inoculation of the pasteurized compost is carried out with ± 0.7 percent of spawn (Figure 9).

Incubation (Figure 10) is done at a temperature of 25°C and a relative humidity of 100 percent. After two weeks the "casing" mushroom beds are covered with peat and chalk for 21 days.

To induce fructification, the CO_2 is removed quickly at a relative humidity of 90-95 percent and a temperature of 16°C. Initiation of fruiting bodies requires eight days (Figure 11).

The production of fruiting bodies proceeds at a temperature of 14° to 20°C. Harvest of the first flux is done on the 21st day and harvesting continues until the 49th day. In a total of seven weeks, one tray produces 17 kg. of mushrooms (Figure 12).

The microbiology of compost preparation. Nowadays it is possible to use different raw materials enriched by composting due to

Figure 9. Inoculation.

Figure 10. Incubation.

Figure 11. Induction to fructification.

Figure 12. Production of fruiting bodies.

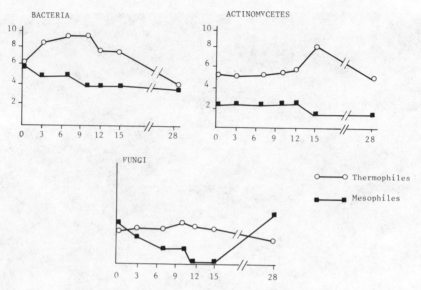

Figure 13. Changes in the population of aerobic bacteria. Actinomycetes and fungi during composting (day 0-15) and the establishment of A. bisporus in compost during incubation (day 16-28). (After Hayes 1969, 1977.)

the knowledge obtained from the composting of horse manure. Studies by Waksman 1930; Lambert and Davies 1934, 1941; Pizer and Thompson 1938; and more recently Hayes 1969 and Imbersion and Lepdae 1972 have contributed considerably to the knowledge of biochemical and microbiological mechanisms of composting (Figure 13).

During composting, and in the early stages, the dominant mesophile flora starts to decrease and there is a rise in thermophiles and thermotolerants. The bacterial population increases aiding the temperature to rise, but his dominating tendency only appears in the central aerobic areas while in the others there is a mixed flora.

A prolonged stage follows dominated by the thermophiles at 50-60°C in which first the bacteria are prevalent, followed later by the Actinomycetes. Fungi, in relatively small numbers, accompany the process and continue to be active. At the end of composting the temperature reduced to 25-30°C.

Cultivation of Lentinus edodes. Lentinus edodes (Berk) Sing. was first cultivated in China about 800 years ago. This mushroom, also known as the Shiitake mushroom, grows naturally on the trunk and branches of the Castanopsis cuspidata shiitree, and Quercus gambelli in Japan, China, Indonesia and New Guinea. In 1971 Japan produced 110,000 tons of these mushrooms most of which were exported to America, England, West Germany and France.

Forest cultivation is carried out on the dead logs of the Quercus sp. which are about 1 m x 5 to 15 cm in diameter. They are inoculated with cultures of Lentinus edodes and remain in the forest at 24-28°C for two years. After that, the logs are transferred to cooler, shady areas of the forest where the temperature is about 12-20°C. Fruiting bodies are formed for 3 to 7 years in the spring and autumn each year.

About 50 kg of mushrooms are obtained at each harvest from 1 cord of wood. This edible fungi requires very precise conditions of humidity, pH (3.5-4.5), and light (10^{-2} -10^{-4} lux) (Ishikawa 1967).

Lentinus edodes utilizes lignin, cellulose and hemicellulose. For this reason experiments have been done using forest and agricultural residues in order to produce these mushrooms in greenhouses. Under appropriate conditions Shiitake mushrooms grow in two weeks yielding about 260 g/Kg of dry waste. It should be added that the forest or agricultural residues that receive this biological pretreatment are better prepared for hydrolysis for later fermentation.

Cultivation of Volvariella volvacea. Volvariella volvacea (Bull ex Fr.) Sing, also known as the Straw mushroom, is a tropical or subtropical species grown in China. The optimum temperature for growth varies from 30 to 35°C. It does not use lignin as a nutrient (Hayes and Lim, 1979) and little is known about its metabolism. It is generally grown on rice straw, but other substrates have been used such as sorghum and wheat straw, maize residues, banana leaves, bagasse, tobacco stems, water hayacinth, sawdust and cotton wastes. Unlike A. bisporus there is no need of a casing layer to induce carpophores and light is indispensable for fructification. Several methods are in use for producing Volvariella volvacea, the traditional outdoor method in Hong-Kong, and two indoor growing methods are used in Malaysia and Hong-Kong as well.

One meter high stacks are made from soaked rice straw bundles. Rice bran or distillery wastes are added. Thermophiles and such species make the temperature rise. Three to four days later the spawn is inoculated. Primordia and fruiting bodies appear in 8 to 14 days. Yields of straw mushrooms are very variable depending on the method used. Indoor production is four times greater than outdoor production.

OTHER EDIBLE FUNGI

The use of biomass for conversion to food has been occupying research workers such as Steinkraus, Zadrazil, Schliemann, and

Figure 14. <u>Pleurotus ostreatus</u>.

Table 11. Mushroom composition percent. (From Vedder 1974).

Water	88 - 90%
Protein	2.7 - 4%
Carbohydrates	3.5 - 5%
Fat	0.2 - 8%
Fiber	0.8 %
Ash	0.9 - 1.1%

Table 12. Minerals in mushrooms per 100 grams of product. (From Carpentier 1971.)

Potassium	400 mg
Phosphorus	130 mg
Chlorine	80 mg
Sodium	20 mg
Calcium	25 mg
Iron	1 mg

Table 13. Vitamin content of mushrooms per 100 gm of product. (From Carpentier 1971.)

B_1	0.12 mg
B_2	0.52 mg
C	8.60 mg
PP	5.85 mg

Kurtzman et al. in the attempt to cultivate other edible mushrooms and utilize different agro-industrial wastes.

The oyster mushroom <u>Pleurotus ostreatus</u>, attacks the straw producing 1 kg of mass from 1 kg of dry straw in two to three months. <u>Strofaria rugoso-annulata</u> gives variable yields on straw with about 600 g per kg of dry straw in 2 months.

NUTRITIVE VALUE OF EDIBLE FUNGI

Like most vegetables, mushrooms have a high H_2O content (88-90 percent). They are rich in proteins and poor in fatty materials. They are also rich in vitamin B_1, B_2, C and K while potassium and phosphorus are the main minerals. Their caloric value is about 30 cal/100g, so they are an excellent diet food (Tables 11, 12, 13).

VITAMINS AND AMINO ACIDS

Certain organisms can synthesize all the growth factors from the elements present in the media on which they are grown. Such organisms rarely give high yields and studies are being carried out in attempt to make genetic changes that will result in greater production.

The commercial production of vitamins and amino acids, essential for metabolism in the higher animals, is needed to complement certain foods that are deficient in these growth factors. Some vitamins and amino acids are produced commercially by fermentation. Examples of this are vitamin B_{12} (cyanocobalamine) and the precursors of the vitamins A, C and B_2 (Table 14).

Lack of vitamin B_{12} results in malnutrition and anemia. The U.S. Recommended Allowance is 3.0g per day. The vitamin B_{12} market is dominated by Rhone-Poulenc.

Commercially produced Vitamin B_{12} is essentially of microbial origin. Strains of <u>Propiniobacterium schermanii</u> give yields of 23 mg/l, this being the highest yield obtained so far. Cobalt is essential for this biosynthesis (5 ppm), but if used in excess it can inhibit the process. Vitamin B_{12} is intracellular, or following fermentation, the cells are collected by centrifugation. Extraction is carried out by acidification or thermal shock. The addition of cyanate changes the coenzyme structure of the vitamin B_{12} (oxyadenosyl) which appears as cyanocobalamine and if offered in this form on the market (Table 15).

The major supplemental amino acids, lysine and glutamic acid, are produced by fermentation. Methionine is produced by chemical

Table 14. Production of vitamins by fermentation. (From Rivière 1975).

Vitamin	Strain	Medium	Conditions Fermentations	Extraction	Yield
Carotène (Vitamin A Precursor)	Bakleslea	Molasses	72h at 30°C	Solvent	1
		Soya oil - Ionone			
	Mycobacterium Smegmatis	Thiamin Hexadecane	Aérobiose		0.0007
Riboflavin	Ashbya	Glucose	6 days at 36°C	In steaming at 120°C	4.25
	Gossypii	Collagène		Reducers	
		Soya oil	Aérobiose	Adjunction to precipitate the riboflavine	
		Glycine			
L-Sorbose	Acetobacter suboxydans	D-Sorbitol 30%	45h at 30°C	Filtration & concentration	7%
		Corn Steep	Aérobiose		
Acid 5-Cétogluconique (Vitamin C Precursor)	Acetobacter suboxydans	Glucose	33h at 30°C		100%
		CO₃Ca Corn steep	Aérobiose		

synthesis. Lysine and glutamic acid are produced industrially for food purposes, with world production being dominated by the Japanese at present.

Thirty-thousand tons of lysine, the essential amino acid in which cereal grains are deficient, are produced annually. Seventy-five percent of this production comes from the Japanese firms Kyowa, Hakko and Ajinomoto.

Glutamic acid, sold commercially in the form of monosodium glutamate, is produced industrially by a strain of Corynebacterium glutamicum that yields as much as 60 grams per liter. Fermentation lasts 40 hours and is carried out at a temperature of 30°C and the pH is slightly alkaline, the limiting factor being biotin (1 to 5 μg/l). At present the substrate being used is molasses and annual production reaches 100 to 200 thousand tons. It is used mainly as a flavouring agent in the food industry.

S.C.O. SINGLE CELL OIL

In general the world suffers from a shortage of oils and fats. Their uses are in soaps, paints, textiles and cosmetics, but mainly as food.

In the U.S.A., 65 percent of the total production of oils and fats are utilized directly or indirectly in food industry.

In the West, 40 percent of the caloric values are satisfied by oils and fats. Microorganisms especially moulds and yeasts, are able to produce lipids. The fat content must be as high as possible and the relation fatty acids/triglyceride composition well known.

The saturated acids present are mainly palmitic and stearic acids and the unsaturated ones, oleic and linolenic acids.

The strains with a high fatty acid content must be able to grow at a reasonable rate in submerged culture, preferably in continuous culture but not necessarily. Studies are underway at the experimental stage into the kinetics of fat formation (Figure 15).

Figure 15 presents the pattern seen usually in batch culture using the strain Candida 107.

The production phase of fat occurs after growth has ceased, with the nitrogen as a limiting factor. Sometimes HPO_4^{-2} or Mg^+ are also limiting nutrients.

In continuous culture Candida 107 gives a good production of oil. The cells have the same lipids content (40 percent) and the yield of conversion is higher at a low dilution rate $<0,1^{-1}$ (Table 16.

However, looking to the quality of the oil produced, Aspergillus terreus has the highest polyunsaturated acids content of Table 16. pH, temperature and O_2 conditions influence the quantity of fats produced, the parameters to be used being a function of the micro-organisms being tested.

Temperature seems to produce larger changes in lipid composition. Various types of substrate should eventually be studied in accordance with local interests. The future will no doubt bring the answer to the technological aspects of microbial fat production. However, even after the process becomes economically viable it will

Table 15. Production of vitamin B_{12} by fermentation. (From Riviere 1975).

Strain	Medium	Aeration	Temp. °C	Incubation Time	Yield
Bacillus megaterium	Molasses Mineral salts Cobalt	Aerobiose	30	18	0.45
Propionibacterium shermanii	Glucose Corn steep Cobalt Ammonia pH = 7	Anerobiose 3 days Aerobiose 4 days	30	150	23
Streptomyces Olivaceus	Glucose Soy flour Cobalt Mineral salts	Aerobiose	28	96	5.7
Bacillus coagulans	Citric acid Triethanolamin Cobalt Corn steep	Aerobiose	55	18	6.0
Pseudomonas	Oxalic acid Betaine Cobalt Mineral salts	Aerobiose			10

Table 16. Fat contents and fatty acids of some oleaginous yeasts and moulds. (From Ratledge 1976.)

Organism	Fat (% W/W)	Fat Coeff.[a]	Relative %(W/W) of principal fatty acids in lipid					
			16:0	16:1	18:0	18:1	18:2	18:3
Candida 107	42	22	37	1	14	36	8	-
Crytoccus terricolus	65	21	Not given					-
Lipomyces lipofer	38	-	17	4	10	48	16	3
Lipomyces starkeyi	38	15	40	6	5	44	4	-
Rhodotorula gracilis	64	15	20	2	1	42	21	8
Aspergillus terreus	57	13	23	TR	TR	14	40	21
Chaetomium glefsim	54	-	58	3	8	27	-	-
Gibberella fujikuroi	45	8	Not given					
Mucor circinelloides	65	14	Not given					
Pythium ultimum	48	-	23	9	7	22	15	2

[a] Fat coefficient = g, total fat formed/100 g substrate utilised.

FOOD FROM BIOMASS

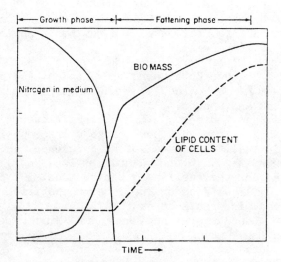

Figure 15. Idealized diagram showing course of lipid formation in an oleaginous micro-organism growing in batch culture. A nutrient, such as nitrogen, is allowed to become exhausted and the excess carbon is then converted to fat. (From Ratledge 1976.)

be necessary to carry out a careful study into the toxicology and complete nutritional value of S.C.O. before launching it commercially for human consumption.

NUTRITIONAL, TOXICOLOGICAL AND MICROBIOLOGICAL EVALUATION OF NEW FOODS

The Protein-Calorie Advisory Group (PAG) of the United Nations Organization published guidelines with the objective of ensuring the safety of new products for human consumption. Complete studies must be carried out on Toxilogical Safety, Nutritional Value, Sanitation, Acceptability and Technological properties before the product can be accepted for human consumption.

ACKNOWLEDGEMENTS

We are very grateful to the authors for authorization of the publication of the tables and for "TOCAN food plant" slides.

REFERENCES

Advi. Com. on Technology Innovation. 1981. Food, Fuel and Fertilizer from Organic Wastes. National Academy Press. Washington, D.C. 154 pp.

Agrela, J. and Nascimento, J. 1981. Cogumelos lenhivoros e substractos de cultura. Alimento para humanos e poligastricos. Actas NOPROT/81 (In Press).

Allaby, Michael. 1977. World Food Resources Actual and Potential. Applied Science Publishers Ltd. London. 418 pp.

Altschul, Aaron M. 1974. New Protein Foods. Academic Press Inc., New York and London. 511 pp.

Betschart, A.A. and Saunders, R.M. 1978. Safflower protein isolates: Influence of recovery conditions upon composition, yield and protein quality. Journal of Food Science. 43:964-968.

Birch, G.G., Parker, K.J. and Worgan, J.T. 1976. Food from Waste. Applied Science Publishers Ltd. London. 301 pp.

Brown, Lester R, 1980. Food or Fuel: New competition for the world's cropland. Interscience 6:365-372.

Caldas, Eugenio de Castro and Conceicao, Manuel de Santos. 1963. Niveis de Desenvolvimento Agricola no Continente Portugues. CEEA. Fundacao Calouste Gulbenkian.

Cluskey, J.E., et al. 1978. Cereal Proteins From Grain Processing. In: Protein Resources and Technology Status and Research Needs (Max Milner, Nevin S. Scrimshaw, Daniel I.C. Wang, (Eds.) pp. 256-277. Avi Publishing Company Inc. Westport, Connecticut.

Collaço, M.T.A. 1981. Producao de SCP a partir do Repiso do Tomate e Bagaco de Ulva com Geotrichum candidum. Pre-tratemento e Hildrolise Enzimatica. Actas NOPROT 81 (In Press).

Couffin, Herve. 1981. Pour une bio-industrie Francaise. Annales des Mines. Janvier: 11-24.

Ghose, T.K. 1969. Foods of the future. Process Biochemistry. December: 43-46.

Godet, Michel and Ruyssen Olivier. 1980. L'Europe en Mutation. Commission des Communautés Europeennes-perspectives europeennes: 117-155.

Hayes, W.A. and Nair, N.G. 1974. The Cultivation of Agaricus bisporus and other edible mushrooms. In: The Filamentous Fungi (J.E. Smith and D.R. Berry, Eds.) Vol. 1, pp. 212-248. Edward Arnold, London.

Hesseltine, C.W. 1965. A millennium of fungi, food and fermentation. Mycologia. 2:149-197.

Hesseltine, C.W. and Wang Hwa L. 1980. The importance of traditional fermented foods. Bio Science, 6:402-404.

Jay, James M. 1978. Fermented Foods and Related Products of Fermentation. In: Modern Food Microbiology (D. Van Nostrand Company Ed.), pp. 253-290.

Lebautl, J.M. 1980. La biotechnologie, cadre pour l'innovation dans les industries agro-alimentaires. Industries Alimentaires et Agricoles. 1045-1049 pp.

Mathur, B.N. and Shahani, K.M. 1979. Use of total whey constituents for human food. J. Dairy Sci. 62: 99-105.

Nicol, B.M. 1975. Manioc Leaves - Leaf Protein. In: The Pag Compendium (Vol. C2). (Moshe Y. Sachs, Ed.). pp. C2007-2012. Worldmark Press, Ltd, New York.

Odell, A.D. 1975. Meat Analogs. Potential in Developing Countries. In: The Pag Compendium (Vol. C2). (Moshe Y. Sachs, ed.), pp. C.2273 - C.2282. Worldmark Press, Ltd. New York.

Parpia, H.A.B. 1966. A Report on Leaf Protein Work at C.F.T.R.I., Mysore: Studies on production of protein for food from green vegetation. In: The Pag Compendium (Vol. C2). (Moshe Y. Sachs, Ed.) pp. C.2021 - 2031. Worldmark Press, Ltd. New York.

Peers, F.G. and Linsell, C.A. 1975. Dietary Aflatoxins and Liver Cancer - A Population based Study in Kenya. In: The Pag Compendium (Vol. D). (Moshe Y. Sachs, Ed.), pp. D.507 - D518. Worldmark Press, Ltd. New York.

Platt, B.S. 1975. The Protein Value of Food for Man. In: The Pag Compedium (Vol. F). (Moshe Y. Sachs, Ed.), pp.F37 - F44. Worldmark Press, Ltd. New York.

Protein Advisory Group of the United Nations System (PAG). 1972. Pag. Guideline No. 11. LaProduccion y Utilizacion Higienicas de Alimentos Proteinicos Desecados.

Protein Calorie Advisory Group of the United Nations Systems (PAG). 1972. Pag Guideline No. 7, Human Testing of Supplementary Food Mixtures.

Protein Calorie Advisory Group of the United Nations System (PAG). 1972. Pag Guideline No. 6, Perchimical Testing of Novel Sources of Protein.

Reed, Gerald. 1981. Use of microbial cultures: Yeast products. Food Technology. January: 89-94.

Rivière, J. 1975. Les Applications Industrielles de la Microbiologie. Masson et Cie., Paris. 203 pp.

Sampayo, M.A., and Martins, M.F.G. 1981. Proteinas de Microalgas crescendo em efluente de fabrica de farinha de peixe. Actas NOPROT/81 (In Press).

Scrimshaw, Nevin S. 1977. Unconventional food: Menu for tomorrow. Technology Review. October/November: 52-57.

Shurtleff, William and Aoyagi Akiko. 1979. The Book of Tempeh. Harper & Row. Publishers. New York, Hagerstown, San Francisco, London, 158 pp.

Siclet, Gerard, 1981. Quelques aspects de l'industrie microbiologique en Europe, aux Etats-Unis et au Japon. Annales des Mines. Janvier: 25-32.

Sinden. J.W. 1981. Strain adaptability. The Mushroom Journal. May: 153-165.

Sorenson, W.G. and Hesseltine, C.W. 1966. Carbon and nitrogen utilization by *Rhizopus oligosporus*. Mycologia. 5:681-689.

Steinkraus, Keith H. 1980. Special analysis: Food from microbes. Bio-Science. 6:384-386.

Tannenbaum, Steven R. and Wang, Daniel I.C. 1975. Single-Cell Protein II. The MIT Press. Cambridge, Massachusetts, London. 707 pp.

Truchot, E. 1979. Principales Sources de Proteines Alimentaires et Procedes d'obtention, C.D.I.I.U.P.A., Massy. 194 pp.

Vedder, P.J.C. 1974. Culture Moderne des Champignons. Stam/Robijns B.V. Culemborg. 384 pp.

Viesturs, V.E., Apsite, A.F., Laukevics, J.J., Ose, V.P. and Bekers, M.J. 1981. Solid-state fermentation of wheat straw with *Chaetomium cellulolyticum* and *Trichoderma lignorum*. Biotechnology and Bioengineering Symp. 11:359-369. Wiley and Sons, New York.

Vlitos, A.J. 1979. Creative botany opportunities for the future. Sugar y Azucar. July: 25-29.

Winarno, F.G. 1979. Fermented vegetale protein and related foods of Southeast Asia with special reference to Indonesia. J.A. Oil Chemists Soc. 56:363-366.

Wogan, G.N. and Shank, Robert E. 1975. Mycotoxins. In: The Pag Compedium (Vol. D). (Moshe Y. Sachs Ed.), pp. D.483 - D.488. Worldmark Press Ltd., New York.

POTENTIAL SUBSTRATES FOR

SINGLE CELL PROTEIN PRODUCTION

J. C. Royer and J. P. Nakas

State University of New York
College of Environmental Science and Forestry
Syracuse, New York 13210
U.S.A.

INTRODUCTION

Due to a rapidly increasing world population, and frequent food shortages, considerable interest has developed in the use of microbial biomass as a food supplement for animal or human consumption. The cultivation of microorganisms has great potential as a food source because of high nutritional value (Table I), high growth rate (Table II), ability to grow on a wide variety of relatively inexpensive carbon and nitrogen sources, and amenability to genetic engineering. Controlled fermentations have an additional advantage over traditional agriculture in that they are not subject to variability in weather conditions and can be implemented in most geographical locations (Moo-Young et al. 1977). Major international conferences on the use of microbial biomass for food were held at MIT in 1967 and 1973, and it was in 1967 that the term "Single Cell Protein" (SCP) was coined to describe microbial biomass of any origin.

HISTORICAL PERSPECTIVES

The use of microbial biomass as a food source is not a novel concept. Mushrooms, baker's and brewer's yeasts, molds on cheese, yogurt, tempeh and many other forms of microbial food have been routinely consumed for centuries. Efforts at mass production of single cell protein, however, began in Germany during World War I, when approximately 60 percent of imported protein was replaced with surplus brewer's yeast and yeast grown specifically for animal consumption (Rose 1979). Interest in microbial protein was renewed in Germany in 1936 when cultivation for human consumption began.

Strains of Candida arborea and Candida utilis were grown on sulfite waste liquor (a paper-making by-product) and incorporated into soups and sausages. Food yeast was reportedly produced from molasses in Jamaica at approximately the same time. Considerable research was devoted to the production of food yeast in the United States and Britain following World War II.

Interest in the production of SCP was revived in the late 1950s and 1960s when several major oil companies began researching the feasibility of producing microbial protein from petroleum fractions and derivatives. British Petroleum pioneered this research, and in 1976 finished construction of a $100 million plant to produce yeast from n-alkanes in Italy. The plant was never operated due to public concern over product safety.

MICROORGANISMS AND NUTRITIONAL VALUE

Although yeasts have received the most attention as SCP organisms, studies have been conducted with molds, bacteria and algae. Each group of microorganisms prossesses specific advantages and disadvantages. Bacteria generally have the highest growth rates and yield products of extremely high protein content. Algae have always been very attractive as organisms for SCP production due to their capacity to utilize the least expensive of all substrates, CO_2, as a source of carbon. Filamentous fungi, while possessing the slowest growth rate of the four general groups, are able to utilize a wide range of growth substrates, particularly polymers such as starch and cellulose which are not easily metabolized by most yeasts and bacteria. The mycelial form of molds provides a distinct physical advantage as it allows for easier harvesting and may impart a more desirable texture when human consumption is considered (Worgan 1976).

Characteristics desirable in most industrial microorganisms are tolerance to high temperature and extreme pH. These characteristics allow operation of fermentations under conditions which lessen contamination problems. In addition, tolerance to high temperatures alleviates cooling requirements which are a major economic concern in any large-scale fermentation.

Despite the fact that yeasts are in some ways inferior choices as SCP organisms, they have been the subject of most intensive studies, particularly large scale cultivation. This phenomenon is most likely a result of their traditional role in wine and breadmaking, and in food production in the earlier part of this century. The influence of tradition and social acceptance in shaping the course of research should not be underestimated, particularly where novel sources of food are concerned.

Table III shows the proximate composite of various inexpensive protein sources and a few examples of SCP products. The total pro-

Table I. Average compositions of cells of the major groups of microorganisms (percent dry weight), from Riviere 1977).

	Filamentous fungi	Algae	Yeasts	Bacteria
Nitrogen	5- 8	7.5-10	7.5-8.5	11.5-12.5
Lipid	2- 8	7-20	2-6	1.5-3
Ash	9-14	8-10	5-9.5	3-7
Nucleic acids	--	3-8	6-12	8-16

Table II. Daily production rates for various forms of food protein (from Riviere 1977).

Organism (1000 kg)	Protein produced (kg/day)	Daily Yield (Percent)
Beef cattle	1	0.1
Soya	10	1
Yeast	10^5	10^4
Bacteria	10^{11}	10^{10}

Table III. Composition (weight percent) of various SCP and other protein sources (from Laskin 1977b).

	Crude Protein	Nucleic Acid	Lysine	Methionine	Fat
Fish Meal	60-65	--	7.0	2.6	5
Soy Flour	45-50	--	6.5	1.4	2
N-paraffin Yeast	55-60	7	7.4	1.8	9
Ethanol Grown Yeast	53	9	6.6	1.4	7
Ethanol Grown Bacterium	78	16	6.4	2.5	6

tein balance of microorganisms, particularly bacteria, is quite favorable with respect to more traditional protein sources. These data must be treated with some caution since the protein content listed is actually "crude protein," which is a protein estimate calculated by multiplying Kjeldahl nitrogen by 6.25. This value can be misleading, since the non-protein nitrogen (nucleic acids, chitin, etc.), which can be quite variable, is measured as protein.

The essential amino acid content of a product is more indicative of its value as a food source. In general, yeasts are fairly high in lysine, but low in methionine compared to bacteria. Since cereal proteins are deficient in lysine, it has been suggested they may be mixed with SCP to create a more favorable amino acid composition.

Table IV lists the vitamin content of some food microorganisms, indicating the high level of B vitamins found in many yeasts.

Favorable nutritional composition is only valuable if the food source in question is digestible, palatable, an economic reality, and is socially acceptable. Feeding trials have shown that the digestibility (percentage of consumed nitrogen absorbed from the alimentary tract) is rather low in algae (50-75 percent) and reaches 95 percent in some yeasts. The digestibility of bacteria tested seems to be quite variable, ranging from 50-90 percent (Rivière 1977). Where whole microbial cells are consumed, various physical, chemical and biological methods have been successfully used to improve digestibility. Another approach has been to completely remove the cell wall by physical means, resulting in a cell extract of very high protein composition and digestibility (Maltz 1981).

Table IV. Vitamin content of some food microorganisms (mg/100g dry weight, from Rivière 1977).

Vitamin	Morchella hortensis	Candida utilis	Saccharomyces cerevisiae	Methylomonas methanica
Thiamine	0.52	0.53	5-36	1.81
Riboflavin	1.31	4.50	3.6-4.2	4.82
Niacin	12.4	41.73	32-100	15.9
Pyridoxine	2.62	3.34	2.5-10	14.3
Pantothenic acid	12.6	3.72	10	2.42
Choline	4.61	--	--	968.0
Folic acid	1.09	2.15	1.5-8	--
Inositol	1.78	--	--	--
Biotin	0.015	0.23	0.5-1.8	--
Vitamin B_{12}	0	0	0	0.96
p-aminobenzoic acid	--	1.7	0.9-10	--

Mycotoxins, which are produced by many filamentous fungi, and endotoxins, which are associated with the lipopolysaccharide layer of many bacteria, are two obvious factors that rule out the use of certain microorganisms as food sources. A less likely problem involves the high levels of nucleic acids produced by actively growing cells. Consumption by humans of high levels of nucleic acids, leads to high blood levels of nearly insoluble uric acid. This can lead to precipitates of uriates in tissues and joints, giving rise to symptoms similar to gout. The Protein Advisory Group (PAG) of the United Nations recommends less than 2 grams of nucleic acid be consumed per day (Scrimshaw 1975). Since the nucleic acid content of actively growing microbes is generally between 12-18 percent by weight, considerable reductions are necessary before an SCP product can be acceptable as a major source of protein for human consumption. Several chemical and biological methods for reducing nucleic acid content have recently been patented (Maltz 1981).

SAFETY AND SOCIAL ACCEPTANCE

While yeasts grown on such substrates as molasses, starchy waste and sulfite waste liquor have been consumed by animals and humans for decades, the safety of SCP products must be determined on an individual basis. Possible toxicity associated with the organism itself (and possible contaminants) and the growth substrate must be considered. Many research projects concerned with SCP production have included crude protein determinations and small-scale feeding studies which provide some information on the nutritional value and safety of the microbial protein. However, before commercial production and animal or human consumption, each SCP product should be subjected to rigorous testing and approval by appropriate regulatory agencies. The Protein Advisory Group (PAG) of the United nations established guidelines for analyses of novel protein sources including preclinical testing and food specifications for both humans and animals. (The PAG ceased activities in 1977, and its functions on SCP were assumed by the World Hunger Program of the United Nations University (Venkatachalam 1979).

There has been much concern over the safety of petroleum-produced SCP. However, at two international symposia (Brussels in 1976 and Milan in 1977), the PAG approved the use of "Toprina" and "Liquipron" (yeasts grown on n-alkanes) and "Pruteen" (a bacterium produced on methanol) as animal feed (Scrimshaw 1979). It seems likely that the number of microbial proteins found suitable for animal and human consumption will continue to expand. The main barriers to the use of SCP will most likely be social or economic.

SUBSTRATES FOR SCP PRODUCTION

Substrates can be classified into three broad categories: (1) high energy yielding materials or substances derived from such materials; (2) recyclable waste materials, and (3) renewable plant materials, Table V (Humphrey 1975). Cultivation of plant materials specifically for SCP substrates is not likely to occur in the near future but recyclable waste materials are, directly or indirectly, renewable plant materials as well as wastes.

HIGH ENERGY SUBSTRATES

High energy yielding materials considered as substrates for SCP production include liqud n-alkanes, ethanol, methanol and methane. n-alkanes (paraffins) are saturated, straight chain hydrocarbons present in varying amounts (0-30 percent) in crude oils. Their presence lowers the calorific value of fuels, and thus they can be considered to have a nuisance value. It has been estimated that approximately 50 million tons per year could be available as substrate for SCP production. The interest in microbial protein production from n-alkanes over the past 20 years is evident from the number of companies that have patented organisms or techniques (Levi et al. 1979). Most of these processes have used yeasts, particularly _Candida lipolytica_. Interest in SCP production from n-alkanes peaked in the mid 1970's with the construction of a $100 million plant by British Petroleum in Italy for production of "Toprina" (_Candida lipolytica_). The failure of the Italian authorities to grant approval for continued operation of the plant due to concerns over toxicity and carcinogenicity of the product coupled with similar failures in Japan signaled the decline in n-alkane-based SCP production (Laskin 1977). While it seems clear that the technology for a safe, high quality product, suitable at least for animal feed is available today (a number of plants based on n-alkanes are now operating in the Soviet Union) the recent escalation in oil prices and the "curious inhibition" on commercial development in Italy and also Japan "appear to have dealt economic and political deathblows to the utilization in Western countries of this new branch of fermentation technology" (p. 362, Levi et al. 1979). It is possible that with new economic and social developments, conversion of n-alkanes to SCP may regain popularity in the future.

The advantages of using ethanol as a substrate for microbial protein production include its availability through synthetic processes or fermentation, non-toxicity, public acceptance, miscibility with water, and the relatively low aeration and cooling requirements during fermentation due to its partially oxidized state (Laskin 1977b). The main barrier to ethanol utilizaton as an SCP substrate remains high cost (Laskin 1977b; Rose 1979). Table VI lists some well-known ethanol-based SCP projects as of 1975. "Torutein," a torula yeast product of Amoco Corporation in the United States, has received clearance as a human food ingredient and is being used as a flavor enhancer and nutritional supplement in a variety of processed foods (Laskin 1977b). A new process developed by Exxon-Nestle in the United States and Switzerland uses the bacterium Acinetobacter calcoaceticus, and yields a product of over 80 percent crude protein and only 3 percent nucleic acid after processing (Laskin 1977b), making it attractive as a potential human food source.

Table V. Possible substrate for SCP (From Humphrey 1975).

Substrate	Classification
Natural Gas	Energy Source Material
n-Alkanes	" " "
Gas-Oil	" " "
Methanol	" " "
Ethanol	" " "
Acetic Acid	Energy Source or Waste Material
Bagasse	Waste Material
Citrus Waste	" "
Whey	" "
Sulfite Waste Liquor	" "
Molasses	" "
Animal Manure	" "
Sewage	" "
Carbon Dioxide	Waste or Renewable Source Material
Starch	Renewable Source Material
Sugar	" " "
Cellulose	" " "

Table VI. Some ethanol-SCP Projects (From Laskin 1977b).

AMOCO (U.S.)	Torula Yeast (Candida utilis)	15 x 10^6 lb/y Hutchinson, Minn.
Mitsubishi Petrochem. (Japan)	Candida Ethanothermophilum	100 t/y
Slovnaft (Czechoslovakia)	Yeast	1000 t/y Kojetin, N. Moravia
Exxon-Nestle (U.S. and Switzerland)	Acinetobacter Calcoaceticus	Pilot Plant

Methane and methanol derived from methane have both renewable and non-renewable origins. While principally obtained from natural gas, methane can also be produced from coal, oil shale, petroleum and by anaerobic degradation of organic materials. Because of this variety of sources, methane and methanol are less subject to environmental and political restrictions than petroleum products and may be considered as potential substrates for SCP production (Cooney and Makiguchi 1977).

Several major international companies have been involved with production of yeast or bacterial protein from methane or methanol. Due to the potential for explosions when methane and oxygen are combined, many industrial processes first convert methane to methanol before cultivation of microorganisms. Interest in SCP production from these substrates has faded considerably in the 1970's, but at least two companies; Hoechse A G in West Germany and Imperial Chemical Industries in Britain are still actively involved. ICI recently began operation of a plant capable of producing 75,000 tons of SCP per year from methanol, ammonia and salts using the bacterium Methylophilus methylotrophus Rose 1981). The product, called Pruteen, is 72 percent protein and has an amino acid profile which compares favorably to that of fish meal (Riviere 1977). Recently, a strain of M. methylotrophus was developed which is more efficient in the conversion of ammonia to protein and should eventually improve yields (Hopwood 1981, Walgate 1980).

WASTE MATERIALS

Waste carbohydrate sources for SCP production can be classified in decreasing order of digestibility as free sugars, starches, and cellulose (including hemicelluloses and lignocelluloses). Cultivation of yeasts for SCP from the free sugars found in molasses, whey, and sulfite waste liquors has been practiced commercially for many years, and yields a well accepted SCP product.

Simple Sugars

Molasses, the byproduct of cane and beet refining, is a traditional carbon source for Candida utilus and Saccharomyces cerevisiae production. It consists of over 50 percent free sugars and requires little or no pretreatment before yeast cultivation. In developed countries, the main product is bakers yeast, while considerable commercial SCP production occurs in Russia, Taiwan, Cuba and South Africa (Forage and Righelato 1979).

Sulfite waste liquor, the effluent from sulfite pulp mills contains about 22 percent free sugars (Romantshuk 1975) and has been used for production of Candida utilus since World War II. Approximately 100 million tons of spent sulfie liquor with a BOD of 25,000 to 50,000 mg per liter is annually produced (Forage and Righelato 1979). A process was developed recently in Finland using the mold Paeci-lomyces varioti, which is capable of utilizing not only the free sugars, but also the acetic acid which is present in the liquor. The product, "Pekilo," contains 55-60 percent protein, 87 percent of which is digestible, and in appearance resembles animal more than vegetable protein. It has been estimated that between 10,000 and 15,000 tons of Pekilo could be produced annually from an average sized sulfite mill (Romantshuk 1975). Use of the product as an animal feed has been approved by the Finnish authorities. The emergence of Kraft and Sulfite pulping processes may limit the future expansion of SCP production from this pulping effluent (Forage and Righelato 1979).

Whey, the liquid effluent of cheese and casein manufacturing, has been used as a substrate for yeast protein since 1940. Whey is approximately 70 percent lactose and 9-15 percent protein on a dry weight basis, and contains a sufficient complement of minerals and vitamins (Meyrath and Bayer 1979). However, the low protein content precludes widespread use as a food source. Approximately 30 billion pounds of liquid whey was produced in the United States in 1974, only half of which was used for human or animal food or purified for lactose (Bernstein et al. 1977). The BOD of whey is approximately 60,000 - 70,000 mg per liter (Forage and Righelato 1979) and it thus represents a considerable disposal problem.

Due to the nature of the growth medium, yeast produced from whey is rather well accepted as a food source for both animals and humans. Kluyveromyces fragilis (formerly Saccharomyces fragilis) is capable of utilizing lactose, and is the most commonly cultivated microorganism. At least three industrial processes for conversion of whey to SCP were operating in Europe as of 1979 (Meyrath and Bayer 1979). Amber Laboratories, a division of Milbrew in Juneau, Wisconsin, U.S.A. has developed a closed loop system which can be run as batch or continuous culture and produces no effluent. By varying environmental conditions, ethanol production can be increased at the expense of cell yield. The 1977 cost estimate for a plant producing 5,000 to 10,000 tons of yeast annually was between 11 and 15.5 cents per pound (Berstein et al. 1977).

Starch

The cost of food grade starch renders it economically unrealistic as a substrate for SCP production under current market conditions (Forage and Righelato 1979). Another problem with starch materials is the high viscosity of concentrations greater than 2 percent which hinders aeration and mixing. Despite these problems, Rank, Hovis and MacDougal Research Limited have investigated the production of microfungi from wheat, potato and corn starch. The organism chosen, Fusarium graminareum, grown on a pilot plant scale, has been well characterized with respect to growth rate and nutritional composition and has been successfully tested in animal feeding trials (Duthie 1975; Anderson et al. 1975).

The Symba Process, developed by the Swedish Sugar company and Chemap of Switzerland, makes use of a coculture of yeasts to convert starch-containing water to SCP. The starch-containing effluent is first saccharified by Endomycopsis fibuliger and then converted to microbial biomass by Candida utilis. A BOD reduction of 90 percent has been obtained in a commercially operated Symba plant utilizing potato processing wastes, and it has been suggested that similar results could be obtained on a variety of potato, corn, rice, wheat, vegetable and bakery wastes. Feeding trials have been conducted in a number of animals by the Swedish University of Agriculture and indicate no toxicity problems (Skogman 1976).

A process for anaerobic production of SCP from starch potato waste was recently developed at Louisiana State University and is now operating at the pilot plant stage. An anaerobic fermentation for SCP production appears to be unique and is advantageous because of lower aeration and harvesting costs. A commercial plant capable of converting 80 tons of starchy waste per day to both methane and SCP is now under construction (Callihan 1982, personal communication).

Cellulose

Cellulose is the most abundant renewable, natural polymer, and because it frequently constitutes a waste material, it has been extensively studied as a growth substrate for SCP production. Unfortunately, the crystalline nature of cellulose, and its intimate association with lignin makes it relatively resistant to microbial attack. The total cellulase enzyme system, which is composed of at least three enzymes is necessary for the utilization of native cellulose. A complete cellulase system, which is a potentially valuable product in itself, is possessed by relatively few microorganisms. Because of this, pretreatment methods are often employed to improve degradability before cultivation of food microorganisms.

Cellulosic agricultural wastes hold tremendous potential for use as SCP. In the U.S., 562 million tons of lignocellulosic wastes are annually produced from crops (Chahal et al. 1981). It has been estimated that approximately 335 million tons of sugar cane bagasse are annually produced and available for SCP production (Callihan and Clemmer 1979). Alkali pre-treated bagasse has, in fact, been tested as a growth substrate for a variety of microbes including Aspergillus terreus in India (Garg and Neelakantan, 1981, 1982), Trichoderma longibrachiatum in India (Sidhu and Sandhu 1980) and Cellulomonas sp. in Cuba (Enriquez 1981). The potential for protein production on alkali-treated bagasse by Cellulomonas sp. and coculture of

Cellulomonas sp. with either Alcaligines faecalis or an unidentified yeast has been thoroughly investigated at Louisiana State University (Rockwell 1976 and Callihan and Clemmer 1979). As a result of eight years of research and $1,500,000 invested, a process was designed which could produce SCP on a commercial scale at an estimated cost of 23.5¢ per lb. This is somewhat higher than the cost of soybean meal, but the higher protein content of the LSU product (55 percent) compared to that of soybean meal (43 percent) may offset this (Callihan and Clemmer 1979). Large scale conversion of bagasse to SCP has been abandoned at LSU in favor of the anaerobic process converting starch waste to methane and SCP.

A process was developed at General Electric Company in Schenectady, N.Y. which converts feedlot wastes to SCP using a thermophilic actinomycete (Bellamy 1974). The organism used, Thermoactinomyces sp., is relatively high in lysine, tryptophan- and sulfur-containing amino acids, and thus could be very valuable for augmenting soy meal protein for animal feed (Humphrey et al. 1977). A demonstration plant built by G.E. at Casa Grande, Arizona was operated briefly before closing for process modifications and has not reopened. Despite the favorable nutritional composition and high growth rates on a number of cellulosic wastes, cultivation of Thermoactinomyces for SCP production ceased in 1975 due to a "general lack of interest" in SCP at the time (Phillips 1982, personal communication). Research has shifted instead towards increasing the production of cellulase enzyme.

Paper waste is the largest single material in municipal solid waste (Updegraff, 1971). The proportion of lignin, and hence susceptibility to microbial decay of paper, depends on the pulping process utilized to produce it. Newsprint, which is produced by mechanical pulping does not alter the percentages of wood components, contains from 17-35 percent lignin, while higher quality papers produced by chemical pulping have much of the lignin removed. Ball-milled newsprint was tested as a growth substrate for Myrothecium verrucaria by Updegraff (1971), and yielded a product of only 10 percent protein. Cellulomonas flavigena grown on domestic refuse paper produced a 53 percent protein product (after separation from utilized paper) which compared favorably with soybean meal when fed to chickens (Kim and Wimpenny 1981).

Brown and Fitzpatrick (1976) suggested that a two step process involving enzymatic hydrolysis with Trichoderma viride cellulase followed by biomass production is more efficient than direct growth of biomass on waste paper. Considerable work has been done on cellulose hydrolysis by T. viride cellulose at the United States Army Laboratory at Natick, Massachusetts (Mandels et al. 1974).

It has been estimated that approximately 220 million tons of timber harvesting and wood processing waste are produced each year in the United States (Chahal et al. 1981). A major contributor to lignocellulose wastes is the pulp and paper industry. Waste materials from pulp and paper mills are often funneled together and dewatered to produce a sludge which is composed of varying amounts of lignin, cellulose, hemicellulose, solubilized sugars, clay, boiler ash and paper additives. This sludge represents a significant disposal problem which is frequently disposed of as land fill. The extensive delignification and fiber modification which results from chemical pulping greatly increases the degradability of these cellulosic wastes, and renders them suitable substrates for SCP production with little or no pretreatment required.

Protein production by Thermomonospora fusca, a thermophilic actinomycete, was tested on a variety of pulps and yields of up to 35 percent protein were obtained on aspen sulfite pulp fines (Crawford et al. 1973; Harkin et al. 1974). Eriksson and Larsson (1975) tested the protein production of Sporotrichum pulverulentum on highly lignified waste fibers from a newsprint mill. Crude protein yields of only 14 percent indicated these fibers were a poor substrate for large scale SCP production.

More recently, a continuous process was developed that allows direct conversion of waste waters from a fiber board factory using S. pulverulentum (Ek and Eriksson 1980). The process is designed for both protein production and water purification, and differs from the Pekilo process in that polymeric materials, along with monomers, are converted to protein. Nutritional studies (Thomke et al. 1980) showed that metabolizable energy and amino acid digestibility of the resultant product in piglets and sheep were considerably lower than that for soybean meal. Such low values are no doubt due to the cell wall structure of the fungus, and it was suggested that a different method of product harvesting may increase digestibility.

Considerable work has been done by Moo-Young and others at the University of Waterloo in Canada and a group at Colorado State University using the newly isolated fungus Chaetomium cellulolyticum. The fungus produces low levels of cellulase enzyme, but high protein relative to Trichoderma viride (Moo-Young et al. 1977). A variety of cellulosic materials have been tested as growth substrates including pretreated wood and sawdust (Chahal et al. 1981 and Pamment et al. 1978), mixes of swine manure and straw (Moo-Young et al. 1981), wheat straw (Viestur et al. 1981), steam treated feedlot waste fibers (Ulmer et al. 1981), and pulp and paper wastes (Pamment et al. 1979).

Solid state cultivation of C. cellulolyticum (Pamment et al. 1978) and co-cultures of C. cellulolyticum and Candida lipolytica (Viesturs et al. 1981) grown on cellulosic materials were found to produce slightly lower yields of protein than traditional submerged cultivation. However, the advantages of solid state fermentation, which include greatly decreased fermentor volume, elimination of foaming, and reduced harvesting costs may offset slightly lower yields.

A prototype plant producing SCP from cellulosic wastes using Chaetomium cellulolyticum recently began operation in Vancouver, Canada. The plant, which is operated by Envirocon Ltd. is capable of producing two tons of SCP per day from pulp and paper mill sludges and other available materials. The protein content of the product varies with the growth substrate used, reaching 40-45 percent when grown on de-ashed pulp sludge. Patents have been obtained on the bioreactor design and the pretreatment process (Moo-Young 1982, personal communication).

Work at our laboratory has involved the screening of cellulolytic fungi on a variety of pulp and paper wastes for production of protein and cellulase enzymes. Sludges from paper mills utilizing mechanically produced ground wood pulp and chemically produced Kraft and Sulfite pulps were analyzed and tested. The highly delignified Kraft mill sludge proved to the the best substrate for protein production, as several fungi were able to produce a product of nearly 20 percent crude protein after seven days growth in shake flasks. Removal of the ash present in the sludge (which exceeds 20 percent by weight), as is done in the Waterloo process, could substantially improve yield. Trichoderma reesei produced the highest protein and enzyme levels, yielding a product of 25 percent protein in four days and cellulase activities of 4.2 μmoles reducing sugar as glucose/ml/min. against carboxymethylcellulose, and 16.9 μmoles reducing sugar as glucose/ml/hour against Whatman No. 1 filter paper (Figure 1).

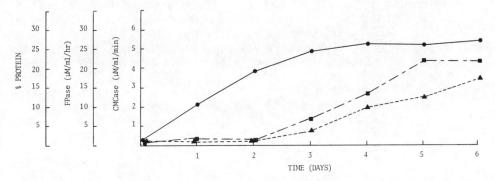

Figure 1. Production of crude protein and cellulase enzyme over 6 days. ●——● = % crude protein (6.25 x Kjeldahl N), ▲---▲ = filter paperase activity (μM/ml/hr.), ■--■ = carboxymethylcellulase activity (μM/ml/min.).

SUMMARY

Interest in large scale SCP production among western countries appears to have faded considerably since the 1960's and early 1970's. This is largely due to the failure of hydrocarbon based SCP ventures which succumbed to increasing oil prices and social and political pressures. The outlook for microbial protein cultivation on renewable waste materials is much brighter. However the abandonment of two major cellulose-to-SCP projects in the United States at General Electric and Louisana State University indicates that the decline has not been restricted to petroleum based processes. The root of the problem is failure to compete economically with other sources of protein such as soy and fish meal, and failure to gain widespread acceptance as a human food source. It seems likely that favorable economic conditions for some application of SCP will exist in the future. Some major oil companies may have overestimated the potential for SCP in an unfavorable economic climate. However, these corporations were responsible for conducting the basic research necessary for future application. A potential for more immediate use of SCP exists in developing countries. These applications are dependent on cheap and available growth substrates and sufficient capital for initial investment. The development of low technology fermentations such as the Waterloo Process, which utilizes a wide variety of universally produced wastes as substrates, may enhance the gradual acceptance of SCP as a major food source.

REFERENCES

Anderson, C., J. Longton, C. Maddix, G.W. Scammell, and G.L. Solomons. 1975. The Growth of Microfungi on Carboyhydrates. In: Single Cell Protein II. Tannenbaum, S.R., D.I.C. Wang, Eds. MIT Press, Cambridge, 1975. pp. 314-329.

Bellamy, W.D. 1974. Biotechnology Report - Single cell proteins from cellulosic wastes. Biotechnology and Bioengineering, 16:869-880.

Bernstein, S., C.M. Tzeng, and D. Sisson. 1977. The commercial fermentation of cheese whey for the production of protein and/or alcohol. In: Single Cell Protein from Renewable and Nonrenewable Rosources. Gaden, Jr., E.L., and A. Humphrey, Eds. John Wiley and Sons, New York, pp. 1-9.

Brown, D.E., and S.W. Fitzpatrick. 1976. Food from waste paper. In: Food From Waste. Birch, G.G., K.J. Parker, and J.T. Worgan, Eds. Applied Science Publishers Ltd, London, pp. 139-155.

Callihan, Clayton D. and J.E. Clemmer. 1979. Biomass from cellulosic materials. In: Economic Microbiology Vol. 4, Microbial Biomass, Rose, A.H., Ed. Academic Press, New York, pp.271-288.

Callihan, Clayton D. 1982. Personal Communication.

Chahal, D.S., M. Moo-Young and D. Vlach. 1981. Effect of physical and physiochemical pretreatments of wood for SCP production with Chaetomium cellulolyticum. Biotechnology and Bioengineering. 23:2417-2420.

Cooney, C.L. and N. Makiguchi. 1977. As assessment of single cell protein from methanol-grown yeast. In: Single Cell Protein from Renewable and Nonrenewable Resources. Gaden, Jr., E.L., and A. Humphrey, Eds. John Wiley and Sons, New York. p. 65-76.

Crawford, D.L., E. McCoy, J.M. Harkin, and P. Jones. 1973. Production of microbial protein from waste cellulose by Thermomonospora fusca, a thermophilic actinomycete. Biotechnology and Bioengineering 15:833-843.

Duthie, I.F. 1975. Animal feeding trials with a microfungal protein. In: Single Cell Protein. II. Tannenbaum, S.R., and D.I.C. Wang, Eds. MIT Press, Cambridge, pp. 505-544.

Ek, M. and K-E. Eriksson. 1980. Utilization of the white-rot fungus Sporotrichum pulverulentum for water purification and protein production on mixed lignocellulosic waste waters. Biotechnology and Bioengineering 22:2273-2284.

Enriquez, A. 1981. Growth of cellulolytic bacteria on sugarcane bagasse. Biotechnology and Bioengineering. 23:1423-1429.

Eriksson, Karl-Erik and Kjell Larsson. 1975. Fermentation of waste mechanical fibers from a newsprint mill by the rot fungus Sporotrichum pulverulentum. Biotechnology and Bioengineering 17:327-348.

Forage, A.J. and R.C. Righelato. 1979. Biomass from carbohydrates. In: Economic Microbiology Vol. 4, Microbial Biomass, Rose, A.H., Ed. Academic Press, New York, pp. 289-313.

Garg, S.K. and S. Neelakantan. 1981. Effect of cultural factors on cellulase activity and protein production by Aspergillus terreus. Biotechnology and Bioengineering 23:1653-1659.

Garg, S.K. and S. Neelakantan. 1982. Effect of nutritional factors on cellulase enzyme and microbial protein production by Aspergillus terreus and its evaluation. Biotechnolgy and Bioengineering 24:109-125.

Hamer, G. 1979. Biomass from natural gas. In: Economic Microbiology Vol. 4, Microbial Biomass, Rose, A.H., Ed. Academic Press, New York. pp.315-360.

Harkin, J.M., D.L. Crawford and E. McCoy. 1974. Bacterial protein from pulps and papermill sludge. TAPPI 57:131-134.

Hopwood, D.A. 1981. The Genetic Programming of Industrial Microorganisms. Scientific American 245:90-102.

Humphrey, A.E. 1975. Product outlook and technical feasibility of SCP. In: Single Cell Protein II. Tannenbaum, S.R., and D.I.C. Wang, Eds. MIT Press, Cambridge, pp.1-23.

Humphrey, A.E., A. Moreira, W. Armiger and D. Zabriskie. 1977. Production of single cell protein from cellulose wastes. In: Single Cell Protein from Renewable and Nonrenewable Resources. Gaden, Jr., E.L. and A. Humphrey, Eds. John Wiley and Sons, new York. pp.45-64.

Kim, B.H. and J.W.T. Wimpenny. 1981. SCP production from domestic refuse paper fractions using the cellulolytic bacterium Cellulomonas flavigena. J. Fermentaton Technol. 59:275-280.
Laskin, A.I. 1977a. Single cell protein. In: Annual Reports on Fermentation Processes. D. Perlman, Ed. Academic Press, New York, Vol. 1: 51-180.
Laskin, A.I., 1977b. Ethanol as a substrate for single cell protein. In: Single Cell Protein from Renewable and Nonrenewable Resources. Gaden, Jr., E.L. and A. Humphrey, Eds. John Wiley and Sons, New York, 1977. pp. 91-103.
Levi, J.D., J.L. Shennan and G.P. Ebbon. 1979. Biomass from liquid n-alkanes. In: Economic Microbiology, Vol. 4, Microbial Biomass, Rose, A.H., Ed. Academic Press, New York. pp.361-419.
Maltz, M.A. 1981. Protein food supplements. Recent Advances. Noyes Data Corporation, Park Ridge, New Jersey. 404 pp.
Mandels, M., L. Hontz, and J. Nystrom. 1974. Enzymatic hydrolysis of waste cellulose. Biotechnology and Bioengineering 16:1471-1493.
Meyrath, J. and K. Bayer. 1979. Biomass from whey. In: Economic Microbiology, Vol. 4, Microbial Biomass, Rose, A.H., Ed. Academic Press, New York, Vol. 4:207-267.
Moo-Young, M. 1982. Personal Communication.
Moo-Young, M., D.S. Chahal, J.E. Swan and C.W. Robinson. 1977. SCP production by Chaetomium cellulolyticum, a new thermotolerant cellulolytic fungus. Biotechnology and Bioengineering 19:527-538.
Moo-Young, M., D.S. Chahal and B. Stickney. 1981. Pollution control of swine manure and straw by conversion to Chaetomium cellulolyticum SCP feed. Biotechnology and Bioengineering 23:2407-2415.
Pamment, N., C.W. Robinson, J. Hilton and M. Moo-Young. 1978. Solid state cultivation of Chaetomium cellulolyticum on alkali-pretreated sawdust. Biotechnology and Bioengineering 20: 1735-1744.
Pamment, N., C.W. Robinson, and M. Moo-Young. 1979. Pulp and paper mill solid wastes as substrates for single-cell protein production. Biotechnology and Bioengineering 21:561-573.
Phillips, J.A. 1982. Personal Communication.
Rivière, J. 1977. Industrial applications of microbiology. Halsted Press, New York, 248 pp.
Rockwell, P.J. 1976. Single cell proteins from cellulose and hydrocarbons. Noyes Data Corporation, Park Ridge, New Jersey. 337 p.
Romantschuk, H. 1975. The Pekilo process: protein from spent sulfite liquor. In: Single Cell Protein II. Tanenbaum, S.R. and D.I.C. Wang, Eds. MIT Press, Cambridge. pp. 344-356.
Rose, A.H. 1979. History and scientific basis of large-scale production of microbial biomass. In: Economic Microbiology, Vol.4, Microbial Biomass, Rose, A.H., Ed. Academic Press, New York. pp. 1-29.

Scrimshaw, N.S. 1979. Summary of the PAG Symposium: Investigations on single-cell protein. In: Single Cell Protein-Safety for Animal and Human Feeding. Garattini, S., S. Paglialunga and N.S. Scrimshaw, Eds. PAG International Symposium on Investigations on Single-Cell Protein, Milan, 1977. Pergammon Press New York, PP. 189-193.

Scrimshaw, N.S. 1975. Single-cell protein for human consumption - an overview. In: Single Cell Protein II. Tannenbaum, S.R. and D.I.C. Wang, Eds. MIT Press, Cambridge pp. 24-45.

Sidhu, M.S. and D.K. Sandhu. 1980. Single cell protein production by Trichoderma longibrachiatum on treated sugar cane bagasse. Biotechnology and Bioengineering 22:689-692.

Skogman, H. 1976. Production of symba-yeast from potato wastes. In: Food from Waste. Birch, G.G., K.J. Parker, and J.T. Worgan, Eds. Applied Science Publishers Ltd. pp. 167-179.

Thomke, S., M. Rundgren and S. Eriksson. 1980 Nutritional evaluation of the white-rot fungus Sporotrichum pulverulentum as a feedstuff to rats, pigs and sheep. Biotechnology and Bioengineering 22:2285-2303.

Ulmer, D.C., R.P. Tengerdy and V.G. Murphy. 1981. Solid-state fermentation of steam-treated feedlot waste fibers with Chaetomium cellulolyticum. Biotechnology and Bioengineering Symp., No. 11, J. Wiley and Sons, New York, pp. 449-461.

Updegraff, D.M. 1971. Utilization of cellulose from waste paper by Myrothecium verrucaria. Biotechnology and Bioengineering 13: 77-97.

Venkatachalam, P.S. 1979. In: Single Cell Protein-Safety for Animal and Human Feeding. Garattini, S., S. Paglialunga and N.S. Scrimshaw, Eds. PAG International Symposium on Investigations on Single Cell Protein, Milan, 1977. Pergammon Press, New York. pp. vii-viii.

Viesturs, U.E., A.F. Apsite, J.J. Laukevics, V.P. Ose, M.J. Bekers and R.P. Tengerdy. 1981. Solid-state fermentation of wheat straw with Chaetomium cellulolyticum and Trichoderma lignorum. Biotechnology and Bioengineering Symp. No. 11, J. Wiley and Sons, New York. pp. 359-369.

Walgate, R. 1980. Single-cell protein organism improved. Nature 284: 503.

Worgan, J.T. 1976. Wastes from crop plants as raw materials for conversion by fungi to food or livestock feed. In: Food From Waste. Birch, G.G., K.J. Parker and J.T. Worgan, Eds. Applied Science Publishers Ltd., London. pp. 23-41.

ENZYMATIC HYDROLYSIS OF STARCH-CONTAINING CROPS -- STATUS AND POSSIBILITIES IN THE COMING YEARS

Poul B. Poulsen

Novo Industri A/S
DK-2880 Bagsvaerd
Denmark

INTRODUCTION

Production of biomass is the condition of all life on this planet. The basic components in biomass are first of all carbohydrates, proteins, and fats. Carbohydrates have the very important task of being both the energy storage and, at the same time, perhaps the most important part of the plant structure.

When plants convert the sun energy through photosynthesis, three types of carbohydrates are formed:

- Mono- and disaccharides (e.g. glucose, fructose, and sucrose);
- Starch and inulin (alpha 1-4 glucose polymer and beta 1-2 fructose polymer);
- Cellulose (beta 1-4 glucose polymer).

The two first groups serve as storage for energy and the last one, the cellulose, is found in the structural system of the plant.

Here, it should be mentioned that there is a fundamental difference between cellulose on one side and starch, inulin, mono- and disaccharides on the other side. In nature, cellulose serves as something which is long-lasting, rigid, and impossible to attack. Starch, inulin, mono- and disaccharides all serve as stored energy which must be convertible more or less quickly to usable energy/action.

The rule -- i.e., the shorter the chain length, the easier it is to convert -- seems to work.

Glucose	(Grape sugar/dextrose)	Grapes
Fructose	(Fruit sugar)	Honey
Sucrose	(Sugar)	Cane and beets
Starch		Grains, potatoes
Inulin		Jerusalem artichokes
Cellulose	(Wood)	Grass, trees

Figure 1. Sources of Carbohydrates.

Thousands of years ago, man found that grapes and honey were delicious foods with a nice taste. Later, experience taught man to make wine (alcohol) from these raw materials. Even the Phoenicians brought grapes with them and planted grape yards in the south of France, Sicily, and in Spain. The Vikings, besides using the honey as a sweetener, also fermented it to mead. Starch was used as the main component of bread. And cellulose, of course, for house-building material and firewood.

This was more or less the situation until around 175 years ago. In the beginning of the 1800's, Europe was completely dependent on the sugar imported from the Caribbean (for sweetening and alcohol making). Then, Napoleon closed all continent harbors to the British. The situation became critical, but then a new invention was made. It is common knowledge that, if you keep a piece of bread in your mouth for a certain period, then it will begin to taste sweet. What Kirchhoff found in 1811, without knowing what starch really was, was that starch treated with acid turned into something sweet. This was the beginning of a whole industry which still exists to a small degree, viz. the production of substitutes for sugar by the conversion of starch. Later, the growing of sugar beets was intensified and, very soon, the sugar imports from America lost its importance.

For many years, only acid was used for the breakdown of starch. However, around 100 years ago, beer science began to develop, and it was found that the very old beer brewing technology could also be used for other purposes. Malt was used to increase the maltose content of an acid-hydrolyzed starch syrup. This was also the beginning of the industrial enzyme technology which will be one of the main subjects of this paper.

In recent years, the use of enzymes for conversion of starch into dextrose and a number of specialty syrups has undergone a tremendous development. The first significant step toward this was the introduction of enzymes produced by Aspergillus niger, capable of hydrolyzing starch and dextrins completely to glucose. This led to a process where the initial starch breakdown, the liquefaction, was done by acid, but where the subsequent breakdown to glucose was done enzymatically.

This process offered the dextrose producers considerable advantages over the straight acid process, the main one being a yield increase, but also with respect to the need for purification and demand on apparatus, great improvements were obtained. Later, the introduction into this industry of the bacterial amylase from Bacillus subtilis for the liquefaction increased the above mentioned advantages still further, and these fully enzymatic methods for hydrolysis of starch are today dominant processes in the sweetener industry.

Dextrose, however, is not as sweet as sucrose and is thus not always ideal as a substitute for sugar. Furthermore, in a concentrated solution, dextrose will crystallize, making the handling of such a product difficult and impractical. Therefore, the finding of enzymes which could transform (isomerize) glucose into fructose stirred a considerable interest in the dextrose industry. If a process could be designed by which approximately half of the glucose in a solution could be converted by means of such an enzyme to fructose, one would have something similar to sucrose in sweetness. This historical background brings us up to the present time.

RAW MATERIALS

Glucose (monosaccharide)

Glucose has the following formula:

Technical grade glucose is called dextrose.

Fructose (monosaccharide)

Fructose has the following formula:

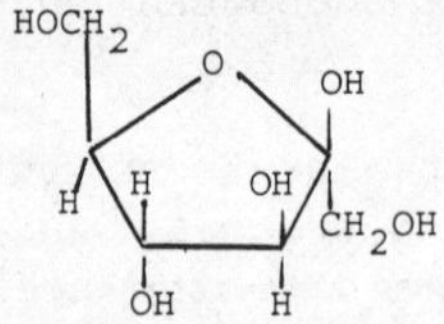

Sucrose (disaccharide)

Sucrose has the following formula:

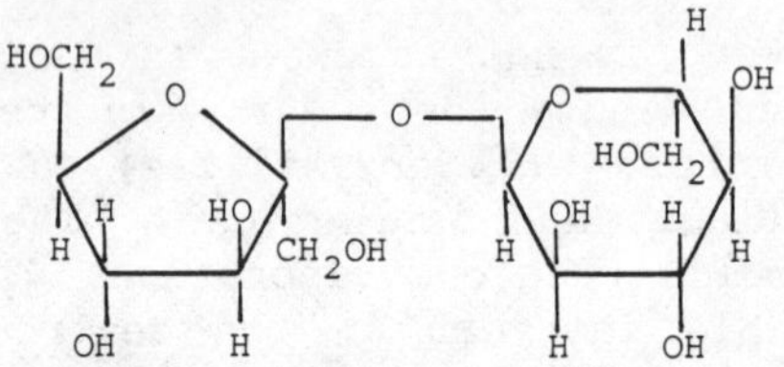

Sucrose (also called saccharose) can be seen to be formed by a combination of glucose and fructose.

Starch (polymer)

Structure. Starch consists of two components that are both polymers of glucose: amylose and amylopectin.

Amylose, which normally makes up 20 to 25 percent of the starch weight is a linear polymer of glucose units joined by 1,4-alpha-linkages (Figure 2). The number of glucose units contained in one amylose molecule typically varies between 500 and 1500, corresponding to a molecular weight of 80,000 to 240,000.

Amylopectin makes up 75 to 80 percent of most starch types. It is a branched polymer (Figure 2) in which the glucose units are joined by 1,4-alpha-linkages in the linear molecule sections and by 1,6-alpha-linkages at the branching points which typically occur at every 20 to 30 glucose units. Molecules of amylopectin can be very large, consisting of 5 to 40 thousand glucose units -- corresponding to molecular weights from around 800,000 and up to several millions.

Gelatinization. In its native state, starch consists of microscopic, partly crystalline granules in which the amylose and amylo-

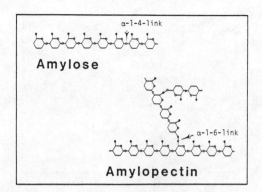

Figure 2. Starch Molecules.

pectin molecules are arranged in a complex folded and stratified manner. At ambient temperature, these granules are practically insoluble in water and not very susceptible to enzymatic hydrolysis. However, when treated with hot water, the starch granules swell and gradually rupture, the amylose and amylopectin molecules unfolding and dispersing into solution. This process is referred to as "gelatinization" of starch.

The high viscosity found in a concentrated solution of gelatinized starch is caused mainly by the unfolded large amylopectin molecules. When cooling the solution towards ambient temperature, these molecules gradually combine to form a gel, whereas the smaller, linear amylose molecules tend to form a crystalline precipitate. The amylose precipitate is difficult to hydrolyze or re-dissolve and is referred to as retrograded starch.

Inulin (polymer)

Inulin is a linear beta-1,2-linked fructose polymer terminated by a sucrose unit residue (Figure 3). Its molecular weight is approximately 6,000. Inulin occurs as an energy reserve in various plants, particularly in those of Compositae family; good sources are, e.g., Chicory, Jerusalem artichoke, and Dahlia.

Pure inulin is almost insoluble in cold water but can be dissolved in hot water. In plant material, part of the inulin is present as fructose polymers with lower chain length which are more soluble in water. These polymers have glucose content of between 5 to 15 percent compared with approximately 3 percent glucose in pure inulin.

Inulin can be extracted from sliced plant material by diffusion at elevated temperatures. The raw juice can be purified by liming/carbonization as in sugar beet processing and then decolourized by carbon.

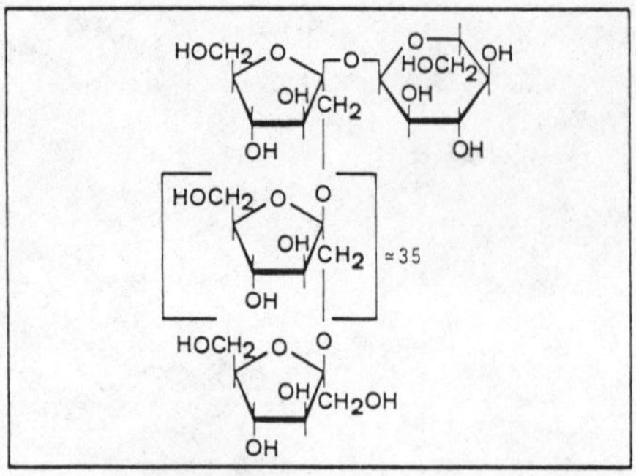

Figure 3. Inulin.

Inulin can be hydrolyzed by acid under relatively mild conditions: pH 1-2, 1-2 hours at 80-100°C. However, as fructose is easily degraded at low pH values, this process gives rise to colour and byproduct formation, e.g., formation of approximately 5 percent difructosedianhydride and thereby to a reduction in fructose yield.

Cellulose (polymer)

Cellulose is an unbranched polymer containing glucose molecules linked by 1,4-beta-glucosidic bonds. The glucose molecules are turned 180° in relation to each other, and the repeating unit in the glucan is therefore not glucose, but cellobiose (Figure 4). All of the glucose molecules are in the chair conformation and, because of the beta configuration, all of the hydroxyl groups are in an equatorial position. This explains why cellulose has a layered structure where the single chains in a layer are bonded together by intermolecular hydrogen bonds. The forces holding the layers together are of van der Waal character.

In pure native cellulose, e.g., cotton fibres, the chains lie parallel so that the reducing ends are adjacent to each other, but by chemical treatment, rearrangement can take place, allowing a reducing end of one chain to be adjacent to a non-reducing end of another chain, a so-called antiparallel configuration (see Figure 5).

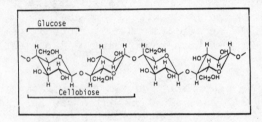

Figure 4. Conformation of the Cellulose Molecule.

The 1,4-beta-glucan chain does not exist as such in nature, the smallest unit being the elementary fibril with a diameter of approximately 3 nm. The elementary fibrils are bonded into microfibrils with a diameter of 10-40 nm, and these are again bonded into macrofibrils which are visible in a light microscope. Further, the structure of the micro- and the macrofibrils is strengthened by a matrix of hemicelluloses and lignin.

In native cellulose, the major part of the fibrils are in a crystalline state. The exact degree of crystallinity is difficult to estimate. It may vary with the source and the treatment of the material.

Several models have been proposed to explain the occurrence of amorphous parts in otherwise totally crystalline cellulose. The basis of most of these models is the view that the glucan chains are folded.

It is quite obvious that the very dense structure of cellulose and the presence of the hemicellulose and lignin matrix make diffusion of cellulolytic enzymes into the fibrils extremely difficult. The hydrophobic character of the 1,4-beta-glucan chains may also contribute to the very low reaction rate observed. With regard to diffusion resistance, it should be mentioned that the distance between parallel layers of the glucan chains is of the same order of magnitude as the average diameter of most cellulolytic enzymes.

ENZYMES

Enzymes are protein substances that catalyze specific chemical reactions. They are found in plants as well as in animals and micro-organisms, and play an important role in the function of any living cell.

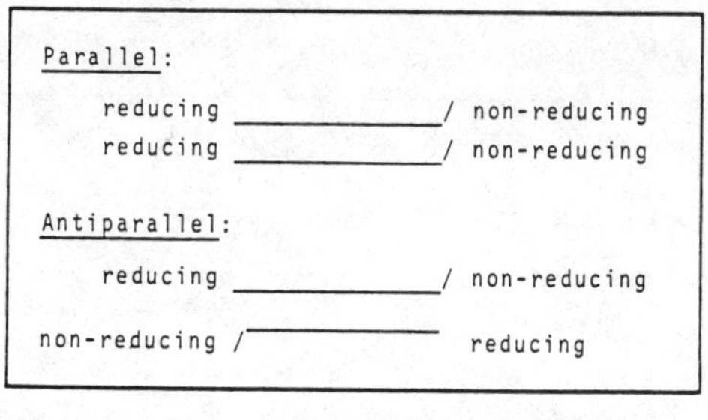

```
Parallel:
    reducing _____/ non-reducing
    reducing _____/ non-reducing
Antiparallel:
    reducing _____/ non-reducing
non-reducing /_____ reducing
```

Figure 5. Conformation of the Cellulose Fibril.

Like other proteins, enzymes can be denatured by heat or chemical treatment, thereby losing their catalytic activity. In consequence, their application is limited to certain temperature and pH ranges.

Alpha-amylases

In principle, 3 types of alpha-amylases exist.

Fungal alpha-amylase. Fungal alpha-amylase can be produced from, e.g., Aspergillus oryzae. The enzyme is an endo-amylase, i.e. it hydrolyzes 1,4-alpha-linkages in amylose and amylopectin almost at random. Prolonged reaction results in the formation of maltotriose and large amounts of maltose (1,4-alpha glucose disaccharide).

Bacterial alpha-amylase (low temperature). Can be produced from, e.g., Bacillus subtilis. This enzyme is also an endo-amylase. The break-down products formed are mainly soluble dextrins and oligosaccharides.

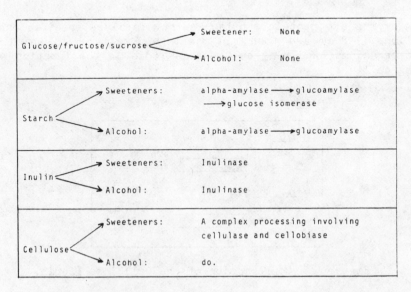

Figure 6. Enzymes Needed For Conversion of Carbohydrates.

Table 1.

	Alpha-amylase		
	Fungal source	Bact. source	Bact. source
Application temperature	Up to 55°C	Up to 75°C	Up to 105-110°C

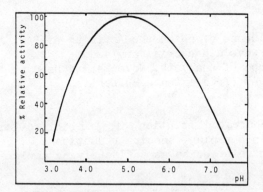

Figure 7. Influence of pH on the Activity of Fungamyl. Fungamyl: Trade name for Novo's fungal amylase.
Method of Analysis: Novo's Standard Method Used at Appropriate pH Values.

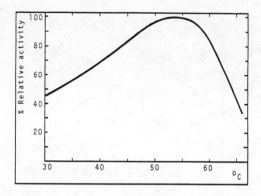

Figure 8. Influence of Temperature on the Activity of Fungamyl. Method of Analysis: Novo Method at Various Reaction Temperatures.

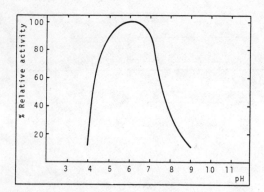

Figure 9. Influence of pH on the Activity of BAN. BAN: Trade name for Novo's bacterial amylase.
Method of Analysis: Novo Method Used at Appropriate pH Values (Tris-maleate Buffer).

In a concentrated solution of gelatinized starch, the hydrolysis results in a rapid viscosity reduction and, consequently, this enzyme is often referred to as a "liquefying amylase".

<u>Bacterial alpha-amylase (high temperature)</u>. Can be produced from, e.g., <u>Bacillus licheniformis</u>. This enzyme is also an endoamylase. In this case too, the break-down products are mainly soluble dextrins and oligosaccharides. This enzyme is also a "liquefying amylase".

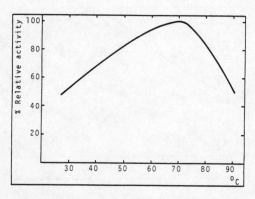

Figure 10. Influence of Temperature on the Activity of BAN.
Method of Analysis: Novo Method Used at Various Reaction Temperatures.

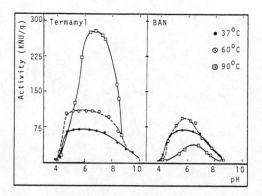

Figure 11. Influence of pH on the Activity of Termamyl At Various Temperatures. Termamyl: Trade name for Novo's high temperature bacterial alpha-amylase.
(Activity Curves For Conventional Alpha-Amylase Ban Shown For Comparison).
Substrate: 0.5 Percent Soluble Starch
Stabilizer: 30-60 ppm Calcium

Termamyl Stability -- Influence of Calcium

In a starch slurry, Termamyl is satisfactorily stabilized in the presence of 50-70 ppm Ca^{++}. In Table 2, figures for the Termamyl stability in a 30 percent starch slurry are shown as a function of pH and temperature for three different levels of Ca^{++} (ppm). The data are considered valid for DE values in the range of 0-12.

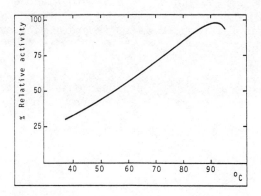

Figure 12. Influence of Temperature on the Activity of Termamyl.
Substrate: 0.5 Percent Soluble Starch
Stabilizer: 30-60 ppm Calcium
pH: 5.7

Table 2. Termamyl Stability (Enzyme Halflives in Minutes).

	93°C	98°C	103°C	107°C
Ca^{++} 70 ppm				
pH 6.5	1500	400	100	40
pH 6.0	800	200	75	20
pH 5.5	300	75	25	10
Ca^{++} 20 ppm				
pH 6.5	450	125	40	10
pH 6.0	250	75	20	5
pH 5.5	100	25	5	2
Ca^{++} 5 ppm				
pH 6.5	150	40	10	4
pH 6.0	75	20	5	2

Glucoamylase (amyloglucosidase/AMG)

This enzyme can be produced from, e.g., Aspergillus niger. It is an exo-amylase hydrolyzing 1,4- as well as 1,6-alpha-linkages in amylose, amylopectin, and oligosaccharides.

The hydrolysis proceeds in a stepwise manner, the break-down product, glucose, being split off from the non-reducing end of the substrate molecule. Maltotriose and, in particular, maltose are hydrolyzed at a lower rate than higher saccharides, and 1,6-linkages are broken down more slowly than 1,4-linkages. Eventually, a practically complete conversion of starch into glucose is obtained.

AMG is often referred to as a "saccharifying amylase".

Glucose isomerase

Glucose isomerase can be produced from, e.g., Bacillus coagulans. This enzyme catalyzes the conversion of glucose to fructose and is normally used in an immobilized form (insoluble) which allows a continuous processing through fixed bed columns.

Inulinase

Inulinase can be produced from, e.g., Aspergilli and Kluyveromyces fragilis. The enzyme product can contain both endo- and exo-activities and thereby the resulting products by the enzymatic action are both monosaccharides and oligosaccharides (which disappear after prolonged action).

STARCH-DERIVED PRODUCTS

The products which are produced from starch will be dealt with in three sections:

- High fructose syrups
- Alcohol (beverage and fuel)
- Other products

High Fructose Syrups

The production of fructose syrups is of a rather new date. For a long time, the price of sugar was relatively low and the production of sugar could satisfy the consumption. However, in the late sixties, things began to change. Because of the low price, the sugar-growing areas were not expanding according to the increasing consumption. And the consumption increased because of larger per capita consumption in third countries, larger softdrink consumption globally, etc.

After a couple of years with decreasing stocks, the sugar price began to go up (from 3 to 5 c/lb), and suddenly there was a boom (it reached a peak of 60 c/lb). One result was that the consumption dropped but, more important, a new substitute for sugar appeared: high fructose syrup.

Fructose syrups have been known and used during decades, however, because the processing only contained an alkali/heating step, only around 35 percent fructose was obtained: And such a syrup could easily be distinguished from sucrose- and invert syrups.

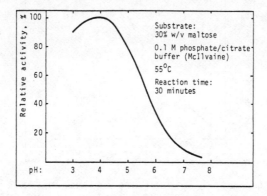

Figure 13. Influence of pH on the Activity of AMG.

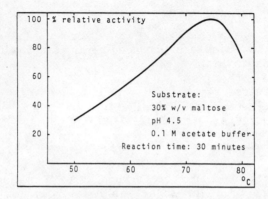

Figure 14. Influence of Temperature on the Activity of AMG.

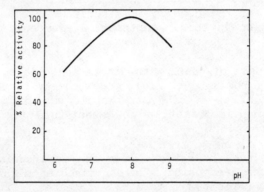

Figure 15. Influence of pH on the Initial Activity of Sweetzyme* Type Q. Sweetzyme: Trade name for Novo's immobilized glucose isomerase.

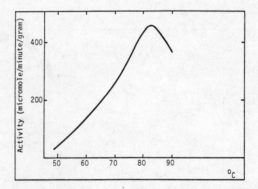

Figure 16. Influence of Temperature on the Initial Activity of Sweetzyme Type Q. The Activity is Measured as the Amount of Fructose Formed Per Minute Per Gram of a 200 IGIC/G Preparation.

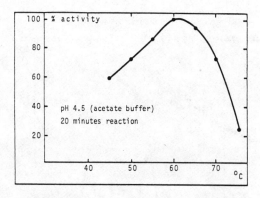

Figure 17. Temperature Profile of <u>Aspergillus</u> Inulinase.

At the beginning of the seventies, more or less when the sugar price boomed, an enzymatic process making 42 percent fructose available appeared. And the 42 percent fructose was so similar to sugar syrups that plants to produce such a syrup were soon built.

As seen in Figures 20 and 21, the penetration of HFCS grew soon and fast. During the next price increase period in 1980, a new and better fructose syrup was introduced. This syrup contains 55 percent fructose, and such a 55 percent fructose (43 percent glucose and 2 percent oligosaccharides) is claimed to be similar to sucrose -- and invert syrup. The penetration in the U.S. in 1981 is shown in Table 3.

For the time being, the U.S.A. has the largest production of fructose syrups, Japan is number 2, and smaller productions are found in Canada, Argentina, and in Europe.

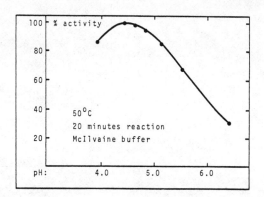

Figure 18. pH Profile of <u>Aspergillus</u> Inulinase.

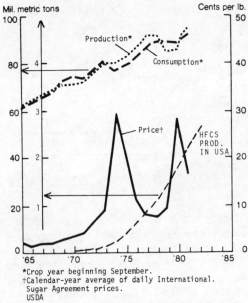

Figure 19.

Table 3. Major Markets For HFCS U.S.A. 1981.

	Sugar 1000 t	HFCS (DS) 1000 t	HFCS pen. %
Baking	1,160	385	25
Confections	873	11	1
Dairy products	408	181	31
Beverage	1,659	1,483	47
Processed foods	204	102	33
Canning	226	308	58
Other food and non-food uses	644	0	-
Total	5,174	2,470	52

Source: McKeany-Flavell co., Inc., Sweeteners Brooker's, San Francisco, June 7, 1982.

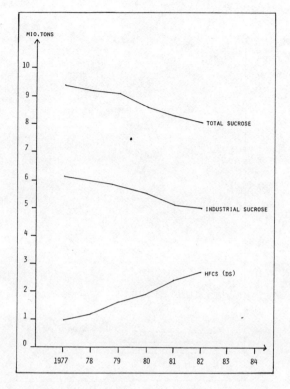

Figure 20. HFCS and Fructose -- U.S.A.

Table 4. HFCS Production in Various Countries in 1981.

	Tons DS
USA	~ 2,400,000
Japan	~ 500,000
Korea	~ 75,000
Europe	~ 200,000
Sugar production worldwide	~ 90,000,000 Tons DS

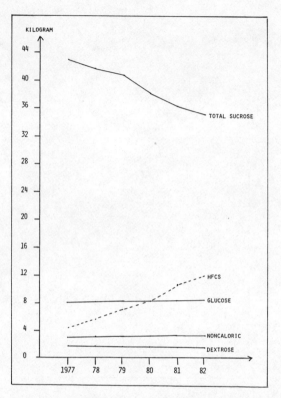

Figure 21. Per Capita Consumption -- Caloric and Non-Caloric Sweet-Eners -- U.S.A.

Table 5. Examples of Potable Alcohol Produced From Sugar-Containing Raw Materials.

Raw materials	Products
Molasses (sugar cane)	Caribbean rum Brazilian cachaca
Wine (grapes)	Cognac Pisco (Peru)
Agave azul tequilana	Tequila (Mexico)
Cherry	Kirsch (Switzerland)
Pear	Pear brandy (Switzerland)
Plums	Slivovice (Balcan)
Palm juice	Ogogoro (Nigeria)

Alcohol (Beverage and Fuel)

Potable alcohol is the term used for all distilled spirits (ethanol content higher than around 20 percent) intended for human consumption. The term covers an enormous number of different spirits produced by many different processes.

Potable alcohol has been produced industrially as well as domestically for many thousand years. In fact, it is one of the oldest industries.

Presumably, the industry has grown from occasional fermentation of sugar-containing juices followed by a natural sun distillation. Examples like the fermentation of cactus juice (has now been developed to tequila a.o. (Mexico)), the fermentation of palm juice (e.g. ogogoro (Nigeria)) can be mentioned. Later, the industry became more sophisticated when it was found that barley could be transformed to malt and thereby be used in the processing of starch-containing crops to alcohol (e.g. whisky (Scotland)). A further improvement was made in China and Japan where special microorganisms were grown on cooked rice in order to produce starch-fermentable agents (koji). Around 1890, the enzymes present in koji were extracted and concentrated by Takamine and sold as Takadiastase. This was the beginning of the modern enzyme industry.

When sugar-containing crops are to be used as raw materials for production of spirits, there is only a need for ethanol-producing agents. However, when starch materials are to be used, it is a must that the starch has been hydrolyzed to fermentable sugars. In some places, this was done by means of enzymes of vegetable origin (malt), and in other places, by means of enzymes of microbial origin (koji). Both methods have survived although there has been a continuous tendency, especially during the last 20 years, towards replacement of malt (and originally koji) by industrially produced enzymes (from microbes). The reason for this replacement has been partly economical and partly the wish for a constant product quality.

Among the sugar-containing and the starch-containing raw materials, a lot of varieties and mixtures are used in the production of potable alcohol.

When trying to determine the basic production of potable alcohol, one runs into a number of statistical problems in addition to the normal statistical uncertainty.

One problem is that, owing to the fact that potable alcohol is an object of taxation, it is tempting to give too low production figures, which may be a problem in some countries. A more serious and frequent problem is the fact that industrial alcohol is some-

Table 6. Examples of Potable Alcohol Produced From Starch-Containing Raw Materials.

Raw materials	Products
Barley	Whisky
Maize and ry	Bourbon whiskey
Potatoes and barley	Aquavit
Potatoes, rye, wheat (etc.)	Vodka
Rice	Chinese brandies

Table 7. Split on Raw Materials Used For Production of Potable Alcohol.

	100% ethanol mill. hl	%
Molasses-based	12	35
Grain-based	10	29
Whisky	5	15
Wine-based	3	9
Potato-based	2.5	7
From fruits, etc.	1.5	5
	34	100

Table 8. Split on Areas of The Production of Potable Alcohol.

	100% ethanol mill. hl	%
Asia and Oceania	4.4	13
Africa	0.2	1
Western Europe	17.2	51
Eastern Europe	4.5	13
North America	6.1	18
Latin America	1.6	5
	34	100

times included in distilled alcoholic beverages. This category is the most heterogeneous in other ways too: it may include beverages with an alcohol content varying from a few percent to 80 percent. This heterogeneity may increase in importance concurrently with the introduction of new types of mixed drinks which are based on distilled alcohol but which may have a very low alcohol content.

Fuel alcohol was a term which began to appear in newspapers after "the energy crisis" in 1973. In Brazil, a National Alcohol Programme was established in 1975 with the aim of replacing imported oil as far as possible by domestically produced renewable resources. The ethanol concept was not new in Brazil, the use of ethanol in automobiles was introduced shortly after World War I and has been alive ever since, especially when sugar prices were low.

A number of factors make large scale biomass cropping and use in place of petroleum a feasibility in Brazil:

- Domestic heating is not required in the country (this accounts for a third of the energy consumption in most developed countries).

- The climate, combined with high levels of rainfall, further contribute to a high growth rate of biomass (cane, cassava, wood, etc.).

- Both technology and political will exist to push the alcohol programme through, even in the face of occasional and inevitable difficulties.

Brazil is, of course, a major agricultural country. This sector of the economy employs 41 percent of the population, and present arable land is twice as big as the area in France, for example. This

Table 9. Alcohol Production in Brazil (Mio Litres).

Year	Hydrated	Unhydrated	Total
1960/61	281	175	456
1965/67	263	314	577
1970/71	385	252	637
1975/76	320	232	552
1976/77	366	302	668
1977/78	286	1161	1447
1978/79	324	2148	2472
1979/80			3798

arable land, nevertheless, represents a mere 4.2 percent of the total land area.

The level of the Brazilian production of hydrated and unhydrated alcohol is shown in Table 9.

The 1978/79 production of alcohol was entirely produced from sugar cane or molasses. It can be broken down as follows:

- Direct cane fermentation: 1.55 mio litres
- Rest from molasses: 0.98 mio litres

Not until 1979/80, was a cassava-based production of alcohol started.

The National Alcohol Programme has now replaced about 20 percent of all gasoline in Brazil by ethanol. The goal is nearly 100 percent substitution of gasoline by hydrated ethanol in 1990 (Table 10).

Table 10.

Year	% ethanol	Gasoline demand (mio l)	Ethanol demand (mio l)
1971	2.4	10,616	254
1977	4.5	14,138	639
1980	20	15,380	3,080
1985	47	19,560	9,200
1990	94	24,950	23,500

Research on ethanol mixtures for diesel engines is presently being carried out by Volkswagen and Mercedes-Benz, the major prodducers of buses and trucks in the country. These two companies are offering systems using 80 percent hydrated ethanol and 20 percent diesel fuel, and the main changes are the injectors and compression ratios. Replacement of diesel fuel would certainly place a considerably greater strain on ethanol and, thereby, on agricultural resources to be developed by 1990.

In 1980, the U.S.A. had a gasoline consumption of 370,000 mio litres (25 times the consumption in Brazil). The increasing oil prices and the increasing dependence on the oil countries have also started a movement in the States for producing gasohol (10 percent ethanol and 90 percent gasoline), from maize especially.

If all the American maize was converted to ethanol, this would correspond to 65,000 mio litres ethanol (or around 18 percent of the total gasoline consumption). This is of course utopia as all the maize cannot be converted to fuel alcohol. However, very significant amounts of fuel ethanol is to be produced in the coming years in the United States. The situation there is this: Only about 10 percent of the maize is going directly to human consumption as maize, corn syrups, etc. The rest is used as fodder and for industrial purposes.

The development of fuel ethanol in the States in the coming years is very difficult to estimate; however, we think the situation that is shown in Table 11 could appear.

Other regions than Brazil and the U.S.A. seem to have a lot of places with ideal conditions for a biomass-based fuel alcohol production. To mention examples, Africa has the natural resources in many countries, Australia has a surplus of wheat, Thailand a surplus of cassava, and the Philippines, Indonesia, etc. have, e.g. sago palms in large numbers where the starch just rots when the palms are in bloom.

Table 11. Anticipated production and consumption of ethanol for motor fuel.

Year	Ethanol production (mio litres)	Percentage substitution of gasoline consumption (1979)
1980	250	0.07%
1981	350	0.09%
1982	575	0.16%
1983	1,100	0.30%
1984	1,500	0.41%
1985	2,000	0.54%

The processing of biomass to ethanol and petrochemical substituents is certainly an industry which can be expected to grow in the coming years.

Other Products

Besides High Fructose Syrup and alcohol, which are by far the largest products made from starch, a few other syrups are also produced: glucose/dextrose syrup, maltose syrups, and "high DE syrups." These products find use especially within the confectionery industry.

Starch, partly degraded starch, and dextrose are also very important raw materials when, e.g., citric acid, mono sodium glutamate (msg), penicillins, and enzymes are fermented.

PROCESSES FOR HYDROLYSIS OF STARCH

If we go back a few centuries, one had two agents for the hydrolysis of starch. In Europe, malt was the preferred agent and, in Asia, koji was used.

Malt

Malt is germinated barley. During the germination, enzymes will be formed or activated. The enzymes formed which are of special interest for the brewing processes are:

- Starch-hydrolyzing enzymes (alpha- and beta-amylases)
- Protein-hydrolyzing enzymes (proteases, peptidases)
- Hemicellulose-hydrolyzing enzymes (cytases)
- Phytin-hydrolyzing enzymes (phytases)

When barley is to be transformed into malt, it has to be brought to germination. Here, a water content of 42-46 percent in the barley is necessary. The malt process is initiated by the barley being soaked (steeped) in water. The steeping takes 2-3 days. After the steeping, the soaked barley is allowed to germinate, a process which takes 6-7 days. During this step, the temperature (10-22°C) is very important in determining which type of malt will be produced. At this stage, the malt is called green malt. The final step is to dry the green malt. The drying step causes: 1) the water content to be reduced from around 45 percent to 1.5-4 percent, 2) the germination and the digestion to be stopped, and 3) the colour and aromatic compounds to be formed.

On an average, 100 kg of barley will yield around 80 kg malt.

Koji

The manufacture of koji can be performed as follows: Dehusked brown rice (unpolished) is pounded slightly with a wooden pestle in order to scratch the surface of the outer epidermis. The rice is then washed thoroughly and soaked in water overnight. The next day, it is cooked in live steam for about one hour. The rice is taken out of the steamer and cooled; burnt wood-ash (2 percent w/w) is then mixed manually and evenly into the rice. When the temperature of the mixture is suitable and not harmful to the microbes, the powdery seed of koji, which is carefully preserved by successive transplantations, is sprinkled over and rubbed on the rice with the fingers so as to distribute the spores as well as to bring them into contact with the surface of the rice. The whole mass is incubated overnight in the warm koji chamber until the temperature of the mass has risen to 35°C. The mass is then divided into small shallow wooden trays which are heaped up in the chamber. The temperature and humidity are controlled by changing the type of heaping and by opening the ceiling window. The propagation of mold as well as the abundant formation of spores can be completed within 5 to 6 days.

The trays are then taken out into the open air, each tray is covered with thin paper and exposed to direct sunlight for one day. It is carefully dried again at 40°C in an indirectly heated drying chamber. The final product is wrapped in a paper bag and stored.

Microbial Enzymes

Especially during the last ten years, the market share of the above-mentioned enzyme products has declined. An increasing number of distilleries have decided to use industrially produced microbial enzyme products. The reasons being economic or technical. The technical reasons are: a) that the microbial enzyme products have a known standardized activity, and b) that the products are much more concentrated, which allows less handling.

The microbial enzymes are produced by means of special, carefully optimized mutants in very large vessels, typically 100-200 m^3.

The enzyme products on the market are either single enzyme products (alpha-amylase, beta-amylase, glucoamylase, protease, etc.) or enzyme mixture products.

MAIN PROCESSES IN DISTILLING

The manufacture of alcohol from starch-containing raw materials is based on the following main processes which are outlined in Figure 22:

- <u>Gelatinization</u>: Dissolution of the raw materials into a mash by steam cooking to make the starch available for enzymatic attack. Normally, a concentration of 15-20 percent starch in the mash is aimed at.

- <u>Hydrolysis</u>: Breakdown of the dissolved starch to fermentable sugars by means of enzymes.

- <u>Fermentation</u>: Conversion of the sugars to alcohol by the action of yeast.

- <u>Distillation</u>: Separation and purification of the alcohol.

The enzymatic hydrolysis consists of two stages:

- <u>Liquefaction</u>: The gelatinized starch is broken down into short molecule fragments (dextrins) by means of alpha-amylase, resulting in a rapid reduction in mash viscosity.

- <u>Saccharification</u>: The dextrins formed during liquefaction are further hydrolyzed to fermentable sugar (glucose) by means of glucoamylase.

German Batch Process

In the German batch process (see flow diagram in Figure 23) the raw material is gelatinized without previous milling by cooking with live steam in a Henze cooker.

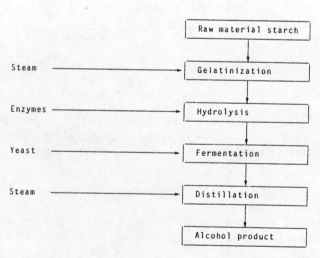

Figure 22. Main Processes in Distilling.

ENZYMATIC HYDROLYSIS OF STARCH-CONTAINING CROPS

No addition of enzyme and no mechanical agitation is necessary in the cooking stage. The cooked mash is blown through a strainer valve into the mash tub where the liquefaction takes place according to one of the two procedures:

1. <u>High temperature liquefaction</u>: The blow-down is carried out within the shortest possible time whereafter the mash is cooled in the mash tub to 80°C.

   At 80°C, alpha-amylase (typically 0.15-0.6 kg per metric ton of starch) is added and the temperature is maintained for 20 minutes before further cooling takes place.

2. <u>Low temperature liquefaction</u>: Before blow-down of the cooker, cold water is filled into the mash tub in an amount sufficient to cover the lowest part of the cooling coil, and alpha-amylase is added.

   The mash is blown into the mash tub under agitation and cooling at such a rate that a temperature between 55°C and 60°C is maintained in the mash tub.

   As soon as the blow-down has been completed, the mash may be cooled further down to the fermentation temperature.

<u>Saccharification</u>: Glucoamylase (typically 1.1-2.0 liter per metric ton of starch) is added to the liquefied mash at 60°C or lower. This will bring about complete saccharification within the

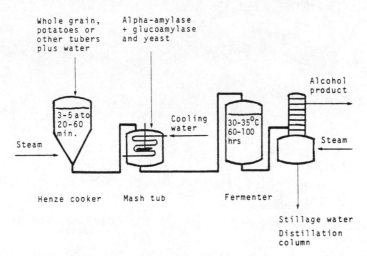

Figure 23. German Batch Process.

normal period of fermentation. Yeast is added to the mash after cooling to 30°C or lower.

Continuous Cooking

Cooking and liquefaction may be carried out continuously, thus giving better process control and more efficient use of equipment. Such a process is shown in Figure 24. A milled corn slurry to which a thermostable bacterial alpha-amylase has been added is heated by direct steam injection to about 150°C in a jet cooker. The mash is then flash-cooled to 80-90°C and the second addition of alpha-amylase made. The cooked mash is held at this temperature for 30-60 minutes to complete liquefaction before it is transferred to the saccharifiation/fermentation tanks.

Continuous cooking followed by continuous fermentation has also been described by Rosen.

MAIN PROCESSES IN HIGH FRUCTOSE PROCESSING

Corn Wet Milling Process

By far the most important starch for high fructose production is maize or corn starch. The "corn wet milling" process usually applied gives, besides high quality starch, steepwater, germ which can be used for oil production, fiber, and gluten which is used as feed (Figure 25). Other important starch sources are potato and cassava (tapioca). Wheat, rice, together with other cereals, are becoming more important as starch raw materials, and in certain geographical areas, more special starches can become of interest, e.g., sago starch in Southeast Asia.

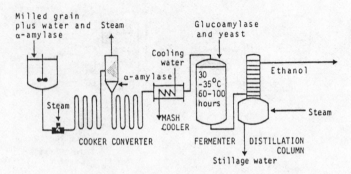

Figure 24. Continuous Cooking Process.

ENZYMATIC HYDROLYSIS OF STARCH-CONTAINING CROPS

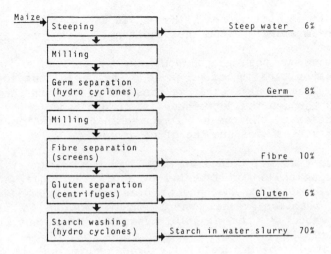

Figure 25. Corn Wet Milling.

Liquefaction Processes

Taking advantage of the heat-stability of B. licheniformis amylase, the so-called single stage liquefaction can be applied (Figure 26).

The process is as follows: a starch slurry containing 30 to 40 percent dry solids is prepared in the feed tank. The pH is adjusted to 6.0-6.5 with sodium hydroxide, and calcium salts may be added if the level of free calcium ions is below about 50 ppm. In a pilot plant, the liquefying enzyme can be added to the feed tank, but in a

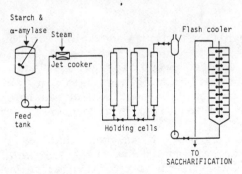

Figure 26. Single Stage Liquefaction - Pilot Plant Layout.

large scale industrial process, the enzyme is often metered directly into the stream emerging from the feed tank.

The slurry is then pumped continuously through a "jet-cooker" where the temperature is raised to 105°C by the direct injection of live steam. Tremendous shearing forces are exerted on the slurry as it is pumped through the jet-cooker, so in addition to the viscosity reduction action of the enzyme, some mechanical thinning also takes place. Peak viscosities are therefore avoided.

The slurry is maintained at this high temperature in the pressurized holding cells for a period of about 5 minutes, after which it is flash-cooled to atmospheric pressure and pumped through a multi-stage reaction vessel where enzyme action is allowed to continue for about 2 hours at 95°C. After this treatment, the liquefied starch will have a dextrose equivalent (DE) of 8-12, depending on the amount of enzyme used. (Dextrose equivalent is defined as reducing sugars expressed as dextrose and calculated as a percentage of dry substance).

When using _B. amyloliquefaciens_ amylase, two enzyme additions are necessary. First, the starch slurry containing the first enzyme dosage is heated to 85°C for one hour. To ensure that all starch granules are gelatinized, the temperature is then raised to 140°C (pressure cooking) for a few minutes. After reducing the temperature to 85°C again, another enzyme dosage is added, and the liquefaction is finished by maintaining this temperature for 30-60 minutes. Although this two-step or "classic" enzyme liquefaction was a big step forward compared to acid liquefaction, the single-step process using heat-stable licheniformis amylase is today the one being most widely used because of its simplicity and energy saving, but also because byproducts are formed to a lesser degree. The dextrin - produced with acid or enzyme -- may be spray dried and used as such, or it may be saccharified.

Saccharification

In the saccharification, the dextrin from the liquefaction is further hydrolyzed to glucose, maltose, and oligosaccharides. Depending on the desired properties of the final syrup, different enzymes can be chosen: amyloglucosidase for glucose production, fungal alpha-amylase or beta-amylase for maltose production, and blends thereof for the so called "high conversion syrup."

An interesting group of enzymes that can be used in the saccharification is called "debranching enzymes," enzymes which specifically hydrolyze the alpha-1,6 linkages or branch points.

Use of glucogenic exo-amylase. The saccharification process is very simple as illustrated where typical conditions are given (Figure 27).

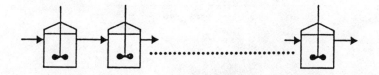

Saccharification process

Dextrin concentration:	30-40% DS
pH:	4.5
Temperature:	60°C
Reaction time:	48 hours - 120 hours
Enzyme dosage:	1.5-1.0 l/ton DS (AMG 150)
Final DE:	98
% dextrose	~ 95

Figure 27. Saccharification Process.

The reaction can be carried out in batch or, as illustrated, continuously in stirred tank reactors in series.

The percentage glucose that can be obtained, called, the DX, is of great concern to the industry. It is depending a.o. on the dry substance at which the saccharification is carried out as illustrated in Figure 28.

This effect is due to the so called reversion reaction; the enzyme itself also catalyzes the re-polymerization of glucose to mainly maltose and isomaltose. The reversion depends also on time,

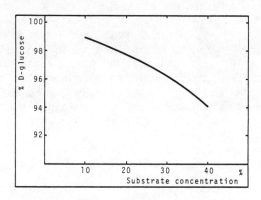

Figure 28. The Influence of Substrate Concentration of Final D-Glucose Level.

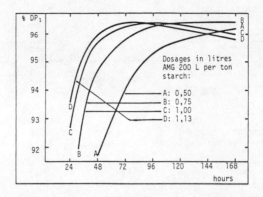

Figure 29. The Effect of Enzyme Dosage on Saccharification.

temperature, and enzyme activity level. In the following figure, the saccharification curves for varying dosages are given as a function of time.

As will be seen, at high dosages, the DX level decreases with time towards the end of the saccharification because of the reversion process.

The long saccharification time, the larger reactor volume, and the technical and commercial success of immobilized glucose isomerases prompted a great deal of research effort in the field of immobilized glucoamylase. It now appears that initial enthusiasm was unfounded because economic benefits are marginal and the technical benefits are questionable.

The obvious place to use an immobilized glucoamylase is in the production of high dextrose syrups in the process outlined above. However, it was soon discovered that the process had a number of severe limitations. The substrate has to be free from particulate matter and soluble protein which might precipitate out and block the column. It therefore has to be filtered and carbon-purified and, in order to facilitate this, it is necessary to liquefy to a DE 15-20 instead of 8-12 which is normal in the standard process. Liquefaction is therefore more costly and complicated.

In order to ensure reasonable enzyme stability, it is necessary to operate at 55°C, which gives severed infection problems even if the substrate is sterile filtered.

The maximum obtainable dextrose levels are limited in both the soluble and immobilized enzyme systems by isomaltose formation resulting from enzyme-catalyzed dextrose polymerization. This phenomenon is more pronounced at high substrate and high enzyme concentrations. The maximum dextrose levels obtained have typically been 2-3 percent lower with immobilized systems. This was thought to be due to two effects: diffusion resistance to the high molecular weight oligosaccharides, leading to a higher residence time in the vicinity of the enzyme to obtain a given degree of hydrolysis; and the slow diffusion of the dextrose formed away from the enzyme, leading to locally high dextrose concentrations in the vicinty of the enzyme.

The only advantages offered are a short reaction time (typically 20-30 minutes) and a considerable reduction in plant size. But because of the problems described above, industrial application is yet to be seen.

Isomerization

Apart from pH and temperature, there are many other factors to be taken into consideration, as illustrated in Figure 30.

As can be seen, optimal utilization of the enzyme depends on how carefully the overall process can be carried out. One very important factor is the substrate quality.

Enzymes are inhibited by certain substances, a classic example being heavy metal ions. Because immobilized enzymes are to be used for long periods of time, the absence of inhibiting substances becomes extremely important. For example: one kg of Sweetzyme which can be used for more than 3,000 hours will be exposed to maybe 10,000 kg of substrate syrup. This means that even very low concentratons of impurities can be harmful. Therefore, in order to get

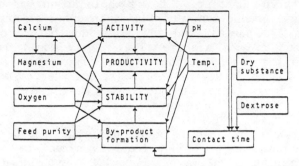

Figure 30. Influence of the Individual Process Parameters on The Activity, The Stability, and Hence The Productivity of Sweetzyme.

good productivity out of the enzyme, impurities which may inhibit or inactivate the enzyme have to be virtually eliminated from the feed syrup before isomerization.

Usually, the syrup is filtered or passed through a centrifugal separator and then filtered to remove particulate matter which may cause clogging of the enzyme. The soluble impurities (peptides, amino acids, ions, etc.) are potential inhibitors and should be removed by carbon treatment and ion-exchange. The efficiency of the carbon treatment and the ion-exchange can be evaluated by measuring the conductivity and the UV-absorbance of the purified syrup. While useful empirical results exist, substrate purification is an area which is being investigated by both enzyme producers and users.

Magnesium ions in the feed syrup activate and stabilize most glucose isomerases. The necessary amount depends on the calcium content as calcium, which is an inhibitor, competes with magnesium. Provided the concentration is below 1 ppm, a magnesium level of 0.0004 M is sufficient for optimum productivity, still using Sweetzyme as an example. A higher level of calcium of up to 15 ppm can be tolerated if the magnesium level is increased to keep the molar ratio magnesium/calcium above 20.

Oxygen in the inlet syrup results in higher byproduct formation, and these byproducts may cause loss of activity, so oxygen should be avoided during isomerization. Low oxygen tension may be obtained by evaporation or vacuum flash.

Under stable process conditions, the activity of immobilized glucose isomerase will decrease exponentially with time. Design lifetime and initial activity will vary from one product to another, but halflives in excess of 50 days are not uncommon.

The syrup flow rate is regulated according to the activity of the enzyme in use in order to maintain the fructose concentration in the effluent syrup constant. If the whole amount of enzyme necessary for a given production were loaded in one reactor, the initial flow would be 230 percent and the flow just before enzyme change only 30 percent of the average flow. This would demand a high flexibility in the other section of the syrup plant, and therefore the enzyme is loaded in several reactors, and the enzyme changes are staggered in time. This is illustrated in Figure 31 where flow variations for one and four reactors are given.

As an example, it can be mentioned that a 100 tons per day plant (dry substance, 42 percent fructose syrup) may consist of 8 reactors, each with a volume of approx. 2.6 m^3 placed in two parallel series, each of four reactors. When designing the reactors, a number of factors must be taken into consideration. These include pressure drop over the enzyme bed, syrup residence time, flow distribution, etc.

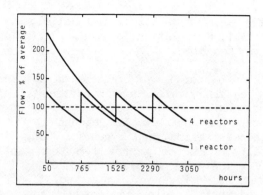

Figure 31. Activity and Flow Variation For One and Four Reactors.

PROCESSING OF INULIN

Although inulin-containing crops are being cultivated, e.g., for chicory salad, coffee extenders, and as vegetables (topinambour) the agricultural basis for industrial utilization of inulin appears to be insufficient at present. However, the processing of inulin is easier than starch.

Inulin can be extracted from sliced plant material by diffusion at elevated temperatures. The raw juice can be purified by liming/ carbonization as in sugar beet processing and then decolourized by carbon. Hydrolysis of the inulin can be accomplished by acid under relatively mild conditions: pH 1-2, 1.2 hours at 80-100°C. However, as fructose is easily degraded at low pH values, this process gives rise to colour- and byproduct formation, e.g., formation of approximately 5 percent difructosedianhydride and thereby to a reduction in fructose yield.

An enzymatic process using pH 4.5 and 50°C would give more than 99 percent hydrolysis within around 1 hour.

PROCESSING OF CELLULOSE

While inulin-containing crops are not available in large commercial quantities, cellulosic raw materials are abundant. Cellulose is estimated to be produced at a rate of 100 billion tons per year. When glucose production from cellulose is not a viable process, at least today, it has to do with the character of the substrate and the enormous complexity of cellulose-containing raw materials as compared to, e.g., starch.

In particular, two problems deserve attention, the presence of lignin and the crystallinity of cellulose. Even if a pure cellulose such as Avicel, a microcrystalline wood cellulose, is used as substrate, the enzymatic degradation is slow, conversion is incomplete, and enzyme consumption high.

The conclusion of our investigations on the processing of cellulose is that, by using high concentrations of the best enzymes developed, either alone or in combinations, about 50 percent degradation of native cellulose can be reached at 50°C, pH 5, 20 percent substrate concentration, and a reaction time of 48 hours. The degree of hydrolysis can be improved by constantly removing the glucose, either by using a membrane reactor or by a simultaneous ethanol fermentation. However, these solutions will hardly be practicable in industrial scale, and thus it appears that an industrial process must include a pretreatment of the lignocellulose in order to make the cellulose more susceptible to enzymatic degradation, a point of view which today is shared by most biotechnologists. But it is still a question whether this leads to an economical process.

It is generally accepted that one of the decisive factors in the economy is the enzyme costs. There is still room for reducing costs of enzyme production by mutations leading to hyper-producers, by optimization of fermentation and enzyme composition, and finally by screening for new and better strains. However, the starting point for optimization is very favourable as one can calculate that, in order to hydrolyze crystalline cellulose, an amount of enzyme protein is required which is approximately 100 times greater than the amount needed for the hydrolysis of starch. Assuming that the production costs for enzyme protein are the same for cellulases and amylases, a tremendous development effort is necessary in order to make enzymatic cellulose hydrolysis economically feasible.

The present market price for fuel alcohol in the U.S. is approximately US$ 0.46-0.48 per litre, and the process economy when using cellulose as a raw material instead of starch will not be competitive with the actual stage of enzyme technology.

It is characteristic that the only processes for the production of ethanol on the basis of lignocellulose on an industrial scale were developed in Germany before the last world war. They were based on a war economy and would not have been contemplated under normal economic conditions. One of the best know processes is the Scholler-Tornesch process for which a flow sheet is shown in Table 12.

We have tried to calculate the process economy without including investments for an acid hydrolysis plant, but as can be seen from Table 13, the variable costs will, even under these favourable conditions, be about US 0.9-1.03 per litre of ethanol. In a future process including a chemical pretreatment as well as an enzymatic hydrolysis, there will be no room at all for enzyme costs.

Table 12. Flow Sheet For The Scholler-Tornesch Process.

Process step	Raw materials	Products formed	Heat consumption, kcal
Hydrolysis 10 ato, 170°C	Wood DS: 1000 kg H_2SO_4: 130-450 kg Water: 12,000 l	Sugars: 530 kg (4% solution)	1,400,000
Neutralization	$CaCO_3$: 133-460 kg	Gypsum, CO_2	
3 stage evaporation		Sugars, 530 kg (20% solution)	2,000,000
Fermenation and Distillation		Ethanol: 240 l	300,000

Our conclusion has been that the utilization of lignocellulose for the production of ethanol or other chemical feedstocks is one of the most difficult tasks encountered in the history of biotechnology. Cellulose has a very dense structure and is shielded by hemicellulose and lignin, so it is very difficult to degrade enzymatically. The enzyme systems to be used must contain at least three enzyme components (endoglucanase, cellobiohydrase, and cellobiase) in adequate amounts which makes production optimization extremely difficult. Therefore, it is doubtful whether the enzyme production costs can be reduced by one or two orders of magnitude. Consequently, it may be necessary to subject the lignocellulose to a pretreatment before the enzymatic hydrolysis, thus increasing the reaction rate to such a degree that the dosage of enzymes can be drastically reduced. Against this background, the outlook for enzymatic cellulose hydrolysis is rather bleak, and only a dramatic breakthrough can change this picture.

NEWER PROCESS LAYOUTS AND FUTURE PROCESS DEVELOPMENTS

Alcohol

What will happen with the technical processes within the potable alcohol industry in the coming years? I think that the processes will be modified, especially with respect to energy consumption and more controlled processing of the raw materials. One should bear in mind that the potable alcohol industry is a very old one where traditions are very important. It is not possible to change raw materials and processing without years of testing in order to prove that the quality of the products has not changed.

In the traditional batch processing, the consumption of energy was approximately 17-24 MJ per liter of ethanol. Of this energy consumption, 7-8 MJ/1 were consumed in the cooking of the raw materials and the rest during the distillation.

Table 13. Economics For The Scholler-Tornesch Process.

	Amount	Unit price US$	Cost US$	Distribution %
Wood dry substance[*)	1000 kg ~ 2,544 x 10^6 kcal	25.5/10^6 kcal	64.9	29-26
Sulphuric acid	130-450 kg	0.08/kg	10.4-36.0	5-14
Calcium carbonate	133-460 kg	0.019/kg	2.5-8.7	1-3
Heat of hydrolysis	1.4 x 10^6 kcal	33.6/10^6 kcal	47.0	21-19
Heat of evaporation	2.0 x 10^6 kcal	33.6/10^6 kcal	67.2	31-26
Fermentation and distillation			28.5	13-11
Cost per 240 l of ethanol			220-254	
Cost per litre of ethanol			0.92-1.05	

[*)Cost calculated on the basis of the combustion energy of cellulose. Value of lignin and hemicellulose not included.

Basis of Calculation

1. 1 ton of coal ~ 8 x 10^6 kcal (25.5 US$/$10^6$ kcal)
2. Coal to steam conversion efficiency: 76 percent
3. Cost of steam: 33.6 US$/$10^6$ kcal
4. Cellulose: 4,000 kcal/kg
5. Costs of hydrolysis equipment not included

When using modern continuous processing, indications seem to show that an energy decrease down to 6-9 MJ/l is obtainable (1-2 MJ/l during cooking and 5-7 MJ/l during distillation) without changing the product quality. In order to obtain this saving, it is necessary to preheat the cooker feed from 20°C to 60°C with recycled surplus process energy, to increase the mash concentration from 20 to 30 percent dry substance, and to reduce the cooking temperature from 150 to about 100°C (the yield loss is limited to a very few percent, dependent on grain quality and particle size).

Recently, it has been reported from Japan (Suntory) that it is possible to reduce the processing temperature to 35°C when using microbial enzymes and a holding time of up to 5 days. Presumably, further optimization will be reported within this special field in the coming years.

The continuous development which is taking place is also important as it allows replacement of at least part of the expensive and troublesome use of malt and koji by industrial enzymes of microbial origin. Furthermore, more and more enzymes are marketed, enzymes which can perform more and more specialized reactions and make the substitution for malt and koji easier so the commercial benefits can be obtained without changing the quality of the potable alcohol.

High Fructose Syrups - Future Process Developments

A minor revolution is presumably under way within this sector. For many years, the idea of using amylopectin debranching enzymes to improve the efficiency of starch conversion has existed.

Processes for the production of dextrose from liquefied starch using either pullulanase or isoamylase together with glucoamylase have been described; however, they have not gained widespread industrial acceptance.

An ideal debranching enzyme for this application should have pH and temperature characteristics which are similar to those of the glycoamylase so that both enzymes can be used together under optimum conditions. Such ideal enzymes have now been found and commercialization will start within the next 12 months. It will then be possible to obtain the following advantages, although not all at the same time:

- Increased dextrose yield
- Considerably reduced glucoamylase requirement
- Reduction in saccharification time
- Saccharification at a higher substrate concentration possible

Other Syrups/Products - Process Developments

Here, the future use of pullulanase and more heatstable acid alpha-amylases are the only changes to be foreseen.

CONCLUSION

Main Changes in Alcohol

- Low energy processes.

Main Changes in HFS

- Introduction of pullulanase.

Main Changes in Other Carbohydrates

- Introduction of pullulanase.

REFERENCES

Christensen, W. 1982. Personal communication. Novo Industri A/S, Bagsvaerd, Denmark.
Hagen, H.A. 1981. Production of ethanol from starch-containing crops -- Various cooking procedures. Paper given at meeting on bio-fuels, Bologna, Italy, June, 1981. Novo Industrias, Bagsvaerd, Denmark.
Lutzen, N.W. 1981. Enzyme technology in the production of ethanol -- Recent process development. Paper given at VI International Fermentation Symposium, London, Ontario, Canada, July 20-25, 1981. Adv. in Biotechnology Vol. II., 161-7, Pergamon Press.
Lutzen, N.W. et al. (To be published). Cellulases and their application in the conversion of lignocellulase to fermentable sugars.
Nakano, M. 1972. A synopsis of Japanese traditional fermenter foodstuffs. Paper given at 4th Int. Fermentation Symposium, Kyoto, Japan, March 19-25, 1972.
Norman, B.E. 1982. A novel debranching enzyme for application in the glucose industry. Paper presented at Starch Days, Detmold, West Germany. Starch 34, 340-346.
Norman, B.E. 1982. The use of debranching enzymes in dextrose syrup production. Paper given at 7th World Cereal and Bread Congress, Prague, June 28 - July 2, 1982.
Norman, B.E. and Lutzen, N.W. 1981. A renewable resource. Paper given at International Symposium on Cereals, Carlsberg Research Center, August 11-14, 1981.
Ostergaard, Jes. 1982. Enzymes in the carbohydrate industry. CNERNA, Versailles, France, May 5, 1982.

Poulsen, P.B. 1981. European and American trends in industrial application of immobilized biocatalysts. Enz. Micro. Tech. 271-273.

Poulsen, P.B. 1982. Industrial practice, potable alcohol. In: Industrial Enzymology, T.G. Godfrey, J. Reichelt, Ed. MacMillan, London.

Poulsen, P.B. and Garberg, P. 1982. Production of alcohol and fructose syrups from cassava. Paper presented at AHARA 82, Bangalore, India, May, 1982.

Rosen, K. 1978. Continuous production of alcohol. Process Biochemistry 5:25-26.

Ueda, S. and Koba, Y. 1982. Alcoholic fermentation of raw starch without cooking by using black-koji amylase. Int. Ferment. Technology 3:237-242.

Zittan, Lena. 1981. Enzymatic hydrolysis of inulin -- An alternative way to fructose production. Paper given at the 32nd Starch Convention of the Association of Cereal Research at Detmold. Starch 33, 373-377.

BIOFILM REACTORS: A REACTION ENGINEERING APPROACH

TO MODELLING

A. E. Rodrigues

Department of Chemical Engineering
University of Porto
Porto, Portugal

INTRODUCTION

Biofilm reactors are often used in water and wastewater treatment and biochemical engineering. In particular, fluidised bed biological reactors (FBBR) have been developed since then high process intensification can be achieved (White 1981). In fact while in a conventional activated sludge tank the biomass concentration is 2-3 g/l, we can get in values of the order of 30-40 g/l, in a FBBR.

Several models for biofilm reactors (Jennings et al. 1976; Grasmick et al. 1981; Mulchay et al. 1980) have been published, both for fixed and fluidised beds, assuming either zero-order kinetics or Monod rate law. Also, diffusional resistances through the liquid boundary layer and biofilm have been considered. In short, we can say that those models do not provide relevant information about each transport mechanism and, sometimes, contain too many parameters which are not independent.

In the following we will show how to apply catalytic reaction engineering fundamentals in order to derive the conversion of substrate in continuous perfectly mixed reactors (CPMR) and plug flow reactors (PFR). The discussion will be limited to zero order kinetics and biofilm mass transfer resistance (Rodrigues et al. 1983).

THE CONCEPT OF BIOFILM EFFECTIVENESS,

Let us assume a slab geometry for the biofilm of thickness, δ, as shown in Figure 1. The biofilm effectiveness is defined as:

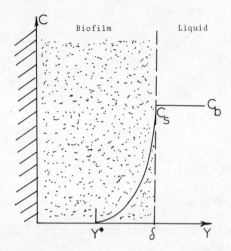

Figure 1. Concentration profile in the biofilm (slab geometry).

$$\eta = \frac{\text{observed rate, } r_{obs}}{\text{intrinsic rate, } r_v} \quad (1)$$

where the intrinsic rate is calculated assuming that in every point of the biofilm the concentration, c, is equal to the surface concentration, c_s.

For zero-order reactions, $r_v = k$ and it turns out that:

$$\eta = 1 - (y^*/\delta) = 1 - x^* \quad (2)$$

where y^* is the point inside the biofilm where the concentration, c and its derivative is zero.

The mass balance for the limiting substrate in a volume element of the biofilm in dimensionless variables is:

$$\frac{d^2 f}{dx^2} = \phi^2 \quad (3)$$

where $f = c/c_s$, $x = y/\delta$ and $\phi^2 = \delta^2 k / D_b c_s$ is the Thiele modulus, with D_b = substrate diffusivity in the biofilm.

The boundary conditions are $f = 1$ at $x = 1$ and $f = 0$, $df/dx = 0$ at $x = x^*$. Integration of Equation (3) leads to the concentration profile inside the biofilm and finally to the biofilm effectiveness:

$$\eta = 1, \quad \phi < \sqrt{2} \quad \text{(chemical controlled regime)} \quad (4a)$$

$$\eta = \sqrt{2}/\phi, \quad \phi > \sqrt{2} \quad \text{(diffusion controlled regime)} \quad (4b)$$

For spherical particles (Figure 2) we can derive the biofilm effectiveness; in fact:

$$\eta = 1 - x^{*3} \tag{5}$$

The calculation of x^* results from the model equation

$$\frac{d^2f}{dx^2} + \frac{2}{x}\frac{df}{dx} = \phi^2 \tag{6}$$

with $f=1$ at $x=1$ and $f=0$, $df/dx=0$ at $x=x^*$. We get x^* from the solution of the cubic equation:

$$x^{*3} - \frac{3}{2}x^{*2} - \frac{6-\phi^2}{2\phi^2} = 0 \tag{7}$$

and then:

$$\eta = 1, \quad \phi < \sqrt{6} \quad \text{(chemical regime)} \tag{8a}$$

The asymptotic expression for high Thiele modulus is

$$\eta = 3\sqrt{2/\phi} \tag{8b}$$

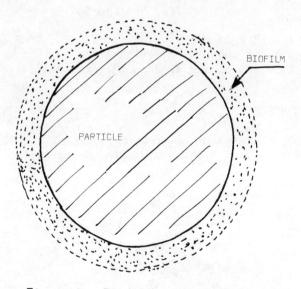

Figure 2. Biofilm: spherical geometry.

CONVERSION IN A CONTINUOUS PERFECTLY MIXED REACTOR (CPMR)

In Figure 3 we sketch a CPMR of volume V fed by a flowrate Q and substrate concentration, c_i. The mass balance over the reactor is:

$$Q c_i = Q c_e + k(1-\varepsilon) V \eta \tag{9}$$

Assuming slab geometry for the biofilm, we can caluclate the conversion of substrate, $X = 1 - (c_e/c_i)$:

$$X_A = N_r \quad \text{(chemical regime)} \tag{10a}$$

$$X_A = \sqrt{\alpha^2 + 2\alpha} - \alpha \quad \text{(diffusional regime)} \tag{10b}$$

with $N_r = (1-\varepsilon)kVc_i/Q$ and $\alpha = N_r N_b$ ($N_b = (1-\varepsilon)D_b V/Q \delta^2$).

CONVERSION IN A PLUG FLOW REACTOR (PFR)

For a plug flow reactor shown in Figure 4 the mass balance of substrate in a volume element dV is:

$$Q dc_s + (1-\varepsilon) k \eta dV = 0 \tag{11}$$

Integrating between the inlet and the outlet of the reactor we get for the substrate conversion:

$$X_p = N_r \quad \text{(chemical regime)} \tag{12a}$$

$$X_p = \sqrt{2\alpha} - \alpha/2 \quad \text{(diffusional regime)} \tag{12b}$$

The conversion in a PFR is always higher than in a CPMR since in the diffusional regime the apparent reaction order is 1/2; moreover, the gap between conversions X_A and X_p does not exceed 25%.

EXPERIMENTAL RESULTS

In Figure 5 we show some results from Grasmick (1978) concerning oxygen depletion in a fixed bed of Biolite particles of 3-5 mm,

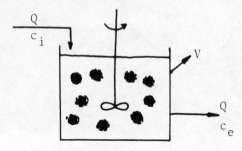

Figure 3. The continuous perfectly mixed reactor.

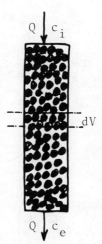

Figure 4. The plug flow reactor.

for organic carbon removal. The results are plotted in terms of $\sqrt{1-X_p}$ as a function of the space time, $\tau = V/Q$ leading to a straight line as predicted by the model, i.e.,

$$\sqrt{1-X_p} = 1 - \frac{\tau}{\tau_c} \quad (13)$$

with $\tau_c = \delta \sqrt{2c_i/kD_b} /(1-\varepsilon)$.

Another set of results is presented from experiments in a BFBR for wastewater denitrification. The reactor is a plexiglass column 135cm high and 9.4 cm in diameter filled with sand particles of mean diameter of 500 and with a bed porosity of 0.42. The substrate contains molasses, sodium nitrate and water. For start-up the final effluent of the sewage plant was used during 10 days, operating in total recycle, at which we add molasses and sodium nitrate. Then the feed was replaced daily at a flowrate of 45 l/h and using low bed expansion 5% (Boaventura et al. 1983).

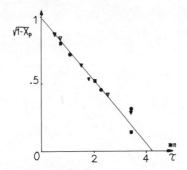

Figure 5. $\sqrt{1-X_p}$ versus space time, for a fixed bed biological reactor.

From the plot of nitrogen removal, E_N (%) versus TOC/N, for different feed nitrogen concentrations, N_o, we get the stoechiometry of denitrification, i.e., TOC/N=1.9. (Figure 6). This is in agreement with the predicted value based on a saccharose formula for molasses, 70% biodegradability of molasses and 30% more molasses

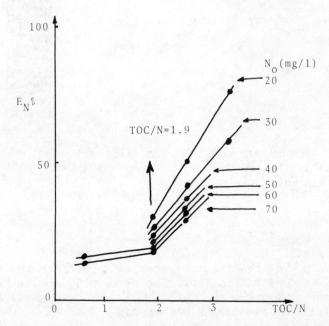

Figure 6. Nitrogen removal, E_N versus TOC/N ratio.

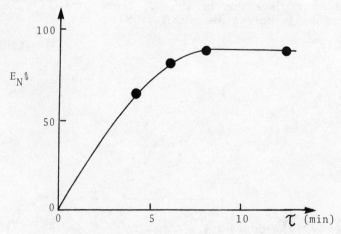

Figure 7. Nitrogen removal, E_N versus space time in a FBBR.

needed for bacterial growth. Finally, nitrogen removal as a function of space time is shown in Figure 7 for the following operating conditions: TOC/N=2.5, C_o=52 mg/l, N_o=220.8 mg/l and Q=0.72 l/min.

REFERENCES

Boaventura, R., Rodrigues, A., Grasmick, A. and S. Elmaleh. 1983. Wastewater denitrification in a fluidized bed reactor. Biotech. and Bioeng. In Press.
Grasmick, A. 1978. La filtration biologique sur colonne garnie. Ph.D. Dissertation, U.S.T.L., Montpellier.
Grasmick, A., Chatib, B., Elmaleh, S. and R. Ben Aim. 1981. Epuration hydrocarbonee en couche fluidise triphasique. Water Research 15: 719.
Jennings, P., Snoeyink, V. and S. Chian. 1976. Theoretical model for a submerged biological filter. Biotech. and Bioeng. 18: 1240.
Mulchay, L., Shieh, W. and E. LaMotta. 1980. Kinetic model of biological denitrification in a fluidized bed biofilm reactor. Progress in Water Technology 12: 143.
Rodrigues, A., Grasmick, A. and S. Elmaleh. 1983. Modelling of biofilm reactors. A.I. Ch. E.J. In Press.
White, M. 1981. BFB's look set for a buoyant future. Processing. January, p. 15.

METHANE GENERATION FROM LIVESTOCK WASTE -- A REVIEW

Sucaattin Kirimhan

Environmental Research Center
Ataturk University
Erzurum, Turkey

INTRODUCTION

In most developing countries, the economic base and majority of the population are still rural, and machinery that requires energy is not heavily utilized. However, the lack of cheap and adequate energy often hampers rural development plans and retards improvement in the quality of rural life. Solving the problems of energy generation and distribution is central to implementation of plans for economic development, especially in rural areas. As solutions dependent on imported fossil fuels become increasingly expensive, the rewards of developing alternative fuel supplies from local materials are bound to grow.

Rural areas usually have large supplies of material-crop residues and animal wastes, theoretically suitable for conversion into a usable source of energy. The process that appears to hold the greatest immediate potential for utilization of these materials as sources of fuel is anaerobic fermentation. This process, also called anaerobic digestion, converts complex organic matter to methane and other gases. It has advantages that recommend it for serious consideration: 1) it is the simplest and most practical method known for treating human and animal wastes to minimize the public health hazard associated with their handling and disposal; and 2) the residue left after removal of the gas is a valuable fertilizer that contains all the essential nutrients present in the raw materials (NAS 1977).

The extraction of energy from wastes using anaerobic digestion to produce bio-gas is not new and general technology is well known. In countries hampered by low natural abundance or inadequate distribution of energy supplies, methane-generating equipment has often been adapted to meet rural needs. Family size methane-generating units have been used in diverse climates and cultures. In India, concern over the loss of cow dung for fertilizer because of its traditional use as fuel, sparked early experiments to develop a system to provide fuel without destroying the dried dung. Nearly 70 percent of India's biogas plants, which now total more than 36,000, were built during the fuel and fertilizer crises of 1975-76. Nearly 27,000 small digesters have been installed in Korea since 1969 through the efforts of the Office of Rural Development. In China, biogas is extensively used for cooking, lighting, fertilizer, and for small internal combustion engines. As of September 1975, over 200,000 family-size digesters were operating in the province of Szechuan. They are built essentially underground, of brick, cement and pebbles, with no moving parts; the gas pressure is kept constant automatically by changing water levels. In the Philippines, fuel is not a major problem as firewood is plentiful. Consequently, interest in biogas stems from its pollution control and public health applications. There are scattered reports of the use of methane generation from waste materials in other countries.

Biogas systems are now receiving attention from several national institutions following the crisis in supply of energy and fertilizer in Turkey. The first demonstration plant was built by the soil and Fertilizer Research Institute in Ankara during 1956-1957. The second research unit on biogas production was built in Eskisehir by the Soil-Water Research Institute during 1962. Research and development are handled by the several research units in the universities. There are now nearly 50 family-sized units based on cattle wastes in Turkey (Kirimhan 1981). The major extension activity is centered at the General Directorate of Soil-Water. The greatest potential seems to be in the digestion of livestock wastes which is estimated to be 150 million tons per year.

Common materials used for methane generation are often defined as "waste" materials, e.g., crop residues, animal wastes, and urban waste including night soil. Some of these materials are already used in developing countries as fuels and/or fertilizers. Use of these materials for methane generation, as illustrated in Figure 1, will allow additional value to be gained from them while the previous benefits are still retained (NAS 1977).

The use of rural wastes for methane generation, rather than directly as fuel or fertilizer, yields three direct benefits:

1) The production of an energy resource that can be stored and used more efficiently,

2) The creation of a stabilized residue (the sludge) that retains the fertilizer value of the original material, and

METHANE GENERATION FROM LIVESTOCK WASTE-A REVIEW 513

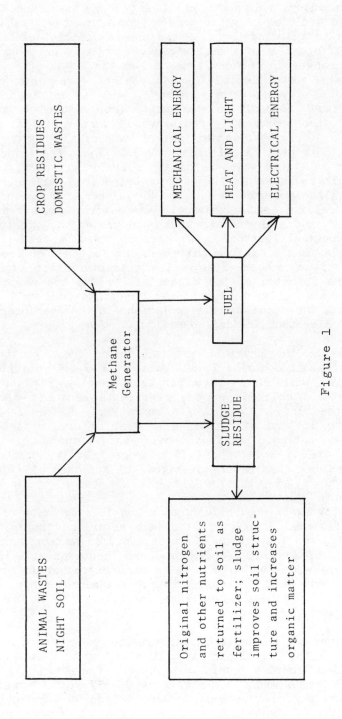

Figure 1

ture and humidity. Evaporation of water may occur under certain conditions. Addition of water occurs from rainwater, wash water, or water added to increase the flow and pumping characteristics of the wastes. Differences in animal waste characteristics also can be a result of changes in the environment and the level of productivity among animals. With both swine and poultry, the diets are highly digestible. The waste from ruminants, such as cattle and horses, has a different composition than waste from simple stomached species. The diet consumed by ruminants is more resistant to digestion. The bacteria that inhabit the stomach of the ruminants enable these animals to utilize cellulosic feeds. These feeds include compounds such as lignin which accompany cellulose in plants and which are difficult to digest in the rumen. Urine from ruminants tends to be more alkaline because their diets are higher in compounds such as potassium, calcium, and magnesium. Wastes from grass fed animals, mature stock, and milk animals will be digested to a greater degree than from growing livestock. As a result these wastes will be less biodegradable than that from animals being fattened or from animals liberally fed on concentrates. The characteristics of livestock wastes are a function of the digestibility and the composition of the feed ration. Livestock wastes can contain feed spilled in the animal pens. There are obvious variations in the characteristics of wastes from livestock feeding operations.

Studies on the physical composition of fresh poultry manure have shown that it contains 75-80% moisture, 15-18% volatile solids, and 5-7% ash with an average particle density of 1.8. Manure excreted from a chicken per day represents about 5% of the body weight of the bird. The nitrogen content of chicken feces ranges from 0.03 to 0.07 gm of nitrogen per gram of dry matter depending upon the feed ration. Combined broiler manure and litter contained an average of 25% moisture, 1.7% nitrogen, 0.81% phosphorus, and 1.25% potassium. Hen manure averaged 40% moisture, 1.3% nitrogen, 1.2 phosphorus, 1.1% potassium.

The daily waste production from a hog is a function of the type and size of animal, the feed, the temperature and humidity within the building, and the amount of water added in washing and leakage. The quantity of feces and urine produced increases with the weight and food intake of the animal. The manure production ranges from 6-8% of the body weight of hog per day. Wet manure can contain 5-9% total solids. Of these 83% may be volatile. The manure averaged 80-85% moisture.

The manure production from dairy cattle ranges from 7-8% of the body weight of the animal per day. Moisture content averages 80-88% in fresh mixed manure and urine. Urine makes up about 30% of weight of manure. Fresh manure contains from 80-85% volatile matter. An average dairy cow will produce between 14 and 18 tons of feces and urine per year. Manure has about 130-180 kg of dry matter per ton.

There are about 4.5 kg of nitrogen, 1.8 kg P_2O_5, 4.1 kg K_2O in each ton of wet manure defecated by the cow.

Waste removed from beef cattle feedlots will have characteristics different from fresh manure. The characteristics change as the manure undergoes drying, microbial action, wetting by precipitation, and mixing and compaction by animal movement. Beef wastes average 80-85% volatile matter. Analysis of a manure slurry from beef cattle manure indicated that each ton of wet slurry contained 2.6, 1.6, and 3.9 kg of nitrogen (N), phosphorus (P_2O_5), and potassium (K_2O), respectively (Loehr 1974).

ANAEROBIC DECOMPOSITION OF LIVESTOCK WASTE

In the anaerobic process, the organic matter consisting of protein, carbohydrates and fat is decomposed by anaerobic bacteria in a manner illustrated in Figure 2. The products of this decomposition are primarily carbon dioxide (CO_2), methane (CH_4), hydrogen sulfide (H_2S), ammonia (NH_3), water and a stabilized effluent. Anaerobic treatment of complex polymeric organic fibers may be considered to be a three-stage process. In each stage, specific bacteria are responsible for the conversion process. In the first stage, a group of facultative microorganisms acts upon the organic substrates. The polymers are converted into soluble monomers that become the substrates for the microorganisms in the second stage, in which the soluble organic compounds are converted into organic acids. These soluble organic acids, primarily acetic acid, are the substrate for the final stage of decomposition accomplished by the methanogenic bacterica. These bacteria are strictly anaerobic and can produce methane. The methanogenic bacteria are very sensitive to changes in environmental conditions such as pH, temperature, and the presence of oxygen (Feddes, McQuitty and Ryan 1980; NAS 1977).

FACTORS AFFECTING THE BIOGAS PROCESS

There are many factors which affect the biogas process. Those which need to be understood in the designing and planning of a gas plant are included here.

Retention Time

An important parameter is the length of time the wastes, especially the solids, remain in the system. In a continuous flow completely mixed anaerobic unit, the liquid retention time and the solids retention time are equal. In units that are not completely mixed, or that employ recycle, the solids retention time is greater than the liquid retention time. The biological solids retention time (SRT) represents the average retention time of microorganisms

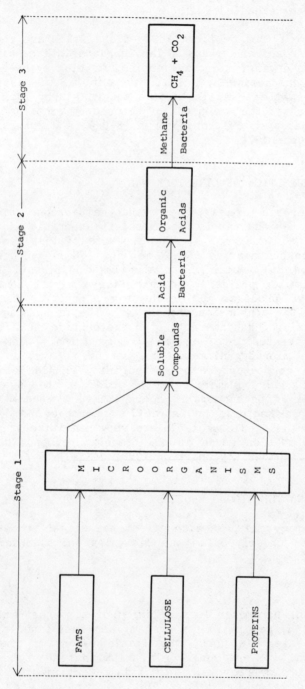

Figure 2. Anaerobic fermentation of organic solids (NAS 1977).

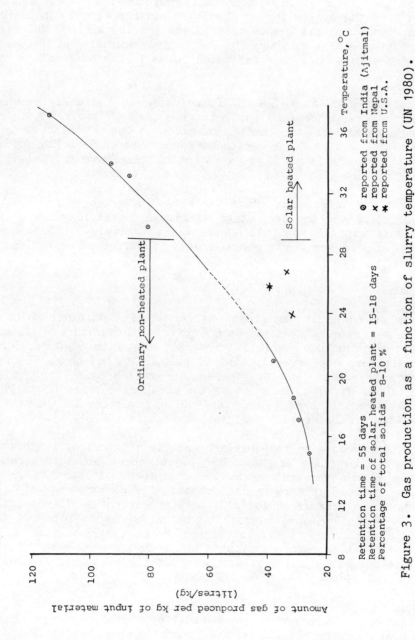

Figure 3. Gas production as a function of slurry temperature (UN 1980).

in the system and can be determined by dividing the amount of volatile solids leaving the system per day. If the SRT of the system is less than the minimum microbial reproduction time, they are removed from the system faster than they can reproduce. In anerobic systems, the microbial growth rate is low and the minimum SRT values are much longer than for aerobic treatment. Minimum SRT values for anaerobic systems have been estimated to be in the range of 2-6 days (Loehr 1974).

The volume of the digester in relation to the daily amount of incoming manure determines the retention time (RT) of the digester and is critical to successful operation. The RT must be great enough to allow time for the methane-formers to convert the acids to biogas. Normally 15 to 30 days retention times are sufficient for manure digestion (BEI 1978).

Three of the most important variables appear to be temperature, retention time and the percentage of solids (specific gravity). The graphs in figures 3 to 6 illustrate these variables. These graphs were produced by various research bodies in different countries (UN 1980).

Temperature

One of the most important environmental conditions which influence the rate of biogas production is temperature. Although biogas production is possible throughout the 0 to 60°C range, the methane-formers are severely limited below 20°C and above 55°C. It is generally agreed that two temperature ranges exist for good biogas production; they are known as the mesaphilic range (20 to 45°C) and the thermophilic range (above 45°C). Investigations have shown that animal manure digests well in both ranges (BEI 1978).

Methane-producing bacteria are greatly affected by temperature. The ideal temperature is about 35°C, just about blood temperature. When the slurry temperature is low the gas production is greatly reduced. At 10°C it more or less stops. The common range of slurry temperature in Asia is 18 to 32°C. A 35°C temperature is not usually possible in an unheated plant. The bacteria cannot tolerate fluctuating temperatures, i.e., hot during the day, cold at night. This will reduce gas production. Plants built in the ground tend to have stable temperatures and keep within the daily allowable fluctuation of $\pm$ 1°C (UN 1980).

Heat losses from a below-ground digester are not zero even in hot climates. There are no theoretical problems in design calculation: heat transfer coefficients, thermal properties, etc. are all known sufficiently accurately. It may also be possible to save on gas use by using solar heaters either to preheat the feed or to run a simple hot-water circulation loop as a thermosyphon. Many prob-

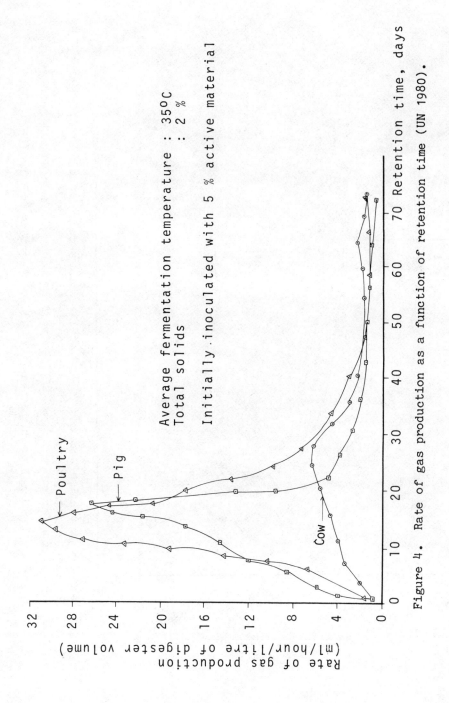

Figure 4. Rate of gas production as a function of retention time (UN 1980).

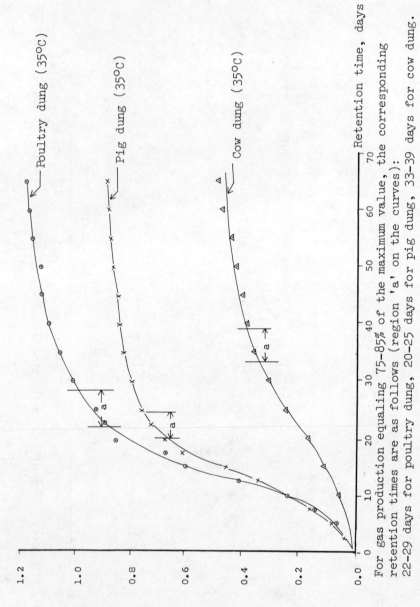

Figure 5. Total gas production as a function of retention time (UN 1980).

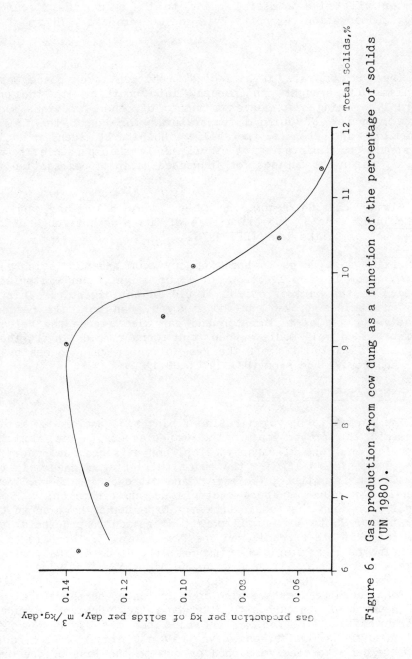

Figure 6. Gas production from cow dung as a function of the percentage of solids (UN 1980).

lems in operating digesters derive from operation with low surrounding temperatures. A simple inventory of the heat loads on the digester will give a useful guide to the most likely sources of energy conservation (Barnett, Pyle and Subramanian 1978).

Mixing

The maximum rate of biological reaction takes place when the organisms are brought continuously into contact with the organic material. Mixing also serves a number of other functions such as: (a) maintenance of uniform temperatures throughout the unit, (b) dispersion of potential metabolic inhibitors, such as volatile acids, and (c) disintegration of coarser organic particles to obtain with a greater net surface for increased rates of degradation (Loehr 1974).

Mixing can be accomplished by mechanical means, by liquid recirculation, or by recycling some of the biogas through diffusers in the bottom of the tank (BEI 1978).

Mixing also helps to break up any scum and to dissipate a bit of it. Scum is a collection of light weight, inert material that collects on the surface of the liquid in the digester. This accumulation, unless broken up and removed, can markedly reduce the effective volume of a digester and can also impede the release of gas from the liquid media. When scum forms a thick layer, the only thing to do is to lift off the gas holder or gain access by other means and remove the scum (BEI 1978; UN 1980).

Carbon Nitrogen Ratio (C/N)

The carbon to nitrogen ratio of biogas digestion raw materials represents the proportion of the two elements. The elements of carbon (in the form of carbohydrates) and nitrogen (as protein, nitrates, ammonia, etc.) are the chief nutrients of anaerobic bacteria. Carbon is utilized for energy and nitrogen for the building of the cell structures. These bacteria use up carbon about 25 to 30 times faster than they use nitrogen. Experiments have shown that a C/N ratio of 25 to 30:1 will permit digestion to proceed at an optimum rate, if other conditions are favourable. If there is too much carbon in the raw materials, nitrogen will be used first, with carbon left over; this will slow down the digestion (UN 1980).

Most fresh manures are also in this range and therefore require no adjustment of the carbon to nitrogen ratio. However, manure with large quantities of bedding, or which has been lying exposed on feedlot surfaces for several days, may not digest well because of the presence of excessive carbon or due to the loss of one or more of the essential nutrients (BEI 1978).

To guarantee the normal biogas production, it is very important to mix the raw materials in accordance with the C/N ratio (UN 1980). Table 1 shows that the contents of C and N and the C/N ratio of each raw material are quite different.

Under normal conditions, equal amounts of methane and carbon dioxide are produced in the digester. However, nitrogen is needed for incorporation into the cell structure; thus, if there is insufficient nitrogen present to permit the bacteria to reproduce themselves, the rate of gas production will be limited by the nitrogen available than is needed to enable the cells to reproduce normally, ammonia will be formed. Its concentration may rise to the point where is inhibits further growth, and the production of methane will slow down or even cease. In summary, if the C/N ratio is too high, the process is limited by the nitrogen availability, and if it is too low, ammonia may be found in quantities large enough to be toxic to the bacterial population (NÂS 1977).

pH

Methane bacteria are very sensitive to changes in pH; the optimal pH range for methane production is between 7.0 and 7.2, although gas production is satisfactory between 6.6 and 7.6. When the pH drops below 6.6 there is a significant inhibition of the methaogenic bacteria, and the acid conditions of a pH of 6.2 are toxic to these bacteria. At this pH, however, acid production will continue, since the acidogenic bacteria will produce acid until the pH drops to 4.5 - 5.0. Under balanced digestion conditions, the biochemical reactions tend to maintain the pH in the proper range automatically (NAS 1977).

Toxic Substances

Since anaerobic digestion is a microbial process any substance that is known to inhibit or kill bacteria is a potential threat to animal manure digestion. Thus when therapeutic dosages of antibiotics or other drugs are administered to a herd, the manure produced during that period should be diverted away from the digester. Similarly, when the barn is disinfected with strong chemicals the wash water should be diverted if possible. It is nearly impossible to predict just how much antibiotics, drugs, cleaning agents, etc. could be tolerated in manure digesters. Certainly most digesters can safely handle manure from animals receiving "normal" low-level dosages. But when disease occurs and heavy dosages are administered it is best not to place manure from those animals in a digester (BEI 1978).

A number of other materials may be toxic to the methane bacteria, and this toxicity results in reduction, sometimes to zero, of gas production. Common toxins include ammonia ($>$1,500 - 3,000 mg/l

Table 1. Approximate carbon and nitrogen contents, and c/n ratios by weight of various raw materials (UN 1980).

Raw Materials	C (%)	N (%)	C/N Ratio
Plant:			
Wheat Straw	46	0.53	87:1
Rice Straw	42	0.63	67:1
Corn Stalks	40	0.75	53:1
Fallen Leaves	41	1.00	41:1
Soybean Stalks	41	1.30	32:1
Weeds	14	0.54	26:1
Peanut Stalks and Leaves	11	0.59	19:1
Dung:			
Sheep	16	0.55	29:1
Cattle	7.3	0.29	25:1
Horse	10	0.42	24:1
Pig	8	0.65	12:1

of total ammonia nitrogen at pH ≻ 7.4); ammonium ion (≻ 3,000 mg/l of total ammonia nitrogen at any pH); soluble sulfides (≻ 50 - 100 mg/l, possibly ≻ 200 mg/l); and soluble salts of metals such as cooper, zinc, and nickel. Alkali and alkaline-earth metal salts, such as those of sodium, potassium, calcium, or magnesium, may be either stimulatory or inhibitory, depending on the concentration. These concentrations have been quantified, and the stimulatory and inhibitory ranges for these salts are shown in Table 2. The toxicity is associated with the cation rather than the anion portions of the salt. When combinations of these cations are present, the nature of their combined effect becomes quite complex. Some cations act as antagonists and reduce the toxic effects of other cations while

Table 2. Stimulatory and inhibitory concentrations of alkali and alkaline-earth cations (NAS 1977).

Cation	Concentration in mg/l		
	Stimulatory	Moderately Inhibitory	Strongly Inhibitory
Sodium	100-200	3,500-5,500	8,000
Potassium	200-400	2,500-4,500	12,000
Calcium	100-200	2,500-4,500	8,000
Magnesium	75-150	1,000-1,500	3,000

others act synergistically, increasing the toxicity of other cations. There are a number of organic materials that may inhibit the digestion process. For all of these materials, toxicity is readily apparent as a reduction in gas production (NAS 1977).

Solid Contents

The suitable solid contents (dry matter concentration) of raw materials is 7 to 9 percent. In actual rural situations, the dry matter concentration of raw materials is changed with the change of season. It is lowered in summer, when the gas production is high, and increased in winter (UN 1980).

The loading rate is influenced by the dry matter content of the manure. The dry matter content of fresh manure ranges between 5 and 25 percent. For most types of farm animals, dilution of the fresh manure will be required.

Dilution and retention time are interdependent and the experience of the region varies considerably. India uses a 1:1 dilution and a 50-day retention period though a 30-day retention is thought possible even with existing designs. At Maya Farms (Philippines) a 1:1 dilution is standard, with a 45-day retention soon to be reduced to 23 days; spend slurry starter is used. The University of the Philippines uses a 1:4 dilution and a 21-day retention; the National Institute of Science and Technology (NITS) dilutions are 1:2 - 1:3. The Agricutlural College at Suweon, Korea, normally practices 30-day digestion at 1:5 dilution; this is reduced in their heated digester, which receives feed at 1:2 dilution for a 20-day retention. The National Institute of Animal Husbandry (Japan) dilutes feed to 1:3 for a retention period of 16 days (BEI 1974: Barnett, Pyle and Subramanian 1978).

CHARACTERISTICS OF BIO-GAS

Bio-gas usually contains about 60 to 70% methane (CH_4), 30 to 40% carbon dioxide (CO_2) and other gases including ammonia (NH_3), hydrogen sulfide (H_2S), mercaptans and other noxious gases. It is also saturated with water vapor.

Two realistic options exist for utilization of biogas on livestock farms: (1) use it directly for cooking, lighting, space heating, water heating, grain drying or gas-fired refrigeration and air-conditioning; (2) transform it into electricity by burning it in an engine and turning a generator. The best strategy for biogas use will depend on the amount and type of seasonal energy demand on a particular farm and the relative costs of each form of energy.

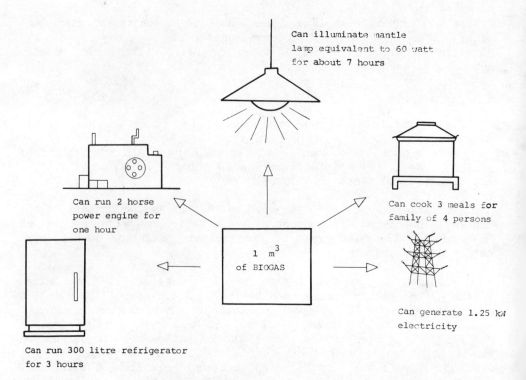

Figure 7. Possible applications of biogas (UN 1980).

If biogas is to be used as a fuel for cars, trucks and tractors, CO_2 and H_2S removal and high pressure storage of the remaining CH_4 has to be considered. Carbon dioxide can be removed by bubbling the gas through lime water. Hydrogen sulfide is commonly removed by the use of iron impregnated wood chips.

The amount of gas required for these applications is given in figure 7.

REFERENCES

Barnett, A., L. Pyle, and S.K. Subramanian. 1978. Biogas Technology in the Third World: A Multidisciplinary Review. International Development Research Center, Ottawa, Canada.
BEI. 1978. Biogas Production From Animal Manure. The Biomass Energy Institute Inc., Winnipeg., Manitoba, Canada.
Feddes, J.J.R., J.B. McQuitty and J.T. Ryan. 1980. Fuel and Fertilier Production From Dairy Manure. Department of Agricultural Engineering, The University of Alberta, Research Bulletin 80-1.
Jewell, W.J. 1974. Energy From Agricultural Waste-Methane Generation. Agricultural Engineering Bulletin No. 397, Ithaca, New York.

Kirimhan, S. 1981. Organik Artiklardan Biyogaz Uretimi. Ataturk Universitesi, Cevre Sorunlari Arastirma Enstitusu, Erzurum, Turkey.

Loehr, R.C. 1974. Agricultural Waste Management. Academic Press, New York.

NAS. 1977. Methane Generation From Human, Animal, and Agricultural Wastes. National Academcy of Sciences; Washington, D.C.

UN. 1980. Guidebook On Biogas Development. Energy Resources Development Series, No. 21, United Nations, New York.

HIGH TEMPERATURE COMPOSTING AS A RESOURCE RECOVERY SYSTEM FOR AGRO-INDUSTRIAL WASTES

J. M. Lopez-Real

Department of Biological Sciences
Wye College
University of London
Wye, Nr. Ashford, Kent, United Kingdom

In a workshop devoted, in principle, to the exploitation of biomass for chemical feedstocks or biofuels, it is perhaps paradoxical to present a paper that urges the return of biomass to that most precious component of our natural resources -- the soil. This paper will discuss the potential use of high temperature composting as a resource recovery system, not only as a technological solution to the ever burgeoning organic wastes of human society, but also as a low technology approach to the maintenance and sustainability of our ecosystems in the future. Whether the biomass being produced is for food, fuel, feedstock or fibre, its continued production is dependent ultimately on the effective management and maintenance of the fertility of the soils utilized. The need for biomass recycling in order to achieve such sustainability is a major theme underlying this presentation.

The composting of organic waste materials has been practiced for centuries by men in many different parts of the world. One of the oldest existing references to the use of "manures" in agriculture is on a set of clay tablets from the Akkadian Empire in the Mesopotamian valley, one thousand years before the birth of Moses. The tribes of Israel were reported to have composted manure with street sweepings and organic refuse outside the city walls. Renaissance literature makes numerous references to composting -- "one aker well compast is worth akers three" (Tusser 1557) and Evelyn (1693), who defined compost as "that rich made mold, compounded with choice mold, rotten dung and other enriching ingredients." A major technological innovation of its time in the 17th and 18th centuries was to capture the heat emitted from composting hotbeds (microbial respiration) for the glasshouses in the winter months. Exotic tropical plants could then be grown in soil placed on the top, thereby exploiting the benefits of the compost heat and nutrients.

That such technological approaches themselves are 'recycled' is apparent today in a research programme in Denmark on the composting of pig manure that is producing hot water for the farmhouse at a temperature only 5°C lower than that of the compost. Chinese agriculture, however, remains the most outstanding example of massive organic recycling through composting and other resource conserving methods (FAO 1977). Interest in composting in the West came largely through the writing of F.H. King (1927), following visits to China, and the work of Sir Albert Howard (1931) in India. The important advances made by Howard in the understanding of the compost process largely paved the way for the exploitation of the system for municipal composting aimed at the treatment of urban organic wastes. Their development was stimulated by a growing awareness that man, with his ever-increasing waste output, needed to find controlled and hygienic processes of disposal. That these problems are even more urgent today will be discussed towards the end of this paper.

Composting may be defined as the decomposition of heterogeneous organic material by a mixed microbial population in a moist, warm aerobic environment (Gray and Sherman 1969). The organic waste itself is normally heavily contaminated with its own populations of micro-organisms (bacteria, fungi and actinomycetes) derived from the atmosphere, water or soil. If moisture is present at the appropriate level, the C : N ratio favourable and the material well aerated, the breakdown activities of the micro-organisms commences. Initially the simple compounds, for example, sugars, exuding from the bruised and broken wastes, will be attacked and, as a result of microbial consumption, heat will be evolved. Providing the heap is of a minimum critical size, the heat evolved as a byproduct of microbial respiration will be trapped, leading in turn to an even greater increase in microbial activity and then more heat evolution. This heat build-up is an essential and basic feature of the composting process leading to development of a specialized microbial population able to rapidly break down normally quite resistant plant materials and to the all-important "pasteurization" of the compost with respect to weed seeds, pests and diseased material.

The temperature and reaction (pH) of the organic wastes follows the classical pattern shown in Figure 1, the duration of each period dependent on how well the variables of the system are controlled. For convenience, the process can be divided into four main phases:

    Phase one   - middle temperature range   (mesophilic)
          two   - high temperature levels   (thermophilic)
          three - cooling down
          four - maturation of compost

Those micro-organisms preferring the middle range of temperature for growth initiate the decomposition cycle, heat is given off and the temperature climbs rapidly. The pH falls as organic acids are

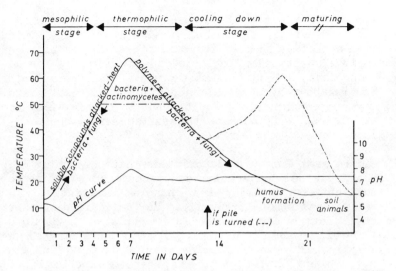

Figure 1. A temperature curve of composting.

produced by microbial action. Above around 45°C, thermophilic (heat loving) micro-organisms begin to predominate and the temperature climbs rapidly to around 60-75°C; this is attained under perfect conditions within 5-7 days of heap construction. The mesophilic populations become inhibited by the time the temperature has passed around 45°C. Though some fungi have been known to remain viable in composts up to 67°C, they do not appear to contribute to the final heating above 70°C which is due to thermophilic bacteria. The more readily degradable substances, sugars, starches, fats and proteins are rapidly broken down. The pH tends towards the alkaline as ammonia is liberated from the proteins present. As the compost cools, a further cycle of activity will begin when the appropriate temperature is attained. Activation of micro-organisms present in the mass occurs, together with a re-invasion from the cooler extremities. These will utilize the remaining materials (e.g., cellulose) compounds that, owing to their structural complexity, are only slowly degraded by micro-organisms and the rate of heat generation cannot balance the heat loss; the compost heap therefore continues to cool down. The types of bacteria and fungi active at this stage are often different from those present at the beginning, reflecting the different types of nutrients that are available during the heating up and cooling down stages. Temperatures at the edges of the heap are considerably lower throughout this process and, if the compost heap is turned so that these sides are brought into the center, a second equivalent cycle then ensues (see Figure 1).

The first three stages of the composting process can be completed within 15-20 days. The final phase, maturation, requires several months. Reactions continue to take place which lead to the stable product of humus or humic acids. Little, if any, heat is generated and the final pH is usually slightly alkaline. During this stage, macro-organisms such as insects, mites, springtails, earwigs, pseudoscorpions, centipedes, millipedes and earthworms invade. This invasion represents a second assault group of decomposers contributing to the breakdown by physical maceration of particles and excretion of organic compounds. Through tunneling activities, they also aerate the compost and increase the surface area where the microorganisms can act.

It can be seen from this description that composting of organic wastes is a truly complex and dynamic process in which temperature, pH and nutrient availability are changing constantly. The time taken for the final product, humus, to be formed is dependent on the interrelated factors of organic waste consistency, particle size, moisture, aeration and heap size. The process has indeed been described aptly in the language of systems dynamics (Finstein 1980). Starting at mesophilic temperature a feedback loop involving heat output and temperature is established. The feedback is positive until approximately 38°C is reached, whereupon it switches to negative. The self-heating then passes to a thermophilic stage and the pattern is repeated. In this stage the feedback switches from positive to negative at approximately 55°C.

High temperature composting is a natural, though probably extremely rare, biological occurrence. With man's intervention and manipulation, it may be carried out as a small-scale garden, market garden or farm scale labour intensive operation. This can be made more or less efficient if the important parameters described in the previous section are controlled to some degree. The process can also be scaled up to a full chemical engineering process for large-scale waste treatment with all the attendant capital costs that are incurred in such operations. Valuable compost as the end product can therefore be produced from both low and high technological systems as deemed appropriate in the context of resources and needs. Since World War II, at least forty different methods for composting have been reviewed extensively in recent years (Gotaas 1956; Gray and Giddlestone 1973; Haug 1980). At the present time, large composting facilities, though rare in North America, are numerous in Europe, with operating examples in Sweden, Belgium, Germany, France, Portugal, Holland, Russia, United Kingdom and Italy. Such silo-type systems use less energy than static pile methods (Easter 1982). Regardless of the plant design or the components of the compost, the basic formula is the same for all systems --proper moisture control, sufficient airflow, steady temperatures and appropriate carbon/nitrogen ratios. The largest silo built to date is the ABV installation at Lanskrona, Sweden, with a diameter of 62 feet (Easter 1982)

Many of these systems were designed initially on a materials handling basis (Gray and Biddlestone 1973), with little consideration of the requirements of the microbial population in terms of nutrients, aeration, moisture and temperature parameters. Extensive research during the mid-1970's in the U.S.A. occurred on the composting of sewage sludge provoked by the scheduled termination in 1981 of dumping at sea. The success of the Beltsville-USDA static pile process attracted attention to the role of composting as a sludge treatment process. In recent years this had led to a more rational design and control of the compost process based on more fundamental scientific considerations (Finstein et al. 1980b). The basic process of composting utilizes the self-heating microbial ecosystem whereby the temperature increases due to the accumulation of metabolic waste heat. Such heat accumulation increases the rate of reaction (i.e., decomposition), but only up to a limiting value, when excessive heat accululation leads to inhibitive temperatures. The central control process was therefore recognized to be heat removal in conjunction with the maintenance of biologically optimal temperatures (Finstein 1980b). This could be achieved by ventilation in response to pile temperature. Ventilation supplies oxygen for aerobic decomposition, removes excess metabolic waste heat thereby maintaining optimal temperatures and, in addition, removes water via vapourization (evaporative cooling). Control was therefore effected through the interrelated parameters of heat output, temperature, ventilation and water removal. Such developments in compost research may lead ultimately to process control comparable to that achieved in the field of biogas production from wastes (Finstein 1980).

This workshop has clearly shown that biomass, whether in the form of a directly grown crop or as a waste product, represents a potentially important resource of chemical feedstock, fuel or food for the future. End product selection from available biomass, in a given locality, will clearly depend on numerous socio-economic factors -- from considerations of health, quantity and quality of wastes, technological and institutional resources, cultural barriers to market demands (National Academy Press 1981). High temperature composting has an important role to play as one approach to the twin problems of increasing organic wastes and the maintenance of soil fertility of intensively cultivated soils.

Recent global projection studies on the world's future population levels and their likely effects on the earth's resource bases (Council on Environmental Qiality 1982; FAO 1982) have highlighted the problem of wastes and the increasing need for the incorporation of organic materials together with fertilizers in order to sustain our food production capabilities. Such projections indicate that by 2000 AD Mexico City may have a population in excess of 30 million people, Calcutta 20 million; Bombay, Cairo, Jakarta and Seoul are expected to be in the 15-20 million range. Some four hundred cities will have passed the million mark. Such rapid urban growth will put extreme pressure on infrastructural services, the treatment of wastes from such vast populations with its attendant sanitary and health implications will undoubtedly become a future major concern. That composting has a potential role to play here has been recognized by the major international development agencies (World Bank 1982).

In addition to the wastes problem, both global studies have recognized that soil losses through desertification, salinization, alkalinization, deforestation, general erosion, humus loss as a consequence of routine agicultural practice are becoming critical throughout the world, affecting both developed and developing nations. The U.S. Soil Conservation Service consider that annual soil losses greater than 1 ton/acre (shallow soils) and 5 tons/acre (deep soils) affect long term productivity. A survey of 283 United States farms (U.S. General Accounting Office 1977) found that 84 percent had annual losses in excess of 5 tons/acre. In Iowa and Illinois 50 percent of the farms surveyed lost between 10-20 tons/acre/year. Higher losses (>20 tons/acre/year) occur on gently sloping lands planted to corn, millet or cotton. Losses of organic matter follows trends similar to those observed for general erosion. Organic matter is also lost if crop residues are burned to protect crops against diseases or if used directly as a fuel source. The burning of such crop residues and dung is a truly disastrous loss for the world's poor, as these organic materials often represent the only source of nutrients needed to maintain soil productivity. Such a dilemma has been succinctly put by the Global 2000 report: "It is the poorest people -- the ones least able to afford chemical fertilizer -- who are being forced to burn their organic fertilizers. The combustion of dung and crop residues is equivalent to burning food."

The recycling of organic wastes through a composting process achieves several important objectives. The process itself results in the conversion of putrescible organic matter to a stabilized end product of considerably lower volume. The compost end product when applied to the soil will trigger off the following benefits:

   (1) increased soil organic matter
   (ii) increased moisture holding capacity
  (iii) improvement in cation-exchange capacity
  (iv) encouragement of beneficial soil fauna
   (v) improved soil aeration
  (vi) improved root penetration
  (vii) provision of plant nutrients
 (viii) suppression of root pathogens

Even small applications of such materials can have marked effects on the soil ecosystem. Addition of as little as 2 tonnes/hectare of organic mulch reduces run-off by 80 percent and erosion by 95 percent (FAO 1981). While inorganic substances exist for organic N, P and K, there is no substitute for organic matter itself. The proteins, cellulose and lignins of plant residues converted to "humus" compost increase porosity and water-holding capacity of soils, preventing erosion and allowing for good root and plant development.

This workshop publication is evidence enough of the vast potential that exists for the conversion and utilization of biomass and biomass products. In a world of finite, rapidly diminishing resources and increasing needs, such approaches to biomass will undoubtedly become more important and more urgent. Biomass production is, however, dependent on the soil resources and our continued success in increasing or maintaining biomass yields will depend on our ability to husband the soil. The ancient art of composting, coupled with our present-day scientific knowledge of the process, may have an increasingly important part to play in this connection in the future.

REFERENCES

Council for Environmental Quality. 1982. The Global 2000 Report to the President. Penguin Books.
Easter, J.M. 1982. Composting silos in Eurpoe. Biocycle May/June 18-21.
Evelyn. 1693. Quoted in The Rodale Guide to Composting. Rodale Press. Emmaus, Pen. 1978.
FAO. 1977. Soils Bulletin No. 40: China: Recycling of organic wastes in agriculture. 107.
FAO. 1981. Economic & Social Development Series No. 23; Agriculture: Toward 2000. rome 1981.
Finstein, M.S. 1980. Microbial ecosystems responsible for anaerobic disgestion and composting. Journal Water Pollution Control 52 (11), 2675-2685.
Finstein, M.S. 1980b. "Sludge composting and utilization: rational aproach to process control." Final report to USEPA, NJDEP, CCMUA.
Gotaas, H.B. 1956. Composting. WHO Monograph No. 31. Geneva.
Gray, K.R. and K. Sherman. 1969. Accelerated composting of organic wastes. Chemical Engineer 20(3). 64-74.
Gray, K.R. and A.J. Biddlestone. 1973. Review of composting -- Part 3: Processes and products. Process Biochemistry 8(10). 11-15, 30.
Haug, R.T. 1980. Compost Engineering Principles and Practice. 655, Ann Arbor Science.
Howard, A. and Y.D. Wad. 1931. The Waste Products of Agriculture. Oxford University Press. London, 1931.
King, F.H. 1927. Farmers of Forty Centuries. Jonathan Cape. London.
National Academy Press. 1981. Food, Fuel and Fertilizer from Organic Wastes. Washington, D.C.
Tusser. 1557. Quoted in The Rodale Guide to Composting. Rodale Press. Emmaus, Penn. 1978.

U.S. General Accounting Office. 1977. "To protect tomorrow's food supply, soil conservation needs priority attention." Report to the Congress by the U.S. Comptroller General, Feb. 14, 1977.

World Bank. 1982. Appropriate technology for water supply and sanitation: Vol. 10. Night soil composting by H. Shuval, C.G. Gunnerson and D.A. Julius.

THERMOCHEMICAL ROUTES TO CHEMICALS, FUELS AND ENERGY

FROM FORESTRY AND AGRICULTURAL RESIDUES

Ed J. Soltes

Professor of Wood Chemistry
Forest Science Laboratory
Texas A & M University
College Station, Texas 77843
U.S.A.

INTRODUCTION

The thermochemical conversion of biomass materials refers to the application of heat resulting in chemical changes. It can be said that the various thermochemical conversion processes developed for biomass are part of a continuum. At one extreme, at relatively low temperatures and in the absence of oxygen, biomass materials are broken down into tars, chars and gases. At the other extreme, at elevated temperatures, and in the presence of at least a stoichiometric amount of oxygen, these primary tar, char and gaseous products are completely oxidized into fully oxidized gaseous products. Man has developed a number of thermochemical processes for biomass, processes not only employing conditions at the two extremes, but also employing conditions resulting in little, partial or near complete oxidation of the feed materials. The processes developed have application in the production of useful products from forestry and agricultural residues, and the purpose of this paper is to examine the potentials for thermochemical conversion of such residues in the generation of chemical, fuel and energy products.

FORESTRY AND AGRICULTURAL RESIDUES

Wood, a product of our forests, is our most important and most widely used biomass material. Wood is remarkably versatile, even the more remarkable because it is renewable. The renewability of wood suggests to many that it is inexhaustible -- that wood will automatically be available in perpetuity. However, our growing wood needs suggest that more and immediate attention must be paid to effective management practices if we are not to find ourselves short, short not only for new uses, but also for traditional uses. The renewability of wood, its versatility, its ubiquitous use on a global level, and its promise in a new age as a source of industrial materials, together suggest concern in potential for overutilization. We must learn how to grow more wood, and to use the available harvest more wisely (Soltes 1980b). Silviculturists, geneticists and pathologists, among others, should be doing what they can to improve the wood supply picture. Our responsibilities as renewable resource conversion scientists are to develop more efficient conversion processes that allow us to meet future wood product demands while using less so-called "merchantable wood," and more so-called "non-merchantable wood." On this basis, we should reject any new wood harvest and utilization schemes or practices which interfere with the supply of wood raw materials to feed existing wood markets, and we should endorse efforts which make better use of our total forest growth, while maintaining or improving environmental qualities. The latter is essential -- our only true renewable resource is productive land.

New uses for wood should then be relegated to residue streams, those parts of our harvest or potential harvest which cannot be used in wood products unless, and only unless, these new products are deemed more valuable than traditional products in some socio-economic context. Wood for energy is the poorest form of wood utilization and should be considered only when other energy forms are not available, or when technology is not available to convert it into more useful forms.

However, in the global perspective, there is need not only for wood for traditional products to improve conditions in less developed nations, and to maintain them in more developed nations; there is need not only for wood for new industrial materials in new national and international strategies for more effective use of both renewable and non-renewable resources; but there is also need for wood for energy, both for basic needs in less developed countries not blessed with fossil resources, and for process energy in more highly developed societies faced with escalating energy costs but producing wood residue streams with currently non-marketable values.

This composite need, coupled with an awareness of renewable resource availabilities, suggests that we should not be too anxious for any major shifts in renewable resources utilization. Instead, the immediate and pressing need is to identify more efficient wood conversion processes to assist our wood-dependent nations to make better use of their wood supply; and for more flexible wood conversion processes, tailored to the use of non-merchanted wood and offering various levels of product sophistication, to assist our more highly industrialized societies to squeeze more value from their wood supply.

We must change also our traditional view of agriculture, especially crop agriculture. U.S. agriculture, blessed with expanses of fertile land, and once seemingly inexhaustible supplies of inexpensive fossil-based resources used for pumping water, for the manufacture of fertilizers, for the planting and harvesting of crops, and for the processing of nature's bounty, soon became energy-wasteful in practices which emphasized crop productivity. Today, in efforts to reduce energy cost inputs to maximize profits, the typical U.S. farmer conserves on the use of energy, often at expense to productivity. The energy requirements for crop agriculture are often found in harvest and process residues, albeit in physical forms which preclude their use in modern agricultural machinery. There is activity in the U.S. in the fermentation of starch- and sugar-containing crops to ethanol, which can serve to extend gasoline supplies. There is activity as well in the use of seed oils for farm diesel usage. However, in both of these, process feeds are again the primary products of agriculture. Agriculture, with few exceptions, continues to suffer from single product practices (Clark, Soltes and Miller 1981). Here too, as in forestry, there is an immedaite and pressing need for conversion processes which can use harvest and process residues in the generation of useable energy products.

An earlier paper in this Advanced Study Institute on Biomass Utilization (Soltes 1982a) states that increasing the accessibility of lignocellulosics, as found in forestry and agricultural residues, is the most important and challenging agricultural research problem today. The cellulose accessibility problem is pervasive, and continues to capture the imagination and ingenuity of researchers around the globe in efforts towards its resolution. The approaches that are being taken to increase plant cell wall accessiblities are analagous to the use of tweezers to carefully and selectively remove, or make more accessible, individual plant cell wall components for utilization. In contrast, thermochemical conversion processes can be likened to hammers, which partially destroy first and target utilization opportunities later. In much the same way that a solution to the plant cell wall accessibility problem would have applicability to both forestry and agricultural residues because the common problem is the cell wall, thermochemical conversion processes do not "see" corncobs, sawdust, or wheat straw, but rather the cellulose, hemicellulose and lignin compositions of plant residues. Products of thermochemical conversion reflect these similar chemical compositions and not dissimilar physical appearances. The heat product of biomass combustion or the gaseous fuel of biomass gasification do not vary with the biomass form fed to the process. Similarly, as will be seen, chemical and fuel product opportunities through thermochemical conversion are also functions of chemical component processing, and are thus applicable to most biomass forms.

DIRECT COMBUSTION

It will be useful in our discussion of thermochemical processing to start with complete combustion of biomass, then follow the thermochemical processing continuum through various stages of incomplete combustion.

The objective in direct combustion is heat (Figure 1). Air is supplied in excess to afford combustion of biomass in the production of fully oxidized, permanent gases. The exothermic reactions involved in this process produce heat. Since wood and agricultural materials have been burned since before recorded time to provide heat, one might assume that there are few, if any, remaining problems in direct combustion of biomass materials. This is not so.

In order to understand the problems involved in burning biomass, one must first understand the combustion process. The most important combustible elements in wood or any agricultural material are carbon and hydrogen which, on combination with oxygen, form carbon dioxide, water and heat. In most cases, air is used as a source of oxygen, and the associated nitrogen must also enter the furnace. In order to insure complete combustion, somewhat more than the minimum theoretical, or stoichiometric, air must be supplied for combination with carbon and hydrogen. This excess air may run from 15 to 150 percent. Excess air and unreacted nitrogen absorb heat with such heat lost to stack gases, and contribute to particulate carry-over by increasing gas velocities and decreasing residence time. Solid fuels such as wood and coal normally require larger amounts of excess air than do liquid or gaseous fuels because of problems in air-fuel mixing. At 50 percent excess air, for example, about 11 percent of the heat produced in the combustion of hogged fuel can be lost to stack gases (Soltes and Wiley 1978).

Water is still an even more serious problem in the combustion of green biomass material. The wet basis moisture content of green southern pine wood and bark is generally about 50 percent. About 1200 BTU is required for the evaporation of water in each pound of wood, or 13 percent of the available energy. Furthermore, excess air must normally be increased under the fuel pile to facilitate evaporation. Water vapor at 500F occupies about 2000 times the

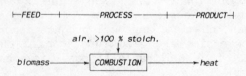

Figure 1. Schematic diagram -- direct combustion.

volume of liquid water. The net result is an increase in stack gas volume with an associated increase in gas velocities creating particulate carryover and losses of heat to stack gases.

This is not to say that green biomass cannot be burned efficently. For example, particulate carryover in the form of unburned combustibles is reduced in the application of the three T's of efficient combustion. These are time, temperature and turbulence. Time is required for the combustion process to be completed before particles can be blown out of the stack. High temperatures are needed to initiate and promote the oxidation reactions. Turbulence gives the needed air-fuel contact. There are a relatively large number of improved conbustion systems now available for bomass, which incorporte these principles (Levelton and O'Connor 1978; Soltes and Wiley 1978). A review of such equipment is outside the scope of this paper.

Water in green biomass creates additional problems. Pound for pound, green wood has a gross heating value about one-fourth that of fuel oil. On a volumetric basis, green wood chips have about one-tenth the heating value of fuel oil. The transportation problems incurred are such that green chips are ten times more costly than fuel oil to transport by truck on a dollar per unit of energy basis. For this reason, there is development activity in the densification and drying of biomass materials destined for fuel usage, at the point of production prior to transportation.

There are yet additional problems in the combustion of agricultural materials, which often contain large amounts of chemicals resulting in high ash production. The present author is part of a multidisciplinary research team at Texas A&M University, now encompassing seven academic departments, created for the purpose of identifying, researching and demonstrating biomass as an alternative energy resource for Texas agriculture. Emphasis on adaptable and reliable off-the-shelf technology suggested the use of fluidized-bed combustion to convert biomass to useable heat. Fluidized-bed technology was selected because of its inherent virtues in high energy conversion efficiency. To our surprise, combustion trials using cotton gin trash and sorghum bagasse experienced several problems. Serious slagging and fouling problems were observed at temperatures as low as 760°C. As well, metal coupons placed in the hot exhaust gas stream experienced high rates of corrosion and erosion. On one occasion, the sand bed used in the reactor agglomerated to the point that the bed could not be salvaged. These problems were traced to the high content of low fusion temperature ash in the feeds to combustion, and the formation of a volatile potassium calcium phosphate compound (Soltes, LePori and Pollock 1982).

As stated earlier, the object in direct combustion is heat. Heat can be utilized in a number of ways in forestry and agriculture

-- directly for drying needs, or in the production of steam for drying needs, or in the production of steam for process energy, or, through cogeneration, both steam and electricity. Despite the problems associated with direct combustion of biomass, it is currently the most direct and most widely used method for converting biomass into energy.

Problems with biomass combustion are related to its direct use as a fuel. Concerns about maximizing the recovery of energy from biomass materials now suggest the development of thermochemical conversion systems for biomass which produce not heat but fuel, gaseous and liquid fuel that can be more easily stored and transported, more easily combusted with fewer operational problems and with higher efficiencies, and more readily metered into high efficiency combustion equipment designed for gaseous and liquid fossil fuels. In other words, several problems in the generation of energy from biomass via direct combustion are avoided in prior conversion of biomass, via gasification or pyrolysis, into better fuels.

It is instructive to look at the combustion of biomass as involving a series of conversion steps (see Figure 2). In the present of heat and air, biomass is first converted to combustible gases, liquid tar and solid char. Anyone who has burned wood in a fireplace is familiar with the semi-solid "creosote" problems experienced when wood is incompletely combusted. These primary conversion products, if given sufficient air and temperature, will eventually undergo secondary conversion towards complete combustion -- a good draft will create the required conditions in a fireplace. Conditions can be found, and processes using these conditions can be developed, which deliberately favor incomplete combustion such that the primary products do not undergo secondary conversion reactions, and the primary products themselves are products which can be recovered directly. Further, process conditions can be identified which favor either the gaseous, liquid or solid fuel product of biomass primary conversion, and these conditions and proceses are what we normally refer to as gasification, pyrolysis or carbonization. The following figure (Figure 3) illustrates the effects of vapor residence time, heating rate, temperature and pressure on the formation of gases, tars and chars in thermochemical conversion of wood (Soltes and Elder 1981).

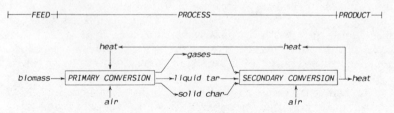

Figure 2. Conversion steps in combustion of biomass.

ENERGY FROM FORESTRY AND AGRICULTURAL RESIDUES

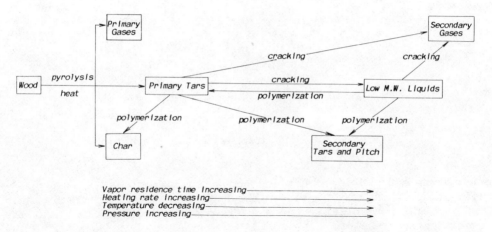

Figure 3. Effects of vapor residence time, heating rate, temperature and pressure of gases, tars and chars in thermochemical conversion of wood.

GASIFICATION

At elevated temperatures just short of those required for combustion, but in the presence of a limited amount of air, biomass will be converted primarily into a mixture of carbon monoxide, hydrogen, and volatile hydrocarbons. This process is called gasification. In gasification of biomass, the objective is gaseous fuel, as shown in Figure 4. To maximize the production of gaseous fuel product, and to minimize that of the other two primary conversion products tar and char, sufficient residence time is required at conversion temperatures such that tar and char undergo secondary reactions towards gaseous product (see Figure 3). This is difficult to accomplish, and unless extraordinary precautions are taken, some tar and/or char is always produced (Soltes 1982b). To a large extent, the solid natures of char, and ash if it is also produced, are such that handling and disposition problems of these materials can be relatively easily solved. Tar, however, is a sticky semi-liquid with corrosive properties and creates troublesome handling problems, not to mention predictable gumming and corrosion problems when entrained

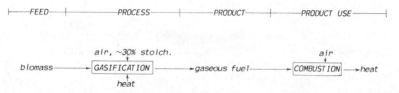

Figure 4. Schematic diagram -- gasification.

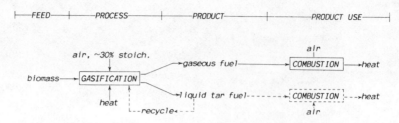

Figure 5. Gasification with tar production.

in the gaseous fuel product and combusted in engines. Biomass gasification, then, may be better shown as depicted in Figure 5, again unless added sophistication and expense is incorporated into the gasification reactor.

GASIFICATION FOR CHEMICALS

A number of factors suggest that methanol, now manufactured principally from natural gas, may soon be manufactured from synthesis gas derived from coal gasification. The carbon monoxide/hydrogen product of wood gasification, similar to that obtained in the gasification of coal, can also be used as a synthesis gas (Soltes 1982a). Methanol which can be synthesized from this gas mixture can be used as a liquid vehicular fuel, or be converted via the Mobil process into gasoline hydrocarbons. Alternately, gas produced via biomass gasification can be used as synthesis gas in Fischer-Tropsch processes to produce paraffinic diesel-like hydrocarbons, or following reforming of these paraffins, gasoline-like hydrocarbons (Kuester 1981).

There have been a number of excellent reviews of biomass gasification in recent years (Klausmeier 1982; Reed 1981; Robinson 1980; Levelton and O'Connor 1978; Sabadell and Mendis 1982), and the reader is referred to these reviews for details of gasification processes. Of particular interest to biomass conversion, it appears that wood, for example, may offer several technical advantages over coal in gasification, because of lower oxygen or steam requirements, and lower shift and desulfurization requirements. Coal contains little oxygen and requires an oxygen-donor, usually large amounts of water, for successful production of gaseous fuel. Despite these and several other technical advantages (Klausmeier 1982), and despite growing use especially by the forest products industry in efforts to replace natural gas fuels with available wood process residue fuels, large-scale wood gasification systems, at least for chemicals, may not come about. The quantities of wood needed to even approach the gas production scales contemplated in coal gasification are simply not available (Soltes, Massey and Murphey 1981), and even if they were, wood as a biomass resource suffers high transportation costs which limit the effective harvesting radius around a conversion plant, thus limiting conversion plant size. Additionally, it is difficult to scale wood handling systems beyond the 5000 or so tons per day used currently in our largest wood conversion facilities. Even if technically and economically feasible, it is difficult to justify the use of wood for gasification for the purpose of generating chemicals in direct competition with coal, especially in the U.S. where coal is a readily available resource. As a general principle in the prudent use of all of our resources, we should avoid direct competition with coal in large-scale, forest products industry-divorced, chemical operations. In the final analysis, the forest products industry, in a raw material pinch, is not likely to dedicate or sell any significant percentage of its wood supply to produce products of little interest and use to its own operations (Soltes 1980b).

Staged Gasification/Combustion

Staged gasification/combustion systems (see Figure 6) are a relatively new application of some old principles (Sabadell and Mendis 1982). Biomass fuel is first gasified using about 70 percent the stoichiometric amount of air required for complete combustion, but here the gasification reactor is close-coupled with a combustion furnace. Gaseous products of gasification while hot and containing vapors and tars in a volatile state come in contact with a secondary source of air for combustion. The sensible heat of the product gases and the subsequent heat produced from combustion of the tars and vapors in the gas provide high conversion efficiency in the production of heat. These systems hold promise for retrofitting onto existing or new oil-or gas-fired systems while using solid biomass fuel feeds, attractive features for process energy needs of the forest products industry.

Uses of Wood Combustion and Gasification

There are many isolated cases of successful implementation of wood combustion and gasification. Wood residues at forest products mill sites are being gasified to provide clean fuel gas for boiler use in the production of needed process steam, and to provide high

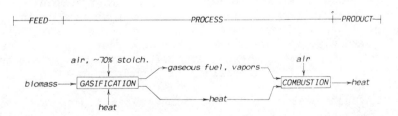

Figure 6. Staged gasification/combustion.

percentages of the energy requirements of dual-fuelled gas/diesel engines in applications requiring shaft power; wood residues are being combusted or gasified to fuel turbines for electrical power generation, both for public utilities, and in cogeneration modes to provide electrical power and process steam requirements for the forest products industry.

These processes, however, are not yet economically competitive for use on broad scales (Soltes 1980b). However, as such technologies develop, as fossil fuel pricing dictates the use of less expensive fuels, as industries are requested to cut back on fossil fuel usage through government dictate (Soltes, Massey and Murphey 1981), and as the forest products industry continues its trends toward materials and energy self-sufficiency, more of these processes will become desirable.

Uses of Agricultural Residue Combustion and Gasification

Agriculture in general is a large consumer of energy, and energy pricing pressures are forcing farmers to adopt less energy-intensive cultural practices and to consider growing their own fuels. A very significant challenge for U.S. agriculture is to maintain agricultural production while attempting to satisfy fuel needs from the same land base. Energy requirements for agriculture are additionally decentralized, that is, needs are for reliable, small-scale biomass energy conversion systems. Although traditional resource recovery systems, such as for example used to process municipal waste, derive large benefits from economies of scale, biomass conversion systems in agriculture are required to be competitive in the range of some 2 to 200 tons per day (Soltes, LePori and Pollock 1982). In Texas alone, it is inconceivable that 73000 internal combustion engines and some 67000 electrical motors used state-wide to pump irrigation water would be displaced by an equal number of biomass-fed engines. Answers to this problem are not yet apparent. One possible solution, again however difficult to implement, would be a farm- or coop-sized energy producer, such as one generating electricity, which could satisfy pumping energy requirements through a power grid.

Other applications for the combustion and gasification of agricultural residues relate to satisfying the energy requirements of various agricultural processes, such as the drying of grain, or the ginning of cotton. These would still be relatively small compared with, for example, an electrical generation plant sized for a forest products operation. The challenge then for energy from and for agriculture is the identification and development of technologies which are at the same time efficient, small-scale, safe and reliable. If such systems could be built, uses would extend beyond agricultural needs.

PYROLYSIS

We now come to pyrolysis. Pyrolysis is at the same time man's oldest and newest thermochemical conversion process: old in the sense of pyrolysis for pine tar, pitch and charcoal fuels; old in the sense of the wood destructive distillation processes used at the turn of the century for organic chemicals production; but also new in the sense of some of its promise (Soltes and Elder 1981).

If thermochemical conversion of biomass is conducted at temperatures still lower (generally below 600C) than those used for gasification, and in the absence of air or under air-starve conditions, all three primary products -- gas, tar and char -- can be recovered (Soltes, Wiley and Lin 1981). Using suitable process conditions, either the char or tar product can be favored.

Pyrolysis for Carbonization

The most common form of biomass pyrolysis is carbonization, that is to produce char. Char is superior as a fuel compared to wood or agricultural residues, and its production can be afforded in very simple systems (Soltes and Elder 1981). A number of options for the production and use of char are given in Figure 7. In primitive carbonization systems, the gaseous and tar products are partially combusted to provide heat as the driving force for the conversion process but to some extent are allowed to escape into the atmosphere. Even under such conditions, chars can be produced in 20 percent mass yield, but because of the higher calorific value of char compared with biomass, are in reality produced in some 30 percent energy yield. Chars are then easily produced, are more energy-intense than their parent biomass forms, have considerable advantages in being essentially smokeless fuels, and thus are heavily used as fuels for home cooking and heating purposes in developing countries not blessed with, or without the resources to purchase, fossil fuels. Heavy usage of wood in charcoal-based economies in many parts of the world, coupled with unsound agricultural practices, are responsible for losses, sometimes irreparable, of forested areas (Soltes 1980b). It is not difficult to increase char yields in more structured carbonization retorts, and it is tempting to suggest that efforts to assist, educate, such societies in the use of more efficient processes should be expedited. However, there are a number of site-specific socio-economic factors in rural areas of underdeveloped countries in charcoal-based economies which make this transition difficult.

In other countries blessed with abundant biomass resources, notably Brazil, charcoal is produced in relatively sophisticated processes as a primary fuel for industrial energy needs, as well as in relatively primitive processes for cottage industries and home use, especially in remote rural areas. In more highly industrial-

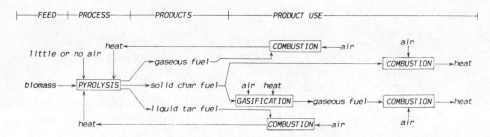

Figure 7. Pyrolysis for carbonization: Carbonization/combustion; carbonization/gasification/combustion.

ized societies, using still more sophistication, carbonization can be an efficient process for the conversion of biomass into a more useful solid fuel form. In the U. S., however, carbonization is relegated primarily to the production of charcoal briquets for home barbeque use. Wood charcoal has some interesting and useful properties which make it a unique raw material for the production of metallurgical and chemical products (Soltes and Elder 1981).

Pyrolysis for Tarification

Pyrolysis has had problems in definition. It is a process term which has been used for both char and tar processes (Soltes, Wiley and Lin 1981). Char is produced in carbonization processes under conditions of long residence time (hours to weeks) at lower temperatures (generally not greater than 400°C) while restricting air supply. Tar product is favored under conditions of short residence time (fractions of a second to minutes) at higher temperatures (generally greater than 400°C) under oxygen-deficient or inert atmospheres. The meanings of gasification and carbonization, both product oriented terms, are clear. There is no good term for processes which favor tar production, but we have suggested the term "tarification" (Soltes, Wiley and Lin 1981) to distinguish from liquefaction, which has other connotations. By direct parallel to carbonization, we can visualize the following processes for tarification (see Figure 8).

If the main objective in tarification is to produce tars, then the key to commercial implementation is the identification of economically competitive technology for the production of higher-valued tar products. There has been much process research in pyrolysis-for-tarification (Chatterjee 1981; Robinson 1980; Soltes and Elder 1981), but there has been little tar product development work. Desirable as tar utilization research may be, our laboratory is one of very few engaged in such research.

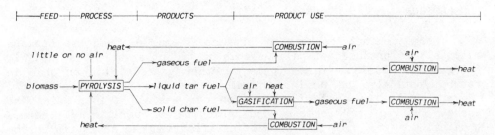

Figure 8. Pyrolysis for tarification: Tarification/combustion; tarification/gasification/combustion.

We have written and presented much lately on the chemical composition of biomass thermochemical conversion tars, on the potentials for biomass pyrolysis as a tarification reaction, and the identification of some higher-valued uses for tars (see References). Of particular interest may be our observation that tars produced from the gasification or pyrolysis of several biomass residues are similar in chemical composition, and that the hydroprocessing (catalytic hydrotreating and hydrocracking) of these tars produces mixtures of hydrocarbons which exhibit similarities to those found in both gasoline and diesel fuels. This suggests that residues from forestry and agricultural activities may be useful for the generation of hydrocarbon fuels for internal combustion engines, and indeed, preliminary engines trials in small stationary engines have been encouraging.

Additionally, variations in the catalytic hydrotreating and hydrocracking processes used to generate these hydrocarbons also produce phenolic compounds which exhibit promise for phenol-formaldehyde adhesive application. Why adhesives? Recent concerns about shrinking inventories of wood in diameters suitable for lumber manufacture (Zobel 1980; Soltes 1980a; 1980b) have suggested that a potentially big market for reconstituted composition wood fiber, flake, and chip products exists, the raw material for which can be logging residue. One doesn't have to cut a 2X4 out of a tree stem; one can also make a 2X4 via extrusion of wood residue particles and the right adhesive. The realization of this desirable development has lagged behind the availability of an inexpensive, effective adhesive. What better use for logging residues can there be than as a raw material not only for the fiber, flake or chip destined for reconstituted products, but also for the production of the adhesive to bind these particles together? An added advantage of the use of pyrolysis in this application is that the other products, gas and char, are fuels that can be used to satisfy much of the energy requirements of the process.

CONCLUSIONS

Thermochemical conversion processes can be used on biomass materials to produce not only useable heat, but a variety of useful, energy-dense fuels. These fuels, whether gaseous, tar or char in nature, can also be used as chemical feedstocks. Thermochemical conversion in the form of combustion can assist us in meeting the energy needs of our forestry and agricultural operations. Thermochemical conversion in the form of gasification can assist in retrofitting bulky biomass fuels to efficient furnaces designed for gaseous fuel usage, and maybe in the generation of a versatile synthesis gas for chemicals needs. Thermochemical conversion in the form of pyrolysis assists the needy without access to fossil fuels by providing charcoal fuels, and may yet provide new approaches to oxychemicals and hydrocarbons for our material and vehicular engine fuel needs. Thermochemical conversion in all forms allows the use of non-merchanted biomass residues in the generation of useful products.

REFERENCES

Brink, D.L. 1981. Gasification. In: Organic Chemicals from Biomass (I.S. Goldstein, Ed.) CRC Press, Boca Raton, FL, 45-62.

Chatterjee, A.K. 1981. State-of-the-art review on pyrolysis of wood and agricultural biomass. Final report, Contract No. 53-319-R-0-206, AC Project PO380, USDA Forest Service, Washington, D.C., 237 pp.

Clark, J.W., E.J. Soltes and F.R. Miller. 1981. Sorghum - a versatile, multi-purpose biomass crop. Biosources Digest 3:36-57.

Elder, T.J. and E.J. Soltes. 1979a. Further investigations into the composition and utility of a commercial wood pyrolysis oil. Paper presented at the American Chemical Society National Meeting, Honolulu, HI, April 1979.

Elder, T.J. and E.J. Soltes. 1979b. Adhesive potentials of some phenolic constituents of pine pyrolytic oil. Paper presented at the American Chemical Society National Meeting, Washington, D.C., Sept. 1979.

Elder, T.J. and E.J. Soltes. 1980. Pyrolysis of lignocellulosic materials: characterization of the phenolic constituents of a pine pyrolytic oil. Wood & Fiber 12:217-26.

Klausmeier, W.H. 1982. Applications for Biomass Gasification: Volume 1. Argonee National Laboratories, Argonne, IL.

Kuester, J.L. 1981. Liquid hydrocarbon fuels from biomass. In: Biomass as a Nonfossil Fuel Source (D.L. Klass, Ed.) American Chemical Society Symp. Series No.144, 163-184.

Levelton, B.H. and D.V. O'Connor. 1978. An Evaluation of Wood-Waste Energy Conversion Systems. Report commissioned by the British Columbia Wood Waste Energy Coordinating Committee, and preparted by B.H. Levelton & Associates, Vancouver, B.C., Canada, 187 pp.

Lin, S.C.K. and E.J. Soltes. 1981. Hydrocarbons via hydrogen treatment of pine pyrolytic oil. Paper presented at the American Chemical Society National Meeting, Atlanta GA, March 1981.

Lin, S.C.K., J.L. Wolfhagen and E.J. Soltes. 1978. Constituents of a commercial wood pyrolysis oil. Paper presented at the American Chemical Society National Meeting, Miami Beach, FL, Sept. 1978.

Reed, T.B. 1981. Biomass Gasification: Principles and Technology (T.B. Reed, Ed.) Noyes Data Corp., Park Ridge, NJ, 401 pp.

Robinson, J.S. 1980. Fuels from Biomass: Technology and Feasibility. (J.S. Robinson, Ed.) Noyes Data Corp., Park Ridge, NJ, 377 pp.

Sabadell, A.J. and M.S. Mendis. 1982. Displacement of industrial oil consumption in developing countries by wood fuel technologies. Paper presented at the First Pan American Congress on Energy, UPADI 82, San Juan, PR, Aug. 1982.

Soltes, E.J. 1980a. Pyrolysis of wood residues: a route to chemical and energy products for the forest products industry? Tappi 63:75-77.

Soltes, E.J. 1980b. Thermal conversion of lignocellulosics: retrospect and prospect. Paper presented at the Intersciencia/JSST Seminar and Workshop on "Materials for the Future: Renewable Organic Resources for Industrial Materials", Kingston, Jamaica, Nov. 1980.

Soltes, E.J. 1982a. Hydrocarbons from lignocellulosic materials. Paper presented at the Ninth Cellulose Conference, Syracuse, NY, May 1982, to be published in: J. Appl. Polymer. Sci.

Soltes, E.J. 1982b. Biomass thermal degradation tars as sources of chemicals and fuel hydrocarbons. Paper presented at the symposium on Feed, Fuels and Chemicals from Wood and Agricultural Residues, E.J. Soltes, org., American Chemical Society National Meeting, Kansas City, MO, Sept. 1982, to be published in: Wood and Agricultural Residues: Research on Use for Feed, Fuels and Chemicals (E.J. Soltes, Ed.) Academic Press, New York.

Soltes, E.J. and T.J. Elder. 1978. Thermal degradation routes to chemicals from wood. A 'special paper' for the Eighth World Forestry Congress, Jakarta, Indonesia.

Soltes, E.J. and T.J. Elder. 1981. Pyrolysis. In: Organic Chemicals from Biomass (I.S. Goldstein, Ed.) CRC Press, Boca Raton, FL., 63-99.

Soltes, E.J., W.A. LePori and T.C. Pollock. 1982. Fluidized-bed energy technology for biomass conversion. Paper presented at the Fourth Symposium on Biotechnology in Energy Production and Conservation, Gatlinburg, TN, May 1982, to be published in: Biotechnol. and Bioeng. Symp. No. 12.

Soltes, E.J., J.G. Massey and W.K. Murphey. 1981. Implications of selected wood use scenarios for the production of energy and industrial materials. Biotechnol. and Bioeng. Symp. No. 11, 3-16.

Soltes, E.J. and A.T. Wiley. 1978. Energy self-sufficiency for the Texas forest products industry: a problem analysis. Paper presented at the annual meeting of the Texas Chapter, Society of American Foresters, Lufkin, TX, April.

Soltes, E.J., A.T. Wiley and S.C.K. Lin. 1981. Biomass pyrolysis - towards an understanding of its versatility and potentials. Biotechnol. and Bioeng. Symp. No. 11, 125-36.

Zaror, C.A. and D.L. Pyle. 1982. The pyrolysis of biomass: a general review. Unpublished manuscript, Dept. of Chemical Eng. and Chem. Tech., Imperial College, London, England.

Zobel, B.J. 1980. Imbalance in the world's conifer timber supply. In: Proceedings of the 1980 TAPPI Annual Meeting, TAPPI, Atlanta, GA, 265-70.

PYROLYSIS OF WOOD WASTES

J. L. Figueiredo, J. M. Orfão,
J. C. Monteiro, S. Alves

Faculdade de Engenharia
Universidade do Porto
Porto, Portugal

INTRODUCTION

Pyrolysis and gasification are thermochemical conversion methods which may be used for the production of energy and chemicals from biomass.

Pyrolysis produces solid, liquid and gaseous fractions by thermal decomposition under inert atmosphere. The yields of these fractions are mostly determined by temperature, heating rate and gas residence time (Soltes and Elder 1981). Gasification involves reaction with an oxidizing agent and consists in a sequence of steps occurring at successively higher temperatures (Antal 1981): -- Pyrolysis (300-500°C) originating volatile matter and char, gas phase reactions of the volatiles (>600°C) and char gasification (>700°C).

Complete gasification of the biomass material is possible by using air, but at the same time introduces unwanted nitrogen in the product gas. An alternative conceptual scheme is shown in Figure 1, where the process is designed to produce fuel gas and char.

The char can be used as a domestic fuel or as raw material for activated carbon manufacture. Better still, it might be possible to produce activated carbon directly in the gasifier. Such an approach is discussed in this communication.

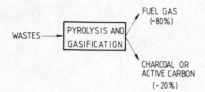

Figure 1. Thermochemical conversion of wood wastes.

LABORATORY TESTS

A small tubular reactor was used to study the pyrolysis of wood wastes such as bark, sawdust, cork and wood shavings (Figueiredo and Órfão 1982). The yields of char, gases and liquid were determined at different temperatures in the range 400-800°C. It was observed that the yield of the liquid fraction goes through a maximum in the temperature range 500-600°C. Operation at higher temperatures favours the production of gases, due to the increasing importance of cracking reactions. Extrapolation of the results shows that the liquid fraction may become negligible at temperatures above 800°C; therefore, a product gas free of tars may be obtained. Char yields decreased continuously to about 20% as the temperature was increased.

Thermograms were obtained for the different materials under nitrogen flow (Figueiredo and Órfão 1982), showing that most of the volatile matter is evolved in the temperature range from 230-380°C. Analyses of the pyrolytic gases were carried out by gas chromatography. As shown in Figure 2, the amount of combustible components (CO, H_2, CH_4) increases with temperature, although the maximum heating value lies around 500°C, due to the higher concentration of methane.

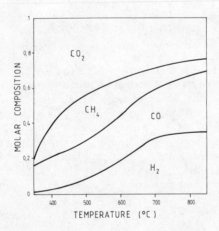

Figure 2. Composition of pyrolytic gases.

The chars obtained by pyrolysis are suitable raw materials for activated carbon manufacture by partial gasification. Gasification experiments with carbon dioxide were carried out in a microbalance flow reactor where the sample weight was continuously monitored (Figueiredo and Orfão 1982). In this way, the rates of gasification could be determined as a function of temperature. When these were plotted in Arrhenius coordinates, a broken line was observed, defining two kinetic regimes: A low temperature regime with high activation energy (111-114 kcal/mole) and a high temperature regime where the activation energy was approximately halved. These results were interpreted in terms of the onset of diffusion limitations in the pores of the char, and they are useful to establish the optimum operating conditions to produce activated carbon. In fact, activation of the char occurs only when the gasification proceeds in the low temperature regime, i.e., in the absence of pore diffusion limitations. In this way, the reactant gas can reach the interior of each particle, opening up its pore structure by carbon removal and increasing the specific surface area. Therefore, the breaking point in the Arrhenius plot defines the maximum gasification temperature to produce an activated carbon. This depends on the material used and on its particle size.

Confirming these results, the specific surface areas of the chars were found to increase with the gasification temperature until a plateau was reached. As an example, an activated carbon with 1084 m^2/g was obtained from sawdust after carbonization at 850°C and gasification with CO_2 to 20% burnoff at 825°C (Figueiredo and Ferraz 1982).

INDUSTRIAL EXPERIENCE

The conversion concept shown in Figure 1 may be attractive for those industries that generate large amounts of wood wastes and that have large energy requirements. A preliminary estimate of the amounts of some cellulosic wastes available in this country is presented in Table 1, showing that there is an adequate supply of raw material for energy conversion. In addition, a "Carbonization and Gasification Chamber" (CGC) operating on this principle has been developed by a local company (Carbotecnica, Lisbon). As shown schematically in Figure 3, it consists of a reactor 6 m long and about 1.5 m in diameter, with no moving parts, and positioned at an angle so that the feedstock can slide down the reactor, in counter-current with the gas. A controlled amount of air is admitted at the bottom for directly heating the reactor, by burning some of the gas.

Each CGC processes about 1000 kg/hr of wood wastes such as pine bark, pine cones, wood chips, rice husks, sawdust or fruit shells and stones, producing about 200 kg/hr of charcoal and a low heating

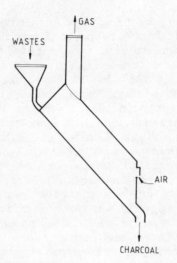

Figure 3. Scheme of the Carbotecnica reactor (CGC).

value gas (1000 kcal/kg gas). At the bottom, the charcoal is extracted after quenching by water injection. Typically, the average solids residence time is 1-1.5 hr and the exit gas temperature is about 850°C.

Such units provide an economically attractive way of replacing conventional fuels. In fact, the waste feed can usually be obtained on site at little or no cost and the gas produced can be cleanly burnt in existing equipment.

In order to exemplify, let us consider one CGC converting 25 t/day of wastes and producing 20 t/day of gas and 5 t/day of char. The product gas can be valued at 1500 Escudos/t by comparison with the cost of fuel (10000 kcal/kg and 15000 Escudos/t) and the char can be sold for briquetting at 6000 Escudos/t. The daily production will then be worth 60000 Escudos, while the maximum cost of the wood wastes would presently be 1000 Escudos/t. Thus, the CGC can be paid

Table 1. Cellulosic wastes available in Portugal (estimated).

INDUSTRY	RESIDUE	AMOUNT (t/d)
Pulp & Paper	Pine bark	150
Cork	Powder	100
Sawmills	Sawdust, shavings	250
Agriculture	Straws, husks, stones, nutshells	?

off in under one year (current installation costs are around 7 million Escudos).

Several of these units are already in operation in Portugal and Spain with very good results, but there is still room for further improvement. So, we are currently engaged in a development programme with the following objectives:

- Obtain better control of the unit;
- Increase the heating value of the gas;
- Investigate the possibility of direct production of an adsorbent carbon.

ACKNOWLEDGEMENTS

This work has been jointly supported by Instituto Nacional de Investigacao Cientifica and by Junta Nacional de Investigacao Cientifica e Tecnologica under research contract No. 410.82.24. We thank CARBOTECNICA for supplying the operating data for the CGC.

REFERENCES

Antal, M.J. Jr. 1981. In D.L. Klass (ed.), "Biomass as a Nonfossil Fuel Source", ACS Symposium Series 144, p. 313.
Figueiredo, J.L. and M.C.A. Ferraz. 1982. In "Adsorption at the Gas-Solid and Liquid-Solid Interface", J. Rouquerol, K.S.W. Sing (eds.), Elsevier Scientific Publishing Co., p. 239.
Figueiredo, J.L. and J.J. Melo Órfão. 1982. Submitted for publication.
Soltes, E.J. and T.J. Elder. 1981. Pyrolysis. In: I.S. Goldstein (ed.), "Organic Chemicals from Biomass", CRC Press, p.63.

HYDROLYSIS OF CELLULOSE BY ACIDS

Irving S. Goldstein

Department of Wood and Paper Science
North Carolina State University
Raleigh, NC 27650

INTRODUCTION

The central role of cellulose in biomass utilization is assured by virtue of its position as the structural polysaccharide which comprises approximately half of all plant material. This abundance coupled with the knowledge that it can be converted back into its glucose precursor by hydrolysis has provided the incentive for research and development efforts on cellulose hydrolysis for over a hundred years. Obviously the production of glucose for use as food or further conversion to liquid fuels or chemicals from inexpensive, renewable and abundant biomass is an attractive goal. Processes for the conversion of cellulose to glucose by acid hydrolysis have reached commercial production, but only under special circumstances. Under free market conditions they have not as yet been able to compete with sugars from specific crops (cane and beets) or from hydrolysis of starch.

In the following discussion I will briefly describe the effect of the structure of cellulose on its hydrolysis by acids and the two classes of processes which have emerged, namely, dilute acid and concentrated acid. Details of individual processes will not be emphasized since these may be found in recent reviews (Goldstein 1980, 1981; Wenzl 1970). Rather an attempt will be made to place the classes of acid hydrolysis in perspective with each other and with enzymatic hydrolysis.

CELLULOSE STRUCTURE AND HYDROLYSIS

Cellulose like other polysaccharides is a polymer of simple sugars in their cyclic form linked by acetal bonds known in carbohydrate chemistry as glycosidic linkages. Simple glycosides are readily hydrolyzed under acid conditions and as might be expected so are such polysaccharides as starch and hemicelluloses in general.

However, cellulose does not follow the same pattern. When exposed to dilute acid there is an initial rapid rate of hydrolysis as measured by weight loss and formation of soluble sugars (glucose and its oligomers). But the extent of this rapid hydrolysis is limited, and the hydrolysis rate soon becomes very much slower. It has been established that the readily hydrolyzed material is amorphous or disordered cellulose. After its removal the residual crystalline cellulose is only slowly hydrolyzed at a rate about 100 times less.

The crystallinity of cellulose is a consequence of the beta configuration of its glycosidic linkages which requires that the cellulose chain assume a linear form. Close packing, high order and intermolecular hydrogen bonds result. The rigid cellulose crystallites resist solution in water and ready hydrolysis, in contrast to the alpha-linked glucose polymer starch which is readily swollen in water and hydrolyzed.

Technical processes for cellulose hydrolysis must overcome this slow hydrolysis of crystalline cellulose in order to effect the conversion in an industrially acceptable period of time. Elevation of temperatures in dilute acid hydrolysis, partial decrystallization by pretreatment before dilute acid hydrolysis, and complete decrystallization by swelling or solution in concentrated acids have been found to be successful techniques in approaching this objective.

DILUTE ACID PROCESSES

As is true of any chemical reaction an increase in temperature will increase the rate of hydrolysis, and complete hydrolysis of cellulose can be effected in minutes or even seconds at elevated temperatures. However, the glucose liberated by hydrolysis is unstable under hot acid conditions and secondary degradation of the glucose becomes significant. Even though the hydrolysis reaction is favored more than the glucose degradation as the temperature is increased, the practical upper limit to glucose yield attainable by dilute acid hydrolysis is about 50 - 60% (Grethlein 1978).

Process technology for dilute acid hydrolysis has been under development since the first decade of this century and is still evolving under the stimulus of the need for alternative liquid fuels. Early process configurations included rotary digesters fuels. Early process configurations included rotary digesters which gave low sugar yields principally from the hemicellulose component of the biomass (Plow et al. 1945), and the percolation processes of the Scholler (Wenzl 1970, Eickemeyer and Henneck 1967) and Madison (Lloyd and Harris 1963) types. Scholler type percolation process plants are still in production in the USSR and one is under construction in Brazil.

More recently continuous processes have been under development in which an acidified slurry of cellulose is heated rapidly to the desired reaction temperature, maintained at temperature for the desired short reaction time and quenched rapidly to give optimum glucose yield. This sequence may be accomplished in a plug flow reactor (Thompson and Grethlein 1979, Church and Wooldridge 1981) or a twin-screw extruder (Rugg and Brenner 1980). The glucose yield limitations are inherent in the kinetics of the cellulose hydrolysis and glucose degradation. Within this framework process configuration can optimize the yield attainable.

An alternative approach to the enhancement of cellulose hydrolysis by dilute acid involves reduction of cellulose crystallinity instead of elevation of temperature. Cellulose substrates of low inherent crystallinity are more amenable to hydrolysis. Reduction of crystallinity can be accomplished by either chemical or physical methods. Swelling of cellulose with alkali (Millett et al. 1954) and liquid ammonia (Tarkow and Feist 1969), or selective dissolution of cellulose followed by precipitation (Ladisch et al. 1978) make the cellulose highly reactive to hydrolysis. Physical methods of decrystallization include high energy radiation (Saeman et al. 1952), vibratory ball milling (Millett et al. 1979) and two roll milling (Tassinari and Macy 1977). Pretreatments of cellulosic substrates to enhance hydrolysis by decrystallization are too expensive for commercial use.

CONCENTRATED ACID PROCESSES

Complete decrystallization of cellulose is effected by concentrated acids which either swell the cellulose or dissolve it and are then capable of bringing about hydrolysis at moderate temperatures in a short time. Under these conditions the yield of glucose can be almost quantitative since degradation of the glucose is minimal.

Decrystallization and subsequent hydrolysis in a single process step may be accomplished by the use of any of a large number of concentrated acids. Most development work has involved hydrochloric and sulfuric acids (Wenzl 1970), but phosphoric (Ekenstam 1936), formic, hydrofluoric (Selke et al. 1983) and trifluoroacetic (Fengel and Wegener 1979) are also effective. Minimum acid concentrations for initiation of decrystallization are critical; for hydrochloric acid it is above 40%, for sulfuric acid above 60% and for trifluoroacetic acid 100%.

The lower cost of hydrochloric and sulfuric acids has led to their having been emphasized in development work, with hydrolchloric acid having been predominant since it is more volatile and more readily recovered than sulfuric acid. The still greater volatility of hydrofluoric acid has attracted recent attention, but the consequent ease of recovery must yet be balanced against its much greater cost.

Hydrolysis of cellulose by concentrated hydrochloric acid has been recognized for 100 years, and by 1932 a technical process using 41% hydrochloric acid recoverable by vacuum distillation was placed into commercial production in Germany (Bergius 1937). These plants operated until the end of World War II when they became uneconomic. Hydrogen chloride gas processes were later studied in Japan (Kusama and Ishii 1966) and the USSR (Chalov and Leshchuk 1962) because of the greater ease of acid recovery, but never progressed beyond the pilot plant.

In Japanese concentrated sulfuric acid processes (Oshima 1965) the acid was either recovered by diffusion dialysis using an ion-exchange membrane or neutralized with lime to produce gypsum for board manufacture. Concentrated sulfuric acid has also been in use in conjunction with mechano-chemical treatments in a screw conveyor (Karlivan 1980), or to selectively dissolve the cellulose from the biomass matrix (Tsao et al. 1979).

Although the technical feasibility has been established of obtaining almost quantitative yields of glucose from cellulose by means of concentrated acid hydrolysis, these processes require extensive recovery of the acid to be economic. This recovery entails considerable cost. It has been estimated (Nguyen et al. 1981) that in the recovery of hydrochloric acid by vacuum distillation in a simulated wood hydrolysis plant that over 35% of the plant operation costs and almost 40% of the total capital costs were related to hydrochloric acid recovery and loss.

COMPARISON OF ACID AND ENZYMATIC HYDROLYSIS PROCESSES

It is apparent from the previous discussion that the two classes of acidic cellulose hydrolysis processes, dilute and concentrated, differ in important ways. It is also useful to compare these differences, and for perspective relate them to cellulose hydrolysis by enzymes (Mandels et al. 1978).

Substrate Susceptibility

Amorphous cellulose is rapidly and completely hydrolyzed by acids of any concentration or enzymes. However, cellulose is ordinarily highly crystalline and its hydrolysis by dilute acids or enzymes is much slower. Cellulose crystallinity does not influence concentrated acid hydrolysis since the reagent brings about complete decrystallization. Dilute acid hydrolysis of crystalline cellulose requires elevated temperatures with concomitant reductions in glucose yield, while enzymatic hydrolysis of crystalline cellulose becomes ever slower with time because of increased inhibition and enzyme inactivation.

An even more important problem in substrate susceptibility involves the presence of lignin. Most of the biomass available and certainly that which is least expensive is highly lignified. The lignin matrix forms a sheath around the cellulose which can completely inhibit enzymatic hydrolysis by preventing access of the enzymes to the cellulose. Some pretreatment is necessary to improve enzyme accessibility, and these are expensive. Acid hydrolysis of cellulose, whether dilute or concentrated, is unaffected by the presence of lignin.

Yield

Dilute acid hydrolysis of cellulose at elevated temperatures can provide a maximum yield of glucose as a practical matter of about 50 - 60% from cellulose with a crystallinity as high as that in wood. Concentrated acid hydrolysis at moderate temperatures can provide almost quantitative yields of glucose. While it is conceptually possible to obtain 100% yield of glucose from cellulose by enzymatic hydrolysis the rapid reduction in the rate of hydrolysis as the reaction proceeds makes it uneconomic to continue to completion.

Rate of Hydrolysis

Dilute acid hydrolysis at elevated temperatures can be completed in seconds and concentrated acid hydrolysis at moderate temperatures in minutes, while enzymatic hydrolysis requires days for completion.

Catalyst Recovery

The use of dilute solutions of inexpensive sulfuric acid permits disposal by neutralization. However, when concentrated acid solutions are used it is essential for the economics of the process that the acid be recovered, but this also entails considerable extra expense. Recycle of the more expensive enzyme catalyst is also of great economic importance.

Equipment

Equipment size in hydrolysis processes is dependent on process configuration, but a generalization about reactor size at a given production level can be made. Obviously reactor size must increase from dilute acid to concentrated acid to enzymatic hydrolysis as the rate of hydrolysis decreases. The change in time scale from seconds and minutes to days requires much greater reactor capacity for the enzyme process. On the other hand the severe corrosion problems and expensive corrosion-resistant materials associated with acid hydrolysis are avoided in enzymatic hydrolysis.

By-Products

The overall economics of a hydrolysis process may be significantly influenced by the value of other products derived from the biomass being processed. Acid hydrolysis of lignified cellulose leaves a lignin residue. That from high temperature hydrolysis conditions is much less reactive and more inert than the residue from low temperature hydrolysis.

SUMMARY

Acid hydrolysis of cellulose to glucose can be accomplished by both dilute acid and concentrated acid processes. Both have been practiced on a commercial scale in the past, but only dilute acid hydrolysis is in production today. The maximum yield of glucose attainable from dilute acid hydrolysis at elevated temperatures is about 50 - 60%. Concentrated acid hydrolysis at moderate temperatures is capable of providing almost quantitative yields of glucose, but these improved yields are offset by the higher operating and capital costs associated with acid recovery.

Acid hydrolysis processes are faster than enzymatic, and require smaller but more corrosion-resistant reactors. They are much more flexible than enzymatic processes in regard to raw material, since they are effective on cellulosic material of any degree of crystallinity or lignification in contrast to the accessiblity barriers encountered with enzymes.

REFERENCES

Bergius, F. 1937. Conversion of wood to carbohyrates and problems in the industrial use of concentrated hydrochloric acid. Ind. Eng. Chem. 29:247.

Chalov, N.V. and Leshchuk, A.E. 1962. Continuous hydrolysis of wood with 46-48% HC1. Izv. Vyssh. Uchebn. Zavad. Lesn. Zh. 5(1):155.

Church, J.A. and Wooldridge, D. 1981. Continuous high-solids acids hydrolysis of biomass in a 1½-inch plug flow reactor. Ind. Eng. Chem. Prod. Res. Dev. 20:371-381.

Eickemeyer, R. and Hennecke, H. 1967. New technical and economic possibilities for wood saccharification. Holz-Zentralblatt 86:1374.

Ekenstam, A. 1936. The behavior of cellulose in mineral acid solutions. Ber. 69:549.

Fengel, D. and Wegener, G. 1979. Hydrolysis of polysaccharides with trifluoroacetic acid and its application to rapid wood and pulp analysis. In: Hydrolysis of Cellulose: Mechanisms of Enzymatic and Acid Catalysis (R.D. Brown, Jr. and L. Jurasek, Eds.) pp. 145-157. American Chemical Society Advances in Chemistry Series 181, Washington, D.C.

Goldstein, I.S. 1980. The hydrolysis of wood. TAPPI 63(9):141-143.

Goldstein, I.S. 1981. Organic Chemicals From Biomass. CRC Press. Boca Raton, Fla. Chapter 6.

Grethlein, H.E. 1978. Chemical breakdown of cellulosic materials. J. Appl. Chem. Biotechnol. 28:296-308.

Karlivan, V.P. 1980. New aspects of the production of chemicals from biomass. In: Future Sources of Organic Raw materials - CHEMRAWN I (L.E. St. Pierre and G.R. Brown, Eds.) pp. 483-494. Pergamon Press Oxford/New York.

Kusama, J. and Isshii, T. 1966. Wood saccharification by gaseous HC1. Kogyo Kagaku Zasshi. 69(3):469.

Ladisch, M.R., Ladisch, C.M. and Tsao, G.T. 1978. Cellulose to sugars: new path gives quantitative yield. Science 201:743.

Lloyd, R.A. and Harris, J.R. 1963. Wood hydrolysis for sugar production Report No. 2029. USDA Forest Products Laboratory, Madison, Wis.

Mandells, M., Dorval, S. and Medeiros, J. 1978. Saccharification of cellulose with Trichoderma cellulase. Proc. 2nd. Annual Fuels from Biomass Symposium, Rensselaer Polytechnic Institute, Troy, N.Y., Vol. II, pp. 627-669.

Millett, M.A., Moore, W.E. and Saeman, J.F. 1954. Preparation and properties of hydrocelluloses. Ind. Eng. Chem. 46:1493.

Millett, M.A., Effland, M.J. and Caulfield, D.F. 1979. Influence of fine grinding on the hydrolysis of cellulosic materials - acid vs. enzymatic. In: Hydrolysis of Cellulose: Mechanisms of Enzymatic and Acid Catalysis (R.D. Brown and L. Jurasek, Eds.) pp. 71-89. American Chemical Society Advances in Chemistry Series 181. Washington, D.C.

Nguygen, X.N., Venkatesh, V., Marsland, D.B. and Goldstein, I.S. 1981. Application of GEMS to preliminary process design and economic analysis for an integrated wood hydrolysis plant. AICHE Symposium Series. 77(207):85-92.

Oshima, M. 1965. Wood Chemistry Process Engineering Aspects. Noyes Development, New York.

Plow, R.H., Saeman, J.F., Turner, H.D. and Sherrard, E.C. 1945. The digester in wood saccharification. Ind. Eng. Chem. 37:36-43.

Rugg, B. and Brenner, W. 1980. Utilization of the New York University continuous acid hydrolysis process for production of ethanol from cellulose. Proceedings Bio-Energy '80 Congress, Atlanta, April 21-24. pp.160-162.

Saeman, J.F., Millett, M.A., and Lawton, E.J. 1952. Effect of high-energy cathode rays on cellulose. Ind. Eng. Chem. 44:2848-2852.

Selke, S., Hawley, M.C. and Lamport, D.T.A. 1983. Reaction rates for liquid phase HF saccharification of wood. In: Wood and Agricultural Residues: Research on Use for Feed, Fuels and Chemicals (E.F. Soltes, Ed.) Academic Press, New York. In press.

Tarkow, H. and Feist, W.C. 1969. A mechanism for improving the digestibility of lignocellulosic materials with dilute alkali and liquid ammonia. In: Celluloses and Their Applications (Hajny, G.J. and Reese, E.T., Eds)., American Chemical Society Advances in Chemistry Series 95, p. 197. Washington, D.C.

Tassinari, T. and Macy, C. 1977. Differential speed two roll mill pretreatment of cellulosic materials for enzymatic hydrolysis. Biotech. Bioeng. 19:1321.

Thompson, D.R. and Grethlein, H.E. 1979. Design and evaluation of a plug flow reactor for acid hydrolysis of cellulose. Ind. Eng. Chem. Prod. Res. Dev. 18:166-169.

Tsao, G.T., Ladisch, M., Ladisch. C. Hsu, T.A., Dale, B., and Chow, T. 1979. In: Annual Reports in Fermentation Processes Vol. 2 (D. Perlman, Ed.) Academic Press, New York.

Wenzl, H.F.J. 1970. The Chemical Technology of Wood. Academic Press. New York. Chapter IV.

EFFICIENT UTILIZATION OF WOODY BIOMASS:

A CELLULOSE-PARTICLEBOARD-SYNFUELS MODEL

        Raymond A. Young and Suminar Achmadi

        Department of Forestry
        University of Wisconsin
        Madison, Wisconsin 53706
        U.S.A.

INTRODUCTION

For efficient utilization of woody biomass, systems must be developed which effectively utilize all components of the woody stem. Since cellulose is probably the most important chemical derivable from wood, we have emphasized schemes which produce cellulose pulp, but also include complete by-product recovery and waste product utilization.

The model which we developed in cooperation with Kenneth Baierl, an engineering consultant, is shown in Figure 1 (Baierl et al. 1980). The scheme emphasizes three major products, <u>cellulose</u> from organosolv pulping, <u>synfuels</u> through recovery of pulping condensates and waste liquor chemicals, including sugar fermentation products, and <u>adhesives</u> from primary paper mill sludge, lignin wastes, and sugar dehydration products. Each segment is discussed in further detail below.

SYNFUELS

A variety of potential synthetic fuels are already produced in conventional pulping as shown in Table 1. Large quantities of methanol and acetic acid (or acetate) are formed in both kraft and sulfite cooks and large amounts of sugars for fermentation are produced in sulfite pulping. Turpentine, phenols, aldehydes and other chemicals are also produced in the pulping of wood, all of which could be utilized for liquid synthetic fuels. It has been estimated that about 5% (10 billion liters) of the gasoline consumed in the United States annually could be derived just from the waste

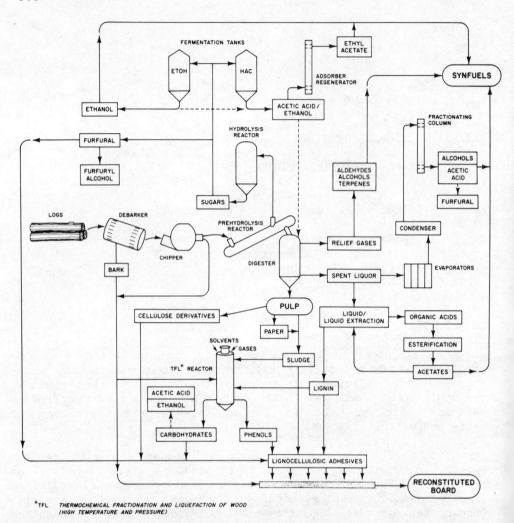

Figure 1. Proposed cellulose-particleboard-synfuels complex model.

chemicals shown in Table 1. Road tests have already proven that a synthetic blend of just pulp mill derived chemicals, such as those shown in Table 1 (acetic acid as ethyl acetate) can power conventional automobiles (Baierl et al. 1981). However, the use of these chemicals as octane boosters is probably more realizable.

The top rated octane boosters include methyltertiarybutyl ether (MTBE), ethanol and ethyl acetate. Obviously, ethanol could be produced by conventional yeast fermentation of pulping waste hexose sugars. Ethyl acetate could also be obtained by esterification of the waste acetic acid with ethanol. The potential for acetic acid

Table 1. Pulp mill chemicals for synthetic fuels.

Chemical	Kraft	Sulfite
	Kg/Ton	
Methanol	5	7-10
Acetic Acid	100-200*	30-90
Sugars	--	200-400
Turpentine	8-10	--

* Calculated as sodium acetate.
≠ Data from Rydholm (1965).

production is even greater than ethanol since it is possible by bacterial fermentation methods to convert by-product sugars to acetic acid with a 100% theoretical yield. This compares with theoretical yields of only 51% from yeast fermentation of hexose sugars. The overall conversion of glucose to ethyl acetate would have a theoretical yield of 58.6% based on fermentation methods (Baierl et al. 1981).

Acidogenic fermentation also has additional advantages. Even with only very mild pretreatment such as dilute alkali swelling, all the major fractions of lignocellulose can be utilized for production of a single class of products. Additional advantages have been outlined by Data (1981).

Clearly, to be able to utilize pulp mill by-product chemicals and bio-conversion products as octane boosters, appropriate methods for separation of the chemicals from the waste product streams must be available. There are a number of possible alternatives for recovery of the organic acids from waste streams. These include: direct and azeotropic distillation, solvent extraction and esterification methods. The distillation methods have to date not shown promise because of high energy costs. However, the Suida process was developed based on azeotropic distillation (Biggs et al. 1961).

Solvent (liquid-liquid) extraction methods have been utilized for many years for recovery of organic acids and a number of processes are available (Biggs et al. 1961). New extractants like trioctylphosphine oxide (TOPO) and several amines have been found to remove acids efficiently from aqueous solutions (Data 1981). The Sonoco Products Co. has used solvent extraction for many years for recovery of acetic and formic acids from neutral sulfite semichemical waste black liquor (Biggs et al. 1961). The process consists of concentration of the liquor to 40-45% solids, acidification with sulfuric acid, and extraction of the acids with 2-butanone.

Probably the most commercially viable method of recovery is through esterification. Data (1981) described a sequential technique where the organic acids are first extracted from dilute aqueous solution with a suitable non-aqueous extractant. Alcohol is then added to the extract layer in the presence of acid catalysts; which results in esterification of the organic acids to esters and expulsion of the product water. The esters, which are now more volatile than the corresponding acids, can be easily separated from the extraction solvent which is regenerated for reuse.

The recovery technique which we have shown in the proposed scheme (Figure 1) is based on the adsorption-regeneration method developed by Baierl (Baierl et al. 1981; Baierl 1979). With this process the waste or product stream is passed over a bed of activated carbon which adsorbs the acids and the water passes through. The carbon bed is then treated with alcohol vapors and heated to yield organic esters at the catalytic sites on the adsorbent. The volatile organic esters are recovered from the vapor phase with regeneration of the column by the alcohol vapors. Baierl (1979) has also demonstrated that ethanol, methanol, acetic acid and furfural can be recovered on similar activated charcoal columns with low process energy. This process is under further development. Additional methods for production of fuels and chemicals are also under investigation as described below.

THERMOCHEMICAL FRACTIONATION AND LIQUEFACTION OF WOOD

To provide an additional source of pulping chemicals and synfuels, a scheme is under evaluation which separates and degrades wood hemicellulose, cellulose and lignin fractions. This reactor (TFL) is an integral part of the model shown in Figure 1. The sequence of proposed reaction steps for thermochemical fractionation and liquefaction of wood is shown in Figure 2.

Waste woody biomass or sludge is the raw material feedstock. Three stages are shown (Figure 2) which represent the sequential fractionation of aspen for subsequent conversion to fuels and chemicals. The conditions become harsher as the material moves from stage I (mildest) through Stage III (harshest).

The first stage (I) is a mild treatment for removal of the hemicellulose fraction. Moderate temperatures and pressures are utilized in this stage with the hemicellulose removal either acid or base assisted. The simultaneous isolation, decrystallization and partial saccharification (SIDS) of cellulose is the second stage (II) of the proposed sequence. The second stage reactions are performed at higher temperatures and pressures with the intent of obtaining a fermentable oil. The third and final stage (III) is designed to convert lignin to a phenolic oil. Critical factors in

EFFICIENT UTILIZATION OF WOODY BIOMASS

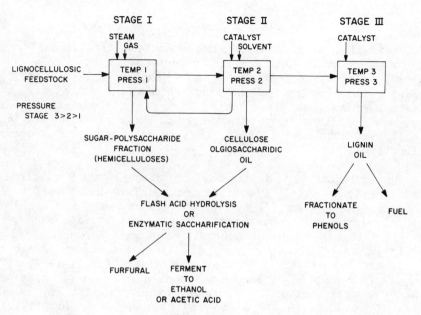

Figure 2. Proposed three stage biomass utilization scheme. Temperature and pressure are increased in the consecutive stages.

this stage are proper choice of catalyst and higher temperatures where lignin is more rapidly degraded. Further details of this process and recent results from trials are discussed in another publication (Young and Davis, 1982).

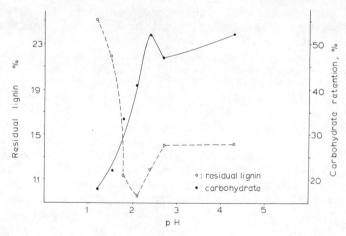

Figure 3. Effect of pH on lignin and carbohydrate content in ethanol-water pulping of aspen.

CELLULOSE

Production of cellulose pulps has been traditionally carried out by the kraft and sulfite processes. Organosolv systems, however, offer the potential for more complete utilization of woody biomass and recovery of pulping chemicals, especially through schemes such as that presented in Figure 1. A variety of solvents have been utilized for organosolv pulping of wood but aqueous alcohol systems appear to hold the strongest promise for commercial application.

The delignifying properties of alcohol-water mixtures were noted already in the 1930's (Kleinert and Tayenthal 1931, 1932) and Aronovsky and Gortner (1936) effectively pulped aspen at 158-186° in butyl alcohol-water mixtures. More recent studies have verified the viability of aqueous alcohol pulping (Katzen et al. 1980; Diebold et al. 1978; Myerly et al. 1981; Bowers and April 1977). Notable are the extensive studies by Kleinert (1971, 1974, 1975) on ethanol-water pulping of spruce and poplar. The use of catalysts to enhance delignification in aqueous ethanol pulping has been demonstrated by Sarkanen (1980) and by Paszner and Chang (1980). Two organosolv pulping systems which we consider compatible with our integrated scheme are the ethanol-water process and acetic acid pulping. An evaluation of aqueous ethanol pulping of aspen is given in this paper and an analysis of acetic acid pulping will follow in a subsequent paper.

Effect of Catalyst

Our initial studies on ethanol-water pulping of aspen chips demonstrated the need for a catalyst in batch-type cooks. We were not able to obtain satisfactory pulps without the aid of an additive. Continuous alcohol-water systems have apparently circumvented this problem (Kleinert 1971; Diebold et al. 1978). A series of experiments were therefore performed to determine the effect of a hydrochloric acid catalyst on delignification and carbohydrate retention.

Figure 3 shows the effect of initial pH on the above two dependent variables for aqueous ethanol pulping of aspen. Although there are certain reservations related to the measurement of pH in aqueous alcohol systems, estimates of the effect of the catalyst can be made on this basis. As the pH is decreased from 4 to about 2 (Figure 3), the delignification is enhanced, with a concomitant loss of carbohydrate. However, when the initial pH is reduced below pH 2 there is a decrease in the extent of delignification. This phenomenon is the result of acid catalyzed condensation of the phenolic lignin fragments onto the pulp at low pH. There is also extensive loss of carbohydrate at these low pH's due to cleavage of polysaccharide glycosidic bonds. A pH of between 2 and 3 appears to be optimum for the HCl catalyzed system.

Oxalic and salicylic acids gave poorer delignification when added to the ethanol-water pulping medium. The role of pH in catalyzed aqueous ethanol pulping is not easy to interpret since different catalysts at the same pH were found to result in different levels of residual lignin content. Also the "hot pH" can vary significantly depending on the system utilized.

Recently Sarkanen (1980) noted that aluminum chloride catalyzed ethanol-water pulping of several wood species at very low levels of addition (0.01 M). This catalyst was utilized throughout the rest of this investigation.

Effect of Liquor-to-Wood Ratio

The liquor-to-wood (L/W) ratio appears to exert a strong influence on delignification in batch-wise ethanol-water pulping of aspen. At a 5:1 L/W ratio it was not possible to obtain a satisfactory pulp at temperatures in range of 155-185°C with up to four hours cooking time. At 185°C the pulp product was dark and yields were low. Sarkanen (1980) reported that at least a 10:1 L/W ratio was necessary in aqueous ethanol pulping and Kleinert (1975) also suggested a high L/W ratio in similar systems.

A series of high L/W ratios were therefore evaluated to determine an optimum ratio for our system. As shown in Figure 4, a high L/W ratio has a significant effect on the extent of delignification. It appears that at least a ratio of 12.5:1 is necessary to achieve sufficient delignification; and, further improvements can be achieved at even higher L/W ratios. Apparently the high dilution is necessary to avoid recondensation of the cleaved lignin back onto the pulp in the acidic solution. The system proposed by Diebold et al. 1978 appears to avoid the lignin recondensation through continuous removal of the pulping liquor which contains the low molecular weight lignin fragments.

Under slightly different pulping conditions, which will be subsequently described, another important effect of dilution was noted. Aqueous ethanol pulping trials were performed at two very high L/W ratios (20:1 and 35:1) as shown in Table 2. It was found that the delignification was somewhat higher at the higher L/W ratio; but more significant, the carbohydrate retention was up to 4% greater (10% based on pulp) with greater dilution (35:1 L/W at same catalyst concentration). It was also noted that only small changes in extent of delignification and percent yield were realized at higher catalyst charges (Table 2) but the selectivity was improved by the dilution as noted above. An increased yield of this magnitude would be important in possible later commercial applications and warrants further investigation.

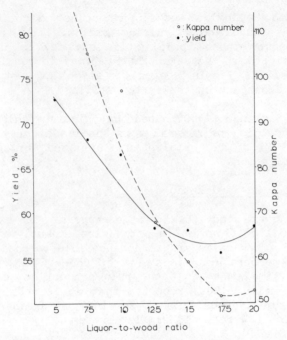

Figure 4. Percent yield and kappa number at different liquor-to-wood ratios.

Table 2. Dilution effect at various amounts of catalyst on residual lignin and carbohydrate retention.

$AlCl_3$ % of dry biomass	L/W 20:1		L/W 35:1	
	Residual Lignin %	Carbohydrate Retention %	Residual Lignin %	Carbohydrate Retention %
4.83	7.3	41.7	6.0	43.9
5.74	7.4	42.1	6.1	43.6
6.64	7.5	40.2	5.2	44.5
7.55	8.4	40.8	4.9	43.6
8.45	7.6	39.0	4.9	42.8

Cooking conditions: 60 minutes preconditioning at 100°C, plus 68 minutes programmed temperature rise. Maximum pulping temperature 165°C.

Effect of Preconditioning

Previous investigators have reported that the selectivity of delignification can also be improved by application of a heated preconditioning stage of 1-2 hours in ethanol-water pulping of wood (Sarkanen 1980). The effect of preconditioning was therefore further evaluated in this investigation. Several different preconditioning methods were tested and monitored on the basis of kappa number and percent yield.

The control experiment (C) was run as described in the experimental section, where the vessel was immersed in an oil bath at 165°C with 30 minutes allowed for the bath to attain equilibrium at this maximum pulping temperature. The vessel was then maintained at the maximum pulping temperature for one hour. A L/W ratio of 35:1 was utilized to insure that lignin recondensation reactions did not occur.

The preconditioning (PC) was performed by immersing the vessel in the bath at 100°C for specified time periods and then raising the temperature to the maximum pulping temperature (165°C). This programmed temperature rise (PTR) of 68 minutes was the time necessary for the oil bath to attain equilibrium at the temperature maximum. Thus, the results shown in Table 3 depict the controls, preconditioning PTR and isothermal preconditioning (PC) with the PTR. For comparison, an additional two hour and 8 minute cook (No. 1) was performed to simulate the time of the PTR and one hour pulping in the preconditioned pulping trial.

The control cook (C) resulted in a higher kappa number compared to all preconditioned cooks (Table 3). Thus, the preconditioning appears to have a beneficial effect on the extent of delignification. However, the prolonged isothermal preconditioning does not offer a further significant advantage over simply using the gradual rise to pulping temperature (PTR, with no isothermal preconditioning).

To determine whether or not the enhanced delignification was due to the longer contact time with the pulping liquor, an additional cook was performed using the standard procedure (30 minutes equilibration at maximum pulping temperature) but prolonging the time of pulping at the maximum temperature. The results from the cook, designated as LC in Table 3, clearly demonstrate that the prolonged cooking at maximum pulping temperature is definitely inferior. Thus, under the conditions employed in this investigation, a preconditioning <u>via</u> gradual temperature rise (PTR) to pulping temperature appears to give the optimum delignification.

To further explore the effect of a PTR to maximum pulping temperature, a series of additional pulping trials were performed, the results of which are shown in Table 4. The preconditioning was started at three different temperatures, 50°, 100° and 150°, and the time to reach maximum pulping temperature was recorded. The wood chips were then cooked at the maximum pulping temperature (165°) for one hour. The results from the cooks indicate that there is probably no advantage to using temperatures below 100° to initiate the preconditioning. The use of initial temperature of 150° also appears to be detrimental to the delignification, apparently due to a too early rise to maximum pulping temperature. Thus, from these results, optimum PTR pulping is an initial temperature of 100° and about a one hour rise to maximum pulping temperature. Presumably, these conditions give the best uniform impregnation of the chips with the cooking liquor prior to the extensive hydrolytic effects which occur at pulping temperatures.

The Effect of Pulping Temperature

The maximum pulping temperature has also been found to be a very significant variable in the ethanol-water system. The data shown in Figure 5 were obtained from cooks with a L/W ratio of 20:1, 0.01 M $AlCl_3$ catalyst, and 60 minutes isothermal preconditioning. The delignification increases dramatically as the maximum pulping temperature is raised, particularly above 165°. The greatest enhancement of delignification is realized between 155° and 165°; however, there is also rapid dissolution of the polysaccharide component as well.

To evaluate the major effects noted in this experiment, a factorial analysis was applied, modified with split plot design for the preconditioning treatments. This allowed a combined analysis of the delignification based on the three variables of temperature (three levels), L/W ratio (two levels) and preconditioning time (four levels). The results of the analysis are shown in Table 5, which gives an analysis of variance of kappa number for ethanol-water pulping of aspen. The magnitude of the F value indicates the relative significance of the independent variables.

Obviously, the temperature has the most significant effect of the three variables, followed by the effect of L/W ratio and of much less consequence is the effect of isothermal preconditioning. In addition, the interaction of temperature and preconditioning was found to be significant. An identical analysis was carried out for total yield and a similar level of significance was noted for the respective variables. The corresponding selectivity curves for ethanol-water pulping at different temperatures, L/W ratios and preconditioning (isothermal) conditions are summarized in Figure 6.

Viscosity

The determination of pulp viscosity can be used to monitor the degree of polymerization (DP) of the pulps and concomitant strength losses in the pulping trials. The effect of the maximum pulping temperature on pulp viscosity is shown in Table 6. For all the conditions shown a low pulp viscosity was obtained, which suggests pulps of low strength. At the highest pulping temperature (185°C) the estimated cellulose DP was only 400. In conventional sulfite pulping the minimum DP is usually about 1400 (Rhydholm 1960).

Myerly et al. 1981 also reported a low DP (540) for ethanol-water pulps. The lower end of the range for dissolving pulp applications is near a DP of 500. However, neither Myerly's nor our investigation was designed for optimization of strength properties.

Experimental

Aspen (Populus tremuloides Michx.) chips (8% moisture content) were pulped in tubular stainless steel vessels of 100 ml capacity. The chips along with the ethanol-water pulping medium (50:50 by volume) and a known weight of catalyst (mainly $AlCl_3$) were introduced directly into the vessel. The vessel was then brought to reaction temperature by immersing in a thermostated polyethylene glycol bath for 30 minutes. In the cases where preconditioning was used, the temperature of the oil bath was raised from the pre-conditioning temperature to pulping temperature after the desired preconditioning time. The cooking time at the maximum pulping temperature was one hour in most cases.

After the completion of the cook, the vessel was removed from the bath and immediately quenched. The chips were mechanically defiberized in a Waring blender for 20 to 30 seconds. The pulps were then washed, dried overnight at 60°C in vacuo. The yield of the cook was calculated as a percent of oven dry wood. The lignin content of the pulp was monitored by kappa number (Tappi Standard UM-236) and the chlorited pulps were subjected to viscosity determination (Tappi Standard 230 so-76).

Residual lignin was calculated by multiplying the kappa number by 0.15 and the carbohydrate retention was obtained from the difference in total yield and residual lignin.

Conclusions - Aqueous Ethanol Pulping

The ethanol-water system is highly dependent on maximum pulping temperature and the liquor-to-wood ratio. A high liquor-to-wood ratio, close to 20 to 1, is preferrable to achieve a low kappa number with reasonable carbohydrate retention. A dilute pulping medium, a two stage cook, or a continuous recycling of the cooking liquor could all be considered as viable alternatives to avoid the secondary lignin condensation reactions.

Table 3. The effect of preconditioning at 100°C in Organosolv Pulping of aspen chips.

Cook No.	Total Cooking Time* PC	+ PTR	+ Cooking	Kappa No.	Residual Lignin %	Carbohydrate Retention %	Total Yield %
	0	30	30	41.4	6.2	46.7	52.9
C	0	30	60	35.3	5.3	44.1	49.4
LC	0	30	98	45.4	6.8	40.4	47.2
1	0	68	60	32.6	4.9	42.8	47.7
2	10	68	60	31.8	4.8	41.2	46.0
3	15	68	60	33.4	5.0	41.6	46.6
4	30	68	60	33.4	5.0	43.0	48.0
5	60	68	60	32.6	4.9	42.8	47.7
6	90	68	60	30.2	4.5	42.4	47.7
7	120	68	60	31.5	4.7	41.7	31.5

* PC: preconditioning time in minutes, PTR: programmed temperature rise from 100° to 165°C which took 68 minutes, C: control cook, no preconditioning, LC: long cooking time, no preconditioning.

Table 4. The effect of preconditioning on delignification in Organosolv pulping.

Temp. °C	Raising Time (minute)	Residual lignin (%)	Carbohydrate retention (%)	Difference in lignin (%)
Control	0	5.3	44.1	--
50-165	103	5.9	43.1	+11.3
100-165	68	4.9	42.8	- 7.5
150-165	23	5.8	44.5	+ 9.4

Cooking conditions: L/W ratio 35:1, and 60 minutes preconditioning.

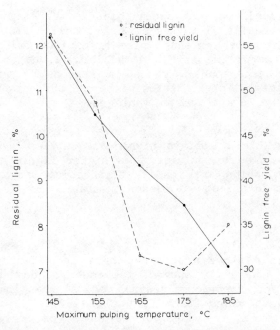

Figure 5. Effect of temperature on hydrolysis of lignin and polysaccharides in acid catalyzed ethanol-water pulping of aspen (l/w ratio 20:1, 60 minutes preconditioning at 100·C.

Table 5. Analysis of variance for kappa number in organosolv pulping.

Source of Variance	df	SS	MS	F
Mean	1	163,617.0		
Temperature (T)	2	6,187.4	3,093.7	40.7**
L/W ratio (R)	1	2,393.2	2,393.2	31.5**
T x R	2	275.9	138.0	1.8
Error	6	455.9	76.0	
Preconditioning (P)	3	158.9	53.0	5.4**
T x P	6	208.9	34.8	3.5*
R x P	3	28.8	9.6	1.0
T x R x P	6	13.5	2.3	0.2
Error	18	177.3	9.9	

Preconditioning of the aspen chips in the ethanol-water liquor appears to be beneficial to the extent of delignification. Probably, the best approach is a gradual rise to maximum pulping temperature from 100° for about one hour. An additional one hour cook at maximum pulping temperature produced a reasonable organosolv pulp. Prolonged isothermal preconditioning beyond the one hour PTR did not offer any further significant advantages.

Although the organosolv pulps were of rather low viscosity, it should be possible to obtain higher viscosities and improved pulp strength properties by pulping to a higher lignin content. Since the organosolv pulps are readily bleachable, the lignin could be easily removed in subsequent bleaching stages.

ADHESIVES

In recent history, adhesives have been derived mainly from petrochemical sources. However, the costs of conventional phenol-formaldehyde and urea-formaldehyde resins have increased dramatically in recent years (Rammon et al. 1982). Therefore, we have emphasized research on adhesives derived from renewable lignocellulosic resources (Kelley et al. 1982, 1982a). Our work on wood surface bonding has been published elsewhere (Young et al. 1982, and above cited references) while our current efforts are on conversion of pulp and paper mill wastes and by-products into durable wood adhesives.

Tremendous quantities of lignocellulosic materials are annually disposed of in the form of pulp and paper mill sludges. Approximately 0.04 metric tons of sludge are produced for every ton of pulp and paper. Thus about 2.5 million metric tons of sludge are produced each year in the United States alone (Young 1982).

Table 7 shows the composition of two primary sludges from a paper mill in Alabama. There is a high content of lignin and, based on the sugar analysis, there is also a considerable amount of cellulose contained in the waste sludge, probably as fiber fragments. Currently we are evaluating methods of derivative formation of both the cellulose and the lignin components of the sludge. The derivative can then be water or solvent extracted for subsequent application in adhesive formulations or other products.

Lignin from waste pulping liquor also has potential adhesive applications. Although conventional kraft lignin has not proven viable for adhesives because of extensive condensation, lignosulfonates from sulfite cooks offer good adhesive properties (Forss and Fuhrmann 1976; Shen and Fung 1979). We anticipate that the organosolv lignins, produced from the integrated scheme shown in Figure 1, will prove even more suitable for adhesives. We feel we

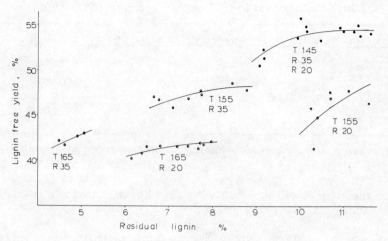

Figure 6. Selectivity curves for organosolv pulping under different conditions (T: maximum pulping temperature, R: liquor-to-wood ratio.

Table 6. The effect of maximum pulping temperature on pulp viscosity.

Temperature	Kappa No.	η(cps)	DP (approx.)
100-145	75.7	9.0	850
100-155	74.2	8.25	825
100-165	51.9	3.7	425
100-175	49.0	3.5	400
100-185	63.80	2.0	< 400

* Cooked with a PTR, one hour at maximum pulping temperature, L/W ratio = 20.

Table 7. Composition of primary paper mill sludges.

Sample	Ash	Lignin	Total Carbohydrate	Glu	Xyl	Gal	Ara	Man
1	33.6	36.5	29.8	63.8	11.2	5.0	1.8	18.2
2	54.2	15.9	28.1	69.3	10.2	3.1	1.2	16.4
3	25.7	33.3	39.5	79.0	13.0	0.0	0.2	7.9

* Given as percentages; sugars as percentage of total carbohydrate; total solids approximately 20%.

can enhance the lignin reactivity by known chemical modification methods (Young et al. 1982; Rammon et al. 1982).

The use of sugars, sugar degradation products and polysaccharides in adhesive formulations is also under investigation. The results shown in Table 8 demonstrate that water resistant adhesives can be produced with just carbohydrate material. The strongest adhesives were obtained when both a sugar and a polysaccharide were cooked in acidic solution for about 15 minutes. Apparently the polysaccharide is necessary for desirable viscoelastic properties and the sugar for production of furfural and phenol compounds, both of which can polymerize to resins. Similar results were obtained by Stofko (1980).

Table 8. Carbohydrate adhesives.

Formulation*		Untreated	Activated	
			Sodium Hydroxide	Sulfuric Acid
Dextrose	Dry	X	X	X
Starch	Wet	-	-	X
Sucrose	Dry	X	X	X
Starch	Wet	-	-	X
Dextrose	Dry	-	-	X
Sucrose	Wet	-	-	X
Dextrose	Dry	-	-	X
Lactose	Wet	-	-	-
Dextrose	Dry	X	X	X
Acacia Gum	Wet	-	-	-
Sucrose	Dry	-	-	X
Oxalic Acid	Wet	-	-	X

X signifies a strong bond was formed and retained.

* 25% aqueous solution in 1N base or acid, heated for 15 minutes at 100 degrees; the wood was activated by dipping the surfaces in 1 N acid or base and heating for one hour at 100 degrees.

Thus, we envision the development of a strong, water resistant adhesive with a formulation based on the following:

a) sludge derived lignocellulosic derivatives
b) sugar and polysaccharide dehydration products
c) modified organosolv lignins

ACKNOWLEDGEMENTS

The assistance of Eric Young with the adhesives evaluation is gratefully acknowledged. The pulping research was supported by the College of Agriculture and Life Sciences and the Graduate School, University of Wisconsin-Madison. The financial support of the MUCIA (Indonesia) program was also greatly appreciated.

REFERENCES

Aronovsky, S.I. and R.A. Gortner. 1936. Ind. Eng. Chem, 28, 1270.
Baierl, K. 1979. Adsorber-Regenerator, U.S. Patent No. 4, 155,849.
Baierl, K., R.A. Young, N. Sell and A. Goldsby. 1981. Production of Synfuels from Waste Biomass Seminars, University of Wisconsin-Green Bay, March 6 and October 6, Green Bay, Wisconsin.
Biggs, W.A., J.T. Wise, W.R. Cook, W.H. Baxley, J.D. Robertson and J.E. Copenhaven. 1961. The Commercial Production of Acetic and Formic Acids from NSSC Black Liquor, Tappi, 44(6), 385.
Bowers, G.H. and G.C. April, 1977. Aqueous n-Butanol Delignification of Southern Yellow Pine, Tappi, 60(8), 102.
Data, R. 1981. Paper presented at the Third Symposium on Biotechnology in Energy Production and Conservation, May 12-15, Gatlinburg, Tennessee.
Diebold, V.B., W.F. Cowan and J.K. Walsh. 1978. Solvent Pulping Process, U.S. Patent 4,100,016.
Forss, K. and A. Fuhrmann. 1976. KARATEX the Lignin Based Adhesive for Plywood, Particleboard and Fiberboard. Paperi Puu, 11, 817.
Katzen, R., R. Fredrickson and B.F. Bush. 1980. The Alcohol Pulping and Recovery Process, Chem. Eng. Progress, February, p.62.
Kelley, S.S., R.A. Young, R.M. Rammon and R.H. Gillespie. 1982. Bond Formation by Wood Surface Reactions: Part III. Parameters Affecting the Bond Strength of Solid Wood Panel, Forest Prod. J., in press.
Kelley, S.S., R.A. Young, R.M. Rammon and R.H. Gillespie. 1982a. Bond Formation by Wood Surface Reactions: Part IV. Analysis of Furfuryl Alcohol, Tannin and Maleic Anhyride Bridging Agents, J. Wood Chem. Technol., 2(3),317.
Kleinert, T. 1974. Organosolv Pulping with Aqueous Alcohol, Tappi, 57(8), 99.
Kleinert, T. 1975. Ethanol-Water Deliginification of Wood-Rate Constants and Activation Energy, Tappi, 58(8), 170.

Kleinert, T. 1971. U.S. Patent 2,037,001.
Kleinert, T. and K. Tayenthal. 1931. Z. Angew. Chem. 44, 788.
Kleinert, T. and K. Tayenthal, 1932. U.S. Patent 1,856,567.
Myerly, R.C., M.O. Nicholson, R. Katzen and J.M. Taylor. 1981. The Forest Refinery, Chemtech, 11, 186.
Paszner, L. and P.C. Chang. 1981. Canadian Patent 1,100,266.
Rammon, R.M., S.S. Kelley, R.A. Young and R.A. Gillespie. 1982. Bond Formation by Wood Surface Reactions: Part II. Chemical Mechanisms of Nitric Acid Activation, J. Adhesion, in press.
Rhydholm, S.A. 1965. Pulping Processes, Wiley-Interscience, New York.
Sarkanen, K.V. 1980. Acid Catalyzed Delignification of Lignocellulosics in Organic Solvents, In: Progress in Biomass Conversion, Vol, 2, K.V. Sarkanen and D.A. Tillman, ed., Academic Press, New York; and private communication.
Shen, K.C. and D.P.C. Fung. 1979. Aspen Particleboards bonded with Spent Sulfite Liquor Powder Treated with Sulfuric Acid, For. Prod. J., 29(3), 34.
Stofko, J. 1980. Bonding of Solid Lignocellulosic Material, U.S. Patent, 4,183,997.
Young, R.A. Ed. 1982. Introduction to Forest Science, John Wiley & Sons, New York.
Young, R.A., R.M. Rammon, S.S. Kelley and R.H. Gillespie. 1982. Bond Formation by Wood Surface Reactions: Part I. Surface Analysis by ESCA, Wood Sci., 14,110.
Young, R.A. and J.L. Davis. 1982. Thermochemical Fractionation and Liquefaction of Wood, Proceedings of the Conference on Fundamentals of Thermochemical Biomass Conversion, October 18-22, Estes Park, Colorado.

METHANOL FROM WOOD, A STATE OF THE ART REVIEW

A. A. C. M. Beenackers and W. P. M. Van Swaaij

Twente University of Technology
Enschede
The Netherlands

INTRODUCTION

The Commission of the European Communities started its pilot plant project "Synthetic Fuel from Wood" early this year. Within this project, four demonstration plants producing 10-60 tons/day of synthesis gas from wood will be constructed and operated before mid-1984.

These plants are all different in concept but aim to produce a similar product: a clean synthesis gas, virtually free of hydrocarbons, low in nitrogen content and suitable as a feed to methanol synthesis plants after passing through conventional shifting and acid gas removal units. See Palz and Grassi (1982) and Beenackers and Van Swaaij (1982) for a description.

This paper reviews the mid-'82 state of the art of producing methanol from wood with the restriction that it is based on open literature only. Furthermore, we particularly focus on the unproven step in this process, the gasification of biomass to a synthesis gas that is low in hydrocarbons and nitrogen. But we start with a short description of the methanol technology because that dictates the specifications for the synthesis gas.

METHANOL PRODUCTION

Already during the previous century methanol was produced from wood though the process was different (dry distillation) and the yields substantially lower than expected for the new "second generation" routes which are the subject of this paper. Present world consumption of methanol is 13×10^6 tons/year (1981) all from synthesis gas which is mainly produced from natural gas though coal-based plants are known to operate in South Africa. Large scale coal-based methanol plants got much attention during the past decade, but actual developments possibly will move towards methanol production at remote natural gas fields rather than to coal conversion (Fifth Int'l. Alcohol Fuel Tech Symp. 1982).

A modern large scale methanol plant may have a capacity of a 1000 tons/day of methanol and is carefully energy integrated with the synthesis gas production plant. Most of the methanol presently produced is used as a chemical ($\pm$ 50% is converted to formaldehyde (Kennedy and Shanks 1981) but it may be introduced on a large scale as an internal combustion engine fuel or fuel extender and possibly also as a peakshaving turbine fuel in electricity generation (e.g. at co-generation of fuel gas and methanol). Furthermore it can be converted to gasoline fractions via the zeolite catalyst process of Mobil Oil.

A few processes are available for methanol production but the processes of ICI (Reed 1980) and Lurgi are most commonly used. These processes were developed during the sixties and early seventies. Both operate at temperatures between 200°C and 300°C, use a copper based catalyst and are low pressure processes typically operating between 50-100 bar thus allowing for centrifugal compressors. Figure 1 gives a schematic view of the reactor section. Because methanol production via:

$$CO + 2H_2 \rightleftharpoons CH_3OH - 90.77 \text{ kJ/mol}$$

with usually some co-production via:

$$CO_2 + 3H_2 \rightleftharpoons CH_3OH + H_2O - 49.52 \text{ kJ/mol}$$

is overall strongly exothermic, heat removal from the conversion/recycle loop and efficient utilization of this heat and of the bleed-stream gas are crucial factors in the process economics. The processes of ICI and Lurgi differ in a few details of which the heat removal from the reactor by direct raising of steam (Lurgi) or by the cold shot technique (ICI) are the most significant ones. This is illustrated in Figures 2 and 3.

One of the possible schemes for wood (or other biomass) based methanol production is given in a simplified form in Figure 4. It is based on an oxygen blown pressurized gasifier, but of course also other types producing a suitable synthesis gas can be used. A major problem with biomass is that, due to the dispersed character of the supply, at least in Europe, wood-based gasifier plants will consequently be small in capacity. A 1000-ton/day plant would require a production area of 1500 km^2. The economy of scale of the whole

METHANOL FROM WOOD, STATE OF THE ART REVIEW

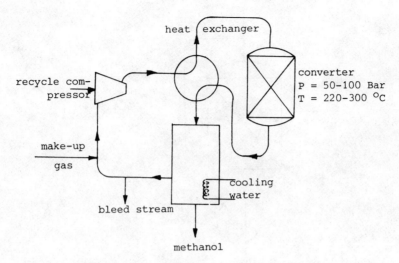

Figure 1. Methanol synthesis loop.

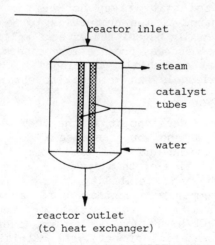

Figure 2. Simplified scheme of the LURGI methanol reactor system.

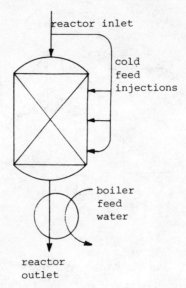

Figure 3. Simplified scheme of a possible methanol synthesis from wood.

methanol complex is relatively important (Beenackers and Van Swaaij 1982) as is illustrated in Figure 5. This is mainly caused by the methanol synthesis plant and the purification section. Therefore addition of the crude synthesis gas to the feedstock of a large scale (e.g. natural gas based) methanol complex could relax the large scale requirement for the biomass gasification. On the other hand, the methanol plant itself could be improved and/or adapted to the wood gasifier or, generally, to small scale operation. First, it should be realized that it may not be necessary to produce a chemical grade methanol if the product is to be used as a fuel and the purification step may be omitted or be simplified. New methanol processes to be developed may produce more economically a fuel grade product. IFP has even specially developed a process to produce a mixture of higher alcohols and methanol to obtain an improved gasoline extender.

A relatively new development is the Chem Systems (Frank 1980) three phase reactor mthanol process (see Figure 6). Here a liquid phase is used to suspend and cool the catalyst. There is also a catalyst entrained version of this system. It is claimed that these processes have a higher flexibility towards fluctuations in the feedstock, better temperature control, a higher conversion per pass (reduced recycle costs) and the possibility of accepting a feed with a wide variety of CO/H_2 ratios. Further, factors like catalyst stability, life expectation of catalyst, attrition/erosion resistance and process economics have to be demonstrated.

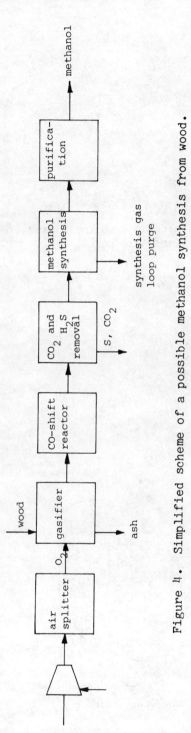

Figure 4. Simplified scheme of a possible methanol synthesis from wood.

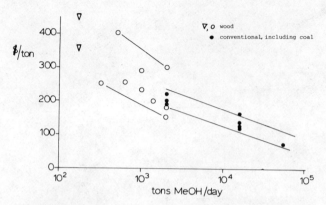

Figure 5. Estimates of methanol production costs (1980 $/ton). From Beenackers and Van Swaaij 1982.

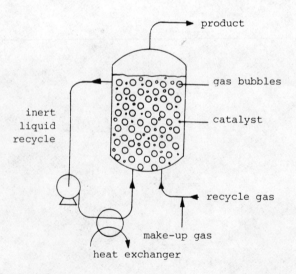

Figure 6. Chem Systems methanol synthesis.

Also by integration of the gasification process with the methanol synthesis process, e.g. via the bleed stream or the unconverted synthesis gas stream, improvements could be obtained especially if a higher conversion per pass in the methanol reactor can be realized. The synthesis gas purification steps can hardly be simplified. Although wood has a very low sulphur content, sulphur still has to be removed with the catalyst presently in use. This requires a separate sulphur removal step although in some cases this removal step can be kept very simple. The amount of CO-shift required (if any) and the need of CO_2 removal depend much on the overall scheme of the process and the amount of CO_2 that can be tolerated in the methanol synthesis.

The economy of methanol via biomass gasification has been the subject of a number of desk studies (Beenackers & Van Swaaij 1982). Generally it shows methanol production costs, by this route, still to be on the high side both due to the smaller scale of operation of a realistic biomass gasifier as compared to a natural gas based plant and also because of the possible lower energy efficiency. These two factors are often not yet balanced by the difference in feedstock price.

The factors mentioned above together with a relative change in feedstock prices could alter this situation in the future. We now turn to the gasification processes for synthesis gas from wood.

AUTOTHERMAL AND ALLOTHERMAL GASIFICATION ROUTES

Gasification of wood is a complex process often including pyrolysis and combustion, either parallel or more or less in series (see Table 1) (Reed 1980).

Gasification processes can be operated in an autothermal or allothermal way. It can be made overall autothermic by combusting part of the feed. Several methods have been proposed to operate autothermally (see Table 2). Methods to add the heat required by other processes have been proposed as well (see Table 3). The advantage of adding heat is a higher yield of CO and H_2 per kg of wood (see Table 15) but locations where external heat is cheaper than the heat from the biomass will remain rare, if anywhere. In the following we will discuss various methods in more detail.

MOVING BED OXYGEN GASIFICATION

Downdraft

Downdraft gasifiers have been known since early in this century. They were redeveloped with CEC funding over the past five years and are presently commercially available for air gasification at a scale of 300 kg wood/hr (C.E.C. 1982). The main advantage is that the gas is low in tars and methane if the gasifier is properly designed ($\pm$ mg/$\pm$ 500 mg tar/mm^3 m^3 and $\leq$ 1 percent methane). Test runs on a small scale have been undertaken by SERI (Wan et al. 1982). They evaluated this process for methanol production (see Table 20). Though the results of Wan's analysis are relatively favourable for this route, we doubt whether the standard co-current gasifier can be scaled up to capacities relevant for methanol production without losing its most advantageous characteristics i.e. being low in tar and methane yield. This is because throat design becomes increasingly critical with increase in capacity. We consider the concept of Novelerg to offer greater prospects in this respect (Dubois 1982). Here, pyrolysis gases are sucked from the top of the bed, burned externally with the oxygen feed, and introduced into the reduction zone of the gasifier. This way, combustion of tar is well controlled and consequently scale up will be less critical. The status of this process is as follows: an experimental unit operating up to 20 bar with a designed capacity of 500 kg/hr at 1 bar was built, but experimental results have not been published as yet.

Cross flow

Gasification of biomass with oxygen/steam mixtures in a cross flow moving bed has been investigated by Foster Wheeler in a pressurized reactor (up to 30 bar) and a capacity of 40 kg/hr (Wilson 1981). So far, no experimental results have been published.

Counter Current Flow

This type of reactor is commercially proven for coal (Hiller 1975) up to pressures of 35 bar and is in the demonstration stage for municipal solid waste. Municipal solid waste has a high ash content and may cause sintering problems in many gasifiers.

These oxygen or oxygen enriched processes (SFW-Funk process) allow for slagging operation and therefore can handle MSW without expensive pretreatment. Success has been limited so far. The Purox unit of 6.5 ton/hr was started up in 1979, South Charleston, U.S.A., and such large units have not been built since then, probably because of the price of oxygen. There is an interesting experiment with an atmospheric gasification system of 520 kg wood/hr in the SFW-Funk municipal waste process plant in Saarbrucken (Hummelsiep 1981). The data are shown in Table 5.

Data with pure oxygen have not been published. A probable disadvantage of this process for methanol is the large content of tar and methane in the synthesis gas. Technically, tar removal is not a problem. After having passed a wet scrubber and an electrofilter, dust and tar content are reportedly below 10 mg/nm^3.

Table 1. Biomass thermal conversion reactions, (idealized), (Reed 1980).

		ΔH_r (kJ/mol)	Process
$C_6H_{10}O_5$	$\rightarrow 6\ C + 5\ H_2O\ (g)$	-462.8	Charring
"	$\rightarrow 0.8\ C_6H_8O + 1.8\ H_2O\ (g) + 1.2\ CO_2$	-336.0	Pyrolysis
"	$\rightarrow 2\ C_2H_4 + 2\ CO_2 + H_2O\ (g)$	+25.94	Fast pyrolysis
" $+ \frac{1}{2}O_2$	$\rightarrow 6\ CO + 5\ H_2$	+299.16	Partial oxidation
" $+ 6O_2$	$\rightarrow 6\ CO_2 + 5\ H_2O\ (g)$	-2832.64	Combustion
" $+ 6H_2$	$\rightarrow 6"CH_2" + 5\ H_2O\ (g)$	-786.6	Hydrogenation

Table 2. Autothermal gasification routes for synthesis gas from wood.

A.	Gasification + partial combustion in one reactor with oxygen or oxygen steam mixtures	Moving beds Fluidized beds Entrained gasifiers	Novelerg Creusot-Loire MINO Foster Wheeler
B.	Gasification with steam while combusting part of the biomass separately with air for heat	Double fluidized beds (DFB)	Battelle
C.	Gasification with chemically bound oxygen and combusting part of the biomass separately for heat	Double fluidized beds	John Brown/ Wellmann

Table 3. Allothermal gasification routes for synthesis gas from wood.

Gasification by	Process Type	Processes
Steam + electricity	Downdraft moving bed + recycle	Novelerg CESP
Steam + solar heat	Entrained flow	Lab status
Steam + indirect heat supply	Always fluid beds	Werner

surized reactor (up to 30 bar) and a capacity of 40 kg/hr (Wilson 1981). So far, no experimental results have been published.

Counter Current Flow

This type of reactor is commercially proven for coal (Hiller 1975) up to pressures of 35 bar and is in the demonstration stage for municipal solid waste. Municipal solid waste has a high ash content and may cause sintering problems in many gasifiers.

These oxygen or oxygen enriched processes (SFW-Funk process) allow for slagging operation and therefore can handle MSW without expensive pretreatment. Success has been limited so far. The Purox unit of 6.5 ton/hr was started up in 1979, South Charleston, U.S.A., and such large units have not been built since then, probably because of the price of oxygen. There is an interesting experiment with an atmospheric gasification system of 520 kg wood/hr in the SFW-Funk municipal waste process plant in Saarbrucken (Hummelsiep 1981). The data are shown in Table 5.

Data with pure oxygen have not been published. A probable disadvantage of this process for methanol is the large content of tar and methane in the synthesis gas. Technically, tar removal is not a problem. After having passed a wet scrubber and an electrofilter, dust and tar content are reportedly below 10 mg/nm^3.

In Table 6 we summarize our knowledge on the status of oxygen gasification of biomass in moving beds from open literature.

Table 4. MSW oxygen updraft gasification.

Process	Capacity (kg/hr)	Reference
SFW Funk process	800	Hummelsiep 1981
Purox	6500	Fischer 1976
Simplex	80	Arbo 1980

Table 5. Mass balance SFW-funk process on wood.

Input: 516 kg wood/hr

100 m^3 air/hr + 30 m^3 oxygen/hr (200°C)

Output (Vol %)
CO: 32 CO_2: 21
H_2: 27 N_2: 15
CH_4 = 5

Table 6. Status of oxygen gasification of wood in moving beds.

	Capacity (kg wood/hr)	Pressure (bar)	Company	Status
Cocurrent	500	9	Novelerg	Constructed
Cross current	40	30	Foster Wheeler	"
Countercurrent	500	1	SFW-Funk	Trial with enriched air only

FLUIDIZED BED OXYGEN GASIFICATION

This is a simple and flexible omnivorous process which, if operated at 800-900°C, produces a low tar but still methane rich synthesis gas. Much higher temperatures are not possible because of ash sintering and melting problems.

Creusot-Loire (Chrysostome and Lemasle 1982) published experimental results from a 40 cm bed at 800°C and capacity of 100 kg dry wood/hr (see Table 7). A plant for 2500 kg/hr is presently designed. The (Swedish) Royal Institute of Technology in cooperation with Stadsvik Energi Teknik develops a similar process (MINO, see Table 7) (Lindman 1981), though at a smaller scale. Because of its high methane content, the product gas must be reformed to CO and H_2. In the MINO process the gas is cleaned of dust in high temperature filters and then catalytically reacted with oxygen by 900-110°C. High temperature filtering and possible catalyst deactivation by sulphur or dust are critical steps that remain to be proven.

Creusot-Loire (Chrysostome and Lemasle 1982) plans to eliminate CH_4 thermally by introducing additional oxygen above the bed and raise the temperature to 1300°C. Critical points here are gas mixing and soot formation. The reported efficiencies (see Table 7) may suggest a slight advantage for MINO but due to the unproven steps in the processes it is too early to judge.

CIRCULATING (FAST) FLUIDIZED BED OXYGEN GASIFICATION

This is a fluidized bed at extreme high gas velocities causing a high solids entrainment. With a cyclone, the bed material is separated from the product gas and continuously recirculated to the bed. It is claimed to have less ash melting and agglomeration problems than the classical fluidized beds thus allowing for higher operating temperatures. This is important if due to the higher operating temperature the methane content can be reduced so that a separate reformer step can be eliminated.

Table 7. Oxygen-wood gasification in a fluid bed (Chrysostome and Lemasle 1982; Lindman 1981; Nitschke 1981; Gravel 1982.

A. Status

Company		P bar	Capacity kg dry wood/hr	Moisture Wt %	T°C	Status
Creusot-Loire	1	1	100	20-40	850	In oper.
" "	2	10-30	2500	20-40	800-1000	Under constr.
MINO	1	30	10	10	750	
"	2	10-30	300	50	700-850	Under constr.
Inter-Uhde (HTW)		10	11000	30	950	Design study
Biosyn		"High"	10000			Prelim. eng. study

B. Reported dry gas composition

		CO	CO_2	H_2	CH_4	N_2	kg O_2/kg wood
Creusot-Loire	1	35.5	30.3	18.7	10	1.8	0.29
After methane converter					<0.5		0.60
MINO	1	20	30.2	32.5	17	0.3	0.17
After methane converter		34.7	21.5	42.2	0.3	0.3	0.44

C. Reported methanol from biomass efficiency

	Creusot-Loire	MINO
kg MeOH/kg dry wood	0.42	0.56
kg O_2 /kg dry wood	0.6	0.44
Thermal eff (%)	53	54

Lurgi now is executing a demonstration program for a capacity of 500 kg wood/hr at a temperature up to 1100°C. Lurgi anticipates a methane content of less than 1 percent and 0.4 kg methanol per kg wood with an oxygen consumption of 0.36 kg per kg wood only (Lindner and Reimert 1982; Lindner 1981), but experimental results will have to prove this; they are expected within six months.

Entrained Bed Oxygen Gasification

This process is semi-commercially available for coal: Texaco (pressurized with a slurry feed) and Koppers-Krupp (atmospheric with a dry feed in a pilot stage for pressurized dry powder coal gasification). It is probably not suitable for wood because of the high pre-treatment costs (<8% H_2O, d_p <0.1 mm) (Lindman 1981) while we expect it to be expensive to operate at a relatively small scale, typically for wood gasification (<500 t/d). The product can be virtually free of hydrocarbons, though a Koppers-Totzek run on peat (2.5 tons/hr) showed 1.5 percent of methane due to the spread in particle diameter, and 5.7 percent inerts in total. Because of the high H and O content of wood the Texaco process is unsuitable for wood, but of course can be used for char and tars from a pyrolysis unit. This method has been suggested by Foster Wheeler (1982).

DOUBLE FLUIDIZED BED STEAM GASIFICATION

This sytem is more or less comparable with the classical double fluidized bed (catalytic) oil cracking process.

Biomass is fed to a steam fluidized bed. Here pyrolysis and (endothermal) steam gasification take place at typically 800-850°C. Fluidizing carrier (sand or aluminaoxide) and char are transported to a second fluidized bed where the remaining char is exothermally oxidized at typically 900-1000°C. The return flow of the hot carrier to the gasifier provides the heat for the gasification.

This system has been strongly advocated now for over ten years already because of its obvious advantages (Bailie 1981):

- low capital needs because no oxygen plant is necessary;
- low operating expenses because no oxygen is used;
- the gas is low in CO_2 because combustion and gasification are separated, but the commercial development for MSW has been slow and for wood, not yet started.

Initial results for wood have been presented recently by Battelle Columbus for a scale of ± 100 kg/hr only (Feldmann 1981).

Table 8. Municipal solid waste double fluidized beds.

Start-up	Capacity (ton/day)	Company
1976	40	Tsukishima (Kagayama et al. 1980)
1980	100	Ebara (Bailie 1981)
1981 (?)	450	Tsukishima (Kagayama et al. 1980)

Table 8a. Battelle double fluidized bed wood gasifier.

Diameter gasifier	0.15 m
Length gasifier	1.8 m
Wood chips diameter	0.6-3.8 cm
Load	6350 kg/hr m^2
Superficial gas velocity gasifier	6 m/s
Superficial gas velocity combustor	0.6 m/s
Temperature gasifier and combustor	870 and 1040 °C
Entrained phase density	65 kg.m^3

Here two solid phases are present in the gasifier: coarse heavy particles, acting as a stationary fluid bed and a fine particulate phase heat carrier circulating with the char over the top as in a riser or a fast fluidized bed. For typical features see Table 8a.

Table 9 shows that the gas composition is not particularly favourable for methanol synthesis; Batelle proposes to use an active steam reforming catalyst as a dense phase to obtain a good synthesis gas, but additional research on thermally stable dense phase particles is still required. Pressurization of this system is not yet foreseen as far as we know.

A similar concept, called LTH process, but then with three concentric tubes (fluid bed gasifier, down comer, fast fluid bed combustor) was designed and operated on a small scale at Lund Institute of Technology (Lindman 1981). It was closed down because of circulation problems, erosion in the fast fluidized bed and difficulties in pressurization.

STEAM GASIFICATION IN SINGLE FLUIDIZED BED

Steam gasification with recycle of product gas has been tried by Lindman et al. but for various reasons found to be not attractive (KTH process (Lindman 1981). Several suggestions have been made to use steam fluidized bed gasification with indirect transfer of heat through a wall (see Table 10).

Table 9. Gas composition Battelle double bed gasifier (vol %), from Feldmann, 1981.

	700°C	820°C
CO	50.5	49.2
H_2	11.2	13.3
$CH_4+C_2H_4+C_2H_6$	19.8	22.4
CO_2	18.5	15.7

With presently available metals the method is technically possible for such a reactive material as biomass. Possibly it is also less complicated than the double fluid bed concepts, but pressurization up to say 15 bar probably is a difficult problem. Work along these lines is going on in various places but integral concepts are not yet in operation and again, product distribution will favour high hydrocarbon yields thus asking for gas reforming.

FASCINATING AND EXOTIC ROUTES

Several other routes have been proposed (see Table 11). For various reasons the first three methods do not need to be discussed here. Electrothermal gasification, however, requires attention because it probably is the only existing integrated methanol-from-wood pilot plant process presently in operation.

Electrothermal Steam Gasification

This process is operated in updraft gasifiers with recycle of pyrolysis gas both in France and Brasil. Refer to Tables 12 and 13 for details.

Table 14 shows the marginally competitive figures for this method based on the availability of very cheap electricity (0.03 $/kWh). Therefore this method is not feasible in Europe except perhaps in France (Dubois 1982).

Methane Hybrid Option

If methanol from wood and methane are each locally feasible, the hybrid option will be relatively attractive. By balancing the process, both shift and CO_2 removal step can be left out and an additional yield due to the balancing is obtained (Marshall 1979).

Such a hybride concept (with a countercurrent fixed bed gasifier) has been proposed by CMC for a 1100 ton/day fuel methanol plant to be located in Ontario (Rock 1982). We believe this to be an attractive contribution to cope with the methanol from wood scale problem. A 250 tons methanol-from-wood plant this way effectively produces twice that amount of methanol. Obviously, feasibility will increase with CH_4 price.

The Oxygen Donor Process

The oxygen donor process is the only double fluidized bed method which overall is an oxygen instead of a steam gasification process; (see Table 17).

Table 10. Methods of indirect supply of heat from combustion to fluidized beds.

Method	References
- Fluidized bed char combustion	Rensfelt 1978
- Fluidized bed wood combustion	Janesh et al. 1982
- External product gas combustion	Janesh et al. 1982

Table 11. Fascinating and exotic routes.

Process	Feature	Feasibility	Reference
1. Air gasification	Cryogenic removal of N_2 from product gas	Not competitive with oxygen gasification	Rowell et al. 1979
2. Flash pyrolysis and steam gasification in solar furnace	Heating rate 1000 W/m^2 temp rise 140000 °C/s $\tau \leqslant 1$ s $d_p \leqslant 1$ mm	Demonstrated at lab scale only	Antal et al. 1980
3. Hydrogen hybrid option	Oxygen from electrolysis of water; H_2 from water added to the product gas	Probably nowhere competitive	Nitschke 1981 Marshall 1979
4. Electricity hybrid option	Increased methanol wood ratio by adding electricity	Very site specific but integral pilot plant exists	Dubois 1982 Brecheret 1982 Brecheret and Zagato 1981
5. Methane hybrid option	Methane reformed and added to balance hydrogen deficient product gas	Site specific but relatively attractive; planned	Rock 1982 Marshall 1979
6. Oxygen donor process	$4C + CaSO_4$ $CaS + 4 CO$	Uncertain, test rig under construction 20 kg/hr	John Brown-Wellmann 1982
7. CO_2 acceptor process	DFB, only 25% of required heat to be transported by $\Phi_r\, c_p\, \Delta T$	Not proven	Lancet and Curran 1982

Table 12. Electrothermal gasification of wood.

	CESP Brasil	Novelerg
Capacity, kg wood/hr	40	600-900
Pressure, bar	1	10
Methanol production	20 kg/hr	-
Source	(Brecheret 1982; Brecheret and Zagato 1981; Oliveira 1980)	(Dubois 1982)

Table 13. Performance CESP electrothermal gasifier, Brecheret 1982.

CO: 45%
H_2: 48.2%
CO_2: 3.5%
Others: 3.3%
1.22 kg wood/Nm^3 synthesis gas
1.26 kWh/Nm^3 synthesis gas

Table 14. Methanol production cost sensitivity analysis for a 1000 ton/day plant gasifier, US $/ton, Brecheret 1982.

Gasifier	Capital Cost	Operating Cost Excluding Electricity	Electricity	Wood	Methanol US $/ton
Fluid bed	57.55	36.26	21.70	71.32	186.9
Fixed bed	57.55	36.26	37.69	87.17	218.7
Electrothermal	57.55	36.26	63.30	48.34	205.5

Table 15. Power and wood consumption per kg methanol, Brecheret and Zagato 1981; Brecheret 1982.

Method	Wood kg	Electricity
Electrothermal:	1.68	2.1 kWh
Fluid bed:	2.88	0.72 kWh
Electrothermal autarkic:	1.68	1.34 kg wood

Besides the complexity that is characteristic for any double fluidized bed system, the complicated unproven chemistry, the danger of sulphur emission and the extra large heat flows necessary (roasting of CaS is highly exothermic!), add to the development risks. But a prospective feature is the design of the solids recirculation system (see Figure 7).

Another feature is that, if successful it will be the only plant producing medium BTU synthesis gas from wood and air only, while no external steam is added. Therefore we would expect wood-to-methanol ratios above 2 but John Brown-Wellmann (1982) mentions that figure and also expects a product gas low in tar and methane, another surprising feature if it can be proven. The consortium planned to operate an atmoshperic 20 tons/day unit within two years from now.

The Carbon Dioxide Acceptor Process

We know of patents (e.g. Lancet and Curran 1982) on applying the CO_2 acceptor process also for biomass. The main feature of this double fluidized bed steam gasification process is the reduction of the solid recirculating flow (or of the temperature difference necessary between gasifier and oxidiser) because 75 percent of the heat is transported chemically (see Table 18).

Problems met in the development of this process for coal were catalyst deactivation and solids attrition. We do not know whether these have sufficiently been solved now but we have no information on the existence of a demonstration plant either on coal or wood. Moreover, the larger reactivity of wood compared to coal and its low sulphur content make the potential advantages of the CO_2 acceptor process less evident for wood. Reported gas composition is not optimal for methanol: the H_2/CO ratio is 4 and CH_4 = 14 vol percent.

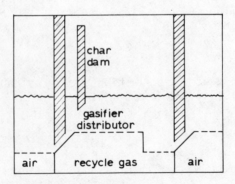

Figure 7. Solids recirculating system of John Brown-Wellmann 1982.

Table 16. The methane hybrid option of CMC, Canada, Rock 1982.

Process	Capacity kg MeOH/d	Capacity gain by hybrid operation
1. $CH_4 \rightarrow MeOH$	p	
2. Wood $\rightarrow$ MeOH	0.24 p	
3. Synthesis gas of 1+2 premixed	1.58 p	27%

Table 17. Oxygen donor process.

Vessel	Principal reactions	$H_r \left(\dfrac{kJ}{mol} \right)$
Combustor	$CaS + air \rightarrow CaSO_4$	-949.4
Gasifier	$CaSO_4 + 4H_2 \rightarrow CaS + 4H_2O$	$+17.05$
	$CaSO_4 + 4CO \rightarrow CaS + 4CO_2$	-186.9
	$C + H_2O \rightarrow CO + H_2$	$+130.5$
	$C + CO_2 \rightarrow 2CO$	$+171$

Table 18. CO_2 Acceptor process for wood.

Gasifier: $C + H_2O \rightarrow CO + H_2$ $+130.5$ kJ/mol
(1000°C) $CO + H_2O \rightarrow CO_2 + H_2$ -41.5 kJ/mol
                 $CaO + CO_2 \rightarrow CaCO_3$ -176.3 kJ/mol

Combustor: $CaCO_3 \rightarrow CaO + CO_2$ 176.3 kJ/mol
(850°C) $C + O_2 \rightarrow CO_2$ -394.51 kJ/mol

Heat flow: 75% chemical; 25% thermal

H_2: 60% CH_4: 14%
CO: 15% CO_2: 9%

Table 19. Optimization to various end products.

Source	Route
Brandon et al. 1982	Coproduction of fuel grade methanol with higher alcohols with a new catalyst
Cahn et al. 1976	One cycle methanol, off-stream to SNG plant
Gluckman and Louks 1981	Power company option
Gluckman and Louks 1981	One cycle methanol to be used for peak shaving + electricity from off-stream

OPTIMIZATION TO VARIOUS END PRODUCTS

Optimization of the wood-to-methanol plant by coproducing other products has been analyzed by several authors. We limit ourselves to mentioning the options proposed, see Table 19.

ROLE OF CATALYSIS IN GASIFICATION

The use of catalyst in the gasification of solid fuels has been studied already for a long time. In fact alkali catalyst activity was already described in a patent letter in 1867 (see Wigmans 1982). It received much attention in the period between the two world wars and recently research and development in this area is intensified (see e.g. Proc. 7th Int. Conf. on Catalysis, Tokyo 1980). An in-depth discussion has not possible here but we will point out a few processes being developed at present.

Catalyst can be used in the gasifier to enhance the reaction rate of char and/or to modify product gas composition. A well known example is the Exxon fluid bed process in which coal is impregnated with K_2CO_3 and gasified with steam. Both the gasification of char and methane formation are favoured. A somewhat similar set-up has been developed by the Pacific Northwest Laboratory (see Figure 8) for wood gasification with steam (Mitchell et al. 1980). Alkali carbonates and nickel or nickel oxides on silica alumina supports have been used as a catalyst in a fluidized bed reactor. Methane formation can be favoured at low temperature but at high temperatures ($\pm$ 850°C) an excellent methanol synthesis gas can be obtained and theoretical yields of 0.86 kg methanol / kg dry wood have been claimed (Mitchell et al. 1980). At these conditions the steam gasification is strongly endothermic and heat will have to be supplied to the fluid bed. This set-up looks very attractive for methanol production because no oxygen is required, no shift reaction is necessary, a low methane content is realized and a high efficiency and conversion are obtained. Possible problem areas could be deactivation of the catalyst (sulphur poisoning), entrainment of catalyst with the ash, and the fact that tar production although reduced remains present. Furthermore, indirect heating would require a large high temperature heat transfer area and the possibility of using a double fluid bed set-up for indirect heating is still uncertain. Nevertheless this process seems to be very attractive. Another possibility is to use a catalyst (e.g. Ni) to improve a secondary gasification step (see Figure 9). In this set-up a tar and methane rich gas is further gasified to produce synthesis gas. This can be carried out in a purely thermal way by introducing O_2 into the reaction chamber or by utilizing a catalyst. The function of the catalyst could be the reduction of the amount of oxygen required, prevention of coke and soot formation and increasing the rate of reforming reactions. Such secondary gasification steps are included in the EC pilot plant program (Palz and Grassi 1982).

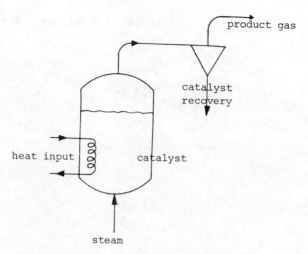

Figure 8. Catalytic gasification Pacific Northwest Lab. (Mitchell et al. 1980).

ECONOMIC EVALUATION

Since the pioneering work of Ader, Bridgwater and Hatt (1981) many additional feasibility studies have been done. We believe however, that with so many points of the total process not proven and with only one single integrated pilot plant operating with a not characteristic process (electrothermal gasification) the data produced are still highly speculative. Some months ago Wan et al. (1982) published an extensive study comparing 4 different routes. The results are summarized in Table 20.

It is seen that predicted production costs agree within 20 percent. Due to the many still unproven steps we conclude that this difference is not significant. Consequently, the simplest, most reliable process has the best chances for industrialization.

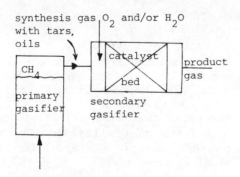

Figure 9. Catalytic secondary gasification/reforming.

Table 20. Feasibility study of methanol from wood, Wan et al. 1982.

Input	Catalytic Steam	Oxygen Downdraft	Entrained Bed	Oxygen Fluid bed
Biomass (25 $/ton)	10% H_2O	10% H_2O	10% H_2O	10% H_2O
Steam kg/kg ODW	0.75	-	0.026	0.180
O_2 kg/kg ODW	-	0.0496	0.559	0.458
p (bar)	10.2	1	1	10
T (°C)	750	600	980	970
Output				
H_2/CO	1.81	0.52	0.81	1.12
CO/CO_2	1.28	3.53	2.32	1.52
Thermal conv. eff.	0.50	0.40	0.44	0.42
Carbon conv. eff.	0.32	0.25	0.29	0.27
Total methanol production costs $/Gal	0.83	0.94	0.98	1.01

CONCLUSION

One small integrated atmospheric 40 kg wood/hr, electrothermal plant is operated in Brazil. Four different processes will be realized under the C.E.C. methanol-from-wood programme. A 400 tons wood/day process, based on the countercurrent moving bed principle, is planned in Canada, integrated with a methanol from methane reforming unit. These plants may give the highly needed data to get more insight into the technical reliability and the relative economical feasibility of the various routes. With the present data available there is no <u>a priori</u> preference for one route over the other but there is a definite incentive to integrate the plant with a methanol from methane process.

REFERENCES

Ader, G., Bridgwater, A.G., and Hatt, B.W. 1981. Techno-economic evaluation of thermal routes for processing biomass to methanol, methane and liquid hydro carbons. Energy from Biomass, 1st E.C. Conf. Palz, W., Chartier, P. and Hall, D.O., Eds. Appl. Sci., Publ., p. 598.

Antal, M.J., Jr., Royere, C. and Vialason, A. 1980. Biomass gasification at the focus of Odeillo. In: Thermal Conversion of Solid Wastes and Biomass. ACS Symp. Ser. 130., J.L. Jones and S. B. Radding, Eds. p. 237.

Arbo, J.C. and Dynecology, Inc. 1980. The Symplex coal and biomass gasification process. In: Energy from biomass and wastes IV. IGT.
Bailie, R.C. 1981. Results from commercial-demonstration pyrolysis facilities (35-45 tons/day refuse) extended to producing synfuels from biomass. In: Energy from biomass and wastes V, IGT, 549.
Beenackers, A.A.C.M and Van Swaaij, W.P.M. 1982. Development of a new method for production of pressurized pure hydrogen from low pressure nitrogen containing producer gas. In: Solar R&D in the European Community, Series E, 3., W. Palz and G. Grassi, Eds. Reidel.
Bickle, R.S., Edwards, A.J. and Moss, G. 1982. Development of the oxygen donor gasifier for conversion of wood to synthesis gas for eventual production of methanol. In: Energy from Biomass, 2., W. Palz and G. Grassi, Eds. Reidel, Dordrecht 41.
Brandon, C.S., Duhl, R. W., Miller, D.R. and Thakker, B.R. 1982. The economics of catalytic coproduction of fuel grade methanol containing higher alcohols. In: Proc. Fifth Int. Alc. Fuel Symp. 1, John McIndoe, Dunedin, N.Z. pp. 3-107.
Brecheret, V. 1982. Proc. Fifth Int. Alc. Fuel Tech. Symp. Vol. I, J. McIndoe, Auckland, 1-107.
Brecheret, V., Zagato, A.J.A. 1981. Methanol from wood in Brazil. In: ACS Symp. Ser. 159, Monohydridic alcohols, 32.
Cahn, R.P. et al. 1976. Concurrent production of methanol and substitute natural gas, U.S. patent 3993467, 23 Nov.
C.E.C. 1982. Energy recovery by gasification of agricultural and forestry wastes in a cocurrent moving bed reactor, EUR 7594 EN.
Chartier, P. and Palz, W. 1981. Energy from biomass, solar energy R&D in the European Community, 1, Reidel.
Chrysostome, G. and Lemasle, J.M. 1982. Gazeification au bois en lit fluidise a l'oxygene et sous pression en vue de produire un gaz utilisable pour la synthese du methanol. In: Energy from biomass, Solar Energy R&D in the European Community. W. Palz and G. Grassi. Eds., Series E, 2. Reidel, p. 28.
Dubois, P. 1982. Design and construction of a pressurized wood gasifier with heat input by oxygen combustion or by electrical heating. In: Energy from biomass, Solar Energy R&D in the European Community, W. Palz, and G. Grassi, Eds., Series E, 2. Reidel, p. 76-88.
Feldmann, H.F. 1981. Steam gasification of wood in multi-solid fluidized bed (MSFB) gasifier, Energy from biomass and wastes V, IGT, p. 529.
Fischer, T.F., 1976. Chem. Eng. Progr., 72, Oct, p. 75.
Frank, M.E. 1980. Chem. Systems liquid phase methanol process, Proc. Intersoc. Energy Convers. Eng. Conf., 15th(2), 1567.
Gehrman, J. (Ed). 1981. Thermochemische Gaserzeugung aus Biomasse, KFA, Jülich, conf. 46, November.

Gluckman, M.Y. and Louks, B.M. 1981. Methanol -- An opportunity for the electric utility industry to produce its own clean liquid fuel, 8th Energy Tech. Conf. march, Washington, DC, Government Inst. Inc., August, p. 920.

Gravel, Guy. 1982. Byosyn, Montreal, Quebec, Personal Communication.

Hiller, H. 1975. Kohle-Druckvergasung, Erdol und Kohle, 28(2).

Hummelsiep, H., SFW-Funk-Verfahren. 1981. In: Thermochemische Gaserzeugung aus Biomasse, KFA, J. Gehrman (ED), Julich, Conf. 46, November, p. 75.

Janesch, R., Werner, F. and DeSanti, D.M. 1982. Synthesis gas obtained from biomass. In: Energy from biomass, Solar Energy R&D in the European Community, Series E, 2. W. Palz and G Grassi, Eds. Reidel. p. 101.

Kagayama, M., Ingarshi, M., Fukuda, J. 1980. Gasification of solid wastes in dual fluidized-bed reactors, ACS Symp. Ser. 130, paper 38.

Kennedy, T.F. and Shanks, D. 1981. Methanol: manufacture and uses, ACS Symp. Ser. 159, p. 19.

Lancet, M.S. and Curran, C.P. 1982. Process producing synthesis gas from wood, U.S. patent 139915, Conoco, June 15.

Lindman, N. 1981. A new synthesis gas process for biomass and peat, Energy from biomass and wastes V, IGT, 571.

Lindner, C. 1981. Vergasung von Biomasse in der Wirbelschicht. In: Thermochemische Gaserzeugung aus Biomasse, KFA, J. Gehrman (Ed), Julich, Conf. 46, November, p. 214.

Lindner, C. and Reimert, R. 1982. Gasification of wood in the circulating fluidized bed methanol production route. In: Energy from biomass, Solar Energy R&D in the European Community, W. Palz and G. Grassi, Eds., Series E, 2, Reidel, p. 115.

Marsden, S.S., Jr. 1982. Six new potential sources for massive quantities of methanol. In: Proc. Fifth Int. Alc. Fuel Symp. John McIndoe, Dunedin, N.Z., Vol. I, p. 21.

Marshall, J.E. 1979. Canadian biomass perspective: A new interest in an old fuel, UK, ISES Conf. (C20) at the Royal Soc.: Biomass for Energy, July, p. 52.

McIndoe, John. 1982. Proc. Fifth Int. Alc. Fuel Symp. 1, Dunedin, N.Z.

Mitchell, L.K., Mudge, R.J., Robertus, R.J., Weber, S.L. and Sealock, L.J. 1980. Methane/methanol by catalytic gasification of biomass, Chem. Eng. Prog., Sept., p. 53.

Nitschke, E. 1981. Die Herstellung von Synthesegas durch Holzvergasung nach dem Rheinbraun HT-Winkler Verfahren. In: Gehrman, J. (Ed.) 1981. Thermochemische Gaserzeuging aus Biomasse, KFA, Julich, Conf46, November.

Oliveira, E.S. 1980. The electro chemical route wood to syngas, Proc. IV Int. Symp. Alc. Fuels, Vol. I, Inst. Pesquisas Technological, Sao Paulo, A38, p. 199.

Palz, W. and Grassi, G., Eds. 1982. Energy from biomass, Solar Energy R&D in the European Community, Series E, 2, Reidel.

Reed, T. 1980. A survey of biomass gasification, Vol. III, SERI, Golden, Colorado, U.S.A.

Rensfelt, E. 1978. Basic gasification studies for development of biomass medium-BTU gasification process. In: Energy from biomass and wastes, IGT, p. 465.

Rock, K.L. 1982. Production of methanol from mixed synthesis gas derived from wood and natural gas. In: Energy from biomass and wastes VI, IGT, paper 31.

Rowell, R.M., A.E. Hokanson. 1979. Methanol from wood: a critical assessment, Progress in biomass conversion, Academic Press, 1, p. 117.

Wan, E.I., J.A. Simmons and J.D. Price. 1982. Economic evaluation of indirect biomass liquefaction processes for production of methanol and gasoline, In: Energy from biomass and wastes VI, paper 37, IGT.

Wigmans, T. 1982. Catalytic gasification of carbon; a mechanistic study, Ph.D. Thesis, Amsterdam.

Wilson, H.T. 1981. Progress in the development of a test facility for biomass gasification studies, In: Energy from biomass, solar energy R&D in the European Community, W. Palz and G. Grassi, Eds. Series E, 2, Reidel, p. 203.

Wilson, H.T., R. Fletcher and R.J. Davies. 1982. Proposed 20 tonnes per day biomass gasification pilot plant, In: Energy from biomass, Solar Energy R&D in the European Community, W. Palz and G. Grassi, Eds., Series E, 2. Reidel, p. 89.

ENERGY BALANCES FOR BIOMASS CONVERSION SYSTEMS

Raphael Katzen

Raphael Katzen Associates International, Inc.
Cincinnati, Ohio
U.S.A.

INTRODUCTION

Biomass conversion systems of any type, irrespective of feedstocks utilized, must be measured on a consistent scale which identifies the energy efficiency of the process and of the overall system. Accurate energy balances, as well as material balances, must be developed so that potential commercialization can be evaluated on the basis of energy efficiency, as well as technical and economic criteria.

Although this premise is becoming accepted as a concomitant of research and development planning and evaluation, there is little consistency in the approaches taken to material and energy balances, and to the definition of energy efficiency. As a result, it is very difficult not only to cross-evaluate different biomass conversion systems, but also to compare such systems with competitive technology based on petroleum or fossil fuel feedstocks with substantially different energy input bases and divergent technologies.

ENERGY EFFICIENCY

The fundamental determination of energy efficiency for a process or for an overall system, including provision for feedstocks and energy supply, and even going on through to distribution of products, is based on a simple ratio of energy output to energy input, as indicated by the equation:

$$\frac{EO}{EI} \times 100 = \%EE$$

where: EO = energy output
        EI = energy input
        EE = energy efficiency

The process energy efficiency ratios the specific output of main product and by-products to the input of feedstocks, chemicals and utilities. However, with the growing emphasis on evaluation of energy consumption as an all-encompassing measure of utility of a technology, it is necessary to go beyond the basic process itself. In the overall consideration, we must include the energy use in steam and power production, and in operation of other utilities such as water supply, waste treatment, etc. Furthermore, particularly in respect to biomass, energy requirements for production and delivery of the biomass feedstocks, or the energy consumption in collecting, processing and transporting biomass wastes, must be taken into consideration.

ENERGY OUTPUTS

Biomass conversion energy system output includes the following key components:

a. Products measured in terms of combustion value (higher heating value), or human or animal food caloric value.
b. By-products measured on the same basis.
c. Stack gases, with heat content calculated above a standard base temperature, utilizing available physical data for sensible and latent heat values.
d. Liquid effluents, with energy values based on sensible heat contents above a standard base temperature. The energy content of dissolved organics in the liquid wastes, if substantial, should also be entered into the accounting.
e. Solid residues will generally have little sensible heat value. If essentially ash, there is no energy value. However, some solid residues contain organics (char), which should be considered in the energy evaluation.

Wastes energy outputs (losses) such as stack gases, liquid effluents, and solid residues have potential for energy recovery through expansive work (gases), heat exchange (gases and liquids) or fuel value (gases, liquids and solids). The extra investment required for energy recovery is often justified by relatively rapid pay-out, and good return on the incremental investment.

ENERGY INPUTS

All input components to a process system must be evaluated for energy content. These inputs may be identified as follows:

ENERGY BALANCES FOR BIOMASS CONVERSION SYSTEMS

a. Feedstocks are evaluated for their equivalent fuel value (higher heating value), particularly when of biomass origin. Since part of the fuel value of the feedstock is contributed by solar energy, an alternative approach is to calculate the amount of energy consumed in growth of the biomass, including fertilizers, fuels for field work and transportation, energy for irrigation, etc.

b. Fuels can be directly evaluated for their energy content (higher heating value). In the case of electrical energy, the true measure must consider the thermal energy requirement for generation of electricity; which in the U.S.A. averages 2,520 kcal per kwh. On the other hand, when electrical energy is of hydro or nuclear origin, it might be argued that the energy equivalent is strictly the conversion equivalent, or 860 kcal per kwh. Here again, a "free ride" is taken for the solar energy utilized to evaporate water from the earth's surface, which is recondensed and collected as rainfall for hydro-power generation.

c. Chemicals and other auxiliary materials should also be evaluated in terms of energy required to produce them. However, in most cases, these energy quantities are too small to be significant.

d. Transport energy requirements need to be taken into account for biomass collection and delivery, and in some cases for product distribution or waste disposal (particularly solid waste that is going to land-fill).

ENERGY EFFICIENCY EXAMPLES

Typical biomass conversion projects are presented in which different approaches to energy efficiency have been utilized. A need for a common basis of evaluation becomes apparent when the different approaches are compared.

a. Methanol from wood.

Energy efficiency was calculated based on the fuel value of the wood process feedstock, also used as a fuel. No account was taken of the energy requirement to either grow the wood (which is negligible, other than solar energy) or to harvest, collect and transport the wood, which can become a sizeable factor.

The process yield is essentially a single product, methanol, with no by-products. It also yields very little in the way of gaseous, liquid or solid wastes (other than a small volume of wood ash). The output energy analysis is therefore fairly simple. The energy input and output are tabulated as follows:

Input	Input Energy
Wood Feedstock, 75.0 OD MT/hr	348.6 MM kcal/hr
Diesel Fuel, 52.1 l/hr	1.0 MM kcal/hr
Electricity, 22,220 kwh/hr	56.0 MM kcal/hr

Output	Output Energy
Methanol, 47,660 l/hr	204.8 MM kcal/hr

Figure 1 depicts the overall energy balance for the wood to methanol process. Process energy efficiency is calculated as 58.8% by the following formula:

$$\text{Process Efficiency, \%} = 100 \times \frac{\text{Methanol Heat Value}}{\text{Wood Heat Value}}$$

An overall thermal efficiency is calculated as 50.5% by the following formula:

$$\text{Overall Thermal Efficiency, \%} = 100 \times \frac{\text{Methanol Heat Value}}{\text{Wood Heat + Electricity + Diesel Fuel Value}}$$

b. Ethanol from cellulosic wastes.

In this case the cellulosic wastes result from operations such as municipal solid waste separation, pulp production, tree harvesting and lumber processing. It may be assumed that energy inputs for the production of these waste raw materials have already been paid for by the primary operation. However, the energy required for collection, processing and transport of these wastes can become a substantial factor in the energy balance. This is often overlooked by simply putting a price for such energy input into the delivered cost of the feedstock. The analysis of a typical process for production of ethanol from cellulosic wastes is based on the following input and output:

Input	Input Energy
Feedstocks, 75.0 OD MT/hr	316.8 MM kcal/hr
Fuel Oil, 1,312 l/hr	12.5 MM kcal/hr
Electricity, 27,420 kwh/hr	69.1 MM kcal/hr

Output	Output Energy
Ethanol, 23,660 l/hr	133.5 MM kcal/hr
Animal Feed, 20.2 OD MT/hr	64.7 MM kcal/hr

Figure 2 is a block diagram representation of the input and output. The overall thermal energy efficiency is 49.8% by the following equation:

Overall Thermal
Efficiency,

$$\% = 100 \times \frac{\text{Ethanol Heat Value} + \text{By-Product Heat Value}}{\text{Cellulosic Waste} + \text{Fuel Oil} + \text{Electricity Heat Value}}$$

The net energy efficiency is 53.4% by the above equation when the effective heat values for each energy input and output are considered. The effective heat values are determined as the net percent of energy recoverable from the materials if combusted; accounting for the inefficiencies associated with the burning of different fuels.

$$\text{Net Energy Efficiency, \%} = 100 \times \frac{\left(.85 \times \text{Ethanol Fuel Value}\right) + \left(.67 \times \text{By-Product Fuel Value}\right)}{\left(.67 \times \text{Cellulosic Waste Fuel Value}\right) + \text{Electricity} + \text{Fuel}}$$

c. Ethanol from corn.

Energy analysis of this operation has been covered by a number of publications, with widely varying results. These variations stem from two factors; differences in treatment of energy values and energy requirements for producing the biomass raw material, and in process technologies having a wide range of energy requirements.

I. Process efficiency, utilizing modern technology for conversion of corn to ethanol, is indicated by the following inputs and outputs for the process.

Input	Input Energy
Corn, 62.3 MT/hr	239.4 MM kcal/hr
Coal, 11.2 MT/hr	75.4 MM kcal/hr
Electricity, 8,313 kwh/hr	20.9 MM kcal/hr

Output	Output Energy
Ethanol, 23,900 l/hr	134.8 MM kcal/hr
DDGS, 20.3 MT/hr	90.2 MM kcal/hr

Figure 3 shows the block energy balance diagram for this process. The process efficiency is indicated to be 94%, based on the heat values of the inputs and outputs. Many representations of efficiency for this process include only the fuel and electric power as energy inputs, which amount to about 3,330 kcal/liter of alcohol, and only the energy value of the alcohol product, not taking into account the by-product or the process feed heat values. This plant thermal efficiency appears to be about 140%.

II. Overall efficiency of the ethanol from corn process yields a better definition of the energy balance. However, even this may be presented on two different bases.

On a fuel value basis, corn is charged to the process at its fuel value, and the ethanol and DDGS by-product are calculated on their fuel value basis. With these and other inputs and outputs, as indicated in item c-I above, an overall thermal energy efficiency of 67% is calculated.

On the other hand, if the calculation is made on the basis of energy required to produce and transport the corn, and a credit is given for the energy value of the DDGS on the same basis, based on the assumption that the DDGS is simply a replacement for part of the corn, an overall growth energy efficiency of 99% is obtained. It may also be argued that since the DDGS has all the protein value of the corn, it has an energy value almost equal to the energy input to produce the corn itself. Such an assumption would redefine the energy efficiency as 125%.

CONCLUSIONS

The sound approach to determine the energy balance and efficiency is to evaluate all input items on the basis of their energy content as fuels, with electric power charged to the process at the energy requirement to produce it. Similarly, all products should be evaluated on the basis of their potential energy value as fuels, with energy losses in unrecoverable form calculated from analysis of waste streams.

If a global analysis is indicated, then the energy values must be taken beyond the battery limits of the process plant into the field where the raw material is being grown or collected. In this case, the energy required to produce, process and transport the raw material should be calculated and utilized in place of its fuel energy value. By similitude, a biomass by-product, such as DDGS, should then also be evaluated in terms of the equivalent energy which would have been required to produce it as a distinct product.

ACKNOWLEDGEMENTS

The author wishes to acknowledge the substantial assistance of Catherine E. Yeats and George D. Moon, Jr. in developing the calculational methods, data and energy efficiency results.

ENERGY BALANCES FOR BIOMASS CONVERSION SYSTEMS

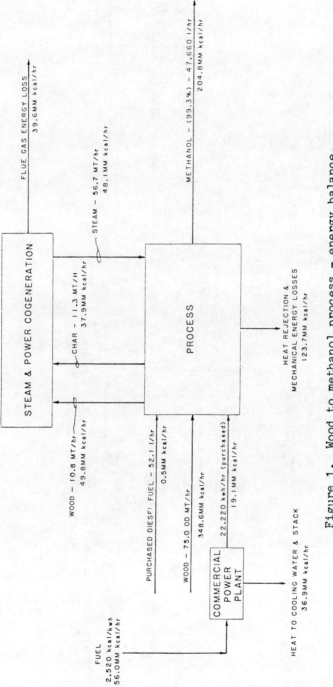

Figure 1. Wood to methanol process - energy balance.

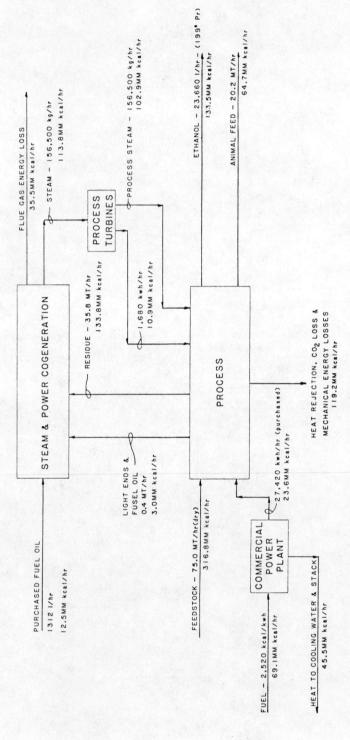

Figure 2. Cellulose alcohol process - energy balance.

ENERGY BALANCES FOR BIOMASS CONVERSION SYSTEMS

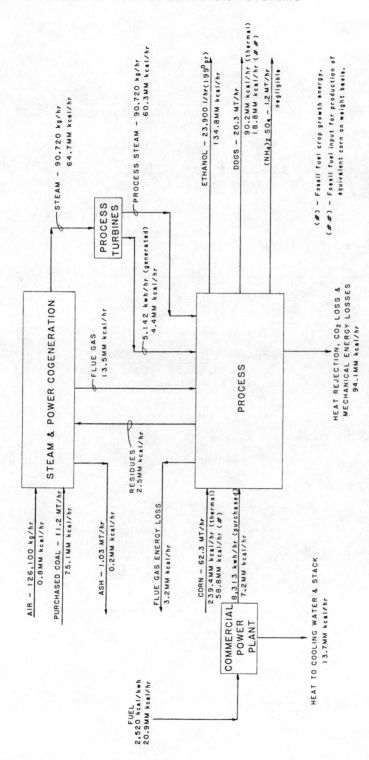

Figure 3. Grain alcohol process - energy balance.

Appendix A. Efficiency calculations and examples

Process Efficiency

$$\text{Efficiency, \%} = 100 \times \frac{\text{Process Products and By-Products Heat Value}}{\text{Process Feedstocks Heat Value}}$$

Example: Wood to Methanol Process Efficiency

$$= 100 \times \frac{\text{Methanol Heat Value}}{\text{Wood Heat Value}}$$

$$= 58.8\%$$

Overall Thermal Efficiency

$$\text{Efficiency, \%} = 100 \times \frac{\text{Process Products and By-Products Heat Value}}{\text{Process Feedstock + Electricity + Fuel Heat Value}}$$

Example: Corn to Ethanol Overall Thermal Efficiency

$$= 100 \times \frac{\text{Ethanol Fuel Value + DDGS Fuel Value}}{\text{Corn Fuel Value + Coal + Electricity}}$$

$$= 67\%$$

Overall Growth Energy Efficiency

$$\text{Efficiency, \%} = 100 \times \frac{\text{Process Products Fuel Value + By-Product Equivalent Growth Energy}}{\text{Feedstock Growth + Electricity + Fuel Energy Requirement}}$$

Example: Corn To Ethanol Overall Growth Energy Efficiency

$$= 100 \times \frac{\text{Ethanol Fuel Value + Equivalent Amount of Corn as DDGS Energy Supplied To Grow}}{\text{Energy Supplied To Grow Corn Feedstock + Electricity + Coal}}$$

$$= 99\%$$

Overall Protein Growth Energy Efficiency

$$\text{Efficiency, \%} = 100 \times \frac{\text{Process Product Fuel Value} + \text{By-Product Protein Fuel Value}}{\text{Feedstock Growth Energy Requirement as Protein} + \text{Electricity} + \text{Coal}}$$

Example: Corn To Ethanol Overall Protein Growth Energy Efficiency

$$= 100 \times \frac{\text{Ethanol Fuel Value} + \text{Energy Supplied To Grow Equivalent Amount of Protein as DDGS}}{\text{Energy Supplied To Grow Protein as Corn} + \text{Electricity} + \text{Coal}}$$

$$= 125\%$$

Net Energy Efficiency

Efficiency, % = 100 × Overall Thermal Efficiency Corrected for Individual Combustion Efficiencies of Products and Inputs.

Example: Cellulose to Ethanol Net Energy Efficiency

$$= 100 \times \frac{(.85 \times \text{Ethanol Fuel Value}) + (.67 \times \text{By-Product Fuel Value})}{(.67 \times \text{Cellulosic Waste Fuel Value}) + \text{Electricity} + \text{Fuel}}$$

$$= 53.4\%$$

Plant Thermal Efficiency

$$\text{Efficiency, \%} = 100 \times \frac{\text{Process Product Heat Value}}{\text{Fossil Fuel Heat Value}}$$

Example: Corn To Ethanol Plant Thermal Efficiency

$$= 100 \times \frac{\text{Ethanol Fuel Value}}{\text{Coal} + \text{Electricity}}$$

$$= 140\%$$

Appendix B. Summary of energy values.

1. Methanol Process — 99.7 MM GPY   BATTELLE
                        377.4 MM LPY

Component	HHV*	
Wood	8,300 Btu/dry lb	4,602 kcal/kg
Methanol	64,575 Btu/gal	4,298 kcal/kg

2. Cellulose-to-Ethanol Process — 49.5 MM GPY   KATZEN '78
                                  187.4 MM LPY

Component	HHV	
Cellulosic Wastes	7,544 Btu/dry lb	4,182 kcal/kg
Animal Feed	5,768 Btu/dry lb	3,198 kcal/kg
Ethanol	84,745 Btu/gal	5,642 kcal/l

3. Corn-to-Ethanol Process — 50 MM GPY   DOE report '78
                             189 MM LPY

Component	HHV	
Corn	6,910 Btu/lb	3,831 kcal/kg
DDGS	8,000 Btu/lb	4,435 kcal/kg
Ethanol	84,745 Btu/gal	5,641 kcal/l
Corn Production	100,000 Btu/bu	944 kcal/kg

4. Fossil Fuels/Electricity (2,520 kcal/kwh)

Component	HHV	
Coal	12,100 Btu/lb	6,708 kcal/kg
Diesel Fuel	140,000 Btu/gal	9,319 kcal/l
Fuel Oil	6 MM Btu/bbl	9,509 kcal/l

* HHV = Higher Heating Value

MUNICIPAL SOLID WASTE: PROCESS TECHNOLOGY

Robert F. Vokes

Consultant
Parsons & Whittemore, Inc.
New York, N. Y.
U.S.A.

INTRODUCTION

MSW may be considered as a useful raw material for many purposes and products. The determination of a course of action is controlled by economic factors which are related to a geographical situation. Therefore, the management of a given source of MSW may involve only the collection, transportation and disposal in a sanitary land fill with a potential of future land reclamation as well as recovery of methane from the biodegradation of the biomass.

However, as the situation changes with time, the given source mentioned above may increase in volume, the landfill site may be consumed and additional process technologies may need to be considered. Our world is replete with examples of such evolution but only recently, as a result of economic factors such as limited oil supplies, environmental concerns, etc., have we concentrated on new solutions to MSW management, including the adaptation of known process technology from many industrial applications.

MSW and other sources of wastes have been combined in some new process technologies. This includes the recovery of organic wastes from municipal and industrial sludges and the refinement of means to isolate cellulose fibers for paper manufacture as well as means to fractionate waste glass according to colors. As the MSW and industrial waste problems increase, it is reasonable to expect that very attractive business opportunities will evolve in the practice of resource recovery.

This treatise on process technology will emphasize the recovery and use of MSW biomass and be further limited to the description of technological developments of the past twelve years.

CRITERIA FOR MSW BIOMASS PROCESS TECHNOLOGY

Certain criteria must be considered prior to the establishment of MSW Biomass process technology. Included are the composition of the MSW, its availability, the market for recoverable materials as well as potential markets for the biomass itself.

The approximate MSW major components as identified by Vokes (1983) are: organic biomass, 55 percent; inorganics, 20 percent; and moisture, 25 percent.

The total biomass generated in the U.S.A. is estimated to be 2 billion tons annually including 140-160 million tons of MSW. Based on 20 percent availability, 400 million tons per year includes about 30 million tons of MSW (Goldstein 1981).

At 55 percent biomass, there are about 18 million tons available annually in the U.S.A., or about 0.5 lb. per capita per day.

At 20 percent availability, 30 million tons per year would produce about 0.15 quads of heat energy as BTU (Goldstein 1981), 75 million tons of steam and about 2000 megawatts of electrical power generation (Dade County Bulletin 1982).

Minerals Market

From 30 million tons per year of MSW, the yield (Dade County Bulletin 1982) would be approximately:

   2 million tons of ferrous material
   100 thousand tons of aluminum
   300 thousand tons of non-ferrous material
   2 million tons of glass

Biomass Potential Markets

Source separation from the municipal and industrial solid waste streams provide resource recovery of metals, paper, plastics, rubber, textiles and wood for specifically related industrial uses. This has been widely practiced as a waste materials industry for many years. Such past practice, new state government regulations (New Jersey Senate 1981) as well as an entirely new newsprint resource recovery industry development (Garden State Paper Co. 1982) seem to have extracted the most useful specific types of biomass to date from the solid waste streams.

Other than energy, the markets for biomass as tipped at landfill sites or processing plants, do not exist to date in the U.S.A.

Some process technology for the resource recovery of paper fiber, (Black Clawson Co. 1972) food for animals (Marsh 1973) and biomass conversion to alcohol (Kidwell-Katzen 1982) has been practiced on both a pilot scale and commercial scale in recent years. The markets for such products have not been developed to permit further expansion.

GENERAL OVERALL CRITERIA

In undertaking the construction of a waste disposal resource recovery plant, first and foremost, a detailed feasibility study must be made of the local situation. Among other things, this should involve an analysis of the nature and type of solid waste pick up and delivery in the area under consideration. Will industrial as well as commercial waste be handled and if so, what is its general makeup and quantity? Is pathological waste to be handled and if so, how much? Is the disposal of hazardous waste involved and if so, what type of hazardous material and what technique should be employed in its handling and ultimate disposal? Are private as well as public waste haulers involved and how many private haulers are involved and what type of equipment is used? Are transfer stations involved? What is the maximum size of packer trucks involved? Are delivery operations union or non-union? What type of truck delivery and queuing times might be involved? How should routing best be arranged? Is pick up and hauling involved in the contractual arrangement? Further, the market for by-products should be thoroughly investigated as well as the local utility rate structure and contract arrangements.

For a more specific and detailed procedure for the design and supply of a resource recovery MSW processing plant, reference should be made to the Dade County (Florida) Resource Recovery Bulletin, 1982.

MSW BIOMASS PROCESSES FOR ENERGY

The National Center for Resource Recovery (NCRR) has published a record of the development of MSW biomass processes for energy since 1974 (NCRR Bulletin 1981). This listing includes the U.S.A. and Canada for a total of 72 MSW processing plants, 19 recovering methane from landfills and 49 planned Resource Recovery facilities.

Of the 72 MSW processing plants, a total of 40 employ mass burning, 27 employ dry shredding for RDF, 2 employ wet shredding for

RDF (pulping, "Hydrasposal") and the balance are either composting or pyrolysis systems.

Most installations tend to incorporate designs which are specific for the "case basis" situation i.e., capacity, MSW composition, plant location, environmental limitations, etc. However, only a very small number of plants can be classified as either employing new technology or operational on a daily basis at rated design capacity and economically viable.

The process technologies described below have been chosen according to the following criteria:

1. Capacity: 2000 tons per day minimum.
2. Integrated: Steam and electric power generation.
3. Non-mass burning.
4. Special biomass recovery potential
5. Operational at full capacity daily.

Burning

Since the late 1800's MSW biomass has been mass burned in Europe. This process has therefore dominated the application of technology in Japan, the U.S.A. and other countries. Mass burning, therefore, does not represent an advance in the management of MSW with respect to resource recovery and the conservation of raw materials and energy.

A major innovation in the early 1900's involved the use of dry shredders to produce a more readily handled and uniform biomass by-product which could be separated from the noncombustibles and become useful as a fuel or feed stock for anaerobic digestion with sewage sludge, or compost or methane gas generation.

This technology, based on the use of shredders for the production of RDF, has met with some serious process control problems which involve the questions of safety and reliability as measured in terms of on-line availability for performance. The problems involve major hazards as a result of impacting explosive components of MSW in the shredders and the further possibility that solid wastes may cause destruction of the shredder because of the size and density of the MSW noncombustible components. Recent designs of such systems have solved some of these problems although effective controls are based upon human surveillance of the infeed MSW materials.

In the late 1960's a new concept of wet shredding was introduced, known as the "Hydrasposal" (Black Clawson Co. 1972 Bulletin; Dade County Bulletin 1982) Technology. This method solved the problem of explosions and produced an RDF which is very uniform and can be completely burned out in standard power boilers for the production of steam and electrical energy.

Along with the dry shredders or hammermills, the use of two types of dry screens was introduced. These are known as the trommel and disc designs borrowed from use in other dry solids fractionating systems.

Resource Recovery (Dade County) Inc.

The system in use by Resource Recovery (Dade County) Inc. is a unique combination of wet and dry shredding technology as shown in the flow diagram -- Figure 1. The capacity is 3000 tons per day of MSW. It is designed and operated for the production of up to 600,000 megawatt hours/year of gross electrical generation.

More details on the design and operation of this commercially operating plant may be found in the Dade County Bulletin 1982.

Other Installations

The NCRR listing includes five other facilities of major commercial size which are designed to produce RDF for energy production. These are:

> Detroit, Michigan - Combustion Engineering, Inc., Stamford, Conn.
> Hempstead, N.Y. - Parsons and Whittemore, Inc., New York, N.Y.
> Niagara Falls, N.Y. - Hooker Energy Corp., Niagara Falls, N.Y.
> Columbus, Ohio - A. E. Stilson and Associates - Columbus, Ohio
> Portsmouth, Va. - Southeastern Public Service Authority, Norfolk, Va.

Note that all of the above plants, except Hempstead, employ dry shredding and that none of these plants are in commercial operation at this writing.

MSW-BIOMASS PROCESSES FOR ORGANIC CHEMICALS AND OTHER USES

Although all of the resource recovery facilities listed by NCRR include the utilization of MSW biomass, mostly for energy production, some of the process technologies are particularly unique. The following brief descriptions of old and new process technology have been chosen on the basis of uniqueness, the availability of information and the author's opinion of feasibility of application for use of MSW biomass as a raw material. New concepts are being disclosed

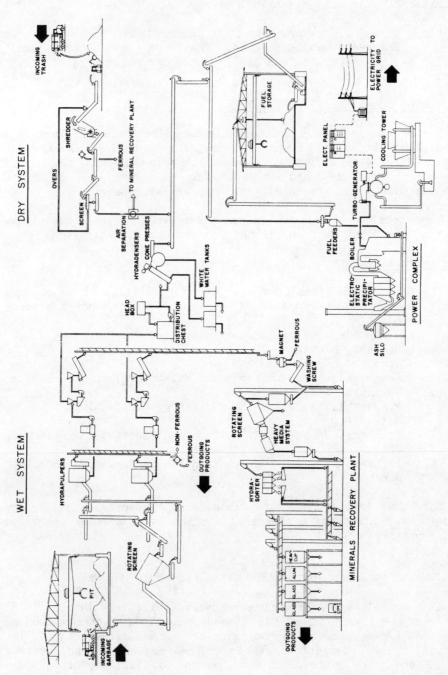

Figure 1. Flow diagram for process in use by Resource Recovery (Dade County), Inc.

frequently, some of which may survive in the market place. Also, in the Western European area, two such processes are in daily commercial operation (Johansson 1982; Marsh 1973).

Western Europe

1) The Sorain-Saar plants in Rome, Italy have been in operation for many years i.e., prior to 1970. The technology was developed by Dr. Silvio Milone and others and includes the unique features of strict control of MSW as collected and the process technologies to produce paper fiber for a local board mill, pelletized food for animals, compost, ferrous metals and energy (Marsh 1973).

Dry separation techniques are used to yield the paper fibers along with the plastics. Wet pulping separates out the plastics for burning and the fiber at about 50% moisture for an adjacent paper board mill.

This process technology may be useful as a means for isolating MSW biomass for other types of utilization.

2) Flakt Inc. has developed new process technology for the production of secondary fibers from MSW for use by the paper industry. Three systems, operating in Europe, also recover metals, plastics and compost (Johansson 1982).

The systems are based upon the use of dry separation techniques and may be used to produce RDF and, of course, biomass in a dry form for other uses. A typical flow diagram is shown in Figure 2.

Flakt Inc. has also developed new technology for bio-fuel preparation, including bark, wood waste, peat or lignite which are part of the industrial solid waste stream and natural resources. This method (Hadenhag, J.G. 1981) also offers a means of processing biomass for other uses.

U.S.A.

1) BC Fibreclaim Inc. initiated the first fiber recovery system from MSW biomass employing the "Hydrasposal" wet shredding technology to produce RDF followed by the "Fibreclaim" technology to isolate the paper fiber in a 50% dry pulp form (Black Clawson Co. 1972). This system was employed in Franklin, Ohio on a commercial basis for about ten years and the fiber was sold to an adjacent roofing felt paper mill. A relatively clean source of cellulose fiber can be produced in this manner for other uses.

2) "ECO Fuel II" (NCRR 1981) by Combustion Equipment Associates and Occidental Petroleum, has produced a unique biomass fuel in a fine dry pulverized form for suspension burning alone or with other

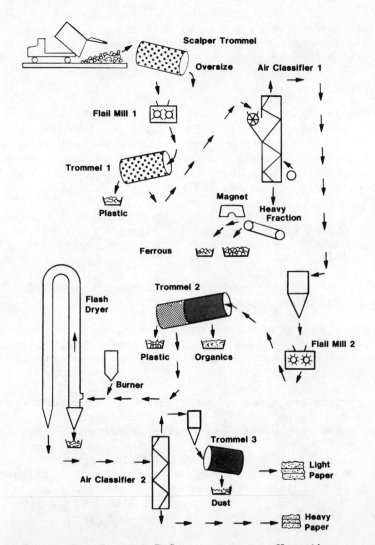

Figure 2. Flakt 3-R System process flow diagram.

fuels. The process and equipment are borrowed from other industries and seem to produce a satisfactory fuel. However, the plants as originally designed, have not proven to be economically viable to date. The biomass recovered is, however, in a form which is readily useful for other products.

3) "News to News" by the Garden State Paper Co., Garfield, N.J., U.S.A. represents a unique industrial development of a resource recovery technology (Snider 1982). The basics for successful operations include source separation and collection of post consumer

newsprint papers and a patented ink and color removal process which produces a fiber composition suitable for high quality newsprint manufacture. Four major plants are in production in the U.S.A. and another in Mexico (Garden State Paper Co. 1982).

Energy conservation results from the recycling of newsprint since the energy requirement of recycled newsprint is only one-half of that in a virgin newsprint mill. Also, the value of newsprint for recycling to newsprint is nearly three times that of its value as a fuel.

Such newsprint recovery technology, however, may be considered as a potential source of MSW biomass in a form which is most useful for producing other products.

BIOMASS CHEMICAL PROCESSES

Pyrolysis

Some experimental pilot plants have been built to pyrolyze the biomass fraction of MSW. The objectives include the production of gas and residual oils for commercial use as fuels or in chemical processes. The advantages claimed include the avoidance of air pollution and a combination of low capital cost and higher energy conversion efficiency.

The only commercial system installed to date is designed according to the Monsanto "Langard" patents and has been operated with some problems which are apparently being solved (Sussman 1974). The process flow diagram is shown in Figure 3 and is self-explanatory since existing technology and standard equipment are employed.

This process represents a unique method of biomass conversion for gas production which could be used as a raw material for other products.

UFBI Ethanol from MSW Biomass

The Gulf Oil Chemicals Corporation, under the direction of Dr. George Emert, developed a simultaneous saccharification and fermentation process and one ton per day pilot plant facility. This patented process has been acquired by the University of Arkansas-Biomass Research Center and a 50-ton-per-day pilot plant has been designed by Raphael Katzen Associates Int'l, Inc. (Kidwell-Katzen 1982).

The United Bio-Fuel Industries Inc. (UFBI) will build the 50-ton-per-day pilot plant and the initial operation will use corn as a raw material with provision for use of MSW biomass alternatively.

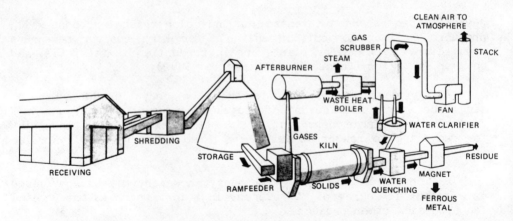

Figure 3. Process flow diagram.

The new technology features the use of MSW biomass as a raw material and the continuous enzyme production and simultaneous saccharification and fermentation. The ultimate objective is to design and build a 2000-ton-per-day commercial facility which would include the flexibility of multiple use of biomass raw materials and the production of multiple end products, i.e. energy, ethanol, etc.

Although this process technology is in the developmental stage, it is presented here as a realistic opportunity for the proposed ethanol production and alternative processing to produce other chemical products.

Acid Hydrolysis and Fermentation

A new technology involving acid hydrolysis and fermentation has developed on an experimental basis by Brenner and Rugg at New York University (Business Week 1980). The concept involves the use of very weak sulphuric acid applied under 500 psi pressure conditions in a continuous reactor such as a multiple screw plastics extruder. The end product is a syrup of sugars which is further converted by enzyme reaction to ethanol.

It is uncertain that this development will prove to be useful as applied to MSW biomass for chemical production. The chemical process is not new, but the technique of using a special extruder reactor vessel may be worthy of consideration.

SUMMARY

Process technology to isolate biomass from MSW is well advanced for the production of RDF for energy. Wet and dry shredding and screening methods borrowed from other industrial processes are employed. The RDF shows promise as a raw material for chemical feed stocks.

New technology such as wet and dry methods of isolating cellulose fibers from biomass show promise as a means of resource recovery for the paper industry and other cellulose using industries.

Pyrolysis and biological as well as acid hydrolysis techniques for conversion of MSW biomass to organic chemical products require further research and development investment.

REFERENCES

Black Clawson Co. 1972. Bulletin: Hydrasposal-Fibreclaim. Franklin, Ohio and New York, N.Y.
Business Week. 1980. 36D, Jan. 14. (See also Chemical Engineering News, Jan, 26, 1981: 53)
Dade County. 1982. Bulletin. Resource Recovery (Dade County), Inc., Miami, Florida
Garden State Paper Co. 1982. Media General - Annual Report, 1982. Richmond, Va.
Goldstein, I.S. 1981. Organic Chemicals from Biomass. CRC Press, Inc. Boca Raton, Fla. 310 pp.
Hadenhag, J.G. 1981. Swedish Pulp and Paper Mission, Portland, Oregon and Flakt, Inc., Old Greenwich, Conn.
Johansson, Per O. 1982. "Secondary Fibers from a New Source-MSW." Flakt, Inc., Old Greenwich, Conn.
Kidwell-Katzen. 1982. Cellulose to Ethanol Pilot Plant Proposal, unpublished. Bio-Fuel Industries, Inc. R. Katzen Associates Int'l Inc., Cincinnatti, Ohio U.S.A.
Marsh, P.G. 1973. Report 118, Soraïn-Saar Solid Waste Process, Rome, Italy. BCF Inc., unpublished. (The Black Clawson Co., New York, N.Y. U.S.A.
Marsh, P.G. 1977. Fibreclaim - A Progress Update. Deutsch Papierwirtschaft. September.
NCRR. 1981. Bulletin,/NCRR, Inc., Washington, D.C. Vol. 11, No. 3
New Jersey Senate. 1981. Public law 1975 amended 22 June 1981, Senate Assembly No. 2283.
Snider, B. Jr. 1982. Paper as Fuel: The Burning Issue. Forest Industries Advisory Council, Boca Raton, Florida.
Sussman, D.B. 1974. Baltimore Demonstrates Gas Pyrolysis. Environmental Protection Agency Report 530/SW-75d.i.
Vokes, R.F. 1983. Municipal Solid Waste: A Raw Material. Article in this volume. Table I.

CHEMICALS FROM BIOMASS

W. J. Sheppard and E. S. Lipinsky

Battelle Columbus Laboratories
505 King Avenue
Columbus, Ohio 43201
U.S.A.

INTRODUCTION

The use of biomass as a source of materials is different from the use of biomass as a source of fuels. First, the scale of production of chemicals is smaller than of fuels. Using the United States as an example, in 1980 total energy use was 79 EJ* (DOE 1982). Chemical feedstock use (which is included in the above) was 3.2 EJ, or 8.4 percent of the total (USITC 1980; C&EN 1981). (OECD data gives a lower number, 6.1 percent for the United States and 6.3 percent for the OECD countries as a whole (IEA).) Another way to compare the chemicals and fuels is to look at ethyl alcohol and gasoline. The United States in 1980 consumed 3.8×10^{11} liters of gasoline, but synthetic ethanol production was only 6.6×10^8 liters. (By way of comparison with biomass production, if all the maize production in the United States were converted to ethanol, it would amount to 8.4×10^{10} liters.)

Second, the market prices for chemicals are higher than for fuels. For synthetic anhydrous ethanol sold in tank truck or tank car lots, the late September 1982 price was $0.48 (U.S.) per liter; gasoline was about $0.26/l at the refinery, excluding gasoline taxes. If the value of the high octane number of ethanol is ignored and the comparison based on heat content alone, ethanol would be about $0.71/l of gasoline equivalent.

At one time, all organic chemicals were made from biomass; for looking at the future role for biomass, the reasons for the predom-

---

* 1 exajoule = 0.95×10^{15} Btu (0.95 quad).

inance of oil and natural gas as feedstocks today is instructive. Starting 150 years ago, high value organic chemicals such as dyes, and later drugs, were manufactured from coal tars and coal oils, which were byproducts of making town gas and metallurgical coke. The availability of low-cost petroleum fractions such as natural gas liquids in the U.S. and naphtha in Europe following World War II and the development of processes and equipment for making fibers, plastics, detergents, etc., on a large scale led to the development of the large industrial complexes familiar to us today. Biomass lost out for several reasons.

Low Cost. By using naphtha in Europe, where it was in surplus, and natural gas liquids in the United States, where natural gas was priced as a waste product, aliphatic raw materials such as ethylene and propylene could be produced cheaply. Aromatic chemicals would be extracted from gasoline precursors since octane needs of the gasoline could be supplied by adding organolead compounds. For a discussion of why naphtha was cheap in Europe and natural gas liquids in the United States, see Appendix A.

Scale. Biomass-based plants are limited in economic size by the cost of collecting, transporting and storing large amounts of material, whereas petrochemical plants can be quite large. (Average transportation costs will increase as the square root of the total amount of feedstock if a uniform distribution is assumed.) The capital investment for two different sized plants usually is related to size ratio raised to the 0.7 power. That is, a plant 10 times as large as another costs only 5 times as much, thus reducing capital charges per unit output to one half. Price ratio - (size ratio)$^{0.7}$; $10^{0.7} = 5$.

Stability of Prices. Agricultural commodities fluctuate greatly in price (see Figures 1-3) depending on season, weather conditions, alternative crops that can be grown on the same land, and alternative uses for the product such as food and animal feed, whereas for many years petroleum prices were stable.

Geography Not Limiting. Oil from many sources can be used to produce the naphtha needed and natural gas liquids can be obtained from any gas field. One is not at the mercy of, say, a typhoon in the Philippines, as one is in the case of coconut oil.

Today many of these factors have changed. Prices for petroleum have risen rapidly and in real (constant dollar) terms have fluctuated (Figure 4). Prices of biomass materials have fallen relative to petroleum as can be seen from the graph of soybean oil prices divided by the cost of Saudi Arabian oil (Figure 5). Although soybean oil has not fallen to the point where it is equivalent in price per unit of heating value, it is a much more attractive feedstock for chemicals than it was 20 years ago. Petroleum is certain to

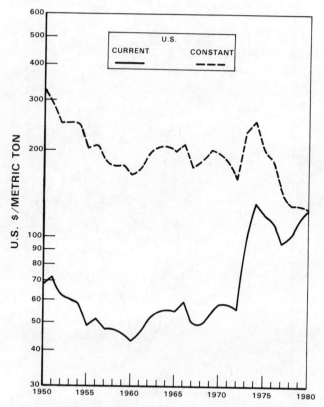

Figure 1. Price of maize landed in Europe.

rise in real terms in the future as a result of depletion of oil fields. Due to the strength of OPEC and the large amount of oil coming from the Persian Gulf, safety of supply through geographic diversity is not what it once was.

CHEMICALS NOW MADE FROM BIOMASS

A large number of chemicals are or could be made from biomass. A review of these will provide useful perspective.

Fermentation Products

Fermentation of sucrose (cane or beet sugar or molasses), glucose (from hydrolysis of starch or cellulose) or fructose (from Jerusalem artichokes) can give rise to ethanol, butanol, acetone, acetic acid, citric acid, xanthan and other gums, and certain amino acids, vitamins, and antibiotics. In the United States fermentation ethanol recently has become cheaper than synthetic ethanol, and in

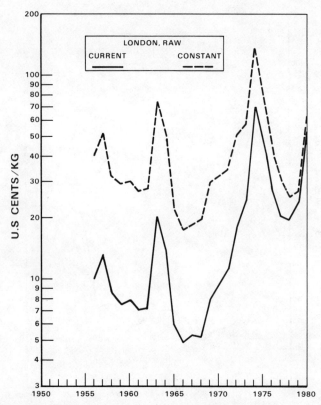

Figure 2. Price of raw sugar, London. (Note: The recent price of sugar has been $0.14 = 0.15/kg.).

South Africa a plant producing butanol and acetone is reported to be operating profitably, although no such plants are known to be operating elsewhere. Citric acid is made commercially by fermentation, but acetic acid is cheaper to produce from ethylene or methanol plus carbon monoxide.

Ethanol is a potentially useful building block for the chemical industry, as shown in Figure 6. Currently, ethylene from cracking of liquefiable petroleum gases (ethane, propane, and butane), naphtha, or heavy distillate oil is a cheaper source of all but ethanol and its esters. However, there was a time (World War II) when ethylene was made from ethanol and, at a high enough price for petroleum feedstocks and given sufficient amounts of ethanol, it could be again.

Lignocellulose

In addition to making ethanol from glucose made by hydrolyzing cellulose, a variety of products can be made from lignocellulose as

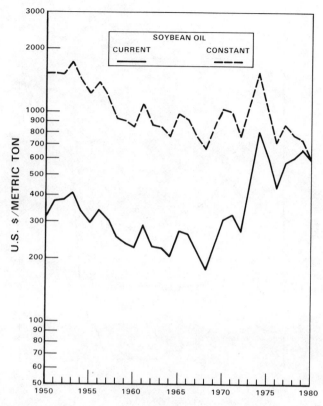

Figure 3. Price of soybean oil, Netherlands ports.

shown in Figure 7. Mild acid hydrolysis dissolves the five carbon sugar polymers (hemicellulose), hydrolyzes them and creates a complex mixture of chemicals in solution. One of these, furfural, steam distills away easily, along with some acetic acid, and can be subsequently purified. Currently, the price of furfural is more than twice the price of phenol, about $1.41/kg for furfural versus $0.55/kg for phenol. If the price of furfural were the same as that for phenol, it could take over many specialty resin markets now served by phenol. Furfural is used today in foundry-core binder resins.

The chemical uses for cellulose include not only the manmade cellulosic materials such as rayon but, of course, cotton fiber, which competes directly with synthetic fibers.

Lignin-derived phenolic polymers are currently reported to be worth the same amount of phenol itself, since they can be used directly in phenol-formaldehyde resins. Potentially, phenol itself can be made from lignin.

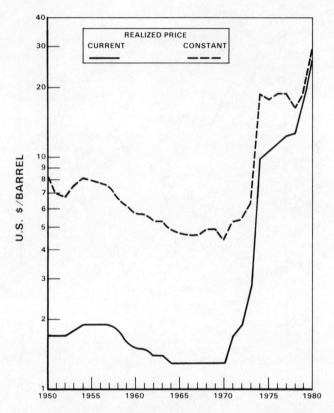

Figure 4. Crude oil, realized price, Saudi Arabian light.

Fats and Oils

Fats and oils are triglycerides; that is, esters of glycerin and fatty acids. Figure 8 gives some of the chemical uses of fats and oils. The structure shown in Figure 8 has several different fatty acids represented, at the top stearic acid (18 carbons), in the center, palmitic acid (16 carbons), and at the bottom, oleic acid (18 carbons with one double bond). If the fatty acid residues are mostly saturated, that is, without double bonds, the triglyceride is solid and called a fat. If there are many unsaturated residues, it is a liquid and is called an oil. If there are several double bonds in each chain, as in linseed oil, it reacts readily with oxygen from the air to form a polymer with triglycerides linked to each other. This is the basis for drying-type paints, which have largely been displaced by latex paints.

Fatty acids are also obtained from the kraft (sulfate) wood pulping process, along with rosin acids. A big use for tall oil fatty acid salts is as flotation agents for ore beneficiation. Rosin acids are used in paper sizing.

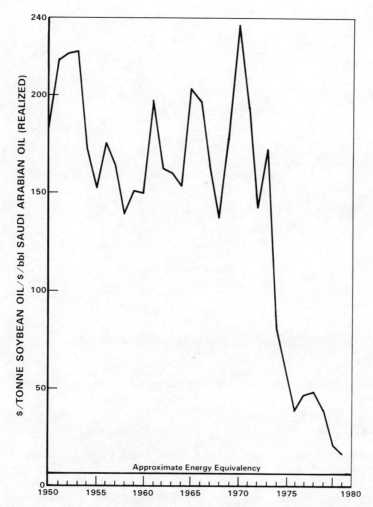

Figure 5. Ratio of soybean oil prices to crude oil prices.

Terpenes

In nature, many compounds are found that appear to be polymers of isoprene (structure A in Figure 9). These are called terpenes (10 carbons), sesquiterpenes (15 carbons), diterpenes (20 carbons), etc. Turpentine, which is obtained from pine trees, pine stumps, and the processing of softwood by the kraft process, is a mixture of 10-carbon hydrocarbons. It is used as a solvent and raw material for making adhesives and specialty resins. The price historically is about $0.40/l, but recently has dropped so low that it has been used as factory fuel.

Structure D is that of cholesterol, which is representative of a variety of materials that nature makes starting with terpenoid

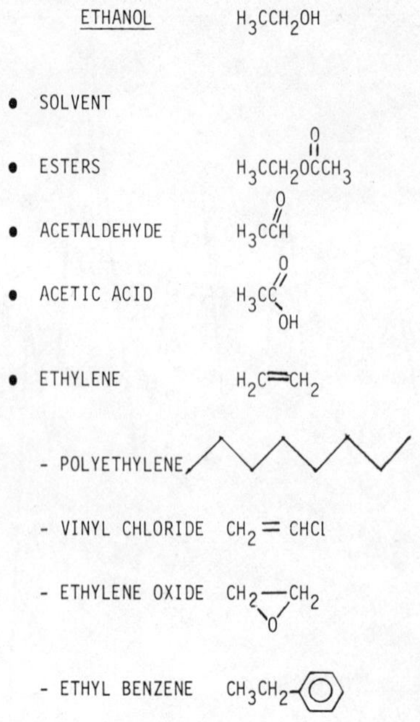

```
ETHANOL           H₃CCH₂OH

• SOLVENT
                               O
                               ‖
• ESTERS          H₃CCH₂OCCH₃
                            O
                            ‖
• ACETALDEHYDE    H₃CCH
                           O
                           ‖
• ACETIC ACID     H₃CC
                           OH

• ETHYLENE        H₂C=CH₂

- POLYETHYLENE  /\/\/\/\

- VINYL CHLORIDE  CH₂=CHCl

- ETHYLENE OXIDE  CH₂—CH₂
                      \O/

- ETHYL BENZENE   CH₃CH₂-⌬
```

Figure 6. Chemicals from ethanol.

compounds, which include sex hormones, Vitamin D, and corticosteriods.

Calcium Carbide

Charcoal from biomass as well as coke from coal can react with lime in an electric furnace to produce calcium carbide. Addition of water generates acetylene, from which many useful chemicals can be made, as shown in Figure 10. The intermediate shown at the bottom, derived from acetone and acetylene, is useful in making terpenoid materials such as Vitamin A. During World War II, Germany used acetylene from coal via calcium carbide to make synthetic rubber and many other materials. Acetylene as a raw material has largely been replaced by cheap ethylene.

Synthesis Gas

Biomass reacted with steam and oxygen or heated with steam in the absence of oxygen will give a mixture of carbon monoxide and hydrogen. This can be used directly as a fuel, or used to make pipeline-quality gas, hydrocarbon liquids (Fischer-Tropsch process), or fuel-grade methanol. Methanol can be used to make gasoline via the Mobil MTG process. Chemical uses of methanol are shown in Figure 11. The synthesis gas can be shifted to all hydrogen by reaction of

CHEMICALS FROM BIOMASS

CELLULOSE

- RAYON, CELLOPHANE
- ACETATE RAYON
- CARBOXYMETHYL-CELLULOSE

HEMICELLULOSE

- C_5 SUGARS $\xrightarrow{H^+}$ FURFURAL

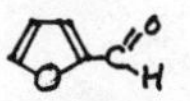

  - ACID-PROOF RESINS
  - LUBE OIL SOLVENT

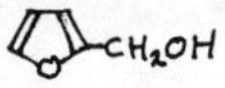

    FURFURYL ALCOHOL

  - FOUNDRY CORE BINDER RESIN

    TETRAHYDROFURAN

  - SOLVENT
  - LYCRA, NYLON, ETC.

LIGNIN

- PHENOLS
- FILLER
- PLASTICS
- VANILLIN

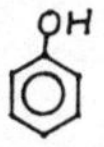

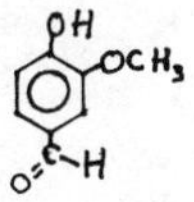

Figure 7. Chemicals from lignocellulose.

carbon monoxide with water over an appropriate catalyst. Hydrogen can be used to make ammonia, which in turn can be used to make nitric acid, ammonium nitrate, organic amines, and urea.

As discussed in the chapter on Biomass -- the Agricultural Perspective in this volume, chemical uses have higher value than fuels but a lower value than food and feed, whereas the volume of fuels is larger than the volume chemicals. We have applied this concept to the ranking of various chemicals. Figure 12 shows the logarithm of volume and the logarithm of unit price for about 140 chemicals, using 1974 U.S. data (Sheppard and Lipinsky 1977). A band of points appear that fits a rough straight line. When the various types of chemicals are grouped by use, it can be seen that the

FATS AND OILS

- LUBRICANTS
- SOAPS
- DETERGENTS
- GLYCERIN
- PAINTS
- PLASTICS
- PLASTICIZERS

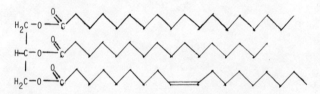

Figure 8. Chemicals from fats and oils.

hierarchy of values and volumes is reflected here. Medicinal chemicals, flavor and fragrances, dyes and pesticides, all things with specific physiological activity, have a high unit value and a low volume. Basic building-block chemicals, on the other hand, have a large volume and low value. This indicates that it is unlikely that

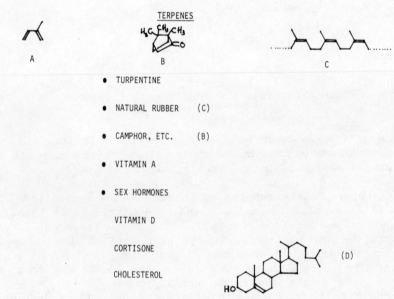

Figure 9. Chemicals from terpenes.

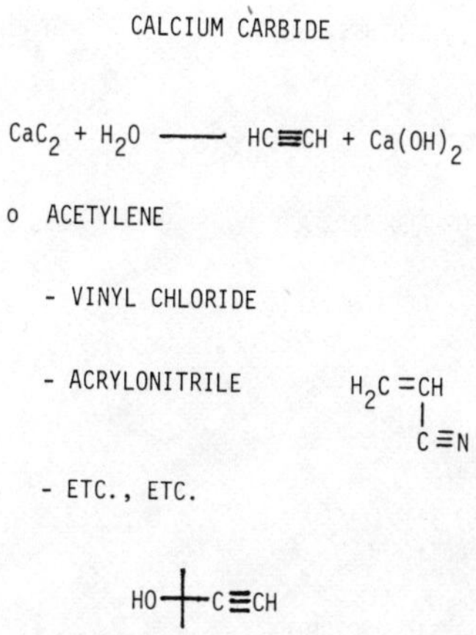

CHEMICALS FROM BIOMASS

CALCIUM CARBIDE

$$CaC_2 + H_2O \longrightarrow HC\equiv CH + Ca(OH)_2$$

o ACETYLENE

- VINYL CHLORIDE

- ACRYLONITRILE $H_2C=CH$
                        $|$
                       $C\equiv N$

- ETC., ETC.

$HO-C\equiv CH$

Figure 10. Chemicals from calcium carbide.

a large volume use will be found for a high priced chemical derived from biomass (or any other source).

It is interesting to see what has happened to the log-log relationship with the changes in oil prices that have occurred in the last decade. Three years were chosen for examination: 1972, the year before the oil embargo; 1976, a year of recovery following the 1975 business recession; and 1980, the most recent year for which consistent data are available. As can be seen in Figure 13, prices rose more for the less expensive chemicals than for the high-price, low-volume chemicals. In other words, a shift in the slope of the trend line in the log-log plot has occurred. Using a larger sample (35 chemicals), the slope in 1972 was -0.46; in 1976, -0.39, and in 1980, -0.36. This might be expected since the petroleum raw material cost is a larger fraction of the final cost of the low cost materials. Suprisingly, the volume shifts did not relate to the price changes. For example, phthalic anhydride, dioctyl phthalate and phthalic-type alkyds increased in price and decreased in volume. Ethylene and polyethylene increased in both price and volume. It appears that the life cycle for the various chemicals -- introduction, growth, maturity, and decline -- continued relatively little affected by the price increases. A second factor emerged -- the 1976-80 crude oil price rise had a smaller proportional impact than the 1972-76 price rise as shown in Table 1. Perhaps the industry was used to shocks, perhaps conservation measures started in 1973 and 1974 had their real impact after 1976. (Between 1972 and 1980 the U.S. chemical industry reduced fuel use by more than 20 percent for each pound produced; no data are available on reduction in feedstock use per pound.)

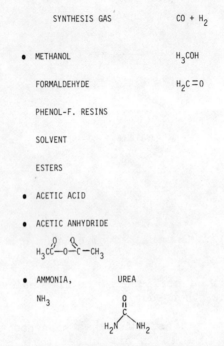

Figure 11. Chemicals from synthesis gas.

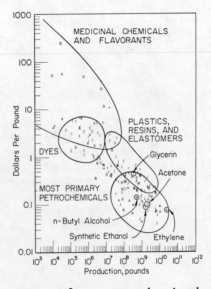

Figure 12. Price versus volume, organic chemicals by type, 1974.

PROSPECTS FOR SOME SPECIFIC BIOMASS-DERIVED CHEMICALS

Up to now, the discussion has focused on families of chemicals and general strategies. Sooner or later, individual companies have to make decisions pertaining to construction of facilities to produce specific chemicals. Which specific biomas-derived chemicals appear to be attractive? The answer to this question involves not only consideration of markets, biomass resources, and available technologies, but the location of and strategic position of the company contemplating the investment. Typical chemicals that may merit consideration in our opinion include two that have been very popular and one that has been largely overlooked:

o Ethanol from site-specifically attractive carbohydrates
o Lactic acid from carbohydrates
o Acetic acid from carbohydrates

Each of these possibilities has its pros and cons. Caveat investor!

Ethanol

The Brazilian Proalcool and United States gasohol programs, which involve the manufacture of fuel ethanol by fermentation of carbohydrates, have tended to obscure the industrial chemical and solvent opportunities implicit in this technology. By use of a somewhat more rigorous distillation, industrial-grade hydrous and anhydrous ethanol can be manufactured rather inexpensively from either simple sugars or starch. It is likely that this technology for ethanol production will displace petrochemical ethanol in most areas of the world.

Table 1. Price increase ratios for selected chemicals and crude oil.

	1976[a] Price 1972 Price	Ratio Relative to Crude Oil	1980[a] Price 1976 Price	Ratio Relative to Crude Oil
Crude Oil (U.S. Average Price)	2.46		1.93	
Ethylene	2.80	1.14	1.46	0.76
Vinyl Chloride	1.88	0.76	1.34	0.69
Polyvinyl chloride	1.50	0.61	0.92	0.48

[a] In constant dollar (inflation adjusted) terms

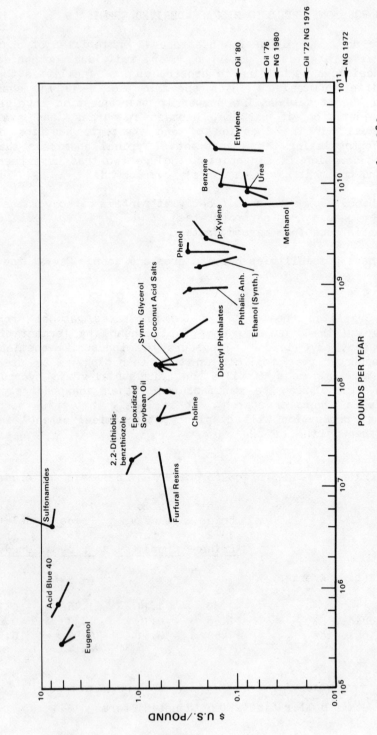

Figure 13. Prices versus volume, organic chemicals, 1972, 1976, 1980.

Lactic Acid

The mass balance aspects of lactic acid production are considably more favorable than for the production of ethanol. Fermentation of carbohydrates yields roughly equal quantities of ethanol and carbon dioxide, the latter having a very low value. Lactic acid reduction from carbohydrates does not entail the loss of carbon dioxide, therefore, almost twice as much salable product per unit weight of carbohydrate is produced. This difference is illustrated in Figure 14. Note that the energy contents, represented by the areas, are about the same for all three chemicals. Despite this considerable advantage of lactic acid technology over ethanol technology, very little interest has been shown in lactic acid via fermentation by the chemical industry. This relative lack of interest reflects a serious lack of development of markets for lactic acid. The chemical structure of lactic acid would appear to be conducive to the development of large markets because the molecule is bifunctional. The small market size may be partially attributed to the high selling price associated with the small production of this chemical at the present time. Other causes include problems in purifying lactic acid when it is made by traditional fermentation processes, difficulties in polymerizing lactic acid by condensation methods and difficulties in dehydrating lactic acid to acrylic products. Successful achievement of dehydration to yield acrylic acid would open the way to the billion pound per year markets that are associated with acrylics. Biochemical approaches are discussed in the paper by D.E. Eveleigh. This is an area worthy of R&D effort.

Direct polymerization of lactic acid or its esters to high molecular weight polymers is difficult to achieve. However, high

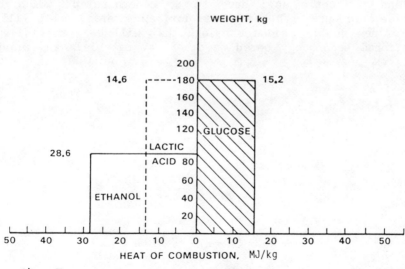

Figure 14. Energy comparison of glucose, ethanol, and lactic acid.

molecular weight polymers can be made from lactide because this internal ester can be polymerized by an addition reaction, instead of a condensation reaction (see chart below). Two examples of products that have been derived from lactide are sutures that are extremely strong biodegradable fibers and the films and molding compounds made by Sinclair (1977). Sinclair's lactide polymerization method makes use of comonomers to tailor the end products. By using this approach, biodegradable products resembling either polystyrene or polyvinyl chloride can be made.

Those who choose to commercialize lactic acid can adopt either the strategy of direct substitution for petrochemicals now made from propylene by mastering the dehydration chemistry leading to acrylics or can attempt to displace polystyrene and polyvinyl chloride by simulating its properties via lactic acid copolymers. Either or both of these approaches may prove feasible.

Acetic Acid

Researchers at several organizations have found microorganisms that are capable of producing acetic acid from carbohydrates more effectively than do the microbes that make vinegar from glucose via ethanol. Vinegar is associated with the loss of carbon dioxide and a poor mass balance. The use of <u>Clostridium</u> <u>thermoaceticum</u> makes possible the production of approximately 0.9 grams of acetic acid per gram of carbohydrate consumed. In a country with low sugar costs, the raw material cost for acetic acid, which sells for approximately $0.55 - 0.66 per kilogram, could be approximately $0.11 - 0.22 per kilogram. Such a low raw material cost provides lots of economic room for processing costs and potential profit. However, these costs are likely to considerable because very dilute solutions of acetic acid need to be concentrated, which is an expensive procedure. Whether these low raw material costs will make possible new acetic acid technology that will be competitive with conventional technology based on other raw materials remains to be seen.

CONCLUSIONS

o Many biomass products have dropped in price relative to crude oil, but most do not compete on an energy content basis. Wood for direct combustion can be an exception at specific sites.

o Chemicals have higher prices and lower volumes than fuels.

o The chemical market chosen should be matched to the biomass source in terms of price of raw material and volume easily obtained. In general, this will be more easily achieved than producing substantial amounts of fuel from biomass.

o High-price, low-volume chemicals from biomass will not profit as much from an increase in oil prices as low-price, high-volume chemicals.

o The best opportunity may lie in the middle of the price range, if one can find a chemical made easily from biomass that competes with the same or a substitute chemical made in several steps from petroleum.

REFERENCES

Chemical and Engineering News, May 4, 1981, p. 41; June 8, 1981, p. 67.
Department of Energy, "Monthly Energy Review," DOE/EIA 0035 (82/02).
Sheppard, W.J. and E.S. Lipinsky, "Can Sucrose Compete With Hydrocarbons as a Chemical Feedstock?" Sucrochemistry, J. L. Hickson, ed. Am. Chem. Soc. Symposium Series No. 41 (1977) pp. 336-50.
Sinclair, R.G., U.S. Patents 4,045,418 and 4,057,537 (1977).
U.S. International Trade Commission, Synthetic Organic Chemicals, 1980, USITC 1183.

APPENDIX A

FEEDSTOCKS FOR PETROCHEMICALS

From Natural Gas

Natural gas as it comes from the well is a complex mixture of hydrocarbons and other gases. Before it can be shipped in a pipeline, typically at a pressure of about 70 atmospheres, it must undergo treatment to remove materials that are corrosive, non-combustible, or that will form liquids or solids in the pipe. The gas as obtained at the well-head is termed wet gas; Table A-1 gives a typical composition of the hydrocarbons in wet natural gas. In addition, water vapor, hydrogen sulfide, carbon dioxide, nitrogen, and helium can be present. In the treatment process the butanes, propane, and heavier hydrocarbons are almost completely removed since they can condense if present in even moderate amounts. The hydrocarbons with five or more carbons, termed natural gasoline, find a ready market in gasoline. The propane and butane, called liquefiable petroleum gases (LPG), can be used as a fuel or cracked in the presence of steam to make a mixture of ethylene, propylene, butadiene, and a heavier fraction called pyrolysis gasoline. In the United States, the rapid expansion of transcontinental gas pipelines during the 1950s led to the availabiity of large amounts of LPG in the producing areas and low prices for it. The petrochemical industry used this low-cost feedstock to develop an industry based on ethylene.

After the industry was established, it was found that it was economic to remove a large fraction of the ethane as well from the wet gas for use as a chemical feedstock, even though this requires more expensive equipment. A typical composition of the hydrocarbons in dry (pipeline quality) natural gas is shown in Table A-1, along with the amount of hydrocarbons removed.

Petroleum refineries are net consumers of butanes, especially in the winter, and net producers of propane. Refiners sell liqefied refinery gases (LRG) that is mostly propane, which can also be used as a petrochemical feedstock or fuel.

From Refineries

To appreciate the relationship of the refining process to petrochemical feedstock production, it is convenient to examine what goes on in a variety of refineries, from the simplest to the most complex.

The simplest configuration, often derisively referred to as a tea kettle or coffee pot, involves only atmospheric distillation of the crude oil and yields a few fractions, as shown in Figure A-1.

Table A-1. Typical composition of wet and dry natural gas (cubic meters per 1000 cubic meters of wet gas).

Component	Formula	Wet	Dry	Removed
Methane	CH_4	880	880	
Ethane	C_2H_6	70	14	56
Propane	C_3H_8	30	2	28
Butanes	C_4H_{10}	20		20

CHEMICALS FROM BIOMASS 653

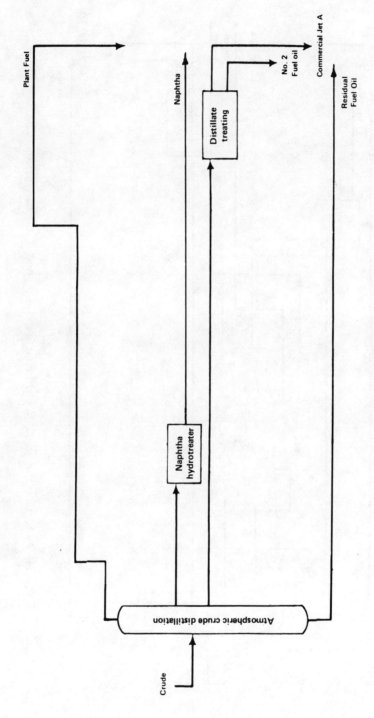

Figure A-1. Simple refinery.

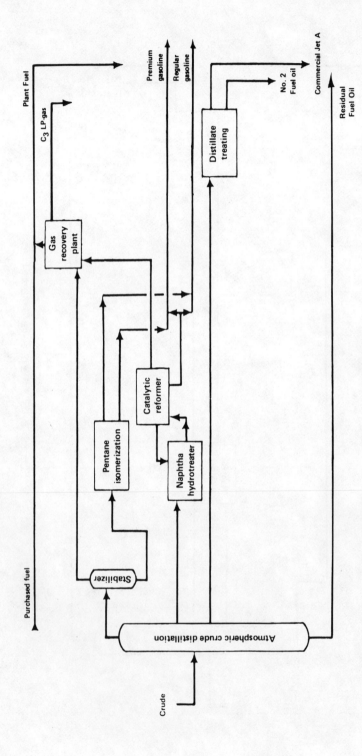

Figure A-2. Refinery with balanced gasoline and fuel oil production.

CHEMICALS FROM BIOMASS 655

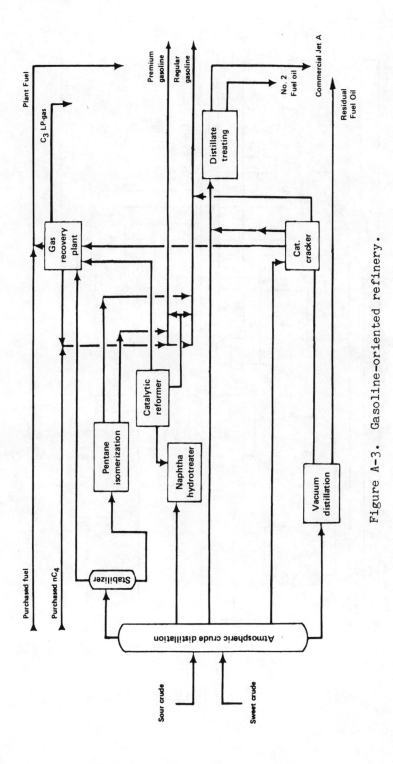

Figure A-3. Gasoline-oriented refinery.

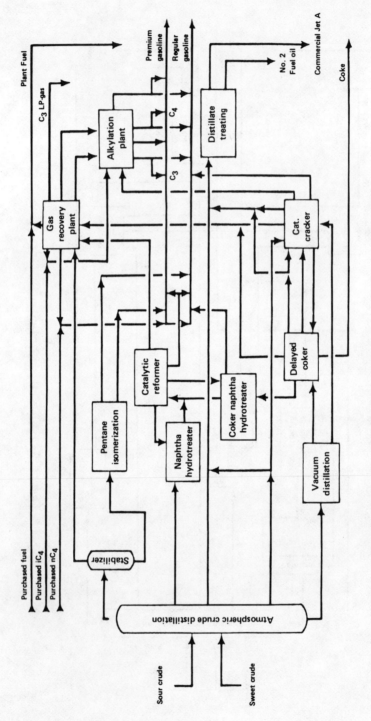

Figure A-4. Refinery maximizing gasoline production.

The pentanes and lighter are used as fuel. The next fraction, which boils in the gasoline range, is too low in octane to be used directly as gasoline. Before sale it is usually treated with hydrogen to reduce the formation of gums and to remove sulfur compounds. It can be used as a solvent, upgraded to gasoline, or cracked in the presence of steam to give ethylene, propylene, butadiene, etc.

The middle distillate fraction can be used for fuel and some of it can be used for jet fuel, depending on the boiling range. Like the naphtha, it is usually treated with hydrogen before sale.

The material that does not distill is sold for fuel use, usually at a low price. This simple refinery is typical of those found in an underdeveloped country with abundant oil production.

If there is a moderate demand for gasoline and local or export markets for all the heavier products, the naphtha can be upgraded to gasoline by reforming as shown in Figure A-2. This process transforms straight chain hydrocarbons into branched chain compounds and benzene derivatives ("aromatic" compounds). The new compounds have a much higher octane number than the straight chain hydrocargons. (The addition of lead alkyl compounds also will increase the octane number.) In addition, the pentanes from the still gases are extracted, isomerized to convert straight chain compounds to branched chain, and blended.

If selling heavy distillate fuel oil is more profitable than selling residual fuel oil because of its lower sulfur and ash content, the bottoms from atmospheric still can be sent to a vacuum still. The vacuum "resid" from some crudes is suitable for proccessing to asphalt. Refineries with gasoline manufacturing capabilities and a vacuum still are characteristic of European areas that lack local natural gas supplies. In areas where the demand for fuel oil is much greater than for gasoline, there is a surplus of naphtha, which depresses it price. In these areas the naphtha is used as the feedstock for the making of olefins (ethylene, etc.) by steam cracking. If fuel oil demand is low, the distillate from the vacuum still, called vacuum gas oil, can be used for olefin manufacture, assuming no cheap LPG is available.

If gasoline demand is very high relative to fuel oil needs, the heavy distillate fraction can be cracked using a catalyst to produce lighter hydrocarbons as shown in Figure A-3. The gasoline fraction produced by cracking has the advantage of having a high octane number.

The cracking operations as well as the reformer and atmospheric still provide an abundance of hydrocarbon gases. If the refinery needs more high octane material, isobutane and olefins such as propylene and isobutylene from the gas streams can be reacted to form so-called alkylate gasoline, as shown in the upper right corner of Figure A-4. Refineries with alkylate production have a gasoline yield equal to about 50 percent of total product and are typical of the large refineries in the United States.

Environmental restrictions, competition from coal, and increasing use of heavy crude oils has led to a glut of high sulfur residual fuel oil. A solution to this problem is to use as much as possible of the residual oil in making distilled products. Figure A-4 shows at the lower left the flow of the "resid" to a delayed coker, where it is heated and slowly decomposes into hydrogen-rich products that vaporize and hydrogen-poor coke. The coke is usable in making electrodes for aluminum smelting and for fuel use.

To obtain aromatic chemicals, such as benzene, toluene and xylene (BTX), the naphtha fraction of the catalytic cracker output or reformer output can also be extracted to yield BTX. In addition, pyrolysis gasoline from the cracking of LPG to olefins can be extracted with a solvent. Aromatic extraction usually gives more toluene and less benzene than the petrochemical market demands; therefore, some toluene is treated with hydrogen over a catalyst to remove the methyl group and yield benzene, a process called hydrodealkylation. Since aromatics have a very high octane value, there is a competition between the petrochemical market and gasoline market for these materials. Regulations limiting the amount of lead that can be used in gasoline has increased the pressure to leave the aromatics in the gasoline pool. Ethanol has been proposed as a gasoline additive to replace lead alkyls, freeing the aromatics for sale as chemicals; however, the current recession has both reduced sales of gasoline and sales of aromatic chemicals for the time being.

ENERGY EFFICIENCY IN THE PROCESS OF ETHANOL PRODUCTION FROM MOLASSES

Ali Beba

Department of Chemical Engineering
Ege University
Izmir, Turkey

INTRODUCTION

Production of ethanol (ethyl alcohol) by fermentation is one of the biochemical methods that has been practiced in converting biomass into liquid fuel for thousands of years. The basic process is well-known, but the conversion paths with different cultures are still being investigated. Recently important research and developmental activities have been created in the following areas:

1. Continuous fermentation technology in order to achieve higher ethanol concentrations at elevated temperatures.

2. More energy efficient chemical engineering unit operations including the process of azeotropic distillation.

3. Different ethanol dehydration routes to obtain absolute (water-free) ethanol by minimizing the energy need of the process.

The major objective of these studies is to determine the technological feasibility of different conversion methods for producing liquid fuels. Ethanol, at the present, is produced from carbohydrates of the following compositions:

Sugar (sugarbeet, sugarcane, molasses, etc.)
Starch (corn, potato, cassava, wild palm tree, etc.)
Cellulose (wood, industrial and agricultural wastes, etc.)

The availability and cost of land, labour, raw material and know-how affect the economics of ethanol production. Economic justification of the conversion technology chosen rests firmly upon the circumstances of the countrys' agricultural, industrial and energy sectors. Figure 1 is adopted from a World Bank study (Kohli 1980) to show the energy and agricultural self sufficiency of several counties. In this figure quadrant I represents countries which have surpluses of energy and agricultural products. Quadrant II shows the countries which have agricultural surplus but not sufficient energy. The countries which have deficiency both in energy and agriculture are shown in quadrant III. The last portion of the chart, quadrant IV, shows the countries which are energy self sufficient but agriculturally deficient.

Some developing countries such as Brazil and Turkey do not have any constraints on agricultural land and have surplus in agricultural products, but they both have deficit in energy supply. Countries with similar conditions have great potential for developing massive biomass programs.

The world grain production, however, has been changing to the disadvantage of the developing countries. Table 1 reflects the changing pattern of world grain trade. As may be seen from this table, the developing countries of Latin America, Asia and Africa have become net importers of grain at the present, whereas these countries were net exporters about a decade ago.

A shift to alternative energy sources is a must for the non-OPEC developing world. Turkey represents a typical example. This country is paying more than 80 percent of her export income for the importation of energy (mostly petroleum) annually. The deficit in the balance of payments of Turkey is mostly contributed to the import of liquid hydrocarbon fuels. Large scale programs utilizing renewable energy sources are most important for countries like Turkey in terms of increasing the availability of energy and allowing a greater portion of these nations' efforts to be devoted to raising their standard of living.

DISCUSSION

Sugar bearing materials are most easily converted into ethanol according to the following reaction:

$$C_6H_{12}O_6 \longrightarrow 2C_2H_5OH + 2CO_2$$

$$(180.1) \qquad (92.1) \qquad (88)$$

Here, it is shown that one mole of saccharose yields two moles of ethanol and two moles of carbon dioxide. The actual conversion mechanism follows a rather complex route which will not be repeated here (Bennet and Friden 1973).

ENERGY EFFICIENCY IN PROCESS OF ETHANOL PRODUCTION

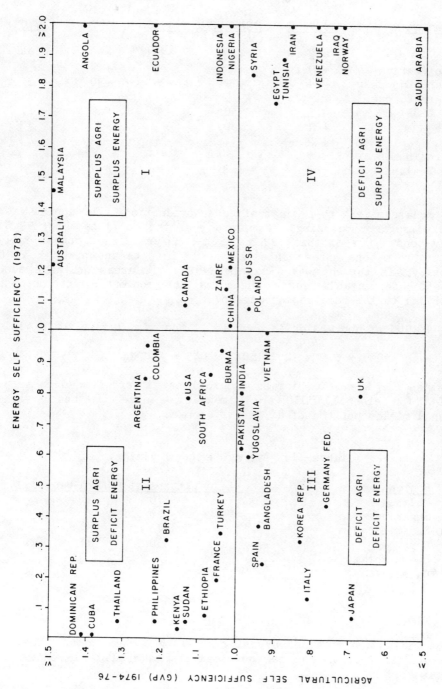

Figure 1. Energy and agricultural self sufficiency.

Table 1. The changing pattern of world grain trade.

Region	1934-38	1948-52	1960	1970	1978
North America	+ 5	+23	+39	+56	+104
Latin America	+ 9	+ 1	0	+ 4	0
Western Europe	-24	-22	-25	-30	- 21
Eastern Europe and USSR	+ 5	0	0	0	- 27
Africa	+ 1	0	- 2	- 5	- 12
Asia	+ 2	- 6	-17	-37	- 53
Australia and New Zealand	+ 3	+ 3	+ 6	+12	+ 14

Molasses is a by-product of sugar production. The annual yield (due to sugar processing) of molasses of the world is about 33.5×10^6 m-tons. Out of this, 22×10^6 m-tons/yr is used domestically, 6.6×10^6 m-tons enters into the world markets and about 5×10^6 m-tons/yr is thrown away (Kovarik 1982). Molasses contains about 46 - 50% fermentable sugar which can be converted into ethanol according to the reaction shown above.

If 90% of conversion is assumed, 1 ton of molasses would yield

$$1000 \times 0.5 \times 92.1/180/1 \times 0.9 = 230 \text{ kg}$$

of ethanol. When the biomass (source) is different than molasses, yields for ethanol will vary somewhat. Table II shows average ethanol yields per ton of different biomass.

Table II. Average ethanol yields.

Source	Yield (liters et OH/ton biomass)
Molasses	270
Sugarcane	70
Sweet sorghum	86
Corn	370
Sweet Potato	125
Cassava	180
Babassu	80
Wood	160

These yields are obtained by using existing technologies. They are expected to increase as new developments take place in conversion processes.

When ethanol is produced to meet the demands of industry (alcoholic beverages, raw or intermediate chemical input) production costs and process efficiencies are not considered as important parameters. On the other hand, if it is produced for its energy content these parameters become very important. Therefore there is a need to have a total understanding of several criteria concerning process economy and efficiency before launching ethanol programs.

Energy Efficiency in Ethanol Production

Since the question of "where the energy balance lies in ethanol production" is still a controversial one, the rest of this paper will focus on this issue.

Due to the differences in the concept of energy efficiency and computational methods applied to determine its value, some researchers claim that the process of converting biomass into ethanol takes place at very low, in some cases negative, efficiencies (Bolt 1964; Lawrason 1964; Mears 1978; Keller 1979; Besorak 1978); whereas some others totally disagree with this view (Radovich 1979; Stallings 1979; Lanouette 1980). Energy efficiency, in general terms, can be defined by the relationship of the energy content of a unit of ethanol produced at a defined concentration and the energy consumed for this production.

The problem usually is in the content of energy consumed during the process. Some take this as the energy utilized in the ethanol plant only, and some others argue that energy consumption should include energy spent in cultivating, harvesting and transporting the crop, manufacturing fertilizers and pesticides, cooking the grain to convert its starch to sugar, fermenting the sugar, and finally distilling the alcohol (Anon. 1980). If the residual from the distillation is to be sold, it must be dried at additional energy cost.

These opposing views must be delicately handled by taking into account the specific countries' circumstances, its agricultural, industrial and energy sectors. The clarification of this concept is very important both from the engineering and economic point-of-view

Mathematical Analysis of Energy Efficiency

The best approach in answering the question of energy efficiency is to use mathematics as a tool. Mathematical analysis can be easily applied to most conversion processes after making the following definitions (Bechtel 1977):

1. Net Thermal Efficiency (NTE)   Indicates the thermal efficiency of the conversion process.
2. Energy Benefit Ratio (EBR)   Shows the amount of the high quality energy produced.

3. State of Technology — Describes how well developed the process is.
4. Availability of Raw Material — Indicates the type and amount of available raw material.

NTE and EBR are quantifiable concepts. In order to clarify their meanings Figure 2 is prepared. This figure illustrates the streams used in these definitions and outlines the legends used in efficiency evaluations (Radovich 1979).

In symbolic form Net Thermal Efficiency (NTE) can be shown as:

$$NTE = Q_5 / Q_1$$

and since

$$Q_2 = Q_3 - Q_4 - Q_5 \quad\quad Q_4 = 0.$$

$$NTE = (Q_2 - Q_3) / Q_1$$

As will be noticed, this definition is based on quantity (content) rather than quality. It assumes the quality of the raw feed to be the same as the quality of the product.

In some conversion processes, such as production of ethanol from molasses, this may not be the case; since product and feed have different characteristics and they can not be compared for their energy values. Therefore (NTE) all by itself is not sufficient to analyze the energy efficiency of every biomass conversion process. It is however, a useful definition in comparing different methods of conversion.

Energy Benefit Ratio (EBR) can be defined as the ratio of the energy content of the fuel product to the energy consumed in the conversion process. By using the legends of Figure 2 it can be expressed as:

$$EBR = Q_2 / Q_3$$

Here it must be pointed out that the conversion processes usually utilize high quality energy, such as fuel oil, coal, and electricity. Since product fuel is also of high quality energy, EBR indicates the gain in usable energy. This concept is applicable to every conversion process. Its value is (must be) greater than 1, i.e., EBR is always above 100%.

Other important parameters in the energy efficiency analysis are the state of technology and the availability of raw material. Technical know-how and domestic resources must be taken into account in choosing the proper technology. Same quality of energy can be obtained from rather different applications. For example nuclear power stations and thermal power stations produce the same quality of energy, namely electricity and heat, but they use very different

ENERGY EFFICIENCY IN PROCESS OF ETHANOL PRODUCTION

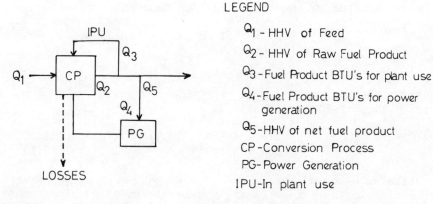

Figure 2. Conceptual biomass conversion process for efficiency evaluations.

know-how levels, technologies and resources. Their complexities are incomparable. Similar differences exist in producing liquid fuels. Oil refineries versus alcohol plants (via fermentation) are typical examples for this case.

As was stated earlier, a total understanding of all the above stated criteria is necessary before making an energy analysis of any conversion system. Besides the concepts of (NTE) and (EBR) other relevant information must be stated clearly in the analysis.

Energy Parameters in Ethanol Production

Several studies have been made for defining the values of NTE and EBR in the process of ethanol production via fermentation and distillation. One study (Klauss 1973) takes the heating value of the feed (Q_1) as 9000 Btu/lb and states the following assumptions:

a) Efficiency of conversion is limited to the plant, i.e., energy used in growing and bringing the raw feed to the plant is <u>not</u> taken into account.

b) Conversion products include ethanol and some methanol + furfural.

c) The residue of conversion is available to generate the required process steam at 50 - 60% efficiency.

The values of streams (see Figure 2) for this case are:

$$Q_2 = 8561 \text{ Btu/lb}$$
$$Q_3 = 6354 \text{ "}$$
$$Q_5 = 2207 \text{ "}$$

When these values are used to determine the values of NTE and EBR the following results are obtained:

$$\text{NTE} = (9561 - 6354) / 9000 \quad (24\%)$$
$$\text{EBR} = 8561/6354 \quad (134\%)$$

State of technology is well developed in small, medium and large scale plants. It is easily applicable in developing countries (with agricultural surplus and energy deficit) for different biomass resources. Product (ethanol) is a prime quality liquid fuel and it is environmentally accetpable.

The values of NTE and EBR for the above example may vary with different raw material and processes. Another study (Beba 1980) investigated the same efficiency parameters for a plant producing 12.5 million liters/yr of 95% ethanol from mollasses. The author concluded the following:

NTE - Not defined due to the unconvertible energy content of raw feed, i.e., Q_1 has no meaning.

$$\text{EBR} = Q_2 / Q_3 = 24.3 \times 10^6 / 8.38 \times 10^6 \quad (290\%)$$

This study also assumes that conversion products include mostly ethanol and that energy analysis is limited to the plant.

The same work showed that in a pilot plant which produced 1.224 liters/hr of 100% (anhydrous) ethanol, EBR is about 116%.

Energy Parameters in Other Conversion Processes

The methods which convert biomass into fuel can be classified as:

1. Biochemical   ----   a) Fermentation
                            b) Anaerobic digestion

2. Thermochemical   ----   a) Pyrolysis
                            b) Gasification
                            c) Liquefaction

Existing methods utilizing the above mentioned conversion technologies can be compared by their overall energy efficiency values. This is possible when efficiency parameters such as NTE and EBR are taken as the basis for comparison. Table III is prepared for this purpose.

As may be seen from this table, fermentation occupies the last row. Biogas ranks as number one (both in NTE and EBR) in this analysis. If the conversion efficiency of a thermal power plant is taken into account, one will notice that even fermentation processes can be considered as high efficiency systems. This is due to the fact that thermal power plants have a maximum efficiency of 36%. This value is easily comparable with NTE value of fermentation.

CONCLUSION

Biomass is a renewable energy source which can be converted into liquid fuels (and/or gas fuels) with different technologies available today. Ethanol production is one such method and it has been practiced for thousands of years.

Fermentation and distillation technology is well known and readily applicable in most developing counties with agricultural surplus. There are several draw-backs concerning this application, however, and intensive research and development activities must be carried out in different phases of ethanol production. One major area of research is dehydration of 95% ethanol-water mixtures.

This study indicates that there is still confusion in the concept of "overall conversion efficiency" in ethanol production via fermentation and distillation. Net Thermal Efficiency (NTE). Energy Benefit Ratio (EBR), state of technology and availability of raw material must all be treated together in order to give a full picture of the conversion process.

Table III. Comparison of energy efficiencies of different processes.

Type of Process	Q_1	Q_2	Q_3	Q_4	Q_5	NTE	EBR
Pyrolysis	9000	3930	450	29	3450	0.34	3.79
Waste-to-oil	9000	5250	NA	NA	NA	-	-
Syngas	9000	6679	549	NA	NA		
Purox (municipal solid waste)	4500	3649	274	1323	2052	0.46	2.28
Biogas	9000	5935	0	635	5300	0.59	9.34
Fermentation	9000	8561	6354	0	2207	0.24	1.34

Streams (Figure 1)

Energy balance in ethanol production is the major issue and it must be resolved before launching ethanol programs (ethanol for energy) in developing countries.

When the efficiency values (NTE values) of ethanol production and thermal power stations are compared, results favour ethanol production rather than thermal power stations.

REFERENCES

Anon. 1980. Gasohol: Does it save energy: Vol. 14:2.
Beba, A. 1980. Investigation of energy efficiency in the process of ethanol production. Thesis, Ege University, Izmir, Turkey.
Bechtel Corp. 1977 (April). Edison Coordinated Joint Regional Solid Waste Energy Recovery Project: Feasibility Investigation.
Bennet, T.P. and Friden, E. 1973. Modern Topics in Biochemistry. Turkish translation. Turkish Ministry of Education. Istanbul.
Besorak, Y. 1978. A new energy Alternative for Turkish highways. Turkish 3rd. Energy National Conference Proceedings, Vol. 2. Ankara. In Turkish.
Bolt, A.J. 1964. Survey of alcohol as motor fuel. Society of Automotive Engineers, SP-254. New York.
Keller, J.L. 1979. Alcohols as motor fuels. Hydrocarbon Processing, May.
Klauss, D.L. 1973. Fuel Gas from Organic Wastes. Chem. Tech. Vol. 3, No. II.
Kohli, H.S. 1980. Renewable Energy: Alcohol from Biomass. Finance and Development. December.
Kovarik, Bill, Ed. 1982. Fuel Alcohol. Earthscan London.
Lanouette, W.J. 1980. Gasohol. National Journal, Washington, D.C.
Lawrason, C.G. 1964. Ethyl alcohol and gasoline as a modern motor fuel. Society of Automotive Engineers, SP-254, New York.
Mears, L.G. 1978. Energy from Agriculture. Environment Vol. 20, No. 10.
Radovich, J.M. 1979. Conversion processes for liquid fuels from biomass. Atlanta International Conference Proceedings, Atlanta, Ga.
Stallings, J.W. 1979. Economics of energy recovery from biomass using fluidized bed combustion. Atlanta International Conference Proceeedings, Atlanta, Ga.

INTEGRATED PROCESSES FOR CHEMICAL UTILIZATION OF BIOMASS

Irving S. Goldstein

Department of Wood and Paper Science
North Carolina State University
Raleigh, NC 27650
U.S.A.

INTRODUCTION

Extraction and process technology of all kinds is faced with the selective isolation of components or conversion products derived from the raw material. The technology can only be economic if the total value of the products exceeds raw material, processing and investment costs. This can occur over the entire range of the total value obtained by multiplying the unit value or values by the yield. For example, the mining of gold, iron, coal and gravel covers the range from very low yields of high value gold to very high yields of low value gravel. Multiple products can also increase total value, and this is the pattern in the petroleum refining and meat packing industries.

Despite this obvious economic truism the history of the conversion of biomass into useful chemicals has more often than not involved single product plants operating at low yields. The failure of such plants to operate profitably is not surprising, for the need to maximize the yield of products from raw materials for greatest economic efficiency is especially important in the case of a multicomponent raw material like biomass. The raw material cost per unit of product can be reduced as much as fivefold when all the components of the biomass are converted into useful products.

Although the major concern of this discussion is chemical utilization of biomass the product mix in an integrated process for biomass conversion does not have to consist exclusively of chemicals. Other valuable applications of biomass, its components or its conversion products are as food, fiber or fuel. Following an analysis of the potential products from total biomass and its individual components it will become apparent from examples how integrated processes using all the raw material can be devised.

POTENTIAL PRODUCTS FROM BIOMASS AND ITS COMPONENTS

Biomass is comprised of the fibrous glucose polymer cellulose embedded in a matrix of other polysaccharides known as hemicelluloses and the three-dimensional aromatic network polymer lignin. These cell wall components constitute by far the largest portion of the biomass. The remainder may be soluble materials known as extractives and the outer material of woody plants known as bark. In certain plant the seeds, roots and sap may contain high concentrations of sugars, starch or hydrocarbons.

Agricultural systems typically have exploited these latter concentrated biomass components for traditional food applications. Needs for liquid fuels have recently focused on such crops as sugar cane, corn (maize) and oilseeds for this alternative use as well.

Total biomass conversion of the mixed components certainly meets the criterion of complete raw material utilization. Such processes are drastic and nonselective, and are the same as those useful for the conversion of coal. Direct combustion, gasification and pyrolysis are applicable to all carbonaceous material. Uses are mainly fuel directed, but some chemicals are obtainable, especially from synthesis gas.

Selective processing of the individual components can provide substantial yields of a wide variety of chemicals and materials. Cellulose can be isolated as a fiber or hydrolyzed to fermentable glucose. Hemicelluloses can be hydrolyzed to sugars which can be fermented or converted to furfural. Lignin can be broken down into aromatics, used as a resin component, or burned.

Glucose and other fermentable sugars can be converted into ethanol for use as fuel or as a building block for a wide range of industrial organic chemicals. Other fermentation or chemical conversion products of biomass hydrolyzates also have broad utility.

INTEGRATED PROCESSES

The combinations of process steps and final product applications apparent from the brief descriptions above are very numerous. They can be grouped into four convenient categories by uses: Chemicals alone, chemicals and fibers, chemicals and food, and chemicals and fuels. Examples of each are presented below, but these should be considered as illustrative rather than exhaustive.

Chemicals Only

In one approach the readily hydrolyzable hemicelluloses are first removed from the biomass and converted into simple sugars under mild conditions that leave the cellulose and lignin essentially unmodified. Subsequent hydrolysis of the cellulose provides almost pure glucose and a lignin residue that can be further processed to phenols and other aromatic compounds. The isolated sugar streams can be converted into various products by fermentation or chemical processing.

Alternatively, the lignin and hemicelluloses can be separated from the biomass before hydrolysis of the cellulose by such techniques as organosolve hydrolysis, autohydrolysis or steaming and extraction. The residual cellulose may then by hydrolyzed by either acid or enzymatic techniques. Obviously such a separation sequence is a variant of conventional pulping, but when chemicals alone are the objective the fiber properties are not important and greater flexibility of process conditions can be tolerated.

Complete gasification to synthesis gas can also be directed entirely to chemicals such as ammonia, methanol, or hydrocarbons.

Chemicals and Fibers

A large number of chemical products are already produced along with pulp in existing conventional pulping operations. These include such materials as tall oil, dimethyl sulfoxide, ethanol and vanillin. There is opportunity for further utilization for chemicals of the lignin and saccharinic acids in kraft black liquor which are not being burned.

New pulping techniques such as organosolve pulping in which the processing conditions would be carefully controlled to prevent extensive cellulose degradation could provide both acceptable pulp and separated and isolated hemicelluloses and lignin in a form suitable for conversion to chemicals.

Chemicals and Food

Systems for coproduction of chemicals and food can be based on such agricultural crops as sugar cane and corn, or on sugar hydrolyzates from cell wall carbohydrates. The biomass of the corn plant has three principal components: grain, stover and cobs. These have well established uses in human and animal nutrition, but all can be readily converted to chemicals as well. By control of product flows between food and chemicals in response to changing needs and conditions increased flexibility and biomass utilization can be attained. Sugar cane yields both juice and bagasse. The sucrose from the juice is utilized in fermentation to rum as well as in the form of crystalline sugar. Hydrolysis of bagasse will also yield sugars convertible to chemicals by fermentation or chemical processing.

Any process scheme involving hydrolysis of cellulose to glucose could provide food grade glucose as well as a chemical raw material. Another option is the conversion of the sugars to single cell protein for supplementing animal rations.

Chemicals and Fuels

Organic materials liberate heat when burned, so biomass and its conversion products, whether solids, liquids or gases, can serve as fuels. This potential source of energy can find application either in external markets or in the internal energy balance of the processing plant.

Ethanol has been widely publicized as an alternative liquid fuel from biomass. An integrated plant producing ethanol by hydrolysis of lignocellulose could produce various chemical coproducts from the other biomass components. Marketing options for ethanol as a chemical raw material or as a fuel would provide potentially useful flexibility.

The gasification of biomass to carbon monoxide and hydrogen may be considered either the production of a low BTU fuel gas or of synthesis gas or of both. All or part may be used as fuel and the remainder converted to ammonia, methanol, methane or higher hydrocarbons. All of these but ammonia can also be considered as fuels. Methanol can be used as a liquid fuel, converted to formaldehyde, or used in the synthesis of other chemicals.

Pyrolysis of biomass yields charcoal, pyrolysis oils and synthesis gas. These too can be used as fuels or some portion of them can be utilized as chemical raw materials to enhance overall process flexibility and profitability.

SUMMARY

The need to maximize the yield of products from raw materials for greatest economic efficiency is especially important in the case of a multicomponent raw material like biomass. Biomass conversion plants should convert all the components of the starting material into useful products. Various integrated schemes have been described involving production of chemicals alone or coproduction of chemicals with fibers, food and fuel.

REFERENCES

Goldstein, I.S., 1980, New technology for new uses of wood, TAPPI 63(2):105-108.
Goldstein, I.S., 1981, Chemicals from biomass: Present status, For. Prod. J. 31(10):63-68.
Goldstein, I.S., 1981, Integrated plants for chemicals from biomass, In: Organic Chemicals From Biomass (I.S. Goldstein, Ed.), pp. 281-285, CRC Press, Inc., Boca Raton, Fla.
Lipinsky, E.S., 1978, Fuels from biomass: Integration with food and materials systems, Science 199:644.
Soltes, E.J. and Elder, T.J., 1981, Pyrolysis, In: Organic Chemicals from Biomass (I.S. Goldstein, Ed.) pp. 63-99, CRC Press, Inc., Boca Raton, Fla.

BIOMASS UTILIZATION: ECONOMIC ANALYSIS

OF A SYSTEMS APPROACH

M. Dicks and J.P. Doll

Department of Agricultural Economics
Department of Economics
University of Missouri-Columbia
U.S.A.

INTRODUCTION

The economic feasibiity of any production process depends not only on prices but also on the possible ranges of physical possibilities and institutional permissibilities. Ideally, the economist would like to obtain from the physical scientist a set of alternative production functions that would permit analysis of all possible alternatives available in any society. Such a general function for agricultural waste systems (utilizing methane production) might be stated as:

(1) $$Y = f(X_1, X_2, \ldots, X_n) - \sum^{n} \lambda_n (rK + wL - C_i)$$

where:

- Y = maximum profit
- $X_1 - X_n$ = mix of desired products
- rK = capital input
- wL = labor input
- C_i = cost vector

This represents an accounting stance which includes the significant benefits and costs that would be consistent with the interests of the decision maker who will use the results in arriving at a decision as to whether the optimal input and output mixes present an economically feasible reality.

Recent research on methane production just completed in Tunisia (Dicks and Blase 1982) demonstrated the need for a systems analysis approach in contrast with the unit approach currently being used (Sanghi 1976; Gaddy 1974; Geller 1976; Fisher 1979). Data collected in China, Kenya, and the United States is currently being analyzed by the methods given in the context of this paper and appear to offer positive benefit-cost ratios in the range of 1.7 to 6.3 depending on amount and characteristic of production facility.

The analysis of systems can be viewed as tripartite in terms of the physical parameters, the economical parameters, and the socio-political goals.

PHYSICAL PARAMETERS

The parameters usually envisioned in any design of a methane digester are temperature (T_D), loading rate (Lr), and retention time (RT). These parameters are controlled by the design of the processing system and will determine the quantity and quality of gas produced, the degree of waste treatment, and the effluent quality.

Each source of raw materials has a unique set of parameters to maximize the full potential of the anaerobic process. Buswell and Mueller (1952) demonstrated this by developing equation (2) to predict the quantity of methane given a known chemical composition of the waste.

(2) $C_nH_aO_b + (n-a/4-b/2)H_2O \rightarrow (n/2-a/8+b/4)CO_2 + (n/2+a/8-b/4)CH_4$

This equation represents the overall equilibrium between substrate (influent) and microbial degradation. The microbial degradation is, chemically, a very complicated process involving hundreds of possible intermediate compounds and reactions, each of which is catalyzed by specific enzymes or catalyst. [See Price and Cheremisinoff 1981 for a complete summary.]

The point to be made here is that the parameters previously mentioned affect the bacterial metabolism and thus their growth rate. A greater rate of metabolic activity precludes a higher rate of gas production and substrate degradation. These parameters will be viewed singly and then the overall kinetics will be discussed correlating the parameters.

Temperature

There is little argument that an increase in temperature (except increase between mesophilic and thermophilic ranges) increases reaction rates and thus gas production. Fair and Moore (1973) indicated that a significant reduction in retention time required to obtain 90 percent digestion could be achieved by increasing digestion temperature. Later work by other researchers (Maly and Fadrus 1971) did not find the same response. Stevens and Schulte (1979) reported that in the 4-25°C range, an increase in temperature increased gas production from 100-400 percent. Loehr (1977) reports that minimum retention time values at 25° and 15°C may be about 1.2 and 1.5 and 2.0 and 3.0, respectively, of the values at 35°C. Several characteristics of increasing temperature can be summarized:

1. Two alternative ranges for microbial activity exist: (a) mesophilic (under 40°C) and (b) thermophilic (45°-65°C). Between the two ranges gas production declines.

2. An increase in temperature reduces retention time.

3. Increased differences between ambient and digester temperatures increases energy requirements of the system.

4. Increases in temperature increase ammonia-nitrogen concentrations.

5. Microbial populations are very sensitive to abrupt changes. A rapid change in temperature of more than 5°C will seriously deplete microbial concentrations.

Loading Rate

The loading rate for anaerobic digesters is generally expressed in terms of weight of volatile solids per unit volume per unit time (kg V.S./m^3/d). These rates vary for different types of digesters from 0.8 to 6.4 kg V.S./m^3/d. Metcalf and Eddy Inc. (1979) recommend loading rates of 0.5-1.6 kg/m^3/d and 1.6-6.4 kg/m^3/d for standard rate (mesophilic) and high rate (thermophilic) digesters, respectively. Optimum rates have been listed between 2.0 and 4.0 kg/m^3/d (Fisher 1979; Slane 1975; Stafford 1980). Several characteristics can be summarized:

1. If the loading rate is too low, the bacteria will exhibit a lower metabolic activity.

2. If the loading rate is too high, an overload situation will result with a buildup of volatile fatty acids (VFA), a decrease in gas production, and an increase in CO_2 content of the gas.

3. An increase in loading rate is achieved by decreasing dilution or detention time.

4. The ability of systems to handle an increase in load rate is dependent on microbial population and metabolic rate.

Retention Time

For systems that do not recycle effluent and where the effluent is a true sample of the reactor mixture, the hydraulic retention time (HRT) and solids retention time (SRT) are synonymous. The solids retention time is the important factor since it represents the average retention time of the microorganisms responsible for digestion. Boundaries of optimum retention time exist where an increase in RT decreases gas production (upper limit) and decrease in retention time decreases gas production (lower limit). Optimum retention times reported in the literature vary from 10-30 days at temperatures 40°-30°C and total solids of 4-8 percent.

PARAMETER CORRELATION

The three parameters just mentioned play a key role in sustaining the necessary levels of microbial concentration. There is a great deal of correlation among these parameters.

A kenetic model [equation (3)] formulated by Ashare, Wentworth, and Wise (1979) offers an attempt at correlating the partial kinetic equations which are based on the interaction of individual or paired parameters.

(3) $$G_o \cdot RT/[VS] = k_1 A(1 - [VS]_2)KRT/1 + KRT$$

where:

- G_o = gas production (volume gas/volume reactor/time)

- RT = hydraulic retention time equal to reactor volume (volume/flow rate)

- $[VS]$ = volatile solids concentration ($[VS]$ /time = Lr)

- k_1 = volume of methane per g COD (a constant dependent on the equation $CH_4 + 2O_2 \rightarrow 2CO_2 + 2H_2O$)

- A = COD to biodegradable vs. ratio of influent (a constant depending on influent source)

- $[VS]_2$ = fraction of refractory volatile solids

- K = rate constant

This equation assumes a continuous stirred tank reactor (CSTR) but may be fitted for other alternatives. Most research has concluded that an equation of the form:

(4) $$K = ae^{bT}$$

BIOMASS UTILIZATION: ECONOMIC ANALYSIS OF SYSTEMS 679

can be used to express the relationship of a reaction rate and temperature T($^\circ$ C) (Loehr 1977). The values of a and b are a function of the specific reaction. Given a known K at T_1 a (K) vector can be arrived at by:

(5) $K_{T_2} = K_{T_1} e^{C(T_2-T_1)}$

where e^C has been given varying values in the literature. A value of 1.072 is given for Arrehnius' hypothesis of a two-fold rate increase for a 10° C change in temperature. Thus as K is dependent on temperature, the model incorporates all of the parameters. To increase the digester temperature (T_D) to levels above ambient temperature, an energy input will be necessary and thus net gas (G_N) may be calculated as:

(6) $G_N = G_o - G_I$

where G_I is calculated as the amount of gas required to maintain T_D and increase temperature of the influent. The value of G_I may be zero for systems utilizing heat exchangers incorporated with generating facilities. In this instance all of the gas is utilized as a consumptive fuel with the waste of consumption (hot water) being a useable by-product. Equation (3) then represents a maximum gas output and defines the parameters necessary to accomplish this task.

ECONOMICS

Equation (6) provides a means for locating the maximal unconstrained gas output located on the production surface described by the physical parameters. This can be illustrated by viewing a set of contour lines representing the shape of one possible surface as in Figure 1.

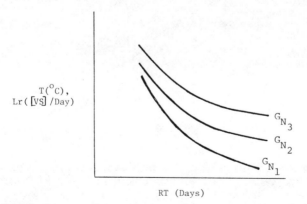

Figure 1. Gas production isoquants as a function of the physical parameters.

As the parameters affect the digester size, the type of technology used to create the influent-effluent flow, it also effects the cost in terms of labor and capital of the system. Most of the literature surveyed, as well as sites visited in several countries, determine the maximum gas production criteria and then compute the cost associated with the technology utilized to provide this situation. This procedure is the primary source of variation in degree of mechanization and cost per unit gas production. An optimal allocation of labor and capital to achieve a maximum level of gas production can be achieved through the use of the Lagrangian given in equation (7).

(7) Max G_N P_1 + $)$ $(rK + wL - C)$

where:

G_N = net gas = $Fn(K,L)$

P_1 = price
r = cost of capital
K = capital
w = wage rate
L = labor
C = total cost

Thus by varying C, a number of solutions can be obtained, as represented by Figure 2.

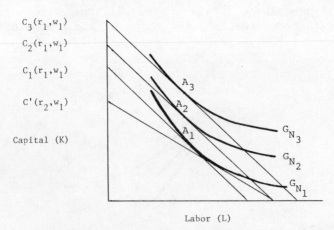

Figure 2. Isocost curves for gas production under constant capital labor ratio (r_1,w_1) and changing capital labor ratio (r_2,w_2) where A_1 represents the optimal K-L mix at each level of production.

BIOMASS UTILIZATION: ECONOMIC ANALYSIS OF SYSTEMS

The value of the effluent has yet to be considered. Again most of the literature either neglects totally the net value (value effluent -- value influent) of the substrate or assigns it a value of zero. The alternatives for use of this effluent vary considerably and most are in the elementary state. These include: (a) use as a fertilizer (value at level of nitrogen); (b) use as an animal feed (value at a level of nitrogen); (c) use as bedding; and (d) use in aquatic fish and/or plant systems. We will assume that these represent a positive value and let (AT) represent any one of an optimal combination of several outputs from the alternative technologies. Then equation (7) becomes:

(8) $\text{Max} \quad \overline{[G_N \cdot P_1 + AT \cdot p_2]} \quad + \quad \overline{[\lambda_1(wL + rK - C_i)]} \; G$
$+ \; \lambda_2 \; \overline{[(wL + rK - C_i)]} \; AT$

Furthermore, there may be benefits to society that have been overlooked in the positive externalities associated with control of odor, nonpoint pollution, and a decrease in health hazards associated with the entire system. Letting A $\overline{[S(t)]}$ represent the value of these positive externalities, equation (8) becomes:

(9) $\text{Max} \quad \overline{[G_N \cdot P_1 + AT \cdot P_2]} \quad + \quad \overline{[\lambda_1(wL + rK - C_i)]} \; G$
$+ \; \lambda_2 \; \overline{[(wL + rK - C_i)]} \; AT \; + \; A \; \overline{[S(t)]}$

Finally, to extract the real value of these systems they should be viewed over a period of time (say 20 years) and thus equation (9) becomes:

(10) $\text{NPV} = B_{20} \; \overline{[G_N \cdot P_1 + AT \cdot P_2]} + B_{20} \; \overline{[\lambda_1(C_i)_G + \lambda_2(C_i)_{AT}]}$
$\quad\quad\quad + B_{20} A \; \overline{[S(t)]}$

where:

$$B_{20} = \sum_{n=1}^{20} 1/(1 + r)^n$$

$$\lambda_1(C_i)_G = \lambda_1(wL + rK - C_i)_G$$

$$\lambda_2(C_i)_{AT} = \lambda_2(wL + rK - C_i)_{AT}$$

ALTERNATIVE TECHNOLOGIES

The number of production unit combinations possible for a biological systems approach is indeed limited only by the demands placed on the system. Our current research in modeling these processes has for the moment focused upon one set of possibilities. The production units to be optimized within the system (an aquasystem) includes methane production, aqua plant production, fish production, and fresh water shrimp production. The possible products from this process include methane gas, fertilizer, high protein animal supplements, fish, shrimp, and clean water.

A flow diagram (Figure 3) can be used to view all of the possible means of collating the units into one system. The factors controlling the decision on which pathways collate the units for optimization are the relative value of outputs and the price of the inputs.

The difficulty in modeling the possible system is inherent in the interconnectedness of the output of one unit with the input needs of the following unit.

The optimization equation for any system, in theory, does not differ from equation (10). However, two sets of equations for each system occur. The first shows the overall accounting for the system (equation 10) and the second is derived through the development of the system based on the single unit optimization. This may be illustrated as:

(11) $\quad MAXNPV = B_{20}(G_N \cdot P_1 - \lambda_1 (C_i)_G) + B_{20}(AT_2 \cdot P_2 \cdot \lambda_2(C^i)AT_2)$

$\quad B_{20}(AT_3 \cdot P_3 - \lambda_3(C_i)AT_3) + \ldots$

Equation (11) represent the basis by which a program will select the best pathway. Moving from unit to unit it will reject the unit into the system if a specified benefit-cost ratio is not reached either for the unit or maintained for the system.

The inter-dynamics of the model deserves more thorough discussion than time and space permit in this paper. Suffice to say that the system is highly complex and offers an enormous array of alternatives. The major limiting factor to the modeling research at this time is the lack of physical data.

SOCIO-POLITICAL IMPLICATIONS

In viewing methane production as a part of a waste management system, the prices set for the products of the alternative technologies (fertilizer, fish, etc.) will greatly affect the system design and feasibility. Policy should be set to promote the development of the systems consistent with the areas factor endowments as well as social acceptability and institutional permissibility.

BIOMASS UTILIZATION: ECONOMIC ANALYSIS OF SYSTEMS

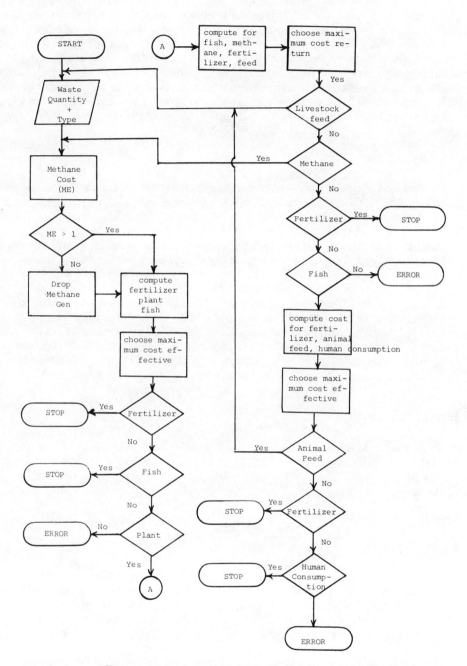

Figure 3. Aqua system flow chart.

The positive externalities, A [S(t)], associated with the system provide a net benefit to society. The cost of achieving these benefits, however, is rarely covered by society. In several countries (notably western countries), the costs associated with government regulation of providing these benefits to society (eliminating the hazards of agricultural wastes) has put a burden on the operator so great that many have been driven out of business. The government needs to reconsider its position and consider subsidizing these operations to the point of pollution abatement.

CONCLUSIONS

The idea of waste management systems is not new in most countries. The idea that agricultural wastes can be a valuable input in a production process is indeed a relatively novel concept. Viewing agricultural wastes and urban wastes as a flow resource with many alternative streams of benefits may well induce innovations which alter the agricultural production systems currently being used.

The benefit-cost approach described is far from complete. Production functions describing all possible alternative uses of the agricultural wastes need to be obtained. However, as stated previously, using several assumptions and data from Kenya, China, Tunisia, and the United States, a linear programming model using A Program Language (APL) has been developed at the University of Missouri. The results of the preliminary analysis look promising.

ACKNOWLEDGEMENTS

A special note of thanks is due Dr. Grant D. Venerable for his assistance with the kinetics in the section on Parameter Correlation.

REFERENCES

Ashare, E., R.L. Wentworth, and D.L. Wise. 1979. Fuel Gas Production from Animal Residue, Part II. An Economic Assessment/Resource Recovery and Conservation, 3, 359-386.
Buswell, A.M. and H.F. Mueller. 1952. Mechanisms of Methane Fermentation. Industry and Engineering Chemistry, 44, 550-552.
Dicks, M. and M. Blase. 1982. Economic Feasibility of Methane Production in Tunisia. M.S. thesis, University of Missouri-Columbia.
Fair, G.M. and E.W. Moore. 1937. Observations on the Digestion of Sewage Sludge Over a Wide Range of Temperatures. Sewage Works Journal, 9, 3.

Gaddy, J.L., E.L. Park, and E.B. Rapp. 1974. Kinetics and Economics of Anaerobic Digestion of Animal Waste. Water, Air, and Soil Pollution, 3, 161-169.

Fisher, J.R. et al. 1979. Economics of a Swine Digester. Paper 79-4530, presented at the 1979 Winter Meeting of the American Society of Agricultural Engineering.

Geller, H. 1976. Economic Feasibility of Methane Generation on Central Mass. Dairy Farms. Science, Technology and Society Review, 2, 45.

Hawkes, D.L. 1979. Factors Affecting Net Energy Production from Mesophilic Anaerobic Digestion. Proceedings of the First International Symposium on Anaerobic Digestion, University College, Cardiff, Wales.

Loehr, R.C. 1977. Pollution Control for Agriculture. New York: Academic Press.

Maly, J. and H. Fadrus. 1971. Influence of Temperature on Anaerobic Digestion. Journal of Water Pollution Federal, 43, 641-650.

Metcalf and Eddy, Inc. 1979. Waste Water Engineering Treatment, Disposal and Reuse. New York: McGraw-Hill Book Co., 445-457, 610-625.

Price, E.C. and P.W. Cheremisinoff. 1981. Biogas Production and Utilization, Chap. 1. Microbiology and Biochemistry Energy Technology Series, Ann Arbor Science Publishers.

Sanghi, A.K. and M. Blase. 1976. An Economic Analysis of Energy Requirements of Alternative Farming Systems for Small Farmers: Some Policy Issues. Indian Journal of Agricultural Economics, 31, 8.

Slane, T.C. et al 1975. An Economic Analysis of Methane Generation Feasibility on Commercial Egg Farms, Managing Livestock Wastes. Third International Symposium on Livestock Wastes.

Stafford, D.A., D.L. Hawkes, and R. Horton. 1980. Methane Production from Waste Organic Matter. CRC Press, Inc.

Stevens, M.A. and D.D. Schulte. 1979. Low Temperature Anaerobic Digestion of Swine Manure. Journal of Environment Engineering Division, ASCE, 105, (EE1).

ECONOMICS OF MUNICIPAL SOLID WASTE

AS A RAW MATERIAL

>Robert F. Vokes
>
>Consultant
>Parsons & Whittemore, Inc.
>New York, N. Y.
>U.S.A.

INTRODUCTION

It is said that "figures don't lie" although "liars can figure" and that data can be chosen to suit the objective of the story teller. Economics is such a discipline. A record of past performance in terms of numbers and currency values can be used as a basis for forecasting and planning future happenings. Unfortunately, the economic scenarios of future operations are influenced by subjective judgments and unforseen events. We have a new term to describe such situations: "sociotechnical systems".

Notwithstanding the inadequacy of our tools to formulate an economic basis for MSW biomass utilization, we have certain very significant economic facts of life to guide us. Hence the following guidelines, observations and scenarios are promulgated to stimulate our thought processes rather than to be proposed as a specific set of objectives.

GUIDELINES

Economic Value of MSW

1. Our "Solid Waste Stream" is a disposal problem which provides an opportunity to practice "resource recovery".

2. In the U.S.A., approximately 200 million tons (EPA 1981) of municipal solid wastes are created annually and the growth rate is in the order of 3 percent per year.

3. Of the 200 million tons, about 150 million tons is created by residential and commercial sources.
4. Of the 150 million tons, about 12 million tons (8 percent) is recovered for re-use in industry. The balance is incinerated (4 percent), or used as landfill (88 percent).

Observations

    a. Except for the 12 million tons of MSW which is recycled,

<u>MSW has no economic value!</u>
<u>MSW has no price tag!</u>

    b. In fact the remaining 138 million tons of MSW (92 percent) creates an economic burden in the order of:

$37/T collection
$8/T disposal
$45/T Total = $6.2 billion/year (EPA 1981)

    c. Note that collection for any purpose is unavoidable at a burden of $5.1 billion/year and that the burden of disposal via landfilling remains at about $1.1 billion/ year.

Economic Values of Incineration Technologies and Resource Recovery

Mass Burning.

1. Mass burning of MSW has been practiced since the late 1800's as a method of disposal. It is essentially a service utility not unlike a municipal landfill site, a sewage disposal or water treatment facility for the benefit of all citizens. Hence, the service may not require an economic, that is, profit motive incentive since the costs of operations may be absorbed via tax levies in the municipalities.

2. Although mass burning is practiced for only 4 percent of the MSW of the U.S.A., it is a commercially proved process based upon world-wide experience. It is the least risky technologically and the economics of scale do not seem to apply (Eberhard 1981).

3. Currently, more than half of the new MSW projects initiated since 1970 in the U. S. A., employ mass burning with energy recovery (NCRR 1981).

Observations.

    a. The process of mass burning with energy recovery, developed as a result of high fossil fuel costs, has a high potential for economic viability which is attractive to the private sector of the economy (NCRR 1981).

b. Air pollution problems do not seem to be insurmountable. A recent EPA decision notes that - emission levels of TCDO (Dioxin) do not present a public health hazard -" (Chrismon 1981).

RDF and Co-generation -- Steam and Electricity

1. MSW is 75 percent combustibles on a dry basis and the energy value of the moist biomass when isolated as RDF is 6000 BTU/lb. vs. oil at 18,300 BTU/lb. (NCRR 1980).

2. On an equivalent heat value basis, one ton of RDF equals 2 bbl. of oil (NCRR 1980).

3. RDF processing cost is not generally known except that it apparently exceeds the income for use as a fuel or the sale of steam or electrical power (NCRR 1981).

4. MSW processing plants producing RDF and operating co-generation electric power facilities are regulated by PURPA (Public Utilities Regulatory Policy Act) and therefore, public utilities are required to purchase electricity at their "avoided" generation cost, that is, the generation cost for less efficient units replaced by RDF fired generation. Hence, "there is little incentive for a utility to participate in a resource recovery project -- other than to provide a community service" (McGowin 1981) and to comply with the law.

Observations.

a. Incineration of RDF with energy recovery, serves the purposes of volume reduction for landfill, the recovery of non-combustibles and may create problems of pollution of the atmosphere.

b. The net economic benefit, if any, seems to be related to the reduction of landfill costs, the possibility that fossil fuel costs would exceed the cost of RDF and its burning separately, or co-fired with other fuels. Incidental economic benefits may also be realized from recovery and sale of metals, glass and ash.

c. RDF production, integrated with co-generation of steam and electricity, avoids the problem of acceptance of RDF by utilities, related to the use of designated boilers and the risks involved with the use of a new type of fuel (McGowin 1981).

d. Resource recovery, RDF co-generation, and minerals recycling represent a developing technology. As a result there is very little data available on the technical performance or economic viability of any of the new processes. In fact, none of the installations to date in the U. S. A. have published information to evaluate these criteria. More research and development investment as well as operating experience is required to determine which processes may be dominant and under what particular set of circumstances.

e. MSW resource recovery and biomass utilization may evolve initially as a public utility type of enterprise, supported and controlled by government or appointed commissions.

INTERRELATIONSHIPS: ECONOMICS OF NATURAL RAW MATERIALS VS. WASTE MATERIALS

In a free market situation, the availability and cost frequently determine the raw material source chosen for industrial uses. This supply-demand relationship applies not only to raw materials but also to the labor and capital required for conversion to marketable products.

However, for other than economical reasons, i.e., political, environmental, social, geographical, etc., all factors, including raw materials, labor, capital, land, etc. may become controlled by the mandates of government. A most recent example is demonstrated by the acts of the State of New Jersey legislature to tax the practice of MSW disposal and use this income to support the development of resource recovery and use of all recyclable materials (New Jersey 1981). This incentive creates a market for MSW biomass raw materials to be recycled for the production of an original product, i.e., newsprint to newsprint, and also creates a market for the use of recycled materials for other products based upon the best, most economically viable use.

Hence, a government mandate to practice resource recycling, for any reason, creates a raw material source for private industrial use and may compete favorably with natural sources of such raw materials, particularly in a controlled economy.

Ultimately, the practice of new technologies for resource recovery of MSW which is presently tipped at landfill sites, may establish MSW as a primary raw material source for stable industrial enterprises.

ECONOMICS OF WASTE AS RAW MATERIAL

PROGNOSIS - MSW BIOMASS UTILIZATION

Scenario I: 1980's - Energy

Visualize the vast existing municipal collection systems as tipping the MSW at industrial sites within the borders of the cities. A total of 5000 Tons/Day per 3 million people is received at each site and processed continuously 24 hours per day and 365 days per year. The tipping fee is established by equating to landfill costs at the time of processing plant construction.

The non-combustibles are recovered and sold to minerals industries at prevailing rates competitive with alternative raw materials.

The 3,750 Tons/Day of biomass is converted to energy as electric power, 135 MW or 1 million plus megawatt hours per year, which is sufficient to provide domestic power requirements for 800,000 people or 250,000 homes (Kohlhepp 1978).

This is adequate electrical power for more than 25 percent of the population who are providing the fuel for generation.

The economics -- who knows today?

Scenario II: 1990's - Chemicals

Visualize Scenario I except that the biomass is converted to ethanol and that 25 percent of the biomass is converted to energy to provide the full requirement of the ethanol production operations.

In this instance, the production of ethanol via the use of micro-organisms (Katzen 1982) would produce in the order of 236,000 gallons of MFG (Motor fuel grade) ethanol per day.

Also, in this instance, the market for MFG ethanol may be mandated by government edict so that "all motor vehicles are required to use 'gasohol' (10-20% ethanol and 80-90% gasoline) for fuel".

This scenario may also apply to the production of methanol.

The economics -- who knows?

Scenario III: Integration - Forest Industries

A forest industry is identified as one which uses trees as a primary raw material for the production of lumber, pulp, paper and allied products. As such, the industrial operations create waste materials in the form of wood, bark, tree branches and leaves, pulp chemical residues (black liquor), waste papers and boards as well as cellulose, plastics, etc. These contributions to the solid waste stream are approximately 75-100 million tons per year (Goldstein 1981).

Visualize an integrated forest industry, lumber and paper, etc., near a large municipality such as Seattle-Tacoma, Washington vs. Weyerhaeuser Co.

Visualize the burning of bark, waste wood, black liquor and MSW biomass in the same industrial park for the purpose of providing electrical energy for the pulp mill, chemical recovery plant, paper mill, and the surrounding municipalities.

The economics -- who knows?

SUMMARY

Each MSW biomass disposal problem represents an opportunity for resource recovery. The case basis may or may not present an economically viable industrial enterprise for private industry, a municipal government authority or a combination of both.

To date, the burning of MSW biomass shows promise to produce electrical energy economically. However, the use of MSW biomass for any other purpose is yet to be demonstrated. The prognosis is that MSW biomass for the energy self sufficient production of chemicals will prove to be economically viable in the event that energy costs continue to rise.

REFERENCES

Chrismon, R. 1981. Solid Waste Report, November 18, p. 175.
Eberhard, John B. 1981. The recovery of energy and materials from solid waste. National Academy Press, Washington, D.C.
EPA. 1981. Solid Waste Data, unpublished, EPA 68 01 6000, Washington, D.C.
Goldstein, I.S. 1981 Organic Chemicals from Biomass. CRC Press, Inc. Boca Raton, Florida.
Katzen, R. 1982. UBFI Project. R. Katzen Associates International, Inc., Cincinnati, Ohio, and UBFI, Inc., Petersvurg, Virginia.
Kohlhepp, D.H. 1978. The dynamics of recycling. National Meeting of Am. I. Che., Salt Lake City, Utah.
McGowin, C.R. 1981. MSW - A Utility Fuel? NCRR Bulletin Vol. II: 4.
NCRR. 1980. Resource Recovery Update No. 5. Washington, D.C.
NCRR. 1981. Bulletin, Vol. 11 No. 3, September. Washington, D.C.

New Jersey Senate. 1981. Public Law 1975 amended 22 June 1981, Senate Assembly No. 2283.
Snider, B., Jr. 1982. Paper as fuel for industry. Adv. Council, Boca Raton, Florida.

TECHNICAL AND ECONOMIC EVALUATION

BIOMASS ENERGY/CHEMICALS INTEGRATED SYSTEMS

Raphael Katzen

Raphael Katzen Associates International, Inc.
Cincinnati, Ohio
U.S.A.

The creation of technically and economically feasible biomass-based energy and chemical systems will require the effective coupling of existing technologies with new scientific concepts. Concentrated interdisciplinary research efforts supported by large-scale funding, from both governmental and industrial sources, are being brought to bear upon the problem of assessing the total biomass potential and how it may be realized.

Engineers are faced with the challenge of coupling research and development results with "best available technology" to produce viable commercial biomass-based energy and chemical systems. The strategy used to accomplish successful commercialization, or possible rejection, of proposed processes is the subject of this paper.

To illustrate the methodology utilized, an example will be given of a grass-roots facility to produce 50 MM gal/yr of motor fuel grade alcohol from corn. These estimates were made in 1978, and should be adjusted for inflation since that time.

To establish the commercial viability of a proposed venture, the engineer must follow a methodical sequence of steps. These steps are categorized as:

o Study of process and technology alternatives
o Development of process design
o Preparation of economic analysis

The first category, study of process and technology alternatives, may involve only a cursory review of existing or proposed technology before selecting the best process. For many evaluations, however, several processes may have to be studied in parallel. Once the best process technologies have been selected, the engineer proceeds with the process design. Table 1 outlines a typical sequence of events.

Table 1. Steps for process evaluation.

1. Establish the design basis
2. Accumulate and assemble all available process data
3. Prepare preliminary block flow diagrams
4. Prepare conceptual flowsheets
5. Prepare general material and energy balances
6. Revise process based on insights gained in Steps 3, 4 and 5
7. Establish final process design including detailed material and energy balances

The design basis must be established first. Next, it is necessary to accumulate all available process data and information, including, if possible, a computer-aided literature search. This prevents disruption of the engineer's thought pattern, or delays in making process decisions due to not having data at his fingertips.

Preliminary block diagrams and conceptual process flowsheets are just that - PRELIMINARY and CONCEPTUAL. This is absolutely necessary in order to establish an optimum design. Flexibility in design is the key at this point. There is nothing more wasteful and demoralizing than to spend time, effort, and money establishing a process flowsheet with detailed material and energy balances, only to discover later that a cursory look at the situation would have shown other options to be more desirable. At this stage of the study, one must decide on such process options as:

- Vapor recompression evaporators versus multiple effect evaporators
- Use of reboilers versus direct steam, or compressors to provide vapor for distillation columns
- Co-generation of electricity versus purchased power
- Staging of distillation towers versus single tower operations

In most situations, these options can be decided by evaluating the unit operations. Sometimes the option may be so complicated that parallel studies have to be performed.

Figure 1 is a block diagram of the basic process modules in the production of motor fuel grade alcohol from grain. From this diagram the engineer sets out to establish the best process under the conditions prevailing at the proposed site. After optimizing

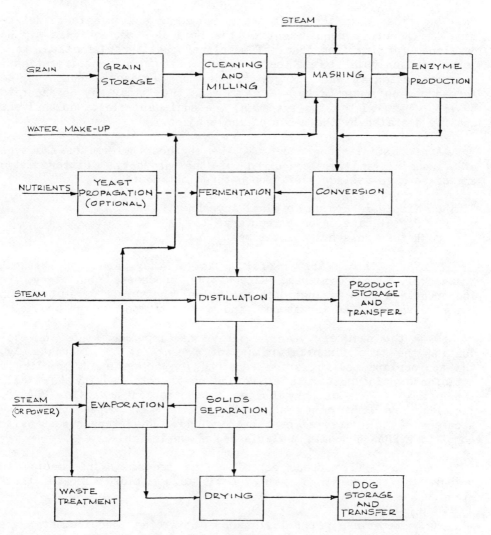

Figure 1. Block diagram of the basic process modules in the production of motor grade fuel from grain.

the overall design, the process flowsheet is finalized. Only after finalizing the process does the engineer prepare a detailed material and energy balance.

At this point, the engineer has established the most desirable process. His process decisions have been based principally on elementary economic rules. The next step is to prepare a complete economic analysis.

The first step in developing an economic analysis is to determine the accuracy requirement of the project. An analysis may be developed ranging from "back of envelope" quality information, to a detailed analysis based on complete drawings, specifications, accurate material and energy balances and estimated costs. The former may be accurate to ± 50%, whereas the latter may be accurate to ±5%. Between the two extremes, the engineer selects an analysis quality dictated by the needs of the project.

After establishing accuracy, the engineer determines the type of financial analysis required of the project. Historically, engineers have relied chiefly upon three methods:

o Simple return on original investment
o Payout time or cash recovery
o Return on average investment

These methods are useful (though they are not without drawbacks), when comparing alternative processes, but they are inappropriate when evaluating a capital expenditure program based on the time value of an investment, and money returned to the project.

Over the past six years, we have developed a method for evaluating the true financial return of a venture; or alternatively, the appropriate selling price required for a venture to realize a satisfactory interest rate of return on investor equity. We call the method a "Preliminary Budget Authorization Analysis (PBAA)". The accuracy is in the range of ±10% to ±25% depending on the accuracy of the data and cost projections. The steps necessary to complete a PBAA economic analysis are shown in Table 2.

The first five steps establish the necessary information to prepare the financial analysis. Let's take a closer look at these steps.

Raw materials, chemicals and utilities requirements are obtained from the detailed material and energy balances. Adjustments must be made for peak loads and process control limitations. The flowsheets provide average usages. Labor requirements are estimated

Table 2. Steps required for economic analysis (PBAA).

1. Establish raw materials and chemicals usage, and labor requirements
2. Prepare listing of total utility requirements
3. Determine cost projections for all purchased items
4. Prepare battery limits investment estimate
5. Determine total plant capital
6. Develop financial analysis

TECHNICAL AND ECONOMIC EVALUATION

from the number of control loops and the extent to which assistance is provided to the operator (use of computer control for example).

The battery limit investment estimate is only a part of the total plant investment. The limits may circumvallate the total plant, or more typically, just the operating units. The procedure for determining the battery limits investment is:

o Size all major equipment
o Determine major equipment purchase price
o Use cost factors to establish the installation cost
o Apply appropriate inflation escalation to construction schedule
o Determine construction cost and contractor fees
o Estimate overhead expenses
o Estimate engineering, start-up and project management cost
o Apply contingency factor

Table 3 shows the investment, by process section, for the Grain Motor Fuel Alcohol plant evaluated in 1978. Table 4 gives a more detailed breakdown for each section of the plant.

The total plant capital required consists of several components. Mistakes can be made in determining the plant investment by

Table 3. Grain motor fuel alcohol plant 50 mm gal/yr plant investment base case.

Section No.	Identification	December 1978 Cost	Percent of Investment
100	Receiving, Storage & Milling	$ 2,086,800	3.96
200	Cooking and Saccharification	2,824,300	5.36
300	Fungal Amylase Production	3,485,900	6.61
400	Fermentation	4,195,600	7.96
500	Distillation	5,123,800	9.72
600	Dark Grain Recovery	13,018,400	24.69
700	Alcohol Storage, Denaturing & By-Product Storage	4,399,900	8.35
800	Utilities	15,090,000	28.62
900	Building, General Services and Land	2,494,000	4.73
		$52,718,700	100.00
	+ 10% Contingency	5,272,300	
	Total Plant Costs	$57,991,000	

Table 4. Grain motor fuel alcohol plant, 50 mm gal/yr - investment breakdown, base case - 1978 cost.

			Equipment	Field Materials	Direct Labor	Construction Overhead
100	Receiving, Storage and Milling	E NE T	$ 725,000 259,300 $ 984,900	182,400	244,000	146,400
200	Cooking and Saccharification	E NE T	32,000 721,600 $ 753,700	302,900	593,000	356,000
300	Fungal Amylase Production	E NE T	$ 0 1,267,300 $ 1,267,300	401,500	665,000	400,000
400	Fermentation	E NE T	$ 1,113,500 649,800 $ 1,763,300	478,000	695,800	417,500
500	Distillation	E NE T	$ 0 1,475,900 $ 1,475,900	801,500	1,182,100	709,200
600	Residue Feed Processing	E NE T	$ 64,200 6,052,300 $ 6,116,500	1,079,400	2,308,000	1,150,000
700	Alcohol Storage, Denaturing and By-Product Storage	E NE T	$ 1,200,000 193,200 $ 1,393,200	580,000	957,000	574,200
800	Utilities	PK	$15,090,000	0	0	0
900	Buildings and General Services		0	0	0	0
	Totals	E NE PK 	$ 3,135,400 10,619,400 15,090,000 $28,844,800	3,825,700	6,644,900	3,753,300

E - Erected PK - Packaged Unit
NE - Not Erected T - Total

Sales Tax & Freight	Constructor Fees	Engineering	Buildings & Structures	Total
49,800	29,300	250,000	200,000	$ 2,086,800
56,700	65,000	247,000	450,000	2,824,300
94,100	83,000	350,000	225,000	3,485,900
107,500	83,500	350,000	300,000	4,195,600
105,000	141,900	518,200	190,000	5,123,800
348,500	216,000	1,200,000	600,000	13,018,400
95,200	114,800	385,500	300,000	4,399,900
0	0	0	0	15,090,000
0	0	0	1,744,000	1,744,000
856,800	733,500	3,300,700	4,009,000	$51,968,700

Land 750,000
Contingency 5,272,300

Total Plant Investment (1978) $57,991,000

not taking into account all cost components. Figure 2 indicates these components.

We have already talked about the battery limit investment. Let us now discuss what the other terms mean.

X-Plant distribution is the investment associated with feedstock receiving or product shipping outside the boundaries of the plant. It includes items such as trucks, garages, and storage depots. Utility pers are investments associated with utilities not included in the battery limits investment. An example is the purchase of steam from a central boiler plant. Investments for general facilities and land must also be included.

Working capital is sometimes omitted or grossly underestimated. Components of working capital are shown in Table 5.

Table 5. Working capital.

1. Raw materials inventory (1 month)
2. Products in stock (1 month)
3. Accounts receivable (12% annual income)
4. Operating cash (2 months)
5. Accounts payable (negative - 12% raw material cost)

The factors associated with each component of working capital will vary depending on the requirements of the project. For example, working capital was minimized by providing for only one week's grain storage. This meant the grain would be purchased from

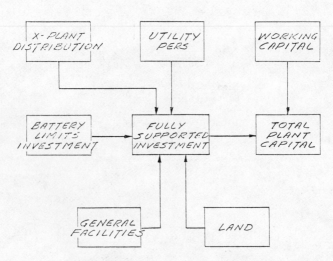

Figure 2. Total plant capital.

TECHNICAL AND ECONOMIC EVALUATION

local grain elevators, probably at a slightly higher price than long-term storage would permit.

After the engineer has completed tasks associated with the first five items listed in Table 2, he is prepared to proceed with the financial analysis.

If the engineer uses one of the simpler analysis techniques mentioned earlier, his remaining work is relatively easy. For example, the steps required to estimate percent return on investment are:

- Develop the single year operating cost
- Determine total income
- Determine gross profit for one year
- Establish percent return on investment after taxes

Table 6 indicates the cost items that must be considered in establishing the single year operating cost. Operating cost items include:

- Fixed charges
- Raw materials
- Utilities
- Labor
- Miscellaneous expenses
- By-product credits

Reference has been made to the financial analysis method called "Preliminary Budget Authorization Analysis (PBAA)". We developed this method because the conventional analyses often do not represent correctly the real life situation. The PBAA analysis has the following specific advantages:

- Time value of money taken into account
- Inflation is included as an input variable
- Each cost item may be inflated at different rates
- Present value of product selling price is established
- Complete profit picture of the venture (both actual and discounted) is established over its projected life
- The effect of leveraged capital can be easily studied
- Tax credits and other governmental incentives can be evaluated

It is apparent that the engineer must have a multi-disciplinary background in order to carry out successfully a financial analysis. In order to reduce the engineer's work load to a manageable level, we developed a computer program to do the tedious but routine computations, prepare annual cost tables, and summarize results. The program compresses several man-weeks of work into a few hours.

LINE DATA   LINE DATA

100 JOB, TITLE   *100 _____,_____
110 TFI, BASIS, DYRS, RINF, TAX   *100 _____,_____,_____,_____,_____
     YRI, YRCA, ROIPC   _____,_____,_____
120 NRM, NBYP, ISP, IDEP, IDEBT,   *120 _____,_____,_____,_____,_____
     ISWCB, IVTFI, IADDC   _____,_____,_____
130 FEES, FYRS, MAINPC, TALPC   *130 _____,_____,_____,_____
140 DEBT, DBYRS, YRL, IROL   *140 _____,_____,_____,_____
150 IOWC   *150 _____
200 RM(1,1), RM(1,2), RM(1,3), RM(1,4)   *200 _____,_____,_____,_____
210 RM(2,1), RM(2,2), RM(2,3), RM(2,4)   210 _____,_____,_____,_____
220 RM(3,1), RM(3,2), RM(3,3), RM(3,4)   220 _____,_____,_____,_____
230 RM(4,1), RM(4,2), RM(4,3), RM(4,4)   230 _____,_____,_____,_____
240 RM(5,1), RM(5,2), RM(5,3), RM(5,4)   240 _____,_____,_____,_____
250 RM(6,1), RM(6,2), RM(6,3), RM(6,4)   250 _____,_____,_____,_____
260 RM(7,1), RM(7,2), RM(7,3), RM(7,4)   260 _____,_____,_____,_____
270 RM(8,1), RM(8,2), RM(8,3), RM(8,4)   270 _____,_____,_____,_____
280 RM(9,1), RM(9,2), RM(9,3), RM(9,4)   280 _____,_____,_____,_____
290 RM910,1), RM(10,2), RM(10,3), RM(10,4)   290 _____,_____,_____,_____
300 STM (1), STM(2), STM(3), STM(4)   *300 _____,_____,_____,_____
310 CW(1), CW(2), CW(3), CW(4)   *310 _____,_____,_____,_____
320 ELC(1), ELC(2), ELC(3), ELC(4)   *320 _____,_____,_____,_____
330 FUEL(1), FUEL(2), FUEL(3), FUEL(4)   *330 _____,_____,_____,_____
340 PW(1), PW(2), PW(3), PW(4)   *340 _____,_____,_____,_____
400 LABR(1,1),LABR(1,2),LABR(1,3),LABR(1,4)   *400 _____,_____,_____,_____
410 LABR(2,1),LABR(2,2),LABR(2,3),LABR(2,4)   *410 _____,_____,_____,_____
420 LABR(3,1),LABR(3,2),LABR(3,3),LABR(3,4)   *420 _____,_____,_____,_____
500 BYP(1,1), BYP(1,2), BYP(1,3), BYP(1,4)   *500 _____,_____,_____,_____
510 BYP(2,1), BYP(2,2), BYP(2,3), BYP(2,4)   510 _____,_____,_____,_____
520 BYP(3,1), BYP(3,2), BYP(3,3), BYP(3,4)   520 _____,_____,_____,_____
530 BYP(4,1), BYP(4,2), BYP(4,4), BYP(4,5)   530 _____,_____,_____,_____
540 BYP(5,1), BYP(5,2), BYP(5,3), BYP(5,4)   540 _____,_____,_____,_____
600 SALEPC, GAOPC, FBR(1), FR8(2), FR8(3)   *600 _____,_____,_____,_____
700 SELP(1), SELP(2), SELP(3)   *700 _____,_____,_____
800 ICON, DTFI, WCR, AITCR, TFIT   *800 _____,_____,_____,_____

*   These lines must be supplied to DOCEWSK as a minimum even
     if you enter zero.

Figure 3.   DOCEWSK data input sheet.

TECHNICAL AND ECONOMIC EVALUATION

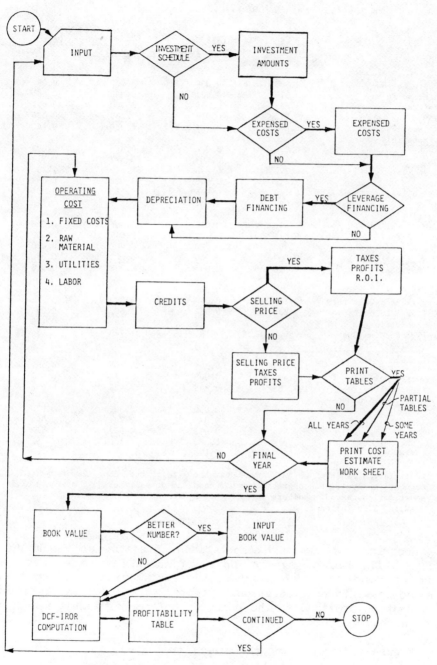

Figure 4. P.B.A.A. program flow logic.

Table 6. Grain motor fuel alcohol plant. Plant operating cost 50 mm gal/yr - base case.

	Equivalent 1978 Cost		1st Year Operation** 1983 Cost	
	Annual, $ MM	$/Gal	Annual, $ MM	$/Gal
Fixed Charges				
Depreciation*	2.900 (5.8)	0.058 (.116)	3.200 (6.4)	0.064 (.128)
License Fees	0.029	0.001	0.040	0.001
Maintenance	1.829	0.036	2.560	0.051
Tax & Insurance	0.914	0.019	1.280	0.026
Subtotal	5.672 (8.6)	0.114 (.172)	7.080 (10.3)	0.142 (.206)
Raw Materials				
Yeast	0.320	0.006	0.449	0.009
NH$_3$	0.373	0.007	0.522	0.010
Corn	44.770	0.896	62.679	1.254
Coal	2.410	0.048	3.374	0.068
Miscellaneous Chemicals	0.180	0.004	0.252	0.005
Subtotal	48.053	0.961	67.276	1.346
Utilities				
Electrical Power	1.646	0.033	2.305	0.046
Diesel Fuel	0.012	0.000	0.017	0.000
Steam (from plant)	0.000	0.000	0.000	0.000
C.W. (from plant)	0.000	0.000	0.000	0.000
Subtotal	1.658	0.033	2.322	0.046
Labor				
Management	0.240	0.005	0.336	0.007
Supervisors/Operators	2.194	0.044	3.072	0.061
Office & Laborers	1.202	0.024	1.683	0.034
Subtotal	3.636	0.073	5.091	0.102
Total Production Cost (TPC)	59.019 (61.9)	1.181 (1.24)	81.769 (85.0)	1.636 (1.70)
By-Products				
Dark Grains	19.175	0.384	26.845	0.537
Ammonium Sulfate	0.413	0.009	0.578	0.011
Subtotal	19.588	0.393	27.423	0.548
Miscellaneous Expenses				
Freight	2.504	0.050	3.506	0.070
Sales	1.930	0.039	2.702	0.054
G&AO	0.644	0.013	0.902	0.018
Subtotal	5.078	0.102	7.110	0.142
Total Operating Cost	44.509 (47.4)	0.890 (.948)	61.456 (64.7)	1.230 (1.29)

* Depreciation is shown as an average over the life of the plant (20 years). In the economic analysis, the depreciation was on a sum-of-year schedule over a 10 year amortization period. Numbers in parentheses are for a 10 year plant life.
** Assumed 7% inflation rate.

The program contains tax computations, credits, etc; but the engineer still has to supply input variables associated with these computations. With the program, the engineer is required to take all factors into consideration. Figure 3 indicates the information that must be provided to the computer before meaningful results can be obtained.

Figure 4 diagrams the computer flow logic.

Basic steps of the program are shown in Table 7. The annual operating cost information developed by the computer is the same as

TECHNICAL AND ECONOMIC EVALUATION

PROFITABILITY, CASH FLOW, AND IROR
FOR SAMPLE PROGRAM

YEAR	TCI	SP	PAT	CUM.P	CUM.D	CSH.FL	CDI	CDCF
1979	21.22	0.	0.	0.	0.	0.	21.22	0.
1980	21.22	0.25	1.79	1.79	1.77	3.55	21.22	2.91
1981	21.22	0.26	2.03	3.82	3.54	7.35	21.22	5.73
1982	21.22	0.28	1.76	5.58	5.31	10.88	21.22	8.10
1983	21.22	0.30	1.45	7.03	7.07	14.11	21.22	10.06
1984	21.22	0.32	1.62	8.65	8.84	17.50	21.22	11.92
1985	21.22	0.35	1.80	10.46	10.61	21.07	21.22	13.70
1986	21.22	0.37	2.00	12.45	12.38	24.83	21.22	15.39
1987	21.22	0.40	2.20	14.66	14.15	28.80	21.22	17.01
1988	21.22	0.42	2.42	17.08	15.91	32.99	21.22	18.56
1989	21.22	0.45	2.66	19.74	17.68	40.96	21.22	21.22
	**					*		

```
INTEREST RATE OF RETURN (IROR) =  10.5%
PAYOUT PERIOD =   6.0 YRS AFTER STARTUP FOR R.O.I.
*INCLUDES SALVAGE VALUE AND RETURN OF WORKING CAPITAL,
      (I.E., SALV=$    3.54 & ROWC=$    0.   ).
**T.F.I. PLUS ANY COMPANY FINANCED WORKING CAPITAL,
      WHERE WORKING CAPITAL ESCALATES AT    7.0% PER YEAR.
```

TCI   - Total Company Investment, $MM

SP   -   Selling Price, $

PAT   - Annual Profit After Tax, $MM

Cum.P   - Cumulative Profit, $MM

Cum.D   - Cumulative Depreciation, $MM

CSH.FL - Cumulative Cash Flow (undiscounted), $MM

CDI   - Cumulative Discounted Investment, $MM

CDCF   - Cumulative Discounted Cash Flow, $MM

Figure 5. Example of summary printout from the P.B.A.A. program.

$$FI* + \frac{WC_{j=1}}{(1+I)^{K-1}} + \sum_{j=2}^{n} \frac{(WC_j - WC_{j-1})}{(1+I)^{K-1}} = \frac{WCE + S}{(1+I)^N} + \sum_{j=1}^{n} \frac{ATN_j}{(1+I)^K}$$

Where:

FI = Plant Investment (w/o working capital)

WC = Working Capital

WCE = Working Capital Last Year

S = Salvage Value

I = Discounted Cash Flow - Interest Rate of Return

j = Year Index Counter; j=1, 1st year plant operation

K = Year Index Counter; K=1, 1st year capital required

N = Total Number Years from Time Capital Required

n = Plant Service Life

ATN = Net After Tax Cash Flow for Operating Years

*FI may be input over several years as

$$FI_{K=1} + \sum_{K=2}^{a} \frac{FI}{(1+I)^{K-1}}$$

Where:
  a = Number of years for plant investment input.

Figure 6. Equation for calculating the discounted cash flow -- interest rate of return (DCF - IROR).

Table 7. P.B.A.A. program flow logic.

1. Input Data File
2. Input Investment Schedule if Required
3. Input Additional Expenses Charged to Project
4. Input Debt or Leveraged Capital Financing
5. Compute Debt Financing Schedule
6. Compute Depreciation Schedule
7. Compute Amount Operating Cost (Table 6)
8. Compute By-Product Credits
9. Establish if Product Selling Price Known or % ROI Required
10. Compute Taxes
11. Compute Tax Credits
12. Print Annual Financial Tables
13. Determine if all Operating Years have been Evaluated
14. Establish Salvage Value of Plant
15. Compute DCF-IROR for Project
16. Print Project Profitability Tables
17. Continue to Another Analysis
18. Stop - End of Analysis

that shown in Table 6. A separate table is prepared for each operating year.

An example of the summary printout from the PBAA program is shown in Figure 5.

In addition to this summary, the program prints all pertinent information for each operating year. The equation for calculating the Discounted Cash Flow - Interest Rate of Return (DCF-IROR) is shown in Figure 6.

The DCF-IROR solution is by trial-and-error. The left hand side of the equation represents the discounted investment and the right hand side represents the discounted cash flows.

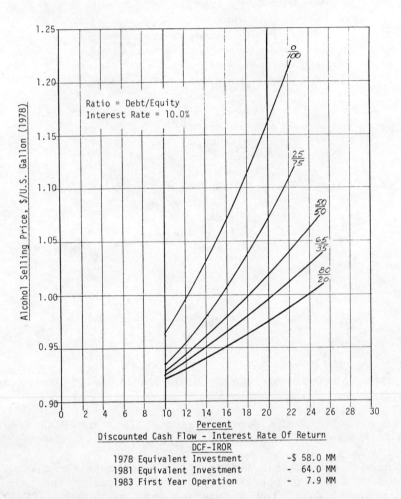

Figure 7. Grain motor fuel alcohol plant. Leveraged capital sensitivity analysis, ethanol selling price vs. DCF - IROR, 20-year plant life, 1978 costs.

It is necessary to evaluate the real effect of inflation, tax credits and leveraged capital over the entire "life cycle" of a venture. For example, the inflation rate of petroleum-based products is substantially different from non-petroleum products. The financial analysis must be able to demonstrate this effect on venture profitability.

Another example is the effect of leveraged capital. One of the U.S. governmental incentives being offered for motor fuel alcohol plants is loan guarantees. An example of how leveraged capital can affect a product's selling price, necessary to achieve an adequate interest rate of return, is indicated in Figure 7.

Note the effect of a 20% DCF-IROR. By increasing the debt to equity ratio from 0/100 to 80/20, the alcohol selling price, at 20% DCF-IROR, is reduced from $1.17/gallon to $0.98/gallon, 1978 prices. Profitability in the leveraged capital case is based on a 20% equity. The leveraged capital case has a lower selling price for a specific DCF-IROR, due primarily to tax laws which permit interest payments to be deducted from the taxable profits; whereas dividends to investors cannot be deducted.

Even with the step-by-step procedure, the determination of commercial feasibility is a difficult task. Not only is a good procedure essential, but the cost factors must be representative of "real life" situations.

ACKNOWLEDGEMENTS

The author wishes to acknowledge the substantial assistance of George Moon, Jr. and Catherine Yeats in assembling the material for this paper, and particularly the development of the computer programmed analysis method by John R. Messick and Kurt F. Kaupisch.

PARTICIPANTS

Maria Vitoria Agra
EMBRABIO - Empresa Brasileira de
  Biotecnologie Representacoes
  Ltda.
Rua Marques Deitu
58 CJ 1101
Sao Paulo, S. P., Brazil

Börge Åhgren
Dept. of Chemical Technology
Royal Institute of Technology
S-100 44 Stockholm, Sweden

Sebastiao M. Silva Alves
Tecnologia Química
Instituto Superior Técnico
Av. Rovisco Pais
1096-Codex, Lisboa, Portugal

J. J. Balatinecz
Faculty of Forestry
University of Toronto
203 College Street
Toronto, Ont., Canada   M5S 1A1

Elchanan S. Bamberger
School of Education of the
  Kibbutz
Department of Biology
University of Haifa
Oranim, Tivon 36000, Israel

Ali Beba
Dept. of Chemical Engineering
Ege University
Bornova, Izmir, Turkey

Jørgen Bech-Andersen
Tecnological Institute Copenhagen
Gregersensvej
DK-2630 Taastrup, Denmark

W. Becker
Institut für Chemische
  Pflanzenphysiologie
Universität Tübingen
7400 Tübingen 1
Corrensstrasse 41
Federal Republic of Germany

A. A. C. M. Beenackers
Twente University of Technology
P. O. Box 217
7500 AE Enschede, The Netherlands

Alda Maria Dos Santos Fidalgo
  Clemente
Biologia/DCTN - LNETI
P. O. Box 21
2686 Sacavém Codex, Portugal

M. Teresa Amaral Collaço
Department of Food Research
Laboratorio Nacional de
  Engenharia E Tecnologia
  Industrial
Rua de Sao Pedro de Alcântara, 79
1200 Lisboa, Portugal

Wilfred A. Côté, Jr.
Renewable Materials Institute
SUNY College of Environmental
  Science and Forestry
Syracuse, New York   13210, USA

Tiberius Cunia
School of Forestry
SUNY College of Environmental
  Science and Forestry
Syracuse, New York 13210, USA

Benoit de Lhoneux
Laboratoire Forestier
Faculté des Sciences Agronomiques
  de l'Université Catholique de
  Louvain
Place Croix du Sud 2, B. P. 4
B-1348 Louvain-la-Neuve, Belgium

José Manuel Cabral de Sousa Dias
EMBRAPA - National Agricultural
  Research Corporation
Super Center Venâncio 2000
Sala 916
70.336 Brasilia, Brazil

Michael R. Dicks
Dept. of Agricultural Economics
University of Missouri
Columbia, Missouri 65211, USA

Jacqueline Doat
Centre Technique Forestier
  Tropical
45 bis, Av. de la Belle Gabrielle
94130 Nogent-sur-Marne, France

José M. C. Duarte
Microbiology Group/DCTN
Institute of Energy/LNETI
P. O. Box 21
2686 Sacavém Codex, Portugal

Ahmet Altan Erarslan
Faculty of Engineering
Division of Applied Chemistry
Middle East Technical University
Gaziantep, Turkey

Douglas E. Eveleigh
Dept. of Biochemistry & Micro-
  biology
Cook College, Rutgers University
P. O. Box 231
New Brunswick, New Jersey 08903
USA

Ingemar Falkehag
Renewing Systems
706 Creekside Drive
Mt. Pleasant, South Carolina
  29464, USA

José Luis Figueiredo
Faculdade de Engenharia
Universidade do Porto
4099 Porto Codex, Portugal

Maria Inez V. S. Florêncio
Laboratório Nacional de
  Engenharia E Tecnologia
  Industrial
R. S. Pedro de Alcantara, 79
1200 Lisboa, Portugal

Kenneth L. Giles
Dept. of Biology & Biotechnology
Worcester Polytechnic Institute
Worcester, Massachusetts 01609
USA

Susan Goldhor
Center for Applied Regional
  Studies
37 South Pleasant Street
Amherst, Massachusetts 01002
USA

Aviv Goldsmith
Graduate Program
SUNY College of Environmental
  Science and Forestry
Syracuse, New York 13210, USA

Irving S. Goldstein
Dept. of Wood & Paper Science
North Carolina State University
P. O. Box 5488
Raleigh, North Carolina 27650
USA

Branca Pinheiro Gonçalves
Dept. of Chemical Engineering
Faculdade de Engenharia
Universidade de Porto
R. dos Bragas
4099 Porto Codex, Portugal

PARTICIPANTS

Maria do Ceu Teixeira Gonçalves
Department of Chemical
 Engineering
Universidade de Porto
R. dos Bragas
4099 Porto Codex, Portugal

Elke Haase
University of Oldenburg, FB 7
Ammelaender Heerstrasse
2900 Oldenburg
Federal Republic of Germany

D. O. Hall
Dept. of Plant Sciences
King's College
68 Half Moon Lane
University of London
London, England SE24 9JF, U. K.

John G. Haygreen
Dept. of Forest Products
University of Minnesota
220 Kaufert Lab.
2004 Folwell Avenue
St. Paul, Minnesota 55108, USA

Merylyn Hedger
Imperial College of Science
Dept. of Social & Economic Studies
53, Princes Gate
Exhibition Road
London SW7 2AZ, England, U. K.

Peter L. M. Heydemann
National Bureau of Standards
10535 Cambridge Ct.
Gaithersburg, Maryland 20879
USA

Erdinç Ikizoglu
Biochemical Engineering Research
 Group
Chemical Engineering Dept.
Faculty of Chemistry
Ege University
Bornova, Izmir, Turkey

Mentz Indergaard
Institute of Marine Biochemistry
University of Trondheim
N-7034 Trondheim, NTH, Norway

Volker Kasche
Fachbereich Biologie
NW 11, Universität Bremen
D-2800 Bremen 33
Federal Republic of Germany

Joseph D. Kasile
Division of Forestry
The Ohio State University
2021 Coffey Road
Columbus, Ohio 43210, USA

Raphael Katzen
Raphael Katzen Associates
 International, Inc.
1050 Delta Avenue
Cincinnati, Ohio 45208, USA

Sücaattin Kirimhan
Environmental Research Center
Atatürk University
Erzurum, Turkey

G. Lakshman
Saskatchewan Research Council
30 Campus Drive
Saskatoon, Saskatchewan,
Canada S7N 0X1

J. M. Lopez-Real
Biological Sciences Department
Wye College (University of
 London)
Wye, Ashford, Kent TN25 5AH
United Kingdom

Bernd Mahro
Department of Biology and
 Chemistry
University of Bremen
P. O. Box 28
Bremen 33, Federal Republic of
 Germany

Luiz Gonsaga Mendes
Escola de Agronomia da
 Universidade Federal da Bahia
44.380 - Cruz das Almas
Bahia, Brazil

Michael M. Micko
Dept. of Agricultural Engineering
Faculty of Agriculture & Forestry
The University of Alberta
Edmonton, Alberta, Canada T6G 2G6

Jette Dahl Møller
Institute of Plant Anatomy and
 Cytology
University of Copenhagen
Sølvgade 83
DK-1307 Copenhagen K, Denmark

Joao Barros Cabral Monteiro
Faculdade de Engenharia
Dept. Eng. Quimica
Universidade de Porto
Rua dos Bragas
4099 Porto Codex, Portugal

James P. Nakas
Environmental & Forest Biology
SUNY College of Environmental
 Science and Forestry
Syracuse, New York 13210, USA

Maria Olsson
Department of Operational
 Efficiency
Swedish University of Agricultural
 Sciences
S-770 73 Garponberg, Sweden

David G. Palmer
Dept. of Forest Engineering
SUNY College of Environmental
 Science and Forestry
Syracuse, New York 13210, USA

John Papadopoulos
Centre Technique de l'Industrie
 des Papiers, Cartons & Celluloses
Domaine Universitaire B. P. 7110
38020 Grenoble Cedex, France

Berhan Pekin
Faculty of Chemistry
Ege University
Bornova, Izmir, Turkey

Maria de Lourdes Ferreira Pocas
Laboratorio Nacional de
 Engenharia E Tecnologia
 Industrial and Instituto
 Superior Técnico
R. Ponta Delgada 65, 1° ESQ
1000 Lisboa, Portugal

Lucelia Pombeiro
Laboratorio Nacional de
 Engenharia E Tecnologia
 Industrial
R. S. Pedro de Alcantara, 79
1200 Lisboa, Portugal

Poul Børge Poulsen
Novo Industri A/S
Novo Allé
DK-2880 Bagsvaerd, Denmark

Maria de Fátima Garcia d'Almeida
 Proença
Department of Food Research
LNETI
Rua Vale Formosa
1 - 1900 Lisboa, Portugal

D. L. Pyle
Department of Chemical
 Engineering and Chemical
 Technology
Imperial College
Prince Consort Road
London, SW7 2BY, England, U.K.

Fernando Henrique da Silva
 Reboredo
INIC - Centro de Engenharia
 Biologica da Univ. de Lisboa
Depart. de Botanica, Faculdade
 Ciencias
1294 Lisboa Codex, Portugal

PARTICIPANTS

Alirio Rodrigues
Department of Chemical
  Engineering
University of Porto
4099 Porto Codex, Portugal

Guy Rollinson
Department of Microbiology
University College
Newport Road
Cardiff, CF2 1TA, S. Wales, U. K.

José Carlos Pereira Roseiro
Dept. of Food Research
LNETI
Rua Vale Formoso
1-1900 Lisboa, Portugal

William J. Sheppard
Battelle Columbus Laboratories
505 King Avenue
Columbus, Ohio 43201, USA

Smiljana Smiljanski
c/o Andrejević
Lole Ribara 37
11000 Beograd, Yugoslavia

Wayne H. Smith
Biomass Energy Systems Programs
3038 McCarty Hall
University of Florida
Gainesville, Florida 32611, USA

Harry Sobel
P. O. Box 5820
Sherman Oaks, California 91403
USA

Ed J. Soltes
Department of Forest Science
Forest Science Laboratory
Texas A&M University
College Station, Texas 77843, USA

Celia A. Spence
Graduate Program
SUNY College of Environmental
  Science and Forestry
Syracuse, New York 13210, USA

S. Suha Sukan
Dept. of Biological Sciences
Middle East Technical University
Ankara, Turkey

Fadime Taner
Temel Bilimler Fakültesi
Kimya Bölümü
Çukurova University
P. K. 171
Adana, Turkey

Fred Terracina
Dept. of Environmental &
  Forest Biology
SUNY College of Environmental
  Science and Forestry
Syracuse, New York 13210, USA

George Tsoumis
School of Forestry and Natural
  Environment
Aristotelian University
Thessaloniki, Greece

Leena Vanhala
Kemira Oy Espoo Research Centre
P. O. Box 44
SF-02271 Espoo 27, Finland

S. Venketeswaran
Biology Department
University of Houston
4800 Calhoun
Houston, Texas 77004, USA

Robert Vokes
Black Clawson Co., New York
Parsons & Whittemore, Inc.,
  New York
412 B Country Club Boulevard
Whispering Pines, North Carolina
  28389, USA

Roscoe F. Ward
United Nations
DC 810
New York, New York 10017, USA

John M. Yavorsky
School of Continuing Education
International Programs
SUNY College of Environmental
   Science and Forestry
Syracuse, New York 13210, USA

Raymond A. Young
Department of Forestry
University of Wisconsin
1630 Linden Drive
Madison, Wisconsin 53706, USA

CONTRIBUTOR

Achmadi, S.
Alves, S.
Andrade, M. E.

Balatinecz, J. J.
Beba, Ali
Becker, E. W.
Beenackers, A. A. C. M.

Clemente, A.
Collaço, M. T. A.
Côté, W. A.
Cunia, T.

Dias, A. T.
Dicks, M. R.
Doll, J. P.
Duarte, J. M. C.

Eveleigh, D. E.

Figueiredo, J. L.

Gandhi, V.
Goldstein, I. S.
Grimme, L. H.

Hall, D. O.

Indergaard, M.

Kasile, J.
Katzen, R.
Kirimhan, S.

Lakshman, G.
Lipinsky, E. S.
Lopez-Real, J. M.

Mahro, B.
Monteiro, J. C.

Nagmani, R.
Nakas, J. P.

Orfao, J. M.

Papadopoulos, J.
Pekin, B.
Poulsen, P. B.

Reboredo, F.
Rodrigues, A.
Romano, E. J.
Royer, J. C.

Sheppard, W. J.
Sobel, H.
Soltes, E. J.

Tsoumis, G.

Van Swaaij, W. P. M.
Venketeswaran, S.
Vokes, R. F.

Ward, R.

Young, B.
Young, R. A.

SUBJECT INDEX

ATP, 323, 323
Accessibility, land, 87
Acid-fermentation, 350
Adhesives from biomass, 567, 580, 582
Aerial photo
  measurements, 105
  volume tables, 103, 105
Aerobic photosynthesis, 231
Aflatoxins, 424
Agar, 143
  production, 149
Agaricus bisporus, 428, 429, 430, 431
Agarophytes, 143
Agarose, 143
Alcohol, 9
  beverage use, 119
  from sugar-cane, 9, 11, 32
  potable, 478, 479, 480, 481, 482
  production from wastes, 345
Alga, blue-green, 145
Algae, chemical composition, 217, 218
  digestibility, 216
  growth rate, 217
  photosynthetic efficiency, 215
  nutritional value, 217, 218
  red, brown, green, 138
  yield, 217
  as a source of energy, 222
  for sewage treatment, 223
Algal biomass, 206
  as feed, 222
  as food, 221

Algal biomass (continued)
  yield of, 217
Algal cultivation, 209
  growth, 207, 212
  production and harvesting, 213
  protein, amino acids in, 220
Algin factories, 151
Alginates, 142, 147
Allothermal gasification, 593
Alpha-amylase
  bacterial, 468, 470
  fungal, 468
Amino acid
  distribution in yeasts, 310
  production, 380
  world production 383
Amylases, 341
Amylopectin, 464
Amylose, 464
Anaerobic microorganisms, 323
Analysis
  multiple discriminant, 100
  of variance, one-way, 104
Anhydrocellobiose, 279
Aquatic plants, 2, 39
Aryl-alkyl ether linkage, 304
Aspergillus terreus, 437

Bacteria
  aerobic, 432
  saccharolytic thermophilic, 379
Bactertial protein from methane or methanol, 450
Bagasse, 41, 130, 335, 375, 376, 672
Bark, tree, 110, 112
Biocatalysts, immobilized, 329, 33 331, 333

Bioconversion processes, 317
Biofilm reactors, 503
Biogas, 16, 312, 348, 512, 515, 525
  alternative use, 17
  characteristics, 525
Biomass
  aquatic, 39, 234, 237
  cellulosic, 366
  combustion, 542
  densification, 183, 184, 186
  forest, 109
  harvested forest, 105, 106
  microbial, 309, 310
  microbial fermentation, 367
  new, 58
  sources, 2, 120
  woody, 249
Biomass conversion
  integrated processes, 669, 670
  thermal reactions, 593
Biomass for energy, 10, 117, 183
Biomass inventory, 66
Biomass plantations, 13
Biomass production, annual, 6
Biomass projects, 9
Biomass from residues, factors affecting supply, 35
Biomass resource, 291
Biomass Resourse base, 33
Biomass steam 291
Biomass table
  one-way tree, 73
  two-way standard tree, 74
Biomass tables
  error of, 98
  regional two-way, 83
Biomass uses, hierarchy of, 120
Biomass utilization, forest, 114
Biomethanation, 42
Biophotolysis, 355, 356, 359, 360
Blue-grass algae, nitrogen-fixing, 223
Bongkrek acid, 424
Bordered pits, 263, 267, 268, 285
Botryococcus branuii, 16
Briquetting of biomass, 185
Bulrush, 234, 237

Bulrush (Continued)
  calorific values of, 239
Burning, mass, 687

CGC - Carbonization and gasification chamber, 555, 556
CO_2 balances, 5
PMR - continuous perfectly mixed reactors, 503, 506
Callus, 194
Calorific values
  bulrush, 239
  cattails, 239
Cambium, 250
Candida, 341
Carbohydrates
  acetic acid from 647, 650
  lactic acid from, 647, 649
  source of, 462
Carbon/nitrogen ratio, C/N, 522, 524
Carburol, 14
Carrageenans, 143, 149, 159
Carrageenans factories, 151
Carrageenophytes, 162
Cattails, 234, 237
  calorific values, 239
  ethanol production from 237
Cell culture techniques, 192
Cell wall
  plant, 275
  sculpturing, 261
  wood, 258, 260, 261, 277
Cellobiase, 497
Cellobiose, 279
Cellobiohydrase, 497
Cellobiohydrolase, 371, 372
Cells, wood, 254
Cellulase, 338, 366, 369
Cellulases, bacterial, 373
Cellulase
  fungal, 370
  price, 368
  production, 327, 328
Cellulolytic bacteria, 375
Cellulolytic microbes, 370
Cellulose, 271, 292, 337, 366, 367, 461, 466, 567, 572, 659
Cellulose I, 278
Cellulose II, 278
Cellulose III, 278

Cellulose (continued)
  accessibility, 274
  acid hydrolysis, 273, 368, 559
  amorphous, 280
  crystalline, 280
  decrystallization, 561
  degradation by enzymes, 371
  derivatives, 275, 287
  enzymatic hydrolysis, 273, 283, 368
  hydrolysis, 337, 369, 538, 560
Cellulose acetate, 275
Cellulose as a chemical feedstock, 272
Cellulose degradation by rumen microorganisms, 353
Cellulosic wastes for SCP production, 452, 455
Char, 547, 553, 555
Charcoal, 111, 547, 642, 672
Chemical feedstocks, 28, 635
Chemicals from biomass, 637, 669
Chemotrophs, 313
Chinese tallow tree, 121, 192, 195
Chippers, whole tree, 103, 107
Chlorella, 206, 207, 337
Chlorella fusca, H_2 photoproduction, 228
Chondrus, 163
Cloning, 192
Clostridium thermocellum, 378
Co-disposal, 171
Co-generation, 171, 689
Combustion, direct, 46, 540, 670
Composting, 432
  high temperature, 529
  temperature curve of, 531
Continuous-culture, 320, 321, 322
Continuous forest inventory (CFI), 85
Conversion
  biomass, 291
  thermochemical, 44, 45, 537, 539, 550, 553
  Conversion processes, energy parameters, 666

Copaiba tree, 192, 201
Copaifera multijuga, 193
Corn wet milling process, 488, 489
Covariance matrix, 79
Crops
  energy, 38
  grain, 121
  hydrocarbon, 121
  oilseed, 121
  oilseed, availability, 122
  oilseed, production, 123
  major grain, prices, 122
Cyanobacteria, 205

d-RDF-densified refuse derived fuel, 170
Densification by compression bailing, 185
Densification of biomass, 183
Densification of biomass by roll-compressing, 185
Densified biomass
  advantages of, 187
  potential uses of, 187
  properties and uses of, 186
Dextrose, 463
Digestion, anaerobic, 312, 350, 351, 352, 511
Disaccharides, 461
Distillation, 485, 486, 487, 488
Double fluidized bed steam gasification, 597
Double bed wood gasifier, 598
Dulse, 162
Dunaliella, 16, 337

EBR - energy benefit ratio, 663, 664
Economic analysis, parameters, 676
Endogluconase, 339, 370, 372, 398, 497
Endotoxins, 447
Effluents, municipal, 233
Energy
  biochemical, 56
  commercial, for cereal output, 28
  commercial, for corn production, 27

Energy (continued)
  from algae, 223
  industrial usage, 56
  production from MSW biomass, 178
Energy analysis, 54
Energy balances, 611, 617, 618, 619, 668
Energy crops, 38
Energy efficiency, 611
  net, 621
Energy Farm Project, U.S. Food, 163
Energy inputs, 612
Energy outputs, 612
Energy from biomass, 13
Energy requirements for agriculture, 25
Energy through direct combustion, 375
Energy use, 4
Enzymatic hydrolysis, 42
Enzymes, 325
  cellulolytic, 280
  classification, 326
  microbial, 485
  production, 380
  world production, 382
Error
  measurement, 76
  sampling, 77
  statistical, 75
Ethanol
  costs from various feedstocks, 33
  dehydration, 659
  fermentation, 40, 348, 365, 659
  high octane number, 635
  production from grain, 119
  production from molasses, 344, 662
  synthetic anhydrous, 635
Ethanol as transportation fuel, 30, 31, 483, 672, 696, 967, 699, 700, 706
Ethanol from biomass, 41, 660
Ethanol from cellulosic wastes, 614, 691

Ethanol from corn, 615, 696, 697
Ethanol production
  energy efficiency in, 663
  energy parameters in, 665
Ethanol-water pulping, 572
Ethanol yields, 662
Eucaryotic green algae, 205
Eucheuma mariculture, 159
Exogluconase, 339
Euphorbia lathyrus, 15, 38, 121

FBBR - fluidised bed biological reactors, 503
Feed stocks, chemical, 118
Feedlot wastes to SCP, 453
Fermentation
  anaerobic, 511
  continuous, 659
  ethanol, 40
  solid state, 312
Fermentation processes, 413
Fiberboard, 112
Fibril, elementary, 467
Fischer-Tropsch process, 642
Fluidized bed char combustion, 600
Fluidized bed oxygen gasification, 595
Fluidized bed wood combustion, 600
Food, 23
  non-conventional, 413
  fermented, 415, 417, 419
Food microorganisms, vitamin content, 446
Forest biomass, 63, 109
Forest inventory by sampling, 65
Forest biomass energy, 52
Fossil fuel reserves, 5
Fructose, 464
"Fuels from Biomass", U.S. programs 376
Fuelwood, 8, 27, 30, 110, 111, 182
Fungi
  edible, 426
  wood-destroying, 282
Furcellaran, 143
Furfural production, 639

Galactans, from red algae, 142
Gasification, 46, 543, 544, 553, 670

Gasification (Continued)
  allothermal, 593
  catalysis in, 604
Gasohol, 11
German batch process, 486, 487
Global environmental monitoring (GEMS), 71, 98
Glucoamylase, 472
  immobilized, 343, 492
Glucose, 272, 299, 463, 559
  fermentation, 272, 670
  yield from cellulose, 563
Glucose isomerase, 472
B-glucosidase, 368, 371, 372
Gracilaria, 160
Green algae as potential conversion system of solar energy, 231
Growth, forms in bacteria, yeast and molds, 316
Growth factor, 314
Guayule, 16, 121, 122
  productivity, 124

H_2 photoproduction, 228
H_2 production, 361
Hardwood, composition, 43
Heartwood, 250, 253
Helical thickenings, 262
Hemicellulose, 282, 284, 366, 375
  mild acid hydrolysis, 639
High fructose syrups, 473, 477, 497
Humus, 532
Hydrocarbon crops, 121
Hydrocarbon liquid, 16
Hydrocarbon plants, 15, 121
Hydrogen from waste, 337
Hydrogen production
  by biophotolysis, 223
  microbial and enzymatic, 351
Hydrogen production by green algae, 227
Hydrogenase, 354
  bacterial, 358
Hydrolysis processes, acid, 564
Hydrolysis
  by concentrated acid, 559, 561
  by dilute acid, 559

Hydrolysis (Continued)
  rate, 583
ICI process for methanol production, 586
Index, site, 103, 104
Industries, fermentation, 365
Integrated forest industry, 692
Inulin, 461, 465, 466
  processing of, 495
Inventory
  biomass, 66
  continuous forest, 85
  error of estimates, 75
  forest, by sampling, 65
  forest, sampling design for, 72
  global forest, 70
  logging, 68
  management, 68, 99
  national forest, 69
  operational, 67, 99
  timber sales, 67
Ipil-ipil, 192
Isomerization, 493

Koji, 485
Koji culture, 327
Koji fermentation, 380
Kombu, 145, 147

LPC
  amino acids in, 416
  extraction, 415
  leaf protein concentrate, 414
Lactase, 344
Lactose, hydrolysis, 346
Laminaria, 153, 155, 161, 163
Landsat, 64, 99
Least squares linear regression functions, 79, 96
Leucaena leucocephala, 193, 198, 199
Life
  expectancy, 53
  quality, 55
Lignin, 259, 281, 283, 288, 299, 366, 580
  distribution, 259
Lignin-cellulose complex, 283

Lignins, acid hydrolysis, 299
Lignocellulose, 271, 580, 638
   chemicals, from, 643
Lignocellulosics
   accessibility, 539
   pre-treatment with sodium hydroxide, 402
Lignocellulosic resources, renewable, 580
Lignocellulosic substrates, microorganisms utilizing, 396
Liquefaction, 47, 350, 463, 489
Liquid fuels from biomass, 667

MSG, production through microbial fermentation, 381
MSW, 170, 687
MSW
   contamination, 179
   as raw material, 623
   secondary fibers from, 629
   source separation of, 630
MSW-biomass, 170
MSW biomass
   acid hydrolysis and fermentation, 632
   burning, 626
   energy recovery, 175
   ethanol from, 631
   pyrolysis, 631
MSW biomass process technology, 624
MSW biomass processes for energy, 625, 690
MSW biomass utilization, 687, 692
MSW cellulose, 368
MSW disposal, tax on, 690
MSW generation, rate, 174
Macrocystis, 158, 163
Macrophytes, marine, 137
Malt, 484
Manure, animal, 133, 512, 514, 522
Marine macrophytes, 137
Marine primary biomass, 165
Material balances, 611
Measurements, aerial photo, 105

Meat analogues, 413
Merchantability, tree, 87
Methane, 313
   biogenic, 52
   from primary marine biomass, 155
   origenic, 52
   production from wastes, 348
Methane bacteria, 518, 523
Methane fermentation, 351
Methane, generation of, 43, 682
Methane hybrid option, 599, 603
Methanol
   production costs, 590
   via indirect liquefaction, 46
Methanol from wood, 585, 606,
Methanol production via ICI process, 586
Methanol synthesis from wood, 588, 589
Microalgae, cultivation, 338
Microbial biomass, 309, 443
   growth, 314, 315
   protein from 311
Microfibril
   cellulose, 259, 278
   elementary, 259
Microfibrils, orientation of, 277
Microorganisms
   mesophilic, 530
   thermophilic, 377, 531
Middle lamella, compound, 282
Molasses, 42, 133
   microbial biomass from, 343
Monosaccharides, 461
Multispectral scanners, 64, 99
Municipal solid waste (MSW), 34, 170
MSW double fluidized beds, 597
MSW gasification, 594
Mushrooms
   minerals in, 434
   vitamin content, 434
Mushroom cultivation, 427, 428
Mycotoxins, 311, 424, 447

National Alcohol Programme, Brazil, 482

Natural compounds in algae, 222
Net thermal efficiency, NET, 663, 664
New biomass, 58
New biomass/solar technology, 59
Newsprint recovery technology, 631
Nonrenewable resources, 57
Nori, 145
Nucleic acids, 310
Nutrient requirements for algal growth, 212

O_2 production, 361
Octane boosters, 568
Octane rating, alcohol, 12
Organization of Petroleum Exporting Countries (OPEC), 170
Organic wastes
  recovery, 623
  recycling, 534
Organogenesis, 192, 194, 201
Organosolv pulping, 572, 578
Osyter mushroom, 434, 435

PAG - Protein-Calorie Advisory Group, UNO, 439, 447
PRF - plugged flow reactors, 503
Palmaria palmata, 162
Parthenium argentatum, 16
Particleboard, 111
Paulownia, 202
"Pekilo", 451
Pelleting, energy requirements for, 188
Pelleting of biomass, 185
Percolation processes, 561
Petrochemicals, feedstocks from, 651
Pharmaceuticals from seaweeds, 152
Photobiology, 17
Photochemistry, 17
Photolithotrophs, 313
Photoorganotrophs, 313
Photophosphorylation, 228
Photosynthesis, 2, 57, 191, 312, 359, 461

Photosynthetic efficiency, 3
  of algae, 215
Phycocolloids, 141
  extraction from seaweeds, 158
  processing seaweeds for, 151
Phycotoxins, lack of, in algae, 219
Phytoplankton
  marine, 137
  monoculturing, 160
Plant cell wall
  accessibility, 182
  rumen digestibility, 287
Plants, wild marine, 137
Poplar, hybrid, fast-growing, 110
Population
  size and rate of increase, 29
  world, 24, 422, 533
Population growth, 53
Porphyra, 147, 161, 163
Poststratification, 89
Power expenditure, 55
Press-to-log, 185
Prestratification, 88
Procaryotic blue-green algae, 205
Process control, 696
Process evaluation, 694
Production costs, algal, 218, 220
Program flow logic, 703
Project financial analysis, 696
Protein
  amino acid composition, 407
  concentrated, 413
  microbial, 443
  soya beans, 414
Protein production on pulps, 454
Protoplasm culture technique, 192
Pulp, 112
  hydrolysis, 286
Pulping
  ethanol-water, 572, 579
  organosolv, 572, 578, 671
Pyrolysis, 46, 114, 157, 547, 553, 555, 670, 672
Pyrolysis for carbonization, 547
Pyrolysis for tarification, 548, 549

Quality of life, 55, 60

RDF - refuse derived fuel, 170
RDF, energy recovery from, 689

RDF production by dry separation techniques, 629
RNA, 310
Rays, wood, 255
Reactor, plug flow, 561
Recombinant DNA, 377, 381
Regression functions
 allometric, 98
 linear, 98
Reserves, fossil fuel, 5
Residues, agricultural, 36, 537
 combustion, 546
 yields at harvest, 32
Residues, Canadian logging and mill 130, 132
Residues, crop, 124, 312, 512, 537
 generation of, 125
 harvest and transport costs, 125
Residues
 field crop, 120
 food processing, 130
 forest, 36, 120, 312, 537
 HC1-lignin, 301
 lignin, 300, 306
 lignin, aldehydes derivable from, 302
 ligning methylation studies, 303
 logging, 109, 110, 127
 logging, generation, 126
 logging, U.S., 124
 manufacturing, 111, 112
 mill, 128
 mill, U.S., 127
 municipal waste, 59
 wood and bark, 129
 wood processing, 109
 wood products industries, generation, 128, 131
Resource recovery, 172
Resources, forest biomass, 71
Retention time, 518, 520
Rice, 393
Rice husks, 391
 acid and enzymatic hydrolysis, 395
 chemical composition, 395
 pre-treatments, 400

Rice husks (Continued)
 pre-treatment with λ-radiation, 405, 406
 pre-treatment with steam, 403
 pyrolsysis, 395
Rice husks as fuel, 394
Rice husks as lignocellulosic material, 396

SCO - Single cell oil, 437
SCP - Single cell protein, 205, 313, 335, 336, 337, 443, 444
 composition, 445
 as major food source, 456
SCP production, high energy substrates for, 448
SCP from rice husk sugars, 395
SSF - saccharification and fermentation, 368
Saccharification, 490, 491
Sample units, permanent, 93
Sampling, double
 for stratification, 89
 second phase, 94
Sampling design
 double with regression, 90
 two-stage cluster, 90, 96
Sapium sebiferum, 193
Sapwood, 250, 253
Scenedesmus, 207, 337
Scholler-Tornesch process, 497, 498
Sculpturing, cell wall, 261
Seagrasses, 137
Seaweed, 138
 cultivation, 158
 giant brown, 145
 harvesting, 150, 157
 as manure and fodder supplement, 152
 microbial conversion, 155
 polysaccharides, 146
 productivity, 138
 resources and chemical composition, 138
 utilization, 138
Seaweed meal, 152, 153
Seaweed utilization, future, 153
Shredders for RDF, 626

INDEX

Shredding technology, wet and dry 627, 629
Sludge
  composition, 582
  treatment, 533
Solar energy, 2, 312, 365
  direct, 59
Solar energy conversion system, 2
Solid culture, 327
Solid waste stream, 171, 172
Spartina maritima, 241
Spirulina, 160, 205, 206, 207, 337
  cultivation, 212
Sporotrichum pulverulentum, 397
  growth, 399
Staged gasification/combustion, 545
Starch, 461, 464, 659
  hydrolysis, 343, 484
Steam, from manufacturing residues, 112
Straw mushroom, cultivation, 433
Structure, bordered pit, 263, 267, 268
Submerged culture, 312, 327, 328
Substrate susceptibility, 563
Sucrose, 119, 464
  fermentation, 637
  hydrolysis, 332
Sugarcane, 130
  processing, 41
Sulfite waste liquor
  Candida sp. grown on, 444
  microbial biomass from, 340
Synfuels, 567
Synthetic fuels from pulp mill chemicals, 569
Synthetic gas, chemicals from, 646
Syringaldehyde, 301, 303

Tables, aerial volume, 90
Tank designs for algal cultivation, 209
Tempeh, 415, 419
  amino acid content, 424
  composition of nutrients, 423

Tempeh (Continued)
  nutritional characteristics, 422
  production, 420, 421
  vitamins and minerals, 425
Terpenes from biomass, 641
Thermal efficiency, plant, 621
Thermal conversion reactions, 593
Thermochemical conversion, 44, 45, 537, 539, 550, 553
Thinnings, 109
Tissue culture techniques, 191, 192, 202
Trace metals, 241, 242, 246
Trace metal concentrations, 242, 246
Triglycerides from biomass, 640
Tukey's pairwise comparison test, 104

Vanillin, 301, 303
Vegetable oils, 15
  price, 34
Vegetative propagation, 192
Vitamin B_{12} production by fermentation, 438
Vitamin production by fermentation, 436
Volume tables, aerial photo, 103
Volvariella volvacea, 433

Wakame, 145, 147
Warty layer, 262
Waste
  agriculture, 366
  forestry, 366
  human, 35, 512
  industrial, 35
  municipal solid, 34, 366
  utilization, 335
Waste management systems, 684
Whey, 131, 313, 335, 451
  production, 133
  sweet and sour, 345
Whole tree utilization, 249
Wild marine plants, 137
Wood
  anatomy, 251
  cross-section, 253
  gasification, 591

Wood (Continued)
  gross structure, 250
  hydrolysis, 299
  liquefaction, 570
  preconditioning for pulping, 575, 578
  radial section, 253
  tangential section, 253
  thermochemical fractionation, 570
  ultrastructure, 258
Wood cell wall, 258
Wood consumption in methanol production, 601
Wood for fuel, 182
Wood wastes, thermochemical conversion, 554

Working capital, 702
World energy use, 4

Xylan, 376
  extraction and hydrolysis, 376
Xylose
  direct fermentation by yeast, 377
  indirect fermentation, 377
  bacterial and fungal catabolism, 378

Yeasts
  aerobic, 346
  anaerobic, 346
Yield tables, 84